Creative Economy

This book series covers research on creative economies based on humanity and spirituality to enhance the competitiveness, sustainability, peace, and fairness of international society. We define a creative economy as a socio-economic system that promotes those creative activities with a high market value and leads to the improvement of society's overall well-being.

As the global economy has developed, we have seen severe competition and polarization in income distribution. With this drastic change in the economic system, creativity with a high market value has come to be considered the main source of competiveness. But in addition to the improvement of competitiveness, we are required to work toward fairness in society.

In the process of developing a mature market, consumers come to understand that what they require most essentially is humanity and spirituality. This cannot be given or bought, but requires sharing with others across cultures and learning and developing further from their richness. Long-term sustainability of a company in this new age also requires building the same values of humanity and spirituality within its own internal organizational culture and practices.

Through this series, we intend to propose various policy recommendations that contribute to the prosperity of international society and improve the well-being of mankind by clarifying the concrete actions that are needed.

More information about this series at https://link.springer.com/bookseries/13627

Kazuo Nishimura · Masatoshi Murase ·
Kazuyoshi Yoshimura
Editors

Creative Complex Systems

Springer

Editors
Kazuo Nishimura
RIEB
Kobe University
Kobe, Japan

Masatoshi Murase
Yukawa Institute for Theoretical Physics
Kyoto University
Kyoto, Japan

Kazuyoshi Yoshimura
Graduate School of Science
Kyoto University
Kyoto, Japan

ISSN 2364-9186 ISSN 2364-9445 (electronic)
Creative Economy
ISBN 978-981-16-4456-6 ISBN 978-981-16-4457-3 (eBook)
https://doi.org/10.1007/978-981-16-4457-3

This Springer imprint is published by the registered company Springer Nature Singapore Pte Ltd.
The registered company address is: 152 Beach Road, #21-01/04 Gateway East, Singapore 189721, Singapore

Preface

In recent years, various problems including environmental problems and economic crises have occurred on a global scale. Globalization brings about benefits, but it can increase the potential risks of "systemic problems" that can lead to system-wide disruptions. The COVID-19 pandemic was declared on March 11, 2020, and has since revealed social disparities in the form of higher risk of death for people with low-socioeconomic status and caused massive destruction of the economy and globalization itself. Our efforts to solve problems often cause further problems, beyond our expectations. What can we do to protect against such emerging problems?

Kyoto University has established joint research units that are not limited to existing disciplines in the humanities and social sciences and in the natural sciences. These interdisciplinary units were formed to solve the increasingly complex and diverse problems of today. In 2010, we organized the International Research Unit of Integrated Complex System Science (IRU-ICSS) with researchers in the fields of mathematics, physics, chemistry, medicine, biology, engineering, informatics, and economics, and started interdisciplinary joint research activities. From 2015 to 2020, such research activities have been further extended through collaboration with the International Research Unit of Advanced Future Studies (IRU-AFS), for which many foreign researchers visited Kyoto University. Some of the visiting researchers have contributed to chapters in this book from transdisciplinary and transcultural perspectives.

A complex system is a generic term for a system composed of numerous interacting elements, ranging from mathematical concepts such as chaos and fractals, evolutionary processes of organisms, ecosystems, organization of organisms, brain mechanisms, and modern economic and even social phenomena, each of which is intricately intertwined with each other. Not surprisingly, a wide range of emergent complex problems can be studied.

The present book is a monograph featuring contributions by selected members of IRU-ICSS and colleagues. This book extends our views from the traditional view of stability, in which global situations are sooner or later stabilized after the impact of destruction, to "creative" complex systems resulting in emergent dynamics.

Indeed, global situations are far more deformed beyond belief. Due to the complex interactions among humans, societies, economies, and many other organizations, our world is more global than ever; thus, it looks like a huge "living" system involving systems nested within systems. Owing to such "self-nested hierarchies" in our world, most of our major problems, such as those related to energy, financial security, the environment, education, and even science issues, cannot be understood in isolation.

Because these problems appear to be very complex, it has been difficult to find ultimate solutions. In other words, extensive efforts to solve these problems have often led to the emergence of additional problems.

A full understanding of our present-day problems now requires nothing less than a new conception of a complex system, what we call a creative complex system. Despite the resulting complexity, once the individual problems are considered as different aspects of a single whole, such seemingly contradictory issues can become totally understandable, because they can be integrated into a single coherent framework. This is the integrationists' approach, or a holistic approach, in contrast to the reductionists' approach. Situations of this kind are truly relevant to understanding the question "What are the creative complex systems?". We try to understand a creative complex system as a "creative" system involving communicative "creative" subsystems.

A "creative complex system" has the function of "creative destruction" in that it causes small-scale subsystems to "fail" as large-scale systems are "created." It also has the aspect of "destructive creation" in that it avoids large-scale catastrophe by actively destroying itself locally, leading to new creation globally. Creative destruction and destructive creation can be seen as two sides of a coin called creative complex systems. This is where the essence of creative complex systems lies, as they behave in ways beyond our understanding.

This book explores the issues of creative complex systems in various disciplines, using a variety of methodologies. Part I, *Toward the Study of Creative Complex Systems*, starts from the essay on the history of IRU-ICSS at Kyoto University. Then, it provides the pathological considerations of complex nature during reproduction and development of humans, a theoretical review concerning mechanisms of coexistence in complex biological networks, and ends with an essay based on science workshops conducted at museums over the years.

Part II, *Creative Complexity in Mathematical Sciences*, discusses the behavior of random walks and diffusions on typical disordered media, and the growth effect of pollution and productive consumption were investigated based on the optimal growth framework in which human and physical capital accumulations are two growth engines. It also provides the temporal topos theoretic formulation for the relationship of the conscious (cognitive) states of an entity for the past, the present, and the future in terms of non-functorially induced mnemonic morphisms. Then, it discusses universality and the role of limitations, the three concepts of "paradox," "analogy," and "fractals," and finally emphasizes the role of paradoxes because there appear common features beyond their fundamental contradictory aspects.

Part III, *Emergent Dynamics in Complex Social and Physical Science*, provides the Chaotic Market Hypothesis (CMH) to capture the essential characteristics of the financial market. It discusses, after a brief review of the history of magnetism, the development of itinerant-electron magnetism in terms of spin fluctuations as well as exotic superconductivity. Then, it reports that quantum size effects (QSEs) can be successfully separated from the surface effects in nanoparticles systems, and that novel magnetic fluctuations can be found related to the QSEs in Pt nanoparticles. It finally discusses the spin-liquid state in organic conductors.

Part IV, *Creative Dynamics of Complex Living Systems*, provides recent evidence supporting the potential of G-quadruplex (G4) structures and telomerase to regulate mitochondrial function and mitochondrial fate. It discusses that during the evolution of living organisms, environmental changes are not only driven by external acting abiotic forces like irradiation but are also influenced by biotic factors originating from communities. It also discusses new horizons in brain science and evolutionary perspectives on suffering.

Part V, *Integrated Complex Science*, provides a survey on computational methods for metabolite identification with the focus on machine learning. It contrasts some of the dominant characteristics of the Western approach to science with those that developed in the Far East and concludes with a plea to integrate the two approaches to improve our understanding of the complex phenomena facing our societies. It finally provides a unified theory and practice of creative complex systems.

Kobe, Japan — Kazuo Nishimura
Kyoto, Japan — Masatoshi Murase
Kyoto, Japan — Kazuyoshi Yoshimura

Contents

Part I
Toward the Study of Creative Complex Systems: From the Foundation of IRU-ICSS

Chapter 1
David Pines and Me

Kazuo Nishimura

Abstract When a large number of elements are interrelated and the entire system exhibits complex behavior, the behavior of the individual elements often does not explain the behavior of the whole. This phenomenon is common not only in life, the global environment, and throughout the universe, but also in many other fields, including economy and society and is called the complexity. The author has been studying the complexity in economics since receiving his Ph.D. from the University of Rochester. This chapter looks back over events linking the Santa Fe Institute, the Institute for Complex Adaptive Matter, and Kyoto University, in addition to establishment and activities of the International Research Unit of Integrated Complex System Science (IRU-ICSS), beginning with his encounter with David Pines. Pines was one of the founders of the Santa Fe Institute, the Mecca of complex systems research. In addition, the author reflects on the activities he has been involved in to improve science and mathematics education with Pines and the complex systems researchers at Kyoto University who are members of IRU-ICSS. The education system is also an example of a complex system.

1.1 Meeting David Pines

There are two American researchers who have acted as my mentors in economics. They are Lionel McKenzie (University of Rochester) and David Pines (Santa Fe Institute). McKenzie was my doctoral advisor when I obtained a Ph.D. in mathematical economics. Pines was a physicist who acknowledged my research and brought me into the world of complex system researchers.

The original version of this chapter was previously published without open access. A correction to this chapter can be found at
https://doi.org/10.1007/978-981-16-4457-3_23

K. Nishimura (✉)
Research Institute for Economics and Business Administration, Kobe University,
2-1 Rokkoudaicho, Nadaku, Kobe 657-8501, Hyogo, Japan
e-mail: nishimura@rieb.kobe-u.ac.jp

K. Nishimura et al. (eds.), *Creative Complex Systems*, Creative Economy,
https://doi.org/10.1007/978-981-16-4457-3_1

The International Research Unit for Integrated Complex System Science (IRU-ICSS) at Kyoto University has been advised by Pines since its inception. In this chapter, I review the activities of the unit from its inception to the present with a focus on the interaction with David Pines.

The Santa Fe Institute emerged from a concept by George Cowan, a physical chemist at the Los Alamos National Laboratory, and was founded with the endorsement of physicists David Pines, Murray Gell-Mann, and Philip Anderson, and economist Kenneth Arrow. It was founded as a research institute in 1984 in Santa Fe, in the state of New Mexico, USA. Both Gell-Mann and Anderson received a Nobel Prize in Physics; Arrow received a Nobel Prize in Economics.

Complex systems are composed of many components that may or may not be mutually interdependent. These systems are a complex interplay of components, in which a part may influence the whole and the whole may affect a part. Such phenomena can be seen in many areas, ranging from nature to human society, in life, the global environment, economics, and other fields. They offer perspectives that differ from and may be unattainable through conventional reductionism. Complex systems have grown into a new field of research, through development of chaos theory and fractal theory and the application of computing power. The book *Complexity* by Waldrop (1992) introduced research being undertaken at the Santa Fe Institute. This publication became a bestseller, and the Santa Fe Institute suddenly rose to a position of renown as a Mecca for the study of complex systems.

The Santa Fe Institute also hosts economists who engage in joint research with physicists. In the early days of the institute, economists included José Scheinkman (Princeton University), who was senior to me at the University of Rochester, and Michele Boldrin (Washington University in St. Louis), junior to me at the same university. We all shared the same advisor, Lionel McKenzie. I too served as an external professor at the Santa Fe Institute, over 9 years from 2008.

In late 1992, I was invited to a complex systems symposium in Tokyo, held in December on the Hongo Campus of the University of Tokyo. Participants from overseas included members of the Santa Fe Institute. Domestic participants included Akito Arima (physicist and President of the University of Tokyo at the time), Minoru Oda (astrophysicist and President of RIKEN at the time), University of Tokyo researchers Kunihiko Kaneko (statistical physics), Hiroshi Shimizu (pharmaceutical sciences), Youichi Kaya (systems engineering), and others. At the time, there were no researchers of complex systems at the Faculty of Economics, University of Tokyo. For that and other reasons, I was invited from Kyoto University to speak.

I talked about research on nonlinear dynamics in the field of economics. Specifically, this included results of joint research undertaken with Jess Benhabib (New York University) (Benhabib & Nishimura, 1979a, 1979b, 1985) and W. Davis Dechert (State University of New York at the time) (Dechert & Nishimura, 1983). My presentation also turned to results of ongoing research then being undertaken by Makoto Yano (Yokohama National University at the time) concerning the oscillations and chaotic behavior of solutions of infinite-dimensional optimization problems.

Brian Arthur, of the Santa Fe Institute, was a discussant of my lecture. Arthur featured prominently in the book *Complexity*, which was released in the USA in

November 1992. He explained the oscillations in the growth path of the economy by comparing it to the path of an airplane and its fluctuations. During the break following my presentation, Arima and Pines asked questions and offered comments. Arima and Pines had been close friends from days long past, and both held my research in high regard.

Minoru Oda, who attended the symposium at the University of Tokyo, was later appointed to the post of President of Tokyo University of Information Sciences. In April 1998, he organized an international conference titled "Human Beings—Inside and Out." I was asked to give a guest lecture at the conference, and was delighted to catch up with Pines, who had of course also been invited. Whereas I spoke on complex systems in economics, Pines gave a presentation titled *Designing a University for the Millennium: A Santa Fe InstitutePerspective*. Oda spoke on *How We have Become Complex Human Beings*, while Hida Takeyuki of Meijo University gave a lecture titled *Fluctuation, Nonlinearity and for Human Beings*. The conference ran over two days, and Pines and I therefore had time to talk at length about each other's research. We discovered common ground and found we were kindred spirits in our thinking on research. Minoru Oda was also a friend of Pines. Later, through activities in complex systems and sciences education, Kyoto University astrophysicists who worked with us would all come under Oda's direct or indirect guidance. Pines would become a cornerstone in the process of forming a complex systems network at Kyoto University.

Born in 1924 in Kansas City, Missouri, Pines graduated from the University of California, Berkeley, in 1944, and then began physics research in the Berkeley graduate division. In 1947, Pines transferred to Princeton University Graduate School where, under the supervision of David Bohm, he continued research. Bohm was a former student of Robert Oppenheimer, who had a leading role in the Manhattan Project at Los Alamos. In 1952, Pines transferred to the University of Illinois at Urbana-Champaign, where he engaged in post-doctoral research under the guidance of John Bardeen. Pines then left the University of Illinois in 1955 to take an appointment as an assistant professor at Princeton University. After the departure of Pines, Leon Cooper took the position of post-doctoral researcher under John Bardeen. Cooper, Bardeen, and Bob Schrieffer (a postgraduate student of Bardeen) further developed the research Pines had engaged in under Bardeen. In 1957, Cooper, Bardeen, and Schrieffer presented a theory that elucidated the phenomenon of superconductivity. For that work, Cooper, Bardeen, and Schrieffer jointly won the Nobel Prize in physics in 1972.

1.2 Complex System Research at Kyoto University

In 1987, I took on a position as professor at the Institute of Economic Research, Kyoto University. Kyoto University had a strong base of complex system research. Yoshisuke Ueda at the time was at the School of Electrical and Electronic Engineering. In 1961, as a doctoral student, Ueda discovered the Strange Attractor while conducting computer simulations of nonlinear ordinary differential equations applied to electrical circuits. This attractor is called the "Ueda Attractor" or "Japanese Attractor." An international conference organized by Ueda was attended by Ralph Abraham, Mitchell J. Feigenbaum, James Yorke, and numerous other prominent people in complex system research outside Japan. In particular, the mathematician Ralph Abraham, a co-author of *Transversal Mappings and Flows* (Abraham & Robbin, 1967) frequently visited Kyoto University to take part in joint research with Ueda. I had read that book while in graduate school in the USA. On one such visit, he gave a seminar at the Institute of Economic Research. Abraham saw my thesis in my office and said that, had he been an economist, he would certainly have wanted to delve into that. Through my relationship with Ueda, I also became acquainted with Tohru Kohda (Faculty of Engineering, Kyushu University) and Yoichiro Takahashi (Research Institute for Mathematical Sciences, Kyoto University). Subsequently, I was invited to join the editorial board for the first issue of *Chaos, Solitons and Fractals* (Elsevier). Ueda taught me much about the way of thinking as a researcher.

In the 1990s, I was invited to serve on the editorial boards of various other international academic journals, in areas including economics, mathematics, and interdisciplinary fields. Those are ongoing.

In 1993, Tadashi Shigoka joined the Institute of Economic Research, shortly after obtaining a Ph.D. from Yale University with his thesis on nonlinear economic dynamics. In 1995, Masahisa Fujita, who had served for many years as a professor at the University of Pennsylvania, took a position at the Institute of Economic Research. Fujita undertook an analysis of "Cities, regions and international trade as complex systems of spatial economies." In the following year, Tomoya Mori, who had obtained a Ph.D. at the University of Pennsylvania under the supervision of Fujita, was brought into the group.

In 1996, I served as a leader in an application for status as a Center of Excellence (COE). The Center of Excellence initiative was sponsored by the Ministry of Education to support mainly international research activities of research groups that had already produced internationally acknowledged yields. Our bid was accepted and in 1997 we launched a project to form a COE for complex economic systems. This would become the second COE at Kyoto University, following the group set up in 1995 by Tasuku Honjyo and Shigetada Nakanishi of the Faculty of Medicine. Our project was completed in 2003, and in 2004 the Center for the Research of Complex Economic Systems was opened within the Institute of Economic Research.

Meanwhile, in 2001, the Institute of Economic Research sponsored an interdisciplinary international conference on complex system science. Not limited to economists, participants in the conference included David Pines and Kunihiko Kaneko in physics, Yoshisuke Ueda and Kazuyuki Aihara (University of Tokyo) in engineering, Christof Koch (Caltech) and Gen Matsumoto (RIKEN) in brain science, and Kazuhiko Aomoto (Nagoya University) and Saber Elaydi (Trinity University) in mathematics.

Ueda spoke on the Strange Attractor he had discovered in 1961, in a presentation titled *Origin of the Broken-Egg Chaotic Attractor*. Christof Koch is renowned as a leading researcher in the field of consciousness. At the time, he was engaged in joint research with Francis Crick, who was famed for discovery of the double-helix structure of the DNA molecule. In 1995, I began looking at measurement of brain activity in trying to explain human behavior. From around 2000, brain science research by economists became established as a field known as neuroeconomics. Saber Elaydi is Editor-in-Chief of the *Journal of Difference Equations and Applications* (JDEA), for which I have been on the editorial board. Through Elaydi, I became acquainted with distinguished mathematicians including James Yorke, who is famed for the paper *Period Three Implies Chaos*, and Alexander Sharkovsky, known for the "Sharkovsky Ordering." James Yorke also took part in the 2007 International Conference on Difference Equations and Applications (ICDEA), which was jointly sponsored by the Mathematical Society of Japan and the Institute of Economic Research, Kyoto University.

When David Pines visited Japan for the 2001 International Conference on Complex System Science, he talked about a concept for a new complex system research institute. At the time, I was not clear on the details of what form it might take. Several years later, while at home in Tokyo in December 2004, I received an unexpected telephone call from the Graduate School of Science at Kyoto University. They told me that David Pines was visiting Kyoto University, and asked whether I could return quickly to Kyoto. The phone call was from Satoru Nakatsuji, a lecturer at the Department of Physics. Nakatsuji (now a professor at the University of Tokyo) and Pines had published a joint paper on the Kondo Lattice in the American Physical Society journal, *Physical Review Letters*.

Returning to Kyoto without delay, I met Pines and engaged in discussion together with several scientist friends of mine. I remember one of those friends was greatly moved on meeting Pines, having read and studied writings by Pines during his days in university. At the time, Pines was accompanied on his visit to Japan by his wife, Suzy. Pines was, needless to say, good friends with Arima and Oda, and well acquainted with many Japanese physicists who could have been regarded as his disciples. Suzy, too, had many friends here, and a visit to Japan was enjoyed as a reunion of old friends.

The network-style Institute for Complex Adaptive Matter (ICAM) had been founded two years earlier in the USA, and Pines had come to Kyoto University with the objective of opening a branch in Japan. Professor Makoto Yao of the Department of Physics and I discussed the matter, and the decision was made to open a Kyoto branch of ICAM, centered on the Institute of Economic Research and the Department of Physics. The branch was named ICAM/Kyoto. Incidentally, another ICAM branch was established in Japan, in the Institute for Solid State Physics at the University of Tokyo.

At the Institute of Economic Research, Kyoto University, projects led by the Center for the Study of Complex Economic Systems were selected for the 21st Century COE Program (2003–2007) and the Global COE Program (2008–2012) of the Ministry of Education. ICAM/Kyoto thus served as another point of contact between complex system researchers at Kyoto University and the world.

To commemorate the establishment of ICAM/Kyoto, the public lecture *Invitation to Chaosand Complex Systems* was held at Kyoto University in October 2005.

Akito Arima, David Pines, Makoto Yao, and I were speakers. After serving as the President of the University of Tokyo, Arima became the Minister of Education from 1998 to 1999, and was in a position to promote a relaxed education policy that is a commonly called *Yutori Kyouiku*. In the midst of this, I published a book entitled *University Students Who Cannot Do Fractions* (Okabe et al., 1999) and criticized relaxed education, which put us in a position of direct conflict. Considering this, it was doubtful whether Arima would have been willing to participate. The people involved, including Yao, were well aware of this. However, it was natural for Arima, a physicist and friend of Pines, to give a lecture at the inaugural public lecture of ICAM/Kyoto. Arima took a mature approach and accepted to give the lecture.

When ICAM was founded, David Pines and Daniel Cox served as co-directors. Cox, a professor with the Department of Physics at the University of California, Davis, visited the Institute for Solid State Physics at the University of Tokyo and the Department of Physics at Kyoto University in 2006.

ICAM was not only conducting research, but also projects to improve the standard of science education, which was in line with my own efforts to improve mathematics education in Japan. In June 2007, I organized an international conference, named the "Interdisciplinary Conference on the Sciences of Complexity and Science Education," and invited James Yorke and Yoichiro Takahashi of chaos, Toshiaki Imada (University of Washington) of brain science, and others, and Pines also gave a presentation. His lecture was chaired by Satoru Nakatsuji, who had transferred from Kyoto University to the University of Tokyo. I had met Yorke on numerous occasions at conferences organized by Elaydi and by that time we were good friends. He had many Japanese friends and disciples; he was a Japanophile, and I too sought out the opportunity to take him to dinner whenever he came to Japan. Although we did not have the opportunity to write a joint paper, conversations with Yorke were often helpful in my research. The same could be said for Pines, and I often got many hints for my research from conversations with him.

Due in part to these efforts, ICAM/Kyoto has steadily promoted interaction with the complex systems group at Kyoto University. Makoto Yao of the Department of Physics, Masatoshi Murase of the Yukawa Institute for Theoretical Physics, Kazuyoshi Yoshimura of the Department of Chemistry, and others have taken part in ICAM international conferences held at the University of California and the Santa Fe Institute.

Meanwhile, I was appointed to the post of External Professor at the Santa Fe Institute (Fig. 1.1) in 2008, and began visiting the institute every year.

In April 2010, I reached retirement age at Kyoto University. Video messages from Pines, Yorke, Elaydi, and from Lionel McKenzie, who had been my advisor for my Ph.D. thesis at the University of Rochester, were replayed during the retirement party. Even after retirement, I have continued my research as a specially appointed professor at the Institute of Economic Research.

Against this background, in 2010, the International Research Unit for Integrated Complex Systems Science (IRU-ICSS) was established at Kyoto University as a cross-disciplinary, interdisciplinary research organization, with members of ICAM/Kyoto playing a central role, and I served as the first Director of the unit.

Fig. 1.1 With David Pines—at the entrance to the Santa Fe Institute

That post was taken up by Kazuo Mino (Institute of Economic Research) in 2013, and then by Kazuyoshi Yoshimura (Graduate School of Science) from 2015.

In December 2010, Jeremy A. Sabloff, President of the Santa Fe Institute, visited the Kyoto University IRU-ICSS. Sabloff returned to Japan in October 2012 and presented a lecture during the Santa Fe/Kyoto Symposium, "An Invitation to Complex System Science," held at Kyoto University Science Seminar House. On the following day, he gave a seminar at Kyoto University Museum, where Terufumi Ohno was at the time serving as museum director.

In June 2013, Luis Bettencourt of the Santa Fe Institute visited the Institute of Economic Research. During an Urban Economics Workshop held by the Institute of Economic Research, Bettencourt gave a report on research he was conducting in relation to a city project being undertaken by the Santa Fe Institute.

Meanwhile, Pines and I would discuss various issues whenever he came to Japan or I visited Santa Fe. Pines was more than 80 years old, but he was constantly actively engaging in new projects. Although his comments during Santa Fe Institute seminars and other occasions were frank and to the point, and he had a severe side when it came to academic matters, Pines was ever warm and welcoming toward his colleagues and juniors. Pines always applied positive thinking in regard to life, and talking with him never failed to stir courage and a positive attitude. When discussing a particular project with him, he said, "There is never any need for stress or anxiety in our work. We are in this because we enjoy doing it." We were in two different fields of work, but the 20 years of interaction since we met at conferences in Tokyo and the Tokyo University of Information Sciences were all due to his broadminded thinking and character.

1.3 Measures to Raise the Level of Science and Mathematics Education

A number of complex system researchers were also members of the Kyoto Prefectural Board of Education/Kyoto University Collaborative Project Committee, which was set up in 2010 to help improve science and mathematics education in public schools in Kyoto Prefecture. This committee was established, centered on Terufumi Ohno, Director of the Kyoto University Museum at the time.

ICAM/Kyoto and the IRU-ICSS united in 2010 to undertake complex system research. Since then, in the area of science education, the Kyoto Prefectural Board of Education/Kyoto University Collaborative Project Committee has worked on raising the science and mathematics abilities of students in public schools. Activities included visiting classes, symposiums, and events at museums and observatories.

From the start, the Santa Fe Institute too has endeavored to promote science and mathematics education, in parallel with complex system research. The Santa Fe Institute provides visiting lectures for junior and senior high schools, and conducts summer schools. Among other activities, there is a High School Prize for Scientific Excellence, awarded to a senior high school that provides outstanding science education. Since 2008, when I became an external professor at the Santa Fe Institute, David Pines, of the Santa Fe Institute and ICAM, and I, at Kyoto University, became mutual points of contact between the institutions, and promoted programs aimed at elevating science education.

In March 2015, Irene Lee, Director of the Learning Lab at the Santa Fe Institute, visited Kyoto University. During the IRU-ICSS Workshop, held in the third-floor lecture room of Kyoto University Museum, she gave a presentation on the topic *Computer Modeling and Simulation in American Education*. She also gave a visiting lecture at a Kyoto Prefectural senior high school. That lecture was arranged by the Kyoto Prefectural Board of Education.

In the area of science and mathematics education, Pines used the ICAM international network to set up the Global Partnership on Science Education through Engagement (GSEE), bringing together activities in science and mathematics education throughout the USA and Europe. I was involved in the endeavor in Japan, where GSEE/Kyoto was established with distinguished advisors. At GSEE/Kyoto, it was Pines' suggestion to ask for help from Akito Arima. I had some hesitation in doing so, since Arima was the one who promoted the idea of a relaxed education policy and I was the one who tried to stop it. However, I thought that Arima would agree with the project to improve the standard of science education, so I decided to ask him. Arima gladly agreed to become an advisor to GSEE/Kyoto. As a result, GSEE/Kyoto started with members including Akito Arima, Michiharu Nakamura (former President of JST), Professor Makoto Kobayashi (winner of a Nobel Prize in Physics), Hiroo Imura (former President of Kyoto University), Kazuo Oike (former President of Kyoto University), and Kazuo Kitahara (Professor at Tokyo University of Science) in the advisory board of GSEE/Kyoto. Such members would not have been obtained without the close relationship shared by Pines and Arima.

I used to visit the office of the President of Musashi Gakuen, which was located in the campus of Musashi University in Ekoda, Tokyo, where Arima was the president at that time, and have frequent meetings with him. Arima seemed to enjoy it, and I was later told by the vice-minister of the Ministry of Education that Arima had happily told him that he was working with me on science education activities.

In May of 2012 and 2013, I attended GSEE meetings held at the University of Chicago. The meetings focused mainly on reporting specific applications to science education in areas of the USA.

In April 2013, a preparatory conference, "Science Education in Kyoto 2013," took place at the Yukawa Institute for Theoretical Physics, Kyoto University. That conference was only for Japanese speakers. Members of the Kyoto University IRU-ICSS, ICAM/Kyoto, the Kyoto Prefectural Board of Education/Kyoto University Collaborative Project Committee, and GSEE/Kyoto advisory board attended. A number of senior high school teachers who have been practicing outstanding science education throughout Japan were also selected to give presentations. Then, in October of that year, researchers involved in science education in the USA, Europe, and Asia took part in the main conference, GSEE/Kyoto Summit 2013, which was also held in Kyoto. During the main conference, we heard reports from some of the senior high school teachers in Japan who had given presentations in the preparatory conference. Just before the main conference, a public lecture, *Let's Enjoy Science*, was held on the Kyoto University campus. With talks by David Pines and Makoto Kobayashi, this event was aimed at senior high school students, and attended by many from Kyoto and other prefectures.

Suzy also accompanied David on that occasion. Both were nearing their 90s, but walked straight and tall, and moved about easily and unaided. During the conference, Suzy caught up with old friends as planned. She had specialized in clinical psychology, and was always interested in discussion relating to my research into brain science and education.

After the Kyoto conference, David and Suzy travelled to Tokyo as planned, where they were able to join my family and me for dinner at a restaurant. I had dined with them many times, but this was the first time my family was included.

In June 2015, the GSEE/Taiwan Summit was held at the National Donghwa University, in Hualien, Taiwan. This was organized by Maw-Kuen Wu, President of the university and a physicist specializing in superconductivity. Maw-Kuen Wu later became the Taiwan Minister of Education. The year from 2015 was a difficult time for Pines, as Suzy succumbed to cancer and passed away in October. He was unable to attend the GSEE/Taiwan Summit.

In January of the following year, 2016, the GSEE/Kyoto Summit was held once more at Kyoto University. It was attended by Jyuichi Yamagiwa (President of Kyoto University) and Akito Arima. People involved in science and mathematics education were invited from Japan, China, Taiwan, Hong Kong, and South Korea.

David Pines attended the GSEE/China Summit, held in Shenzhen, China, in December 2016. The "STEM and Gifted Education Conference and Summit" took place in Hong Kong directly after. The GSEE/China Summit in Shenzhen was sponsored by the physicist Hong Ding (Institute of Physics), whereas Tai-Kai Ng (Hong Kong University of Science and Technology) organized the Hong Kong conference.

1.4 Education in Osaka City

Raising the standard of science and mathematics education in Japan was my motivation in working with Pines on GSEE activities. In many countries, the importance of science and mathematics education has been acknowledged and budgets are being allocated accordingly. In contrast, the current state of science and mathematics education in Japan was troubling. The more conferences I attended, the more apparent it became to me that improving science and mathematics education in senior high school is too late. The work needed to begin with improving teaching of science and arithmetic at the elementary school level.

Since around 2005, I have created self-study textbooks for elementary schools and attempted to raise results of public schools in Tokyo with performance below the Tokyo average to the top level, and to lift results of public elementary schools at the lowest levels in Kyoto Prefecture up to the national average. I was also involved in efforts to raise the grades of children with learning difficulties to the average of their class in a public school in Osaka City.

These attempts produced startling results. However, they were limited to individual schools and individual school years. When a new school principal was appointed, the undertaking was in many cases terminated. Principals are recognized by the Board of Education for their new initiatives. Therefore, when an incoming principal continues a successful project that was started by the previous principal, credit for success continues to go to the previous principal. It does not improve the evaluation of the incoming principal. From that experience, I strongly felt the need to enlist the cooperation of Boards of Education to broaden application of endeavors to improve academic ability. In that regard, collaborative projects being undertaken by the Kyoto Prefectural Board of Education and Kyoto University were invaluable. Additionally, over four years from April 2013, I served as a member of the Osaka City Board of Education. Board member Fujio Ohmori, now a professor at Tohoku University, was appointed as Chairman of the Board in the following year. The Osaka City Board of Education discuss all issues such as the hiring of teachers, the hiring of principals, teacher scandals, and problematic behavior of students, so it was a valuable experience for me to learn about the various issues surrounding schools.

At GSEE international conferences, I spoke about undertakings with Education Boards during presentations. I also reported on the results of a study I had conducted jointly with Tadashi Yagi (Professor at Doshisha University) concerning changes in the productivity of engineers and researchers in Japan. Many of the reports presented by overseas participants concerned guidance in science for senior high students. Almost no other research dealt with arithmetic and science in elementary schools.

Pines and other overseas participants listened intently to my report. That may have been the first time they had encountered such a perspective.

After finishing my four-year term as a member of the Osaka City Board of Education, from 2017 I began working on improving the academic ability and normative awareness of students at public junior high and elementary schools in Osaka City, as an advisor to the Board of Education. This is because I believe that safety and academic achievement are the roles that parents expect of schools. As for normative awareness, it did not take long to see results because the entire city was involved from the beginning. I will give an overview of the results. As for academic improvement, good results have been obtained in the model schools, but it will take time to spread the results to all schools in the city.

First of all, when serving as a member of the Board of Education, I made an effort to emphasize basic morals for preschool education and early elementary school students. In particular, the following four norms were included in the Osaka City Basic Plan for the Promotion of Education: "Be kind to others," "Don't lie," "Follow the rules," and "Study." In an earlier survey I conducted with Professor Yagi and others, we found that these four norms were often taught in early childhood to people who later had achieved social success.

In Osaka City, there were many instances of violent behavior among schoolchildren. Osaka City scored low in nationwide academic achievement tests, and the city has had the worst ranking in a nationwide study of school violence. Until 2014, the rate of student violence per 1000 students was more than three times the national average. So, first, we "pre-declared" the rules in writing, which are things that are natural not to do.

Virtually no schools in Japan have a clear set of rules that specify what children should not do. There are school precepts and school regulations, but those kinds of rules are different. School precepts tend to be abstract. School regulations are not necessarily about specific actions, but about attitude and dress. In extreme cases, school regulations may place restrictions on students such as specifying that underwear should be white, or that hair should be dyed black if it is not already so. In fact, a case had been brought to court about an Osaka Prefecture senior high school that compelled a student with naturally brown hair to dye it black. On the matter of bullying, in many instances discussion focuses on why it happened; that is, after the event. However, students can be clearly shown beforehand, obvious "things they should not do," such as hiding something belonging to someone else, ignoring, or hitting, etc. If the school then takes immediate action upon occurrence of such events, in many cases matters should not develop to the point of bullying.

Of course, some people might say these things obviously should not be done even if they are not codified in writing. However, in law there is the principle of nonretroactivity, which means you should not be punished for an action taken before a law was enacted to make that action illegal. The same thing applies in schools. A child who is judged for an action despite the absence of a rule forbidding it will likely feel indignation.

On 17 November 2015, the Osaka City Board of Education announced a set of "Rules for School Safety" that specified what children obviously should not do, and a

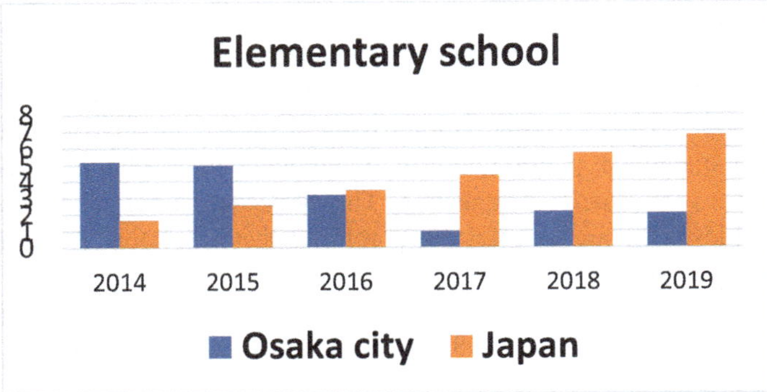

Fig. 1.2 Trends of school violence rates (per 1000 children) in Japanese elementary schools

set of responses to be taken if a child broke a rule. The significance of the responses is that they reinforce children's awareness of what they should not do, thereby reducing problem behavior.

Promoting understanding of the Rules for School Safety among parents and children, raising awareness of the rules with posters and rules tables did reduce cases of violent behavior as the rules were accepted (Fig. 1.2). By the 2017 school year, the number of cases had fallen to 25% of that in 2014. In particular, violent incidents in Osaka City elementary schools fell to 1 child per 1000, in contrast to the national average of 4 children per 1000. Surveys of the children also showed an increase in the number who said they enjoyed school.

The efficacy of the Rules for School Safety stems from the fact that even infraction of rules for small problems have been dealt with in a consistent manner. This has also prevented small issues from developing into major problem behavior. It is understandable that emphasis tends to lean toward dealing with serious problem behavior. However, the functions of the Rules for School Safety are in prior clarification and the willingness to apply them to even seemingly trifling matters, thereby raising children's awareness.

Concerning the improvement of academic ability, we set out to apply what had been learned from implementation in single school years and individual schools in Tokyo, Kyoto, and Osaka, to all schools in Osaka City. We prepared teaching manuals on arithmetic and Japanese language for teaching staff. Implementation of classroom instruction based on the manuals then began at about 80 model schools. Application of the manuals spread from the model schools to all elementary schools in the city from April 2019. As good results were obtained in the subjects' arithmetic and Japanese language, the undertaking was expanded to include sciences from the middle of 2018. This endeavor may be regarded as a GSEE version for elementary schoolchildren. David Pines would no doubt have been very pleased, had I been able to report the results.

1.5 Farewell to David

In the latter half of 2017, I received a message from Pines, saying that he had been diagnosed with cancer and would begin treatment. He would also be suspending his GSEE activities. Pines had lost his wife, Suzy, to cancer two years earlier, so this was deeply disturbing. Then, on 5 May 2018, I received a message titled "David." His family informed me that David Pines had passed away. It was just before his 94th birthday. He had moved from New Mexico to Illinois, where his children lived, and had been undergoing treatment there.

Despite his considerable achievements, Pines missed receiving a Nobel Prize as he had been away from the University of Illinois for four years from 1955. The main media reported his death as a loss to society, and gave high acclaim to the yields of his research and his contribution to society (Chang, 2018).

While working to the last with ICAM and the Santa Fe Institute, Pines continued his research throughout America, of course, and around the world, teaching also at the University of California, Davis, and other institutions. David Pines was active throughout his life, and for me, too, it was a rare honor to have such a role model. His forward-looking attitude sometimes caught me by surprise, such as when I learned that he was preparing a 10-year research plan, just as he turned 90. It is indeed unfortunate that the 10-year plan was terminated prematurely. Having already lost a mentor in the world of economics with the death of Lionel McKenzie in 2012, the departure of yet another mentor in interdisciplinary research was all the more poignant.

Acknowledgements I acknowledge the JSPS Grant-in-Aid for Scientific Research No. 20H05633 and No. 16H03598 for support of our projects. While editing this manuscript, I learned that Akito Arima passed away on 7 December 2020 at the age of 90. I have fond memories of discussing GSEE/Kyoto's activity plan and talking about Pines with him in the Chancellor's office on the campus of Musashi University. I sincerely pray for his soul to rest in peace.

References

Abraham, R., & Robbin, J. (1967). *Transversal mappings and flows*. Benjamin.

Benhabib, J., & Nishimura, K. (1979a). On the uniqueness of steady states in an economy with heterogeneous capital goods. *International Economic Review, 20*(1), 59–82.

Benhabib, J., & Nishimura, K. (1979b). The Hopf bifurcation and the existence and the stability of closed orbits in multi-sector models of economic growth. *Journal of Economic Theory, 21*, 421–444.

Benhabib, J., & Nishimura, K. (1985). Competitive equilibrium cycles. *Journal of Economic Theory, 35*, 284–306.

Chang, K. (May 11, 2018). *David Pines, 93, insightful and influential physicist, dies*. The New York Times.

Dechert, D., & Nishimura, K. (1983). A complete characterization of optimal growth paths in an aggregated model with a non-concave production function. *Journal of Economic Theory, 31*, 332–354.

Okabe, T., Tose, N., & Nishimura, K. (1999). *University students who cannot do fractions* (in Japanese). Toyokeizai Shinposya.
Waldrop, M. (November 1992). *Complexity: The emerging science at the edge of order and chaos.* Simon and Schuster.

Kazuo Nishimura Ph.D., is a Specially Appointed Professor of the Research Institute for Economics and Business Administration at Kobe University in Japan. He is also Professor Emeritus of Kyoto University and a member of the Japan Academy. He received his Ph.D. from the University of Rochester in 1977. Nishimura served as President of The Japanese Economic Association in 2000–2001 and a fellow of the Econometric Society since 1992. He is known for contributions in complexity economics and served as an external professor of the Santa Fe Institute, 2008–2017. Nishimura has done not only mathematical economics, but also research on human capital and education, and brain science, which he has applied to teaching.

Chapter 2
To Err is Human: The Complex Nature of Human Reproduction and Prenatal Development

Kohei Shiota

Abstract Developmental errors occur frequently early in gestation and at least 10% of human embryos are morphologically and/or cytogenetically abnormal around the fifth week of gestation. The causes of developmental anomalies are genetic, environmental, or a combination of multiple factors. It appears that developmental errors and reproductive losses occur exceptionally frequently in the human species, but most abnormal embryos/fetuses die in utero and end in spontaneous abortion. High mutation rates in humans can be deleterious for human health and survival, but mechanisms may exist that favor normal over faulty conceptions and bring about natural elimination of abnormal human conceptuses.

Keywords Reproduction · Prenatal development · Developmental error · Reproductive loss

2.1 Introduction

We are born after an intrauterine life of 9 months. Although most individuals are born healthy, some 3% of newborn babies have abnormalities or diseases of prenatal origin, which are called birth defects or congenital anomalies. Such anomalies include structural abnormalities (malformations), functional disturbances (e.g., cognitive and motor impairments and intellectual problems), and developmental delay (growth retardation). When infants were examined at 2 years, 10% of them were found to have prenatal and perinatal handicaps that required special medical or educational care for physical defects and/or mental retardation (Bierman et al., 1965). In addition, a substantial proportion of human conceptions are lost as spontaneous abortion. Thus, human beings suffer from a high prenatal morbidity and mortality. This is because the prenatal developmental process is controled by complex genetic programs and their disturbance can interfere with the normal developmental process and affect the development and survival of the conceptus in utero. Furthermore, the unborn

---

K. Shiota (✉)
International Research Unit of Integrated Complex System Science, Yukawa Institute for Theoretical Physics, Kyoto University, Kitashirakawa, Kyoto 606-8501, Japan

K. Nishimura et al. (eds.), *Creative Complex Systems*, Creative Economy,
https://doi.org/10.1007/978-981-16-4457-3_2

human is exposed to various intrinsic and extrinsic noxious agents that can disturb morphogenesis and/or functional development.

2.2 Many Fertilized Human Ova Die in Utero

Humans suffer a high incidence of reproductive loss. It has been estimated that 15–20% of human conceptions end in spontaneous abortion, although the abortion rate varies by the method of ascertainment. Warburton and Fraser (1964) analyzed the outcome of 6,835 conceptions using the reproductive histories taken by personal interview and found that 15% of total conceptions ended in spontaneous abortion. When females at the reproductive age were carefully followed, 24% of conceptions ended in spontaneous abortion after the first missed menstrual period (Kauai Pregnancy Study) (Bierman et al., 1965; French & Bierman, 1962). The pregnancy loss rate was highest between the fourth and seventh weeks and decreased as gestation advanced.

In the 1980s, early pregnancy diagnosis soon after implantation became possible by measuring the human chorionic gonadotropin (hCG) level in the female serum and urine. Follow-up studies of hCG-positive pregnancies revealed that 20–62% of implanted human ova are lost by mid-gestation and a significant proportion of those abortions were recognized neither as clinical pregnancy nor as spontaneous abortion, and are referred to as "biochemical pregnancies" or "occult pregnancies" (Edmonds et al., 1982; Miller et al., 1980; Whittaker et al., 1983; Wilcox et al., 1988; Zinaman et al., 1996). For example, Wilcox et al. (1988) reported that the total pregnancy loss rate after implantation was 31% and that 22% of the pregnancies identified by elevated hCG levels ended before pregnancy was detected clinically. Zinaman et al. (1996) revealed that 31% of the pregnancies detected by serum hCG were lost during the first three cycles following detection and 41% of these losses (13% of the total pregnancies detected) were occult and recognized only by elevated hCG levels. Thus, a substantially large proportion of human conceptions are aborted spontaneously from such an early stage that the mother is unaware of pregnancy. The clinical data cited above suggest that the efficiency of human reproduction is not high enough with a maximum of approximately 30% per cycle.

2.3 Developmental Abnormalities Occur Frequently Early in Development

Various studies on aborted human specimens have shown that 40–60% of spontaneous abortuses are associated with morphological and/or chromosomal abnormalities. Poland and her colleagues made a systematic study on a large number of spontaneous abortuses and found structural abnormalities in 84% of aborted embryos and 26% of fetuses (Poland et al., 1981). Chromosomal abnormalities have been found in 30–60% of aborted embryos (Boué et al., 1976; Creasy et al., 1976; Hassold et al., 1978; Kajii et al., 1980). These results suggest that morphological and chromosomal abnormalities occur frequently in early human pregnancies and many of those abnormal embryos die in utero and are eliminated as spontaneous abortion.

A direct observation of early human embryos was conducted by an American gynecologist Arthur Hertig and his colleagues. They examined 34 pre- and postimplantation embryos (2–17 days after fertilization), which had been procured after hysterectomy, and found that more than one third of the embryos were morphologically abnormal or had a sign of intrauterine death (degeneration) (Hertig et al., 1956, 1959). Based on their observation, Hertig (1967) proposed an estimation that out of 100 human ova exposed to sperm in vivo, 16 fail to be fertilized, 15 die during the first week, 27 die during the second week, 8 die between the third and sixth weeks, 3 die during the following gestational months, and only 31 are born alive (including 1 malformed case). In Kyoto, the late Professor Hideo Nishimura conducted a systematic study of human embryos at the period of major organogenesis. Nishimura collected over 30,000 human embryos procured after induced abortion given to pregnant women for socio-economic reasons based on the Maternity Protection Law of Japan. The embryo collection is considered to represent the intrauterine population in Japan because the specimens were procured after therapeutic abortion given to virtually healthy women and their availability was not biased by the condition of the embryos and fetuses (Nishimura et al., 1968; Nishimura, 1975; Shiota, 2018). Nishimura and his coworkers made a systematic survey of the embryos and found that the prevalence rates of various malformations were at least several times higher than the corresponding figures in newborns (Nishimura, 1975). We histologically examined 37 embryos at the early postimplantation period (between 14 and 24 days after fertilization) and found that 5 cases (13.5%) were grossly abnormal and 7 cases (18.9%) were degenerating (Shiota et al., 1987). Gross abnormalities included distorted embryonic disc, disorganized neural groove or tube, and neural tube dysraphism.

Chromosomal abnormalities are found in 6 cases among 1000 newborns, whereas 30–60% of spontaneously aborted embryos/fetuses are chromosomally abnormal (Hook, 1981). Direct chromosome analyses of human sperm, oocytes, and embryos fertilized in vitro have shown that at least 20% of human gametes and zygotes are associated with chromosomal abnormalities (Angell et al., 1983; Plachot et al., 1986, 1987; Rudak et al., 1978, 1985). By analyzing the published data, Rudak et al. (1978) estimated that the frequency of aneuploidy in the sperm of normal males was around 40%. Plachot et al. (1986) examined 55 oocytes recovered in an in vitro fertilization (IVF) program and found chromosomal abnormalities in 22% of the oocytes examined. Angell et al. (1983) examined eight-cell human embryos recovered after IVF and 2 of the 11 embryos examined for ploidy were haploid and 2 of the 3 embryos examined chromosomally showed evidence of nondisjunction (abnormal segregation of chromosomes occurring during meiotic division). Furthermore, Plachot et al. (1987) cytogenetically examined 68 human embryos at the two- to eight-cell stage and found that 23% of them had chromosomal abnormalities, including triploid or haploid compliment and their associated mosaicism. These results suggest that both morphological and chromosomal abnormalities are much more prevalent in human gametes and early embryos than in the newborn population.

2.4 Intrauterine Fate of Normal and Abnormal Human Embryos and Fetuses

As discussed above, human beings suffer from a high incidence of reproductive failure and many of the aborted conceptuses are abnormal. Thus, spontaneous abortion appears to be a screening device for abnormal conceptuses and contributes to eliminating pathological embryos/fetuses before birth. However, it is not clear how abnormal conceptuses die in utero during gestation because it is difficult to obtain the prevalence rate of abnormal embryos/fetuses at each gestational period.

The present author constructed an intrauterine life-table for normal and malformed human embryos and fetuses based on the data for over 20,000 cases in the Kyoto Collection of Human Embryos and analyzed the intrauterine fate of normal and abnormal human conceptuses (Shiota, 2021). It was found that among embryos viable in utero at the start of the fifth week after fertilization, 10% of them were found to have major external malformations or eventually become malformed later in gestation. By selective natural death of defective embryos and fetuses, the prevalence of malformations was estimated to drop to 9.2, 8.5, 7.5, and 5.3% by the end of the fifth, sixth, seventh, and eighth weeks, respectively, and eventually reach 1.0% at term (Table 2.1). The prenatal mortality rate of externally malformed embryos between the fifth week of gestation and term was as high as 93%, which is significantly higher than the corresponding rate for externally normal embryos (25%). The prevalence

Table 2.1 Estimated proportion of malformed embryos/fetuses at the start of each gestational interval

Gestational week/s	Total malformed cases (%)	Neural tube defects (%)	Holoprosencephaly (%)	Cleft lip (%)	Polydactyly (%)
Week 5	9.6	1.3	3.1	–	–
Week 6	9.2	1.1	2.9	–	–
Week 7	8.5	1.0	2.3	1.9	1.0
Week 8	7.5	0.9	1.8	1.8	0.9
Weeks 9–12	5.3	0.3	0.8	1.4	0.9
Weeks 13–16	2.8	0.06	0.3	0.6	0.07
Weeks 17–20	1.5	0.07	0.01	0.2	0.08
Week 21-term	1.0	0.07	0.01	0.17	0.08
Newborn	1.0	0.07	0.01	0.17	0.09

Data modified from Shiota (2021)

Table 2.2 Estimated proportion of human conceptuses with chromosomal abnormalities

Gestational week	Any cytogenetic abnormality (%)	Clinically significant cytogenetic abnormality (%)
Week 5	5.0	4.7
Week 8	4.2	3.8
Week 12	2.4	2.1
Week 16	1.1	0.8
Week 20	0.8	0.5
Week 28	0.7	0.33
Live births	0.6	0.27

Data from Hook (1981)

rates of malformed embryos and fetuses cited above should be considered as minimal estimates because the sample did not include stillborn malformed cases. Hook (1981) compiled the published data of chromosomal abnormalities in human spontaneous abortuses and estimated that the proportion of cytogenetically abnormal conceptuses is 5% at 5 weeks of gestation and drops to 4.2% by 8 weeks, 2.4% by 12 weeks, and 0.6% at term (Table 2.2).

Thus, it is evident that morphological and chromosomal abnormalities occur early in gestation at least several times more frequently than in newborns but only a small proportion of the abnormal embryos survive the intrauterine life and are born alive. Without spontaneous abortion, the prevalence rates of major malformations and chromosomal aberrations among newborns would be 10% or higher.

2.5 Causes of Developmental Anomalies in Humans

The causes of human birth defects can be classified into three categories: genetic, environmental, and unknown (Table 2.3). Genetic causes include gene mutations and chromosomal aberrations, and they account for 15–25% of newborn cases with birth defects (Brent, 2004). Mutant genes can be transmitted from parents to children, but approximately a quarter of the mutations found in clinical cases are known to occur during the formation of gametes (i.e., de novo mutations). Many chromosomal abnormalities are also known to occur during the formation of germline cells. The genetic basis of human birth defects will be further discussed later in this chapter.

Human embryos and fetuses in utero may be exposed to various environmental factors that can interfere with their development. Table 2.4 lists the environmental agents that have been proven to cause congenital anomalies in humans (teratogens). Known human teratogens explain less than 10% of all birth defects, but the birth defects of this category can be prevented by avoiding the exposure or intake of teratogenic agents during pregnancy. The list cannot be used in isolation because various parameters are involved in producing birth defects, such as the timing of

Table 2.3 Causes of human congenital malformations

Cause	Contribution to total malformations (%)
Genetic	15–25
Autosomal and sex-linked inherited genetic disease	
Chromosomal abnormalities	
Environmental	10
Maternal conditions	
Infectious agents	
Mechanical problems (deformations)	
Chemicals, prescription drugs	
Ionizing radiation, hyperthermia	
Unknown	65–75
Polygenic	
Multifactorial (gene–environment interactions)	
Spontaneous errors of development	

Data from Brent (2004)

exposure (gestational age), magnitude (dosage), length of exposure, and individual susceptibility. For example, the embryo is highly susceptible to teratogenic agents during the period of major organogenesis (3–8 weeks after fertilization), which is called the critical period of teratogenesis. Further details of human teratogens and their risks have been discussed by Schardein (2000) and Shepard and Lemire (2010).

In two thirds of the patients with birth defects, however, a single cause cannot be identified but a combination of genetic and environmental factors is assumed to be responsible. This group of birth defects are referred to as multifactorial or polygenic in origin. Included in this category are various common birth defects such as cleft lip and palate, neural tube defects (anencephaly and spina bifida), and congenital heart defects. In the multifactorial/threshold model, the disease "liability" distributes normally in the population and is determined by combined effects of genetic and environmental factors (Carter, 1961; Falconer, 1965). It is assumed that those people for whom the liability exceeds a "threshold" are affected and have a malformation (Fig. 2.1).

In many common birth defects of multifactorial/polygenic origin, contributing factors and the mode of their interaction are obscure and yet to be clarified. Nicotinamide adenine dinucleotide (NAD) is a crucial biomolecule involved in cellular energy metabolism, cell signaling, and numerous cellular processes such as cell division, DNA damage repair, and mitochondrial function (Houtkooper et al., 2010; Yang & Sauve, 2016). NAD deficiency during pregnancy due to genetic disruption of its

Table 2.4 Known human teratogens

Teratogenic agent	Abnormalities induced
Maternal conditions	
Hypothyroidism (cretinism)	Mental retardation, intrauterine death
Diabetes mellitus	Increase various malformations
Phenylketonuria	Mental retardation
Virilizing tumors	Masculinization of female external genital organs
Infections	
Rubella virus	Congenital cataract, congenital deafness, cardiac malformation
Cytomegalovirus	Microcephaly, mental retardation, intrauterine death
Herpes simplex virus	Microcephaly, microphthalmia
Varicella virus	Hydrocephaly, microcephaly, microphthalmia, skin lesions
Venezuelan encephalitis virus	Microcephaly, microphthalmia, hydrocephaly
Parvovirus B19	Hydrops fetalis, intrauterine death
HIV virus	Growth retardation, microcephaly
ZIKA virus	Microcephaly
Toxoplasma	Hydrocephaly, microcephaly, calcification of cerebral cortex, microphthalmia
Syphilis	Hepatosplenomegaly, mental retardation, dysgenesis of bones and teeth
Mechanical factors	
Constriction band, amniotic band	Limb amputation, facial deformation
Uterine malformation	Deformation of limbs and face
Physical agents	
Ionizing radiation	Microcephaly, mental retardation, microphthalmia
Hyperthermia	Neural tube defects, microcephaly, mental retardation
Drugs and environmental agents	
Alcohol (ethanol)	Fetal Alcohol Syndrome (typical facial dysmorphogenesis, mental retardation, microcephaly)
Androgenic hormones	Masculinization of female external genitalia
Anticancer drugs	CNS malformations, skeletal malformations, fetal death

(continued)

Table 2.4 (continued)

Teratogenic agent	Abnormalities induced
Coumarin derivatives (Warfarin)	Hypoplasia of nasal bones, epiphyseal stippling
Antithyroid drugs	Scalp defect, gastrointestinal tract malformations, umbilical hernia
Tetracycline	Bone and teeth staining
Streptomycin	Hearing disturbance, interference of CNVIII
Thalidomide	Phocomelia malformations, cardiac malformations
Antiepileptic drugs (phenytoin, trimethadione, valproic acid, etc.)	Neural tube defects, cardiac malformations, cleft lip and palate
Nonsteroidal anti-inflammatory drugs (NSAIDs)	Persistence of fetal circulation
Angiotensin converting enzyme (ACE) inhibitors	Growth retardation, cranial malformation, pulmonary hypoplasia, fetal hypotension, intrauterine death
Diethylstilbesterol	Malformations of uterus, oviduct, and upper part of vagina, vaginal cancer after puberty
Retinoic acid	Microtia, anotia, cleft palate, cardiac malformations
Methylmercury	Severe CNS dysfunction (Fetal Minamata Disease)
Polychlorobiphenyl (PCB)	Skin pigmentation, growth retardation
Cocaine	Vascular disruption, intrauterine death, postnatal behavioral problems

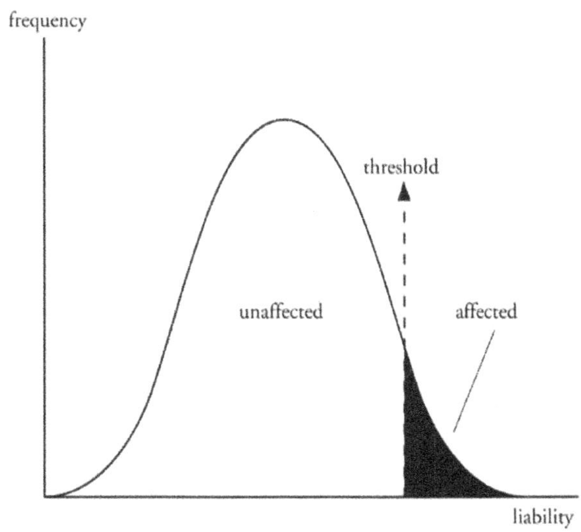

Fig. 2.1 A multifactorial threshold model describing the situation in which a genetically predisposed individual is affected when exceeding a threshold of genetic and/or environmental factors (Falconer, 1965)

synthesis pathway has been identified as a cause of multiple congenital malformations and recurrent miscarriages in humans (Shi et al., 2017). Recently, Cuny et al. (2020) showed that using a mouse model of loss-of-function in NAD synthesis, the teratogenic and embryolethal effects of the genetic variant were exacerbated by maternal malnutrition and mild hypoxia. Their study has provided experimental evidence for the gene–environment interaction as a cause of birth defects and miscarriages.

2.6 Mutation as a Cause of Developmental Abnormalities

Sex cells (sperm and oocyte) have a haploid set of chromosomes, which means each gamete has a half number of chromosomes as compared with normal somatic cells. Therefore, a special process of cell division is required to generate sex cells. During gametogenesis, DNA is replicated and two specific cell divisions (meiosis I and meiosis II) generate four haploid sex cells (four sperm in the male and an ovum and three polar bodies in the female).

In the female, meiosis I is completed and meiosis II starts before birth in all the oocytes in the ovary. In the oocyte-generating cell-lines, there are 22 cell divisions before meiosis and two during meiosis, giving 23 chromosomal replications in total (Vogel & Motulsky, 1997), because only one replication occurs during the two meiotic divisions. It should be noted that all the oocytes that have initiated meiosis II enter a period of meiotic arrest before birth and meiotic division does not resume until after puberty. Meiosis II is completed only when the cell is fertilized. On the other hand, spermatogenetic cells in the male testis (spermatogonia) are produced continuously throughout the reproductive life and therefore the number of cell divisions and chromosome replications increase with age (Crow, 2000) (Table 2.5).

Gene mutations and chromosomal aneuploidies (especially nondisjunction) can occur in the process of gametogenesis. Because germline cell divisions are so different between the sexes, the chances of gene mutations and chromosomal aneuploidies occurring are quite different between the male and female. It is accepted that chromosomal aneuploidies, especially trisomy 21 (Down syndrome) and other trisomies, increase with advancing maternal age (Fig. 2.2). Among women younger than 25 years old, the incidence of having a baby with Down syndrome is 1 in 1500

Table 2.5 Number of male germ-cell divisions

Age (years)	Chromosome replications
15	35
20	150
30	380
40	610
50	840

Data from Crow (2000)

Fig. 2.2 Maternal age-specific estimates of trisomy % among all clinically recognized pregnancies (adopted from Hassold & Hunt, 2001)

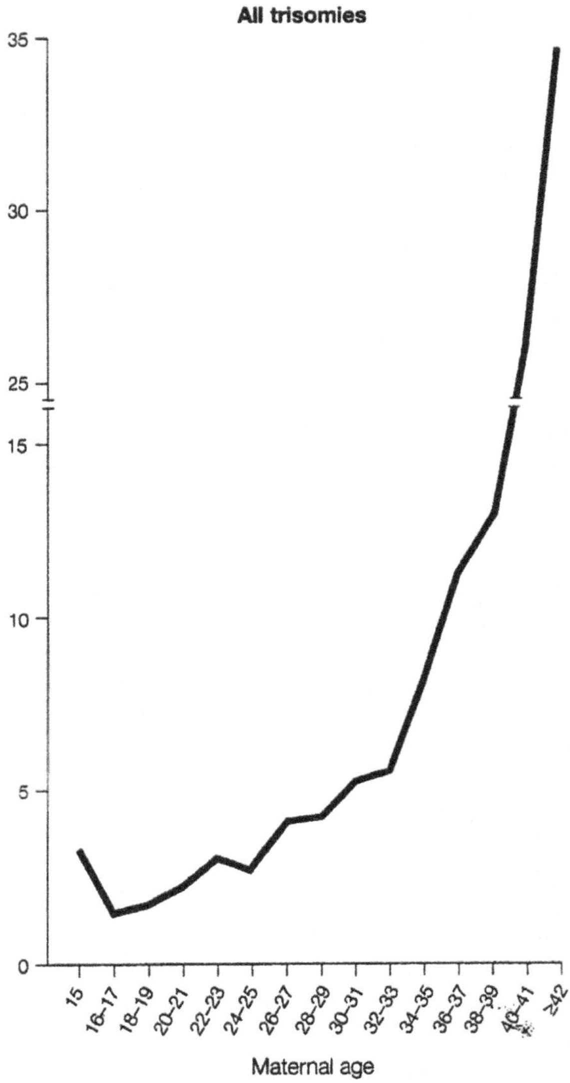

births, while after 35 years of age, it increases exponentially with increasing maternal age and becomes 1 in 80 births or more among mothers over 40 years old. Autosomal trisomies are characterized by an extra number (three) of specific chromosomes in the zygote and is caused mainly by a failure of separation of homologous chromosomes to two daughter cells (a phenomenon called nondisjunction) during meiotic cell divisions. The prolonged arrest of cell division in the ovary may increase the risk of abnormal chromosomal segregation during meiosis II, but the biological basis of maternal age effect on chromosomal nondisjunction is yet to be elucidated (Hassold & Hunt, 2001).

There is little contribution of aging of the father on chromosomal aneuploidies, but paternal age effects have been noted in some congenital anomalies, particularly dominantly inherited diseases such as achondroplasia and Apert syndrome (Risch et al., 1987). A paper was published by Weinberg in 1912, in which he claimed that sporadic cases of achondroplasia (dominantly inherited, short-limbed dwarfism) were most often found in last-born children of a sibship and suggested a possible mutational effect (Weinberg, 1912). Mutation was a vague idea in those days but more than 40 years later, Penrose (1955) showed that paternal age was the main cause of Weinberg's observation.

Risch et al. (1987) reported a strong paternal age effect for 12 syndromes such as Apert, Crouzon, Marfan, and Waardenburg syndromes. Several other traits were also found since, and it became clear that gene mutations may occur disproportionately in males. Analyses of mutations at FGFR2, FGFR3, and RET genes using molecular markers revealed that almost all the new mutations had a paternal origin (Glaser et al., 2000; Moloney et al., 1996; Schuffenecker et al., 1997; Wilkin et al., 1998). Such an increase in mutations with advancing paternal age may be explained as a result of an elevated risk of gene mutation due to the increased number of DNA replications (Table 2.5) as well as reduced fidelity in DNA repair mechanisms in older fathers. A slight paternal age effect has been reported for some other birth defects such as congenital heart defects (Olshan et al., 1994) and CHARGE syndrome (Teller et al., 1998). The parental-age effects on human mutation have been discussed in detail by Crow (2000).

2.7 Spontaneous Mutation and Its Impact on Human Health and Survival

Classically, the natural mutation rate in humans was estimated to be about 10^{-5} per locus per generation (Vogel & Motulsky, 1997). A direct characterization of human genome sequence variation showed that the rate of de novo germline base substitutions is approximately 10^{-8} per base pair per generation (1000 Genomes Project Consortium, 2010).

By measuring the amino acid changes in 46 proteins in the human ancestral line, Eyre-Walker and Keightley (1999) found 143 nonsynonymous substitutions among 41,471 amino acids. Given that 231 neutral mutations were theoretically expected, they estimated the number of deleterious mutations eliminated by natural selection as 88 ($231-143$). This means that about 38% (88/231) of the mutations are spontaneously eliminated. They concluded that 1.6 ± 0.8 deleterious mutations occur per diploid per generation. If each zygote has more than one deleterious mutation as they proposed, how can we stay healthy and survive? Their estimation may be an overestimation because they assumed the number of genes per diploid genome as 60,000. Because the human gene number is 22,000–23,000, the more realistic number of deleterious mutations would be about 0.6 per diploid per generation.

Most of these mutations may possibly be eliminated spontaneously so that they may not be accumulated in the human genome and human beings may not become extinct.

For the removal of deleterious mutations from the population, Muller (1950) proposed the concept of "genetic death," which means that individuals carrying a deleterious mutant gene suffer from reduced fitness or higher mortality so the gene may not persist in the population. Crow (2000) proposed a "quasi-truncation selection model" in which the number of new mutations per person has a normal distribution and those individuals with the mutation number above a threshold (Crow assumed a standard deviation of 15) fail to survive or reproduce. The measures for eliminating harmful mutations from the gene pool may include infertility, illness, and death, but spontaneous abortion may also contribute to this process.

2.8 Conclusion

When compared with other animal species, mutations and developmental errors occur more frequently in humans. One explanation for this observation is that in the human, the life span and the duration of the reproductive age are exceptionally long, during which time germline cells become aged and deleterious mutations accumulate. In the female, oocytes stay at the resting stage for decades, which increase the risk of chromosomal nondisjunction. In the male, spermatogonia repeat cell divisions and the risk of gene mutations increase as the father becomes older. Some of such mutations are unfavorable but, on the other hand, the high mutation rate in humans may possibly have facilitated human evolution. Infertility, reproductive loss, and developmental errors may be important devices for eliminating deleterious mutant genes and preventing their accumulation in the human gene pool.

References

1000 Genomes Project Consortium. (2010). A map of human genome variation from population-scale sequencing. *Nature, 467*, 1061–1073.

Angell, R. R., Aitken, R. J., van Look, P. F., Lumsden, M. A., & Templeton, A. A. (1983). Chromosome abnormalities in human embryos after in vitro fertilization. *Nature, 303*, 336–338.

Bierman, J. M., Siegel, E., French, F. E., & Simonian, K. (1965). Analysis of the outcome of all pregnancies in a community. Kauai pregnancy study. *American Journal of Obstetrics and Gynecology, 91*, 37–45.

Boué, J., Philippe, E., Giroud, A., & Boué, A. (1976). Phenotypic expression of lethal chromosomal anomalies in human abortuses. *Teratology, 14*, 3–20.

Brent, R. L. (2004). Environmental causes of human congenital malformations: The pediatrician's role in dealing with these complex clinical problems caused by a multiplicity of environmental and genetic factors. *Pediatrics, 113*, 957–968.

Carter, C. O. (1961). The inheritance of congenital pyloric stenosis. *British Medical Bulletin, 17*, 251–254.

Creasy, M. R., Crolla, J. A., & Alberman, E. D. (1976). A cytogenetic study of human spontaneous abortions using banding techniques. *Human Genetics, 31*, 177–196.
Crow, J. F. (2000). The origins, patterns and implications of human spontaneous mutation. *Nature Review Genetics, 1*, 40–47.
Cuny, H., Rapadas, M., Gereis, J., Martin, E. M. M. A., Kirk, R. B., Shi, H., & Dunwoodie, S. L. (2020). NAD deficiency due to environmental factors or gene-environment interactions causes congenital malformations and miscarriage in mice. *Proceedings of the National Academy of Sciences of the United States of America, 117*, 3738–3747.
Edmonds, D. K., Lindsay, K. S., Miller, J. F., Williamson, E., & Wood, P. J. (1982). Early embryonic mortality in women. *Fertility and Sterility, 38*, 447–453.
Eyre-Walker, A., & Keightley, P. D. (1999). High genomic deleterious mutation rates in hominids. *Nature, 397*, 344–347.
Falconer, D. S. (1965). The inheritance of liability to certain diseases, estimated from the incidence among relatives. *Annals of Human Genetics, 29*, 51–76.
French, F. E., & Bierman, J. M. (1962). Probabilities of fetal mortality. *Health Reports, 77*, 835–847.
Glaser, R. L., Jiang, W., Boyadjiev, S. A., Tran, A. K., Zachary, A. A., Van Maldergem, L., Johnson, D., Walsh, S., Oldridge, M., Wall, S. A., Wilkie, A. O., & Jabs, E. W. (2000). Paternal origin of FGFR2 mutations in sporadic cases of Crouzon syndrome and Pfeiffer syndrome. *American Journal of Human Genetics, 66*, 768–777.
Hassold, T., & Hunt, P. (2001). To err (meiotically) is human: The genesis of human aneuploidy. *Nature Review Genetics, 2*, 280–291.
Hassold, T. J., Matsuyama, A., Newlands, I. M., Matsuura, J. S., Jacobs, P. A., Manuel, B., & Tsuei, J. (1978). A cytogenetic study of spontaneous abortions in Hawaii. *Annals of Human Genetics, 41*, 443–454.
Hertig, A. T. (1967). The overall problem in man. In K. Benirschke (Ed.), *Comparative aspects of reproductive failure* (pp. 11–41). Springer.
Hertig, A. T., Rock, J., & Adams, E. C. (1956). A description of 34 human ova within the first 17 days of development. *The American Journal of Anatomy, 98*, 435–493.
Hertig, A. T., Rock, J., Adams, E. C., & Menkin, M. C. (1959). Thirty-four fertilized human ova, good, bad and indifferent, recovered from 210 women of known fertility; a study of biologic wastage in early human pregnancy. *Pediatrics, 23*, 202–211.
Hook, E. B. (1981). Prevalence of chromosome abnormalities during human gestation and implications for studies of environmental mutagens. *Lancet, 2*, 169–172.
Houtkooper, R. H., Cantó, C., Wanders, R. J., & Auwerx, J. (2010). The secret life of NAD+: An old metabolite controlling new metabolic signaling pathways. *Endocrinol Rev, 31*, 194–223.
Kajii, T., Ferrier, A., Niikawa, N., Takahara, H., Ohama, K., & Avirachan, S. (1980). Anatomic and chromosomal anomalies in 639 spontaneous abortuses. *Human Genetics, 55*, 87–98.
Moloney, D. M., Slaney, S. F., Oldridge, M., Wall, S. A., Sahlin, P., Stenman, G., & Wilkie, A. O. (1996). Exclusive paternal origin of new mutations in Apert syndrome. *Nature Genetics, 13*, 48–53.
Muller, H. J. (1950). Our load of mutations. *American Journal of Human Genetics, 2*, 111–176.
Miller, J. F., Williamson, E., Glue, J., Gordon, Y. B., Grudzinskas, J. G., & Sykes, A. (1980). Fetal loss after implantation. *Lancet, 316*, 554–556.
Nishimura, H. (1975). Prenatal versus postnatal malformations based on the Japanese experience on induced abortions in the human being. In R. J. Blandau (Ed.), *Aging gametes* (pp. 349–368). S Karger, Basel.
Nishimura, H., Takano, K., Tanimura, T., & Yasuda, M. (1968). Normal and abnormal development of human embryos: First report of the analysis of 1,213 intact embryos. *Teratology, 1*, 281–290.
Olshan, A. F., Schnitzer, P. G., & Baird, P. A. (1994). Paternal age and the risk of congenital heart defects. *Teratology, 50*, 80–84.
Penrose, L. S. (1955). Paternal age and mutation. *Lancet, 269*, 312–313.

Plachot, M., Junca, A. M., Mandelbaum, J., de Grouchy, J., Salat-Baroux, J., & Cohen, J. (1986). Chromosome investigations in early life. I. Human oocytes recovered in an IVF programme. *Human Reproduction, 1*, 547–551.
Plachot, M., Junca, A. M., Mandelbaum, J., de Grouchy, J., Salat-Baroux, J., & Cohen, J. (1987). Chromosome investigations in early life. II. Human preimplantation embryos. *Human Reproduction, 2*, 29–35.
Poland, B. J., Miller, J. R., Harris, M., & Livingston, J. (1981). Spontaneous abortion. A study of 1,961 women and their conceptuses. *Acta Obstetricia et Gynecologica Scandinavica,* suppl 102, 1–32
Risch, N., Reich, E. W., Wishnick, M. M., & McCarthy, J. G. (1987). Spontaneous mutation and parental age in humans. *American Journal of Human Genetics, 41*, 218–248.
Rudak, E., Jacobs, P. A., & Yanagimachi, R. (1978). Direct analysis of the chromosome constitution of human spermatozoa. *Nature, 274*, 911–913.
Rudak, E., Dor, J., Mashiach, S., Nebel, L., & Goldman, B. (1985). Chromosome analysis of human oocytes and embryos fertilized in vitro. *Annals New York Academy of Sciences, 442*, 476–486.
Schardein, J. L. (2000). *Chemically induced birth defects* (3rd ed.). Marcel Dekker Inc.
Schuffenecker, I., Ginet, N., Goldgar, D., Eng, C., Chambe, B., Boneu, A., Houdent, C., Pallo, D., Schlumberger, M., Thivolet, C., & Lenoir, G. M. (1997). Prevalence and parental origin of de novo RET mutations in multiple endocrine neoplasia type 2A and familial medullary thyroid carcinoma. *American Journal of Human Genetics, 60*, 233–237.
Shepard, T. H., & Lemire, R. J. (2010). *Catalog of teratogenic agents* (13th ed.). Johns Hopkins Univ Press.
Shi, H., Enriquez, A., Rapadas, M., Martin, E. M. M. A., Wang, R., Moreau, J., Lim, C. K., Szot, J. O., Ip, E., Hughes, J. N., Sugimoto, K., Humphreys, D. T., McInerney-Leo, A. M., Leo, P. J., Maghzal, G. J., Halliday, J., Smith, J., Colley, A., Mark, P. R., … Dunwoodie, S. L. (2017). NAD deficiency, congenital malformations, and niacin supplementation. *New England Journal of Medicine, 377*, 544–552.
Shiota, K. (2018). Study of normal and abnormal prenatal development using the Kyoto collection of human embryos. *Anatomical Record, 301*, 955–959.
Shiota, K. (2021). A life-table analysis of the intrauterine fate of malformed human embryos and fetuses. *Birth Defects Research, 113*, 623–632.
Shiota, K., Uwabe, C., & Nishimura, H. (1987). High prevalence of defective human embryos at the early postimplantation period. *Teratology, 35*, 309–316.
Teller, A. L., Cormier-Daire, V., Abadie, V., Amiel, J., Sigaudy, S., Bonnet, D., de Lonlay-Debeney, P., Morrisseau-Durand, M. P., Hubert, P., Michel, J. L., Jan, D., Dollfus, H., Baumann, C., Labrune, P., Lacombe, D., Philip, N., LeMerrer, M., Briard, M. L., Munnich, A., & Lyonnet, S. (1998). CHARGE syndromes: Report of 47 cases and review. *American Journal of Medical Genetics, 76*, 402–409.
Vogel, F., & Motulsky, A. G. (1997). *Human genetics. Problems and approaches.* Springer.
Warburton, D., & Fraser, F. C. (1964). Spontaneous abortion risks in man: Data from reproductive histories collected in a medical genetics unit. *American Journal of Human Genetics, 16*, 1–25.
Weinberg, W. (1912). Zur Vererbung des Zwergwuchses. *Arch Rassen-u, Gesel Biology, 9*, 710–718.
Whittaker, P. G., Taylor, A., & Lind, T. (1983). Unsuspected pregnancy loss in healthy women. *Lancet, 1*, 1126–1127.
Wilcox, A. J., Weinberg, C. R., O'Connor, J. F., Baird, D. D., Schlatterer, J. P., Canfield, R. E., Armstrong, E. G., & Nisula, B. C. (1988). Incidence of early loss of pregnancy. *New England Journal of Medicine, 319*, 189–194.
Wilkin, D. J., Szabo, J. K., Cameron, R., Henderson, S., Bellus, G. A., Mack, M. L., Kaitila, I., Loughlin, J., Munnich, A., Sykes, B., Bonaventure, J., & Francomano, C. A. (1998). Mutations in fibroblast growth-factor receptor 3 in sporadic cases of achondroplasia occur exclusively on the paternally derived chromosome. *American Journal of Human Genetics, 63*, 711–716.
Yang, Y., & Sauve, A. A. (2016). NAD(+) metabolism: Bioenergetics, signaling and manipulation for therapy. *Biochimica Et Biophysica Acta, 1864*, 1787–1800.

Zinaman, M. J., Clegg, E. D., Brown, C. C., O'Connor, J., & Selevan, S. G. (1996). Estimates of human fertility and pregnancy loss. *Fertility and Sterility, 65*, 503–509.

Kohei Shiota M.D., Ph.D., is Professor Emeritus of Kyoto University and Shiga University of Medical Science. He is a former Executive Vice President, Kyoto University, and former President of Shiga University of Medical Science. Shiota was also a Visiting Professor at Free University of Berlin. He is well known for his contributions to the studies of normal and abnormal development, genetic epidemiology, and reproductive and developmental toxicology.

Chapter 3
Short Notes on Theories of Species Diversity

Atsushi Yamauchi, Kei Tokita, Toshiyuki Namba, and Tae-Soo Chon

Abstract In the 1970s, a theoretical ecologist deduced that a biological community tends to be unstable with increasing numbers of species and interactions. Because many species coexist and have complex interactions in natural communities, ecologists have investigated the characteristics and underlying mechanisms of species diversity. In this chapter, we review theoretical studies that investigate mechanisms of coexistence in complex biological networks. A variety of hypotheses have been proposed to explain the mechanism of species diversity, each of which is focused on specific processes of species interactions and/or community dynamics. We categorize these theoretical models, describing types of focal mechanisms as either "fitness dependence" or "fitness independence." Despite the diverse hypotheses, no key factor of species diversity has been clarified, mainly due to difficulties in testing the hypotheses. We conclude that more interdisciplinary and integrative studies are needed to understand species diversity in depth.

A. Yamauchi (✉)
Center for Ecological Research, Kyoto University, Hirano 2-509-3, Otsu 520-2113, Shiga, Japan
e-mail: a-yama@ecology.kyoto-u.ac.jp

K. Tokita
Department of Complex Systems Science, Graduate School of Informatics, Nagoya University, Furo-cho, Chikusa-ku 464-8601, Nagoya, Japan
e-mail: tokita@i.nagoya-u.ac.jp

T. Namba
Division of Biological Science, Graduate School of Science, Osaka Prefecture University, 1-1 Gakuen-cho, Naka-ku, Sakai 599-8531, Osaka, Japan
e-mail: tnamba@b.s.osakafu-u.ac.jp

T.-S. Chon
Ecology and Future Research Association (EnFRA), 21 Dusil-ro 45 Beon-gil, Geumjeong-gu, Busan 46228, Republic of Korea

Pusan National University, 2 Busandaehag-ro 63 Beon-gil, Geumjeong-gu, Busan 46241, Republic of Korea
e-mail: tschon@pusan.ac.kr

K. Nishimura et al. (eds.), *Creative Complex Systems*, Creative Economy,
https://doi.org/10.1007/978-981-16-4457-3_3

In this world, many species stably coexist within complex interaction networks. Previously, ecologists believed that the complexity of ecological networks contributed to ecosystem stability. For example, Elton (1958) hypothesized that more diverse communities should be less susceptible to invasion by exotic species (diversity–invasibility hypothesis). In the 1970s, however, May (1972) suggested that complexity destabilizes randomly constructed linear systems and, since then, ecologists have explored mechanisms that result in the stable coexistence of a large number of species. In this chapter, we categorize these theoretical models with summarizing hypotheses of mechanisms of species diversity and ecosystem stability. The models are categorized as either fitness-dependent or fitness-independent based on how they deal with demographic processes, namely, with or without species interactions, respectively.

3.1 Fitness-Dependent Theories (with Species Interactions)

Organisms within an ecological community interact with each other, resulting in positive or negative effects from one species to another species. The interactions can be classified into several categories based on a combination of reciprocal effects among interacting species (Odum, 1953); for example, competition (negative–negative), predation or consumption (negative–positive), and reciprocal mutualism (positive–positive). The interactions influence the reproductive and survival processes of each individual, eventually determining an individual's regeneration ability, which is quantified as an expected offspring number and is referred to as "individual fitness" in ecology. Because individual fitness governs the population growth of a species at a higher scale, it may eventually be responsible for the stability of the ecological community. Therefore, several theoretical studies have focused on the importance of species interactions that affect individual fitness and community stability; this is referred to as fitness-dependent theory in this chapter. In the following, we describe several categories of fitness-dependent theory.

3.1.1 Random Network

One category of fitness-dependent theory is focused on random networks, in which the sign and strength of interactions between species are determined randomly. This assumption was initially adopted by May (1972), in a paper that was a pioneering theoretical study about the relationship between the stability and complexity of an ecological community. He focused on a Jacobian matrix of community dynamics in an equilibrium state. The matrix was called a "community matrix" (Levins, 1968; Vandermeer, 1970) or "interaction matrix" (May, 1972). It should be noted that, in modern usage, an interaction matrix is the coefficient matrix (Vandermeer, 1970) of a Lotka–Volterra system, while a community matrix is Jacobian. The matrix is characterized by three properties, namely, the number of species, interaction strength, and connectance (the probability of an interaction existing between any two species). Importantly, in this model, reciprocal interactions between species (i.e., effects from species A to B and from species B to A in turn) are determined independently of each other. Therefore, the system can include various types of relationships between a pair of species, namely, competition, predation, and mutualism, in a flexible manner. Based on the distribution of the eigenvalues of the matrix, May showed that the system tends to become unstable with increasing species numbers, interaction strength, and connectance. This result may be inconsistent with ecological observations that many species are likely to coexist stably in real ecosystems. After May (1972), theoretical ecologists began investigating factors that contribute to the stability of complex systems [see Tokita and Yasutomi (2003), Namba (2015), and Landi et al. (2018) for reviews]. Those studies tended to focus on effects of various interaction types on system stability.

In the past three decades, researchers have performed global analyses of nonlinear evolutionary dynamics, such as the generalized Lotka–Volterra equations and replicator dynamics (Hofbauer & Sigmund, 1988), with many kinds of species and random interspecies interactions based on statistical mechanics (Galla, 2006; Obuchi et al., 2016a, 2016b; Opper & Diederich, 1992; Tokita, 2004; Yoshino et al., 2007, 2008). They analytically elucidated the diversity (number of coexisting species), level of extinction, and species abundance distributions depending on ecological parameters, such as the level of symmetry of interspecies interactions, ecosystem productivity, resource distribution, degree of interactions, higher-order interactions, and so on.

3.1.2 Food Web

Each animal survives and reproduces by consuming other organisms, which is referred to as predator–prey interaction or consumer–resource interaction. These interactions form a network of trophic flows, a so-called food web, which is an essential structure in an ecological community. Notably, various food web models have different network structures. The simplest model is a random food web, where predator and prey species (or consumer and resource species) are chosen randomly in the community (Erdős & Rényi, 1960). Meanwhile, the cascade model assumes that each species is labeled by a uniformly random value, in which a certain species consumes only species with values less than its own for a certain probability (Cohen & Newman, 1985; Cohen et al., 1990). In the niche model, each species is similarly labeled randomly with a niche value, but a species consumes all prey species within one range of values whose randomly chosen center is less than the consumer's niche value (Williams & Martinez, 2000). These constructed food webs represent characteristic network metrics, which are compared to those of empirically observed communities. Williams and Martinez (2000) emphasized that the niche model fit the data more precisely than the cascade model. Furthermore, the bipartite model was also proposed to describe networks between two layers, assuming an absence of interactions within the same trophic level (Mougi & Kondoh, 2012; Thébault & Fontaine, 2010).

Researchers tend to analyze food web stability by choosing models with structural properties of the random food web, cascade model, niche model, or bipartite model. As described below, however, those structural properties may not significantly influence the functional properties of food web stability.

3.1.2.1 Static Food Web

In the random network model of May (1972), species interactions were considered to be temporally constant, or static. Food web dynamics can be investigated by assuming static interactions, which can be regarded as the static food web. Researchers studied, theoretically, the effects of predator–prey or consumer–resource interactions on community stability by extending the concept of the random network (i.e., introducing a correlation between reciprocal interactions in a pair of species). A positive correlation implies dominance of competition (negative–negative interaction) and/or mutualism (positive–positive interaction) in the community, whereas a negative correlation represents prevalence of predation or consumption (negative–positive interaction).

Galla (2005) and Yoshino et al. (2007) analyzed the stability of systems, with various degrees of symmetry in effect, between interacting species on a spectrum from symmetric (i.e., mutualistic or competitive) to antisymmetric (i.e., prey–predator) relationships. Yoshino et al. (2007) developed a replicator dynamics model with resource competition that formulated frequency dynamics of species in a community by normalizing the dynamics of densities. Their model simultaneously included two types of interspecific interactions, namely, direct interaction and resource competition, in which direct interactions represent various levels of symmetry. They showed that antisymmetric (prey–predator) relationships comparatively promote the diversity and stability of the system when compared to asymmetric and symmetric relationships. Similarly, Allesina and Tang (2012) investigated the effects of correlation between reciprocal interactions, deriving stability conditions with a comparable form to May (1972). They showed that a system tends to be stabilized and destabilized by small and large correlations, respectively. Trophic interactions promote stability due to negative correlations, although cascade and niche models with realistic food web structures are less stable than the corresponding unstructured predator–prey case. To confirm the effects of correlation on stability in real communities, Tang et al. (2014) evaluated the contributions of several factors to the stability of observed food webs by randomizing the network with respect to focal factors. Accordingly, they showed that the effect of topological structure on stability is small compared to that of correlation. Furthermore, Allesina et al. (2015) showed that stability was promoted by substructures of food web models; that is, "broad degree distributions" in the cascade model (the distribution of the number of partners each species interacts with) and "intervality" in the niche model (the tendency that predators prey on consecutive species within an interval in a hierarchy). They also showed that the average interaction strength has little influence on stability per se, compared to the effects of variance and correlation.

Those studies clarified that trophic interaction could contribute to community stability. Despite that, however, these studies still could not specifically reverse May's conclusion that system stability declines with an increasing number of species.

3.1.2.2 Adaptive Food Web

The above models consider static trophic interactions, although predators (or consumers) might modify their predation rates on prey species (or consumption rates on resource species), depending on availability, via behavioral responses or evolutionary adaptations. Kondoh (2003) investigated, theoretically, the effects of adaptive food choice on food web stability based on computer simulations of the random and cascade models. In this study, stability was measured as community persistence, the probability that all species persist for a given time. The model indicated that in adaptive food webs, the stability tends to increase with increasing species number and connectance, which is opposite to the prediction of May (1972). The positive relationship between species number and community persistence is regarded as a unique characteristic of the adaptive food web model. Poderoso and Fontanari (2007) considered replicator dynamics with random mutualistic and competitive interactions, which correspond to the Lotka–Volterra equations with random interactions including mutualistic, competitive, and prey–predator relationships, and analytically showed that temporal changes of species interactions promoted community species diversity and stability.

3.1.3 Mutualism

Biological communities include various types of mutualistic interactions between species, for example, plant–pollinator and plant–disperser relationships, etc. It should be noted that the mutualistic network is characterized by bipartite structure and nestedness (Bascompte et al., 2003). The bipartite network includes two types of nodes, in which interactions occur exclusively between different types. The nestedness is the topology of an ordered bipartite network, where the interactions of the specialists form subsets of those of the generalists. These network properties are considered to influence system stability.

Okuyama and Holland (2008) analyzed, theoretically, properties of mutualistic communities with empirically informed parameters, and found that mutualistic communities with nested structures show a largely positive relationship between stability and number of species, in opposition to predictions of the random network and static food web. Thébault and Fontaine (2010) compared, theoretically, stability between trophic and mutualistic networks, and found that the network architecture favoring stability characteristically differs between those systems. Trophic networks tend to be stable under compartmented and weakly connected architectures, whereas mutualistic networks become stable under highly connected and nested architectures. However, those results are inconsistent with the analytical result of Allesina and Tang (2012) showing that the number of species negatively affects stability even in nested mutualistic communities. Okuyama and Holland (2008) and Thébault and Fontaine (2010) evaluated the magnitude of stability by persistence (the proportion of persisting species once equilibrium is reached) and resilience (the speed at which the community returns to stable equilibrium after a perturbation). On the other hand, Allesina and Tang (2012) evaluated stability by eigenvalues of the community matrix. The difference in the criteria of stability may have caused the inconsistency among the analyses.

3.1.4 Competition

In some systems, interspecific competition might be more significant than trophic interactions. For example, in communities in which plant species dominate, competition for resources among plants could be an important factor. In the 1960s and early 1970s, competitive interactions were considered the preeminent process in determining community structure (Cody & Diamond, 1975; MacArthur, 1972). However, in the late 1970s and 1980s, there emerged questions about the lack of tests to falsify null hypotheses assuming that the observed patterns could be generated by random assemblage without invoking competition between species, as well as about the validity of assumptions in the equilibrium theory of competition (Connor & Simberloff, 1979; Strong et al., 1984). Although the prevalence of interspecific competition was confirmed by a few critical reviews (Connell, 1983; Schoener, 1983), its role in biodiversity may have been overlooked since then and even strongly rejected (Rohde, 2005), because it usually decreases species diversity via a negative effect between species. Indeed, Yoshino et al. (2007) and Allesina and Tang (2012) indicated that the community is destabilized by a positive correlation between reciprocal interactions involving competition with negative–negative relationships. However, some theories propose that multiple species can stably coexist under interspecific competition with some conditions, from either static or fluctuating aspects, as listed below.

3.1.4.1 Static Competitive Interaction

O'Sullivan et al. (2019) studied species diversity, theoretically, in spatially structured competitive communities by focusing on the Lotka–Volterra competitive equations. They proposed a model combining Lotka–Volterra equations, metacommunity structure, and continuous emergence of new species. They revealed that the model can represent high species diversity with characteristic patterns in species abundance distributions (SADs, see below), in agreement with empirical observations.

This analysis may suggest that metacommunity-scale biodiversity regulation can be attained even in competitive communities, although it may depend significantly on the immigration of new species from outside the communities. In each iteration of their simulation, the authors added a new species (with a low biomass) that possessed a positive growth rate in at least one patch, which may be a strong basis for increasing further the number of species. Their model reproduces well-documented patterns in the SADs, although effects of immigration should be considered carefully.

It has been indicated that extinction and colonization processes with a certain type of interspecific competition can result in the coexistence of multiple species even without immigration of new species (Hastings, 1980; Kinzig et al., 1999; Levins & Culver, 1971; Tilman, 1994). The colonization model considered that offspring compete for empty sites to establish their colonies, where empty sites are regenerated by disturbances of the established colonies. Each species is assigned competitiveness, in which more competitive species can replace colonies of less competitive species. If the competitiveness negatively correlates with the colonization rate of the species, this process can result in a steady equilibrium coexistence of multiple species (Kinzig et al., 1999; Lehman, 2000; Tilman, 1994). Under continuous competitiveness, the equilibrium solution of species frequency becomes a smooth continuous distribution of competitiveness, resulting in stable coexistence of an infinite number of species. On the other hand, under discrete competitiveness, the equilibrium shows alternative appearances of high and low frequencies along the order of competitiveness (Kinzig et al., 1999; Yamauchi et al., 2021). In this case, some species represent very low frequency or become extinct, although the system reaches a stable equilibrium. Accordingly, this system generally ensures coexistence of multiple species.

3.1.4.2 Fluctuating Competitive Interaction

Environmental fluctuation may temporally alter the competitive superiority of a species (Hutchinson, 1961). Even in such a case, it is expected that the species with the largest geometric average of fitness dominates the community, excluding other species, in temporally discrete models. However, Chesson and Warner (1981) theoretically showed that the random temporal fluctuation of competitive ability can promote species diversity under lottery competition even in the presence of a variation in average fitness among species. The lottery model assumes that the fecundity of each species is determined by temporal environmental conditions, which causes fluctuation of specific competitiveness. The reproduced offspring compete for a small number of empty sites generated by the deaths of adults, where one individual is chosen to win that site, and the losers die off. In this system, coexistence is achieved by a "storage effect" that is a buffering mechanism against the effects of bad conditions through variable recruitment and less variable high adult survivorship (Chesson, 1983), or covariance between environment and competition (Chesson, 1994, 2000). In the lottery model, relative densities of species continuously change through varying species recruitment because of environmental fluctuations. In more general models of species competition in variable environments, differences in the nonlinear dependence of the growth rates on population densities or a common competitive factor between species (relative nonlinearity) can also be a mechanism of coexistence (Chesson, 1994, 2000). As stated above, there have been statistical mechanics-based analytical studies on the effects of fluctuation of species interactions on community species diversity and stability. Poderoso and Fontanari (2007) considered such a model in which both species abundance and the interspecies interactions change with time, albeit in widely separated timescales, and showed that temporal changes of species interactions promoted species diversity and stability of the community. Yoshino et al. (2007) analytically clarified that resource fluctuation reduces both the diversity and stability of a community.

The formulation of the lottery model is rather similar to that of the colonization model, although there are three critical differences. The colonization model assumes that superior species can replace inferior species, that encounters between offspring and sites are mass–action processes, and that some empty sites always remain after encounters. It is a temporally continuous model and the colonization and extinction rates are assumed constant. On the other hand, the lottery model assumes that new recruits can only occupy empty sites that become available by adult death, and that recruits of each species can fill empty sites in proportion to their frequencies. This model includes a temporal environmental variation, in which reproductive abilities of species change continuously. These differences result in different modes of species composition in coexistence, namely, stable equilibrium under a constant environment in the colonization model and continuous fluctuation under varying environment in the lottery model.

In temporally continuous Lotka–Volterra competitive equations with seasonally fluctuating coefficients, seasonal variation promotes or demotes coexistence depending on seasonal differences between reproduction and intense competition (Namba, 1984). If the per capita competition coefficient of one species against another is large in a season after reproduction when the species is abundant, it imposes strong competitive pressure on others. For the two species to coexist, both species should reduce per capita competitive pressures from each other in seasons when the enemy species is abundant. In this model, alternative stable periodic solutions can appear in any combination of stability of two-boundary single-species solutions, and two species may coexist even if one or both species cannot increase when rare (Namba & Takahashi, 1993). Therefore, the mutual invasibility condition when rare is not a necessary condition for coexistence.

3.1.5 Multiple Types of Interactions

The stability of a community consisting of multiple types of interactions was also analyzed (Fontaine et al., 2011). Ringel et al. (1996) studied, theoretically, population dynamics in a relatively small community (four species) with a mixture of antagonistic and mutualistic interactions. They showed that the mutualisms tend to destabilize equilibrium when feasibility (all population densities are positive) is ignored, although mutualisms can be strongly stabilizing when looking at the stability of only feasible equilibria. Mougi and Kondoh (2012) studied a mixture of prey–predator and mutualistic interactions in large communities by using computer simulations of the cascade and bipartite models, which showed that the moderate mixture of two interactions is likely to stabilize the community dynamics. This result appears inconsistent with that of Allesina and Tang (2012), which showed that mutualism is expected to decrease stability because it is accompanied by positive correlation. Mougi and Kondoh (2012) suggested that the difference between the two models was caused by their assumption of the negative relationship between the number of interactions and the interaction strength. Mougi and Kondoh (2014) extended their previous analysis by including competition in addition to antagonistic and mutualistic interactions. They also showed that interaction-type diversity generally enhances the stability of complex communities.

On the other hand, Melián et al. (2009) focused on plants interacting with both herbivores and pollinators or seed dispersers in a complex community. They investigated the importance of combinations of interaction types in each species for species diversity by analyzing an empirically observed ecological network. Both the ratio of mutualistic to antagonistic interactions per plant and the number of basic modules with an antagonistic and a mutualistic interaction were very heterogeneous across plant species, with a few plant species showing very high values for these parameters. They theorized that the observed correlation between strong interaction strengths and high mutualistic-to-antagonistic ratios in a few plant species significantly increases community diversity. However, it should be noted that their model considered dynamics of plants only, assuming that densities of interacting arthropods, namely, pollinators and herbivores, were relatively constant and could be treated as fixed parameters. Consequently, effects of interaction types on community stability may have been revealed only in part.

There is a series of studies on statistical mechanics of random replicator dynamics, which is mathematically equivalent to Lotka–Volterra equations of many species, with the mixture of competitive, prey–predator, and mutualistic species interactions. They analytically obtained diversity and stability of the replicator dynamics with random interspecies interactions. Some of them considered fully connected random symmetric interactions (de Oliveira & Fontanari, 2000, 2001, 2002; Diederich & Opper, 1989; Tokita, 2004), random asymmetric interactions (Galla, 2005, 2006; Opper & Diederich, 1992; Yoshino et al., 2007, 2008), and sparse random symmetric interactions (Obuchi et al., 2016a, 2016b), which predicts multiple peaks of species abundance distribution.

3.1.6 Multiplex Ecological Networks

Recently, multilayer networks or multislice networks with nodes and edges in separate layers and interlayer connections have received considerable attention and have been studied intensively in physics and network science (Boccaletti et al., 2014;

Kivelä et al., 2014; Mucha et al., 2010). In particular, multilayer networks with different types of edges or interlayer connections are called multiplex networks. Multiplex ecological networks are multilayer networks linked via different types of interactions (Kéfi et al., 2018), but sometimes flattened into a single layer so that interlayer connections disappear (García-Callejas et al., 2018).

Kéfi et al. (2012) incorporated nontrophic interactions, such as habitat modification, predator interference, and facilitation, into a traditional food web linked by trophic interactions and then studied the system as a multiplex network including both positive and negative nontrophic interactions. In a network on the central Chilean coast (Kéfi et al., 2015, 2016), nontrophic interactions were more than twice as abundant as trophic interactions; the structure of those nontrophic links was nonrandom, and competitions for space and habitat/refuge provisioning by sessile and/or basal species were the most abundant nontrophic interactions (Kéfi et al., 2015). Although this was a very complex network including 106 species, 1362 trophic links, 3089 negative nontrophic links, and 172 positive nontrophic links, they could identify only 14 multiplex clusters using network statistical modeling. Those clusters were groups of species differing from each other in at least one of the types of links involved, the pattern of incoming and outgoing links, and the identity of the species they interact with (Kéfi et al., 2016). To simulate dynamics of the network, they used the bioenergetic consumer–resource model with multi-prey Holling type functional responses (Yodzis & Innes, 1992). Nontrophic interactions were incorporated by modifying the growth terms, functional responses, and metabolic rates. Simulation results of this model suggested that the way nontrophic interactions were mapped onto the trophic interactions tended to increase species persistence and the total biomass realized (Kéfi et al., 2016). By analyzing geographic variability in these multiplex networks, it was revealed that some of the changes observed in the network structure across the environmental gradient were mediated by species richness and that different types of interactions responded differently to environmental factors such as sea surface temperature and coastal upwelling (Lurgi et al., 2020).

Layers in ecological multilayer networks can be different habitat patches or time points such as different seasons. Then, interlayer edges represent movement between layers or changes in abundance (Hutchinson et al., 2019; Pilosof et al., 2017). Historically, effects of movement between habitat patches were studied in the framework of metacommunity theory (Holyoak et al., 2005). In a metacommunity model of food webs studied by Mougi and Kondoh (2016), metacommunity complexity, measured by the number of local food webs and their connectedness, elicited a self-regulating, negative-feedback mechanism and thus stabilized food web dynamics. Moreover, the presence of metacommunity complexity could give rise to a positive food web complexity–stability relationship measured by eigenvalues of the Jacobian matrices. Gravel et al. (2016) expanded May's original formulation of local stability (May, 1972) to include an additional matrix of dispersal rates, and analytically and numerically found that dispersal can increase the stability of meta-ecosystems. These approaches can be transferable to some types of multilayer networks if their dispersal matrices are generalized to represent any interlayer edges (Hutchinson et al., 2019).

A multiplex network where one layer represents predator–prey interactions and another represents plant–herbivore interactions can have interlayer edges that describe how the landscape of fear alters prey/herbivore habitat use (Hutchinson et al., 2019). A community of plants interacting with both herbivores and mutualists studied by Melián et al. (2009) can be expressed as a plant–herbivore network and a plant–mutualistic network interconnected via a common set of plant species. Interlayer edges connect each plant species to its counterpart in the other layer and represent the extent to which mutualism affects herbivory and vice versa (Pilosof et al., 2017).

Although a few reviews (García-Callejas et al., 2018; Hutchinson et al., 2019; Pilosof et al., 2017) have already been published, studies of biological communities as multiplex ecological networks remain recent and scarce (Kéfi et al., 2018). Further accumulation of multiplex ecological network data would help investigation of whether the structural regularities observed in the datasets currently available are general and whether they can be extrapolated to different ecosystem types (Kéfi et al., 2018). Cooperation with researchers in other fields studying multilayer networks, for example physicists or network scientists, would undoubtedly promote our understanding of multiplex ecological networks.

3.2 Fitness-Independent Theories (Without Species Interactions)

As described above, there are many theories that take effects of fitness into account. By contrast, theories without such effects have also been proposed, which imply that species are identical in fitness (as a result of survival and reproduction). This type of theory can be referred to as fitness-independent theory. In these theories, species can coexist because there are no fitness differences among species, although species abundance represents characteristic properties, depending on the model assumptions. The fitness-independent theories provide null hypotheses in the patterns of species abundance for a statistical test of fitness effects in comparison to empirical data.

3.2.1 Niche Apportionment Models

Abundance of species may represent the size of the niche (a range of resource utilization) of each species; therefore, some theories focus on niche partition patterns among species, and are called niche apportionment models (Tokeshi, 1990). The pattern of niche partitioning may be simultaneously influenced by both interspecific competition and mechanical processes of community assemblage (e.g., order of species introductions into a community) in reality, although models are unlikely to consider dynamics of competition and effects of fitness explicitly. Therefore, we categorize these models into fitness-independent theory.

Motomura (1932) found that relative abundances of species tended to fit a geometric series, which may correspond to a certain partitioning process. Namely, a first species preempts a fraction k of the total niche, a second species k of the remainder, a third again a fraction k of what remains after the first and the second species have carved out their shares, and so on. On the other hand, a broken-stick model was proposed by MacArthur (1957) in which the niche partitioning process is compared to shares of pieces of a stick that is broken into fragments either sequentially or simultaneously based on certain rules. The size of a stick piece is considered an amount of resources for a species, which determines the abundance of the species. This process imitates a random assignment of a range of niches to each species. The model results in a characteristic distribution of species abundance, depending on the fragmentation process of stick pieces. The broken-stick model has been used in many ways to investigate temporal and geographic trends in species abundance (Sugihara, 1980).

Tokeshi (1990) also developed several variants of niche apportionment models by extending the above models. According to comparisons between patterns of those models and empirical data of a chironomid community, he found that the data fitted well to a random assortment model where abundances of different species are not mutually related at all. For further information on niche apportionment models, see May (1975), Tokeshi (1993, 1996, 1999), Magurran (2004), and McGill et al. (2007).

3.2.2 Neutral Theory

Hubbell (2001) proposed a neutral theory of biodiversity and biogeography, which is named in an analogy to a theory in population genetics, that is, the neutral theory of molecular evolution (Kimura, 1983). This theory assumed that all species have identical fitness, by which they are neutral in force of natural selection. In this community, a species composition dynamically changes through stochastic extinctions of present species and migrations from an external species pool. The process of stochastic changes of species abundance is named ecological drift. This model also represents characteristic SADs, patterns of which fit well to those of empirical data of natural communities (Hubbell, 2001; Volkov et al., 2003).

In this century, studies of nonequilibrium statistical mechanics contributed to an exact solution of Hubbell's neutral model (Alonso & McKane, 2004; Etienne & Alonso, 2005; McKane et al., 2000, 2004; Vallade & Houchmandzadeh, 2003; Volkov et al., 2003). It should be noted that Etienne (2005) and Etienne and Olff (2005) applied the coalescent theory of population genetics to the neutral model and obtained a sampling formula, that is, an exact solution of zero-sum multinomial distribution that can be used for maximum likelihood estimation using real data.

3.3 Tests of Theories

The theories of species diversity have been tested by comparing derived patterns with observed data, although there are various constraints in such treatments. Here, we describe both advantages and limitations in three types of approaches to testing mechanical theories of species diversity.

3.3.1 Evaluation of Metrics of Community Structure

In some cases, theories of ecological community structure can be compared to empirical data directly by focusing on metrics of a modeled community. Williams and Martinez (2000) constructed three types of theoretical food webs, namely, random, cascade, and niche models, by using parameters that are measured in empirical food webs. They examined the predictability of those models by focusing on 12 metrics of communities, for example, fractions of species in various trophic levels, fractions of carnivores and omnivores, food chain lengths, etc. According to the comparison, they concluded that the niche model most represented the properties of natural communities.

The metrics test may be a powerful approach for comparing theories, although is substantially restrictive. As discussed above, there are various types of interspecific interactions in an ecological community. Among those, trophic interactions are relatively easy to detect and quantify because the consequences are clear, to some degree. Therefore, researchers can determine the metrics in a food web when they spend significant effort on empirical research. On the other hand, effects of competition and mutualism are difficult to measure relative to trophic interaction. Consequently, an application of this type of test could be limited.

3.3.2 Goodness-of-Fit Test of SADs

In tests of theories of species diversity, community SADs (species abundance distributions) have been intensively studied and can be plotted in various styles. The simplest SAD plot is a histogram of species number within a given octave of abundance (i.e., density or individual number) or an octave of log-scale abundance. Another type of SAD plot uses logarithms of species abundance and orders them from the most abundant to the rarest species in a rank abundance diagram (RAD), which necessarily becomes a decreasing function. It is known that when SADs are illustrated from empirical community data, they are likely to represent characteristic shapes, namely, a decreasing hollow curve with a long tail in the standard histogram, a lognormal-like distribution in the log-scale histogram, and an accelerated reduction at low rank species in a RAD (Hubbell, 2001; Magurran, 2004; McGill et al., 2007).

Researchers tend to discuss the plausibility of theories based on comparison of a graphical shape without additional predictions, which is referred to as a goodness-of-fit test. This results in difficulty in clarifying which mechanisms are key determinants in a species composition, and whether fitness is a significant factor. McGill et al. (2007) emphasized the need for distinct predictions that can be tested. One possible direction to help overcome this problem is an integrative comparison of model predictions across many different patterns, for example, various SADs, species–area relationships, beta-diversity, etc. (Hubbell, 2001; Levin, 1992; May et al., 2015; McGill et al., 2007). For example, May et al. (2015) tested the neutral theory by comparing model predictions with forest data in several patterns of different aspects of dynamics and biodiversity structure. They indicated that the model was unable to match the species–area relationship and beta-diversity simultaneously, although it correctly predicted each pattern independently and up to five patterns simultaneously.

By using statistical mechanics, Tokita et al. obtained the exact solution of a SAD of random replicators with fully connected symmetric interactions (Tokita, 2004), fully connected asymmetric interactions (Yoshino et al., 2008), and sparse-symmetric interactions (Obuchi et al., 2016a, 2016b). Notably, Obuchi et al. (2016a, 2016b) showed for the first time that sparse interactions give multi-peaked SADs.

3.3.3 Labeled SADs

Problems with goodness-of-fit tests might also be resolved by the extraction of more detailed information from a SAD by referring to species identities, namely, a "labeled" SAD (McGill et al., 2007). In this approach, trends of species abundance are considered by connecting them with species properties, for example, body size (Russo et al., 2003; Wilson, 1991), various life history traits (Murray & Westoby, 2000; Murray et al., 1999; Shipley et al., 2006), ecological niche (resource utilization) (Sugihara et al., 2003), etc. It should be noted that these factors are relevant to resource usage, which potentially includes information about species interactions, typically, competition among species for resources. The analyses of labeled SADs suggested the existence of relationships between species properties and community structures, implying that species interactions could be an important determinant of species diversity.

Recently, Yamauchi et al. (2021) tested a colonization model with species abundance data from the literature by labeling both taxonomy and resource utilization. According to the study, the data significantly support the colonization model with a competition–fecundity trade-off, suggesting that interspecific competition could be a determinant of species diversity and community structure.

3.4 Conclusions

An invader, or a species at low density, can increase if the long-term per capita growth rate is positive, and the long-term low-density growth rate can be used to quantify species coexistence (Chesson, 1994). Chesson (1994, 2000, 2018) divided the per capita growth rate of an invader into two terms: the average fitness difference from the resident and the dependency on niche overlap between the invader and resident. The first term has the opposite sign for the two species but the second term can be positive if the niche overlap is smaller than a threshold value, and the overall growth rates become positive if the niche differences are larger than the relative fitness differences. Chesson identified two mechanisms for coexistence, the stabilizing mechanism, which increases the niche differences, and the equalizing mechanism, which tends to minimize average fitness differences. The latter allows slow declines of inferior species over time. Clearly, in our framework, the former is a fitness-dependent mechanism and the latter is a fitness-independent mechanism resulting in an unstable coexistence such as that in Hubbell's unified theory. This theory, termed modern coexistence theory (Letten et al., 2017; Mayfield & Levine, 2010), states that coexistence depends on the relative magnitude of the stabilizing and fitness difference terms. Stabilizing processes are defined as any mechanism that causes species to limit themselves more than other species and, thus, a sign of stabilizing processes is that species' per capita growth rates decrease as their relative abundance or frequency increases, which is a pattern referred to as negative frequency dependence (Adler et al., 2007).

In empirical studies, it is difficult to estimate (or manipulate) stabilizing niche differences and relative fitness differences independently (HilleRisLambers et al., 2012). Nevertheless, some methods have been proposed to estimate the parameters of a relatively simple model as a function of the densities of target and other component species and decompose them into average fitness and stabilization terms (Adler et al., 2007, HilleRisLambers et al., 2012). Thus, a study found that coexistence occurred when stabilizing effects of niche differences overcame fitness differences in serpentine annual communities in California (Levine and HilleRisLambers, 2009). However, the overall importance of stabilizing niche differences and relative fitness differences is still confusing. Our understanding of community structure and biodiversity will be strengthened by synthesizing complementary perspectives of niche and neutral theories rather than debating which theory provides better explanations (Adler et al., 2007; Leibold & McPeek, 2006).

Using the structure of population genetics theory as a guide, Vellend (2010, 2016) attempted to synthesize the "Theory of Ecological Communities" based on just four fundamental or "high-level" processes: selection, ecological drift, dispersal, and speciation. In an ecological community, selection results from deterministic fitness differences between individuals of different species. Ecological drift is a random process in community dynamics. Drift in a local community causes species' relative abundances to fluctuate at random and, eventually, one species will dominate the community, with all other species fluctuating to extinction. Because selection and drift often result in extinction of species and constrain species diversity, maintaining diversity in a community requires a source of continuous input of new species. A new species may arise via speciation and join the community. A species can also be added to a regional or local community via dispersal (immigration). In this framework, niche-based theory (Chase & Leibold, 2003) including stabilizing mechanisms involves selection and possibly dispersal, whereas Hubbell's neutral theory, assuming the absence of fitness differences, involves speciation, dispersal, and drift. This process-first approach has articulated the widespread importance of local negative frequency-dependent selection and spatially and temporally variable selection, and a possible influence of drift, and so on. However, the details of low-level processes such as competition, predation, stress, disturbance, productivity, etc., still seem necessary to understand community structure and dynamics in depth.

Despite these efforts to integrate a variety of theories and hypotheses to explain the mechanisms maintaining species diversity, a key factor has not yet been clearly revealed. To understand processes and patterns in ecosystems, including those involving human beings, we should progress toward more interdisciplinary and integrative studies, both theoretically and empirically.

References

Adler, P. B., HilleRisLambers, J., & Levine, J. M. (2007). A niche for neutrality. *Ecology Letters, 10*, 95–104.

Allesina, S., & Tang, S. (2012). Stability criteria for complex ecosystems. *Nature, 483*, 205–208.

Allesina, S., Grilli, J., Barabas, R., Tang, S., Aljadeff, J., & Maritan, A. (2015). Predicting the stability of large structured food webs. *Nature Communications, 6*.

Alonso, D., & McKane, A. J. (2004). Sampling Hubbell's neutral theory of biodiversity. *Ecology Letters, 7*, 901–910.

Bascompte, J., Jordano, P., Melián, C. J., & Olesen, J. M. (2003). The nested assembly of plant-animal mutualistic networks. *Proceedings of the National Academy of Sciences of the United States of America, 100*, 9383–9387.

Boccaletti, S., Bianconi, G., Criado, R., del Genio, C. I., Gomez-Gardenes, J., Romance, M., Sendina-Nadal, I., Wang, Z., & Zanin, M. (2014). The structure and dynamics of multilayer networks. *Physics Reports, 544*, 1–122.

Chase, J., & Leibold, M. (2003). *Ecological niches: Linking classical and contemporary approaches*. University of Chicago Press.

Chesson, P. L. (1983). Coexistence of competitors in a stochastic environment: The storage effect. In H. I. Freedman & C. Strobeck (Eds.), *Population biology, lecture notes in biomathematics* (pp. 188–198). Springer.

Chesson, P. (1994). Multispecies competition in variable environments. *Theoretical Population Biology, 45*, 227–276.

Chesson, P. (2000). Mechanisms of maintenance of species diversity. *Annual Review of Ecology and Systematics, 31*, 343–366.

Chesson, P. (2018). Updates on mechanisms of maintenance of species diversity. *Journal of Ecology, 106*, 1773–1794.

Chesson, P. L., & Warner, R. R. (1981). Environmental variability promotes coexistence in lottery competitive systems. *American Naturalist, 117*, 923–943.

Cody, M. L., & Diamond, J. M. (1975). *Ecology and evolution of communities*. Harvard University Press.

Cohen, J. E., & Newman, C. M. (1985). A stochastic theory of community food webs I. Models and aggregated data. *Proceedings of the Royal Society Series B-Biological Sciences, 224*, 421–448.

Cohen, J. E., Briand F., & Newman, C. M. (1990). *Community food webs: Data and theory* (biomathematics vol. 20). Springer.

Connell, J. H. (1983). On the prevalence and relative importance of interspecific competition—Evidence from field experiments. *American Naturalist, 122*, 661–696.

Connor, E. F., & Simberloff, D. (1979). The assembly of species communities—Chance or competition. *Ecology, 60*, 1132–1140.

de Oliveira, V. M., & Fontanari, J. F. (2000). Random replicators with high-order interactions. *Physical Review Letters, 85*, 4984–4987.

de Oliveira, V. M., & Fontanari, J. F. (2001). Extinctions in the random replicator model. *Physical Review E, 64*.

de Oliveira, V. M., & Fontanari, J. F. (2002). Complementarity and diversity in a soluble model ecosystem. *Physical Review Letters, 89*.

Diederich, S., & Opper, M. (1989). Replicators with random interactions—A solvable model. *Physical Review A, 39*, 4333–43336.

Elton, C. (1958). *The ecology of invasions by animals and plants*. Chapman & Hall.

Erdős, P., & Rényi, A. (1960). On the evolution of random graphs. *Publications of the Mathematical Institute of the Hungarian Academy of Sciences, 5*, 17–61.

Etienne, R. S. (2005). A new sampling formula for neutral biodiversity. *Ecology Letters, 8*, 253–260.

Etienne, R. S., & Alonso, D. (2005). A dispersal-limited sampling theory for species and alleles. *Ecology Letters, 8*, 1147–1156.

Etienne, R. S., & Olff, H. (2005). Confronting different models of community structure to species–abundance data: A Bayesian model comparison. *Ecology Letters, 8*, 493–504.

Fontaine, C., Guimarães, P. R., Kéfi, S., Loeuille, N., Memmott, J., van der Putten, W. H., van Veen, F. J. F., & Thébault, E. (2011). The ecological and evolutionary implications of merging different types of networks. *Ecology Letters, 14*, 1170–1181.

Galla, T. (2005). Dynamics of random replicators with Hebbian interactions. *Journal of Statistical Mechanics-Theory and Experiment*, P11005.

Galla, T. (2006). Random replicators with asymmetric couplings. *Journal of Physics A: Mathematical and General, 39*, 3853–3869.

García-Callejas, D., Molowny-Horas, R., & Araújo, M. B. (2018). Multiple interactions networks: Towards more realistic descriptions of the web of life. *Oikos, 127*, 5–22.

Gravel, D., Massol, F., & Leibold, M. A. (2016). Stability and complexity in model meta-ecosystems. *Nature Communications, 7*.

Hastings, A. (1980). Disturbance, coexistence, history, and competition for space. *Theoretical Population Biology, 18*, 363–373.

Hofbauer, J., & Sigmund, K. (1988). *The theory of evolution and dynamical systems: Mathematical aspects of selection*. Cambridge University Press.

Holyoak, M., Leibold, M. A., & Holt, R. D. (2005). *Metacommunities: Spatial dynamics and ecological communities*. University of Chicago Press.

Hubbell, S. P. (2001). *The unified neutral theory of biodiversity and biogeography*. Princeton University Press.

Hutchinson, G. E. (1961). The paradox of the plankton. *American Naturalist, 95*, 137–145.

Hutchinson, M. C., Mora, B. B., Pilosof, S., Barner, A. K., Kéfi, S., Thébault, E., Jordano, P., & Stouffer, D. B. (2019). Seeing the forest for the trees: Putting multilayer networks to work for community ecology. *Functional Ecology, 33*, 206–217.

Kéfi, S., Berlow, E. L., Wieters, E. A., Navarrete, S. A., Petchey, O. L., Wood, S. A., Boit, A., Joppa, L. N., Lafferty, K. D., Williams, R. J., Martinez, N. D., Menge, B. A., Blanchette, C. A., Iles,

A. C., & Brose, U. (2012). More than a meal … Integrating non-feeding interactions into food webs. *Ecology Letters, 15*, 291–300.
Kéfi, S., Berlow, E. L., Wieters, E. A., Joppa, L. N., Wood, S. A., Brose, U., & Navarrete, S. A. (2015). Network structure beyond food webs: Mapping non-trophic and trophic interactions on Chilean rocky shores. *Ecology, 96*, 291–303.
Kéfi, S., Miele, V., Wieters, E. A., Navarrete, S. A., & Berlow, E. L. (2016). How structured is the entangled bank? The surprisingly simple organization of multiplex ecological networks leads to increased persistence and resilience. *PLoS Biology, 14*.
Kéfi, S., Thébault, E., Eklöf, A., Lurgi, M., Davis, A. J., Kondoh, M., & Krumins, J. A. (2018). Toward multiplex ecological networks: Accounting for multiple interaction types to understand community structure and dynamics. In J. C. Moore, P. C. de Ruiter, K. S. McCann, & V. Wolters (Eds.), *Adaptive food webs: Stability and transitions of real and model ecosystems* (pp. 73–87). Cambridge University Press.
Kimura, M. (1983). *The neutral theory of molecular evolution*. Cambridge University Press.
Kinzig, A. P., Levin, S. A., Dushoff, J., & Pacala, S. (1999). Limiting similarity, species packing, and system stability for hierarchical competition-colonization models. *American Naturalist, 153*, 371–383.
Kivelä, M., Arenas, A., Barthelemy, M., Gleeson, J. P., Moreno, Y., & Porter, M. A. (2014). Multilayer networks. *Journal of Complex Networks, 2*, 203–271.
Kondoh, M. (2003). Foraging adaptation and the relationship between food-web complexity and stability. *Science, 299*, 1388–1391.
HilleRisLambers, J., Adler, P. B., Harpole, W. S., Levine, J. M., & Mayfield, M. M. (2012). Rethinking community assembly through the lens of coexistence theory. *Annual Review of Ecology, Evolution, and Systematics, 43*, 227–248.
Landi, P., Minoarivelo, H. O., Brannstrom, A., Hui, C., & Dieckmann, U. (2018). Complexity and stability of ecological networks: A review of the theory. *Population Ecology, 60*, 319–345.
Lehman, C. L. (2000). *The competition–colonization equations: Ecological and evolutionary implications*. University of Minnesota (Ph.D. Thesis).
Leibold, M. A., & McPeek, M. A. (2006). Coexistence of the niche and neutral perspectives in community ecology. *Ecology, 87*, 1399–1410.
Letten, A. D., Ke, P. J., & Fukami, T. (2017). Linking modern coexistence theory and contemporary niche theory. *Ecological Monographs, 87*, 161–177.
Levin, S. A. (1992). The problem of pattern and scale in ecology. *Ecology, 73*, 1943–1967.
Levine, J. M., & HilleRisLambers, J. (2009). The importance of niches for the maintenance of species diversity. *Nature, 461*, 254-U130.
Levins, R. (1968). *Evolution in changing environments*. Princeton University Press.
Levins, R., & Culver, D. (1971). Regional coexistence of species and competition between rare species (mathematical model/habitable patches). *Proceedings of the National Academy of Sciences of the United States of America, 68*, 1246–1248.
Lurgi, M., Galiana, N., Broitman, B. R., Kéfi, S., Wieters, E. A., & Navarrete, S. A. (2020). Geographical variation of multiplex ecological networks in marine intertidal communities. *Ecology, 101*.
MacArthur, R. H. (1957). On the relative abundance of bird species. *Proceedings of the National Academy of Sciences of the United States of America, 43*, 293–295.
MacArthur, R. H. (1972). *Geographical ecology: Patterns in the distribution of species*. Harper & Row.
Magurran, A. E. (2004). *Measuring biological diversity*. Blackwell Science.
May, R. M. (1972). Will a large complex system be stable. *Nature, 238*, 413–414.
May, R. M. (1975). Patterns of species abundance and diversity. In M. L. Cody & J. M. Diamond (Eds.), *Ecology and evolution of communities* (pp. 81–120). Belknap/Harvard University Press.
May, F., Huth, A., & Wiegand, T. (2015). Moving beyond abundance distributions: Neutral theory and spatial patterns in a tropical forest. *Proceedings of the Royal Society B-Biological Sciences, 282*.

Mayfield, M. M., & Levine, J. M. (2010). Opposing effects of competitive exclusion on the phylogenetic structure of communities. *Ecology Letters, 13*, 1085–1093.

McGill, B. J., Etienne, R. S., Gray, J. S., Alonso, D., Anderson, M. J., Benecha, H. K., Dornelas, M., Enquist, B. J., Green, J. L., He, F. L., Hurlbert, A. H., Magurran, A. E., Marquet, P. A., Maurer, B. A., Ostling, A., Soykan, C. U., Ugland, K. I., & White, E. P. (2007). Species abundance distributions: Moving beyond single prediction theories to integration within an ecological framework. *Ecology Letters, 10*, 995–1015.

McKane, A., Alonso, D., & Sole, R. V. (2000). Mean-field stochastic theory for species-rich assembled communities. *Physical Review E, 62*, 8466–8484.

McKane, A. J., Alonso, D., & Sole, R. V. (2004). Analytic solution of Hubbell's model of local community dynamics. *Theoretical Population Biology, 65*, 67–73.

Melián, C. J., Bascompte, J., Jordano, P., & Křivan, V. (2009). Diversity in a complex ecological network with two interaction types. *Oikos, 118*, 122–130.

Motomura, I. (1932). A statistical treatment of animal associations. *Japanese Journal Zoology, 44*, 379–383.

Mougi, A., & Kondoh, M. (2012). Diversity of interaction types and ecological community stability. *Science, 337*, 349–351.

Mougi, A., & Kondoh, M. (2014). Stability of competition–antagonism–mutualism hybrid community and the role of community network structure. *Journal of Theoretical Biology, 360*, 54–58.

Mougi, A., & Kondoh, M. (2016). Food-web complexity, meta-community complexity and community stability. *Scientific Reports, 6*.

Mucha, P. J., Richardson, T., Macon, K., Porter, M. A., & Onnela, J. P. (2010). Community structure in time-dependent, multiscale, and multiplex networks. *Science, 328*, 876–878.

Murray, B. R., & Westoby, M. (2000). Properties of species in the tail of rank-abundance curves: The potential for increase in abundance. *Evolutionary Ecology Research, 2*, 583–592.

Murray, B. R., Rice, B. L., Keith, D. A., Myerscough, P. J., Howell, J., Floyd, A. G., Mills, K., & Westoby, M. (1999). Species in the tail of rank-abundance curves. *Ecology, 80*, 1806–1816.

Namba, T. (1984). Competitive co-existence in a seasonally fluctuating environment. *Journal of Theoretical Biology, 111*, 369–386.

Namba, T. (2015). Multi-faceted approaches toward unravelling complex ecological networks. *Population Ecology, 57*, 3–19.

Namba, T., & Takahashi, S. (1993). Competitive coexistence in a seasonally fluctuating environment II. Multiple stable states and invasion success. *Theoretical Population Biology, 44*, 374–402.

Obuchi, T., Kabashima, Y., & Tokita, K. (2016a). Multiple peaks of species abundance distributions induced by sparse interactions. *Physical Review E, 94*.

Obuchi, T., Kabashima, Y., & Tokita, K. (2016b). Relative species abundance of replicator dynamics with sparse interactions. *Journal of Statistical Mechanics-Theory and Experiment*.

Odum, E. P. (1953). *Fundamentals of ecology*. Saunders.

Okuyama, T., & Holland, J. N. (2008). Network structural properties mediate the stability of mutualistic communities. *Ecology Letters, 11*, 208–216.

Opper, M., & Diederich, S. (1992). Phase-transition and 1/f noise in a game dynamic-model. *Physical Review Letters, 69*, 1616–1619.

O'Sullivan, J. D., Knell, R. J., & Rossberg, A. G. (2019). Metacommunity-scale biodiversity regulation and the self-organised emergence of macroecological patterns. *Ecology Letters, 22*, 1428–1438.

Pilosof, S., Porter, M. A., Pascual, M., & Kéfi, S. (2017). The multilayer nature of ecological networks. *Nature Ecology & Evolution, 1*.

Poderoso, F. C., & Fontanari, J. F. (2007). Model ecosystem with variable interspecies interactions. *Journal of Physics A: Mathematical and Theoretical, 40*, 8723–8738.

Ringel, M. S., Hu, H. H., Anderson, G., & Ringel, M. S. (1996). The stability and persistence of mutualisms embedded in community interactions. *Theoretical Population Biology, 50*, 281–297.

Rohde, K. (2005). *Nonequilibrium ecology*. Cambridge University Press.

Russo, S. E., Robinson, S. K., & Terborgh, J. (2003). Size-abundance relationships in an Amazonian bird community: Implications for the energetic equivalence rule. *American Naturalist, 161*, 267–283.

Schoener, T. W. (1983). Field experiments on interspecific competition. *American Naturalist, 122*, 240–285.

Shipley, B., Vile, D., & Garnier, E. (2006). From plant traits to plant communities: A statistical mechanistic approach to biodiversity. *Science, 314*, 812–814.

Strong, D. R. J., Simberloff, D., Abele, L. G., & Thistle, A. B. (1984). *Ecological communities: Conceptual issues and the evidence*. Princeton University Press.

Sugihara, G. (1980). Minimal community structure—An explanation of species abundance patterns. *American Naturalist, 116*, 770–787.

Sugihara, G., Bersier, L. F., Southwood, T. R. E., Pimm, S. L., & May, R. M. (2003). Predicted correspondence between species abundances and dendrograms of niche similarities. *Proceedings of the National Academy of Sciences of the United States of America, 100*, 5246–5251.

Tang, S., Pawar, S., & Allesina, S. (2014). Correlation between interaction strengths drives stability in large ecological networks. *Ecology Letters, 17*, 1094–1100.

Thébault, E., & Fontaine, C. (2010). Stability of ecological communities and the architecture of mutualistic and trophic networks. *Science, 329*, 853–856.

Tilman, D. (1994). Competition and biodiversity in spatially structured habitats. *Ecology, 75*, 2–16.

Tokeshi, M. (1990). Niche apportionment or random assortment—Species abundance patterns revisited. *Journal of Animal Ecology, 59*, 1129–1146.

Tokeshi, M. (1993). Species abundance patterns and community structure. *Advances in Ecological Research, 24*, 111–186.

Tokeshi, M. (1996). Power fraction: A new explanation of relative abundance patterns in species-rich assemblages. *Oikos, 75*, 543–550.

Tokeshi, M. (1999). *Species coexistence: Ecological and evolutionary perspectives*. Blackwell Science.

Tokita, K. (2004). Species abundance patterns in complex evolutionary dynamics. *Physical Review Letters, 93*.

Tokita, K., & Yasutomi, A. (2003). Emergence of a complex and stable network in a model ecosystem with extinction and mutation. *Theoretical Population Biology, 63*, 131–146.

Vallade, M., & Houchmandzadeh, B. (2003). Analytical solution of a neutral model of biodiversity. *Physical Review E, 68*.

Vandermeer, J. H. (1970). The community matrix and the number of species in a community. *American Naturalist, 104*, 73–83.

Vellend, M. (2010). Conceptual synthesis in community ecology. *Quarterly Review of Biology, 85*, 183–206.

Vellend, M. (2016). *The theory of ecological communities*. Princeton University Press.

Volkov, I., Banavar, J. R., Hubbell, S. P., & Maritan, A. (2003). Neutral theory and relative species abundance in ecology. *Nature, 424*, 1035–1037.

Williams, R. J., & Martinez, N. D. (2000). Simple rules yield complex food webs. *Nature, 404*, 180–183.

Wilson, J. B. (1991). Methods for fitting dominance diversity curves. *Journal of Vegetation Science, 2*, 35–46.

Yamauchi, A., Ito, K., & Shibasaki, S. (2021). Competitive quasi-exclusion under colonization process explains community structure. *Ecology and Evolution, 11*, 4470–4480.

Yodzis, P., & Innes, S. (1992). Body size and consumer-resource dynamics. *American Naturalist, 139*, 1151–1175.

Yoshino, Y., Galla, T., & Tokita, K. (2007). Statistical mechanics and stability of a model eco-system. *Journal of Statistical Mechanics-Theory and Experiment*.

Yoshino, Y., Galla, T., & Tokita, K. (2008). Rank abundance relations in evolutionary dynamics of random replicators. *Physical Review E, 78*.

Atsushi Yamauchi is a professor in the Center for Ecological Research, Kyoto University, studying mathematical ecology. He obtained a Ph.D. in Science at Kyushu University. He was a research associate at the University of Tokyo, and an associate professor at Nagasaki University. He joined Kyoto University as an associate professor in 2001 and was promoted to professor in 2007. His research interests are population dynamics and the evolution of organisms.

Kei Tokita is a professor at the Graduate School of Informatics, Nagoya University. He obtained a Ph.D. in Science at the University of Tokyo. He was an assistant professor and associate professor at Osaka University, and was a visiting researcher at Harvard University during the periods of 1996–1998 and 2003–2004. His research interests are in the statistical mechanics of complex systems and mathematical biology.

Toshiyuki Namba is an emeritus professor at the Graduate School of Science, Osaka Prefecture University. He obtained a Ph.D. at Osaka University. He was a lecturer, associate professor, and professor at Senshu University before moving to Osaka Women's University in 1994, which was united into Osaka Prefecture University in 2005. His research interests lie in structure and dynamics of ecological communities.

Tae-Soo Chon is an emeritus professor at Pusan National University, Busan, Republic of Korea, and currently serves as the chair of Ecology and Future Research Association. After receiving a Ph.D. in Entomology at the University of Hawaii, Honolulu, USA, he was appointed as an assistant professor in 1983, and was promoted to professor in 1992. Broadly based on computational sciences and engineering (e.g., pattern recognition), his interests lie in ecological modeling/informatics and mathematical biology applied to invasion population dynamics, stream community dynamics, and computational behavior for monitoring.

Chapter 4
Museum Workshop: Evolution of Human Intelligence and Education

Terufumi Ohno

Abstract This chapter describes some interesting behaviors common to participants in science workshops conducted at museums, and shows that these behaviors reflect the implicit intelligence that our ancestors acquired and accumulated during the long process of human evolution. Our intellectual activities are heavily influenced by this implicit intelligence, for better or for worse. In evolutionary terms, our language is a very recently acquired intellectual capacity and thus can be considered immature. Recently, a number of lifelong learning approaches have been proposed with emphasis on more rational, explicit, and interactive aspects. However, our intellectual activity is strongly influenced by implicit intelligence and our interactive capacity is immature. We should take these aspects into account if we really wish to construct more effective lifelong learning programs.

Keywords Human intelligence · Evolution · Curiosity · Oversight · False-belief · Dialogue · Lifelong learning

4.1 Introduction

In the past, knowledge and skills, once acquired through formal education and higher education, could be effectively utilized throughout one's life. However, we are living in a world with rapidly changing technology, numerous environmental issues, and economic globalization (OECD, 2018), where knowledge and information acquired in formal education may quickly become useless. Based on this recognition, countries and international organizations such as the OECD (2018) have begun to search for ways to develop learning skills in learners that they can use for the rest of their lives while flexibly adapting to a rapidly changing society, either individually or in collaboration.

T. Ohno (✉)
Kyoto University, Yoshida Honmachi, Sakyo-ku, 606-8501 Kyoto, Japan
e-mail: ohno.terufumi.76u@st.kyoto-u.ac.jp

Takada Junior College, Toyono 195, Isshinnden, Tsu City 514-0115, Mie Prefecture, Japan

K. Nishimura et al. (eds.), *Creative Complex Systems*, Creative Economy,
https://doi.org/10.1007/978-981-16-4457-3_4

In Japan, the Central Council for Education Report in December 2016 set the basic policy for revision of the governmental curriculum guideline with the aim of fostering "zest for living," which is the general intelligent ability to proactively respond to changes in society that are difficult to predict in the midst of a declining birthrate, aging population, and severe economic and financial conditions of this country (Central Council for Education, 2016). Human intelligence is a product of evolution, and therefore its strengths and weaknesses are constrained by the course of that evolution. Therefore, any attempt to develop flexible intellectual abilities in learners cannot be powerful without taking into account the evolution of intelligence.

Since 1997, I have been working in two museums, the Kyoto University Museum and Mie Prefectural Museum. In these museums, I have been intensively involving myself in holding science workshops to stimulate learning motivation of people of all ages from kindergarten children to senior citizens. In these workshops, participants showed several recurring behaviors. These behaviors provided me with plenty of opportunity to get insight into the nature of human intelligence in its evolutionary terms. Here, I report some of their behaviors, describe my hypothesis about their causes, and discuss their implications for education.

1. Children's emotions can be stimulated in a science workshop.
2. Participants make two common mistakes: oversights (not seeing something that does exist), and wrong assumptions or false-belief (seeing something that does not exist).
3. People are poor at dialogue.
4. Participants reflect on their activities in the workshop, whether explicit or implicit.

4.1.1 Children Are Getting Hooked in Science Workshops

My favorite workshop for children is "Let's investigate trilobites", where children were asked to guess what kind of creature trilobites were and how they lived. The aim of the workshop is to make children familiar with the process of science activity; that is, finding new knowledge through the science cycle of observation, hypothesis formulation, and verification.

In the workshop, we first handed each child a trilobite fossil and asked them to observe and sketch the morphological features of the fossil. I tell them that sketching provides hints to answer the questions. Children, then, earnestly observed and sketched the fossil to find hints. Eventually, they intuitively realized that trilobites are closely related to members of arthropods like insects, shrimps, and crabs. These animals are all familiar to children. Then, they began to connect their observations with their previous knowledge and experiences of these creatures. This subconscious connection became a hint for later inferences. Therefore, after sketching, when we asked the first question: "What kind of creatures are trilobites related to?", they immediately named the animals that belong to the arthropods. Often, nearly 20 names were mentioned. When we told them that most of the answers were correct, the children began to feel more confident in their inquiry activities, knowing that they had made a reasonable guess of a creature that has been extinct for 250 million years.

I also asked the children a series of questions and asked them to guess the answers: "What features do trilobites have in their eyes?" (the answer is compound eyes); "How do they protect themselves from natural enemies?" (the answer is, for example, by curling up like a pill bug); "How do trilobites grow?" (the answer is by molting). After the children had guessed, I showed them the fossils to prove the validity of their guesses.

During one of the workshops, in response to the question, "How do trilobites protect themselves from their natural enemies?" three girls presented their guess, "They protect themselves from their natural enemies by rolling up in a ball." When I showed them a trilobite fossil with its body curled up as they had guessed, they proudly held up the specimen as proof that it was as they had thought (Fig. 4.1a).

The next moment, however, something very interesting happened. These children began to gaze at the specimens with fascination (Fig. 4.1b). I interpreted this as a moment when the children experienced new discoveries on their own and became

Fig. 4.1 **a** Discussion and speculation on how trilobites lived. **b** The moment when learning emerges in the museum

aware that they had such abilities: they were moved and felt a sense of self-efficacy. I call this moment "The moment when learning emerges in the museum." They then enjoyed the workshop with increasing curiosity and motivation to learn.

4.1.2 Why Do Science Workshops Stimulate Participants Emotions?

The process of science consists of observation, formulation, and verification of hypothesis as a dispassionate activity. However, as described above, in this dispassionate cycle of the workshop, children became emotional, excited, felt self-efficacy, and became more curious.

Why does the dispassionate, analytical process of science stimulate children's emotions and invite them to learn? As with all evolutionary phenomena, human intelligence must have evolved as an adaptation to increase the chances of survival. Therefore, I believe that the answer lies in the evolution of human intelligence.

4.1.3 Environmental Conditions for Our Ancestors

Like all evolution phenomena, human intelligence must have evolved as adaptation to increase the survival probability in response to the environment where human ancestors lived. Changes in past climates are revealed by the study of fossil plants and pollen, which are indicators of the environment recorded in geological strata, or by the study of oxygen and carbon isotope ratios. The most important characteristic in the Cenozoic climate is that Antarctica has been covered by ice sheets for the past 35 million years and the Arctic has been similarly covered for the past 2 million years. As a result of the associated cold and arid climate (Zachos et al., 2001), the area of forest has decreased and the area of grassland has increased. This has had a major impact on the evolution of human intelligence.

The main scene where important stages of human evolution occurred was the African continent. Fossil records show that human ancestors and apes (including chimpanzees, gorillas, and orangutangs) split from their last common ancestor between 10 and 6 million years ago.

The discovery of well-preserved fossils of the 4.4-million-year-old *Ardipithecus ramidus* in Ethiopia has provided a major clue to the reconstruction of the appearance and ecology of our distant ancestors (Gibbons, 2009). This species, our remote ancestor or its close relative, lived along riversides in Ethiopia about 4.4 million years ago. Its foot thumb faced the other toes and the foot was able to grasp objects (Lovejoy et al., 2009). This suggests that it was adapted to arboreal life among trees of sparse forests. This is in harmony with the inference that the species inhabited sparse forests developed along rivers surrounded by savannas, as revealed by studies of sedimentological and stable isotopic (carbon and oxygen) data from excavation sites (Gani & Gani, 2011).

Then, 4.2 million years ago, the genus *Australopithecus* appeared. Fossil footprints ascribed to *Australopithecus afarensis* have been found on a 3.7-million-year-old volcanic ash layer in Raetoli, Tanzania, suggesting that our ancestors began walking upright and bipedal on the ground at the latest around this time.

Eventually, about 3 million years ago, the area of the rainforest began to shrink and the savanna began to spread (Bonnefille, 2010). Our ancestors were forced to change their way of life from arboreal to terrestrial, and faced the difficult task of surviving on savanna grasslands full of predators and other natural enemies.

4.1.4 Biga (Two-Horse Chariot) Mode of Science Activity

This situation, I believe, has given rise to science consisting of two components: dispassionate and emotional cycles (Fig. 4.2).

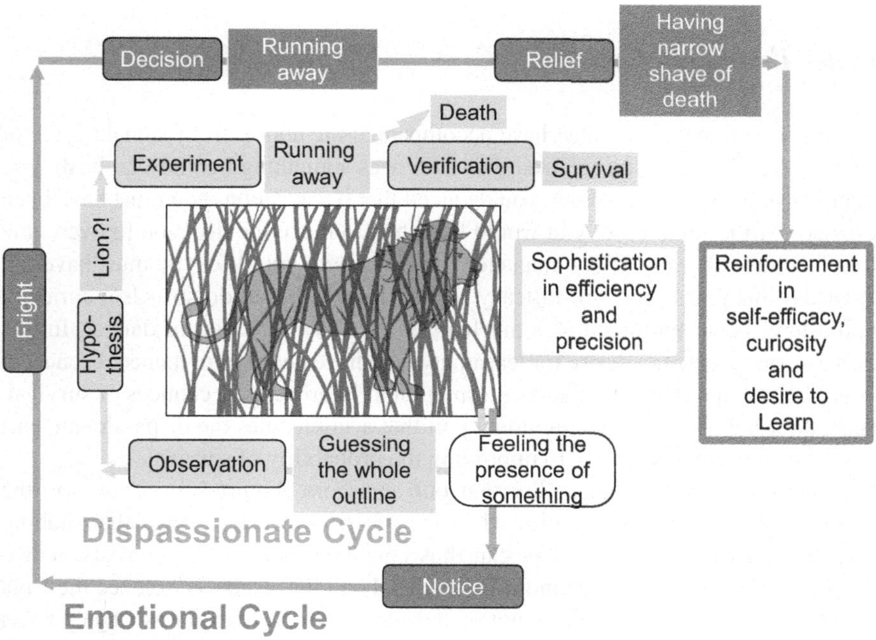

Fig. 4.2 Biga (two-horse chariot) mode of science activity

4.1.5 Dispassionate Cycle

Let us imagine that you were in the middle of the savanna a long time ago. How did you react when you realized that there was something behind that bush (Fig. 4.2)? You would have made observations to fill in the gaps covered by the bush foliage to obtain a complete outline of the hidden object (observation). Then, when the outline was complete, you would have presumed that the silhouette was a lion (making hypothesis). Once you had deduced that it was a lion, you would have conducted your experiment by running away from it. Eventually you would have verified your hypothesis in the form of a successful survival from the lion threat (verification). Because individuals who overlooked the object or made the wrong inference probably died (falsification), it is not difficult to imagine that behind our successful ancestor, many individuals were killed as a result of inappropriate reaction. This led to accumulation of genes of careful individuals. Thus, we are the descendant of those ancestors who increased their survival chance by refining this science cycle of observation, hypothesis, and verification. This is the dispassionate cycle of science activity.

4.1.6 Emotional Cycle

A strong emotional cycle must have accompanied the above dispassionate cycle of science. Again, when you realized that there was something behind that bush, you would have been startled. When you deduced that it was a lion, you must have been terrified. At the same time, you would have had to make the decision to overcome your fear and run away. If you managed to escape with your life, you must have felt a sense of happiness and self-efficacy that you were able to do so. Also, curiosity would have been reinforced as a mechanism for early detection of danger. In this scenario, our ancestor's desire for learning would have been strengthened, because it would have helped them to gather information to improve their chances of survival. The author believes that the emotional part that accompanies the dispassionate part of science contributes greatly to improving the probability of survival.

Scientific analysis is an ability that our ancestors acquired to survive on the savannas of their natural enemies when they were forced to change their habitat from the trees to the ground. We can thus conclude that science consists of two cycles: the dispassionate and emotional cycles. I call this aspect of science the Biga (two-horsed chariot) mode. The emotional cycle of this mode explains why children get emotionally excited in science workshops.

4.1.7 *Seeing is Not Always Believing and Three Men Do Not Make a Wise Man*

One of my science workshops for adults is called "Anatomische Tabellen von Muscheln," or the clam workshop, which is a program to determine the number of adductor muscles of a clam by observing the attachment scars on its shell inner surface (the answer is two). The program consists of two stages. In the first observation stage, each participant observes and sketches the inside of a clam shell and determines the number of muscle scars. In the second stage, the participants form groups of six to eight people and consult each other through dialogue to reach a conclusion on the number of clam shells.

One may think that the task of this workshop is so simple that almost all the participants should give the correct answer after the first stage of observation. For the eight workshops I conducted, however, in this stage, only about 50% of the participants correctly answer that there are two muscle scars (Table 4.1). Observation does not lead to the truth.

In the second dialogue stage, even after the participants have consulted each other through dialogue, about one third of the participants end up with incorrect answers. Dialogue is not as powerful as the proverb says: "Three men make a wise man."

4.1.8 *Common Mistakes in the Observation Stage*

On the inner surface of the clam shell, there are two tear-drop shaped muscle scars with fine growth lines. These scars are connected by a line called the pallial line along which the mantle of the clam body attaches to the shell. The line has an inward bend called the pallial sinus corresponding to the location of clam's inhalant and exhalant syphons (Fig. 4.3).

Many workshop participants guess erroneously the muscle scar number at three. They make two typical failures: oversight and false-belief.

Table 4.1 Participant answers on number of muscle scars in eight clam workshops

Number of participants	Correct answer and percentage		Improvement in percentage points
	Observation stage	Dialogue stage	
352	177 (50.3%)	225–231 (63.9–65.6%[a])	13.6–15.3[a]

[a]Several participants did not report their answers. In such cases, estimated maximum number of positive answers is also given, where participants without an answer are treated as giving correct answer

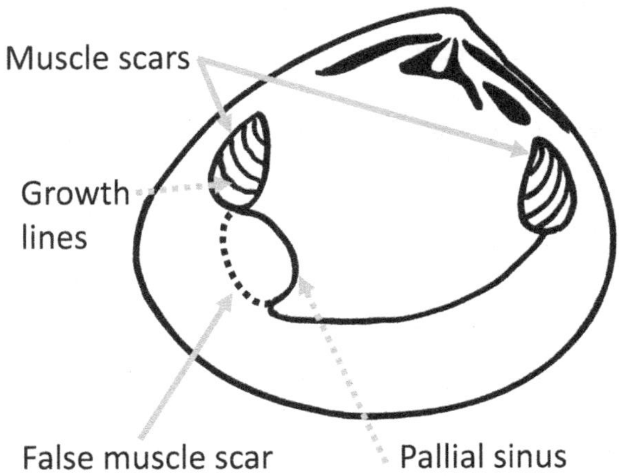

Fig. 4.3 Diagram of inner surface of a clam shell

4.1.9 Oversight

Adductor muscle scar has fine growth lines, which are traces of migrating and growing adductor muscles in accordance with the growth of the clam shell. Thus, they serve as a reliable key in identifying muscle scars. Interestingly, many participants draw growth lines on their sketch, but do not take them into consideration when they infer the number of scars.

The cause of this oversight is probably related to so-called procedural memory. When we learn to ride a bicycle, we consciously step on the left pedal and then on right pedal and so on. After some time, however, we automatically step on two pedals alternatively without conscious awareness. Thus, procedural memory enables us to perform various kinds of activities without conscious attention and makes it possible to allocate attention to other urgent tasks like those useful for survival.

It seems that many of the participants make use of procedural memories and automatically execute sketches without conscious awareness. Consequently, they draw detail of the characteristics of the inner shell surface without an awareness of what they are drawing, which leads to incorrect inference of muscle scar number.

4.1.10 False-Belief

Many participants automatically draw a half circle on the opposite side of the semi-circular pallial sinus to complete the (third) false muscle scar (Fig. 4.3). Explanation for this kind of mistake, or false-belief, might be found in evolutionary adaptation of our ancestors to increase survival chances in the extremely hazardous habitat of the African savanna.

When our ancestors lived on the savannas of Africa, early detection of dangerous natural enemies such as predators must have translated into increased survival rates.

Suppose that one of our ancestors noticed something behind the savanna bush foliage. If the person had the ability to automatically fill in the contours, the person could drastically reduce the time it takes to identify the nature of the object. If the object was a dangerous carnivore like a lion, this early detection would increase the person's chance of survival (Fig. 4.4).

The tendency to automatically fill in the contours may lead to false-belief, but misidentification, for example, a fallen tree being mistaken for a lion, does not risk one's life, although energy would be wasted in running away from a fallen tree. On the other hand, if, by meticulously filling in the contours, one could correctly judge whether it is a lion or a fallen tree, the observer may also increase the probability of falling prey to a lion, because meticulous observation takes more time and may allow a predator to move dangerously close.

As explained above, both oversight and false-belief were very effective survival adaptations in the hostile African savannas where our ancestors once lived. However, now that the natural and social context has changed completely, these traits function as obstacles to correct perception.

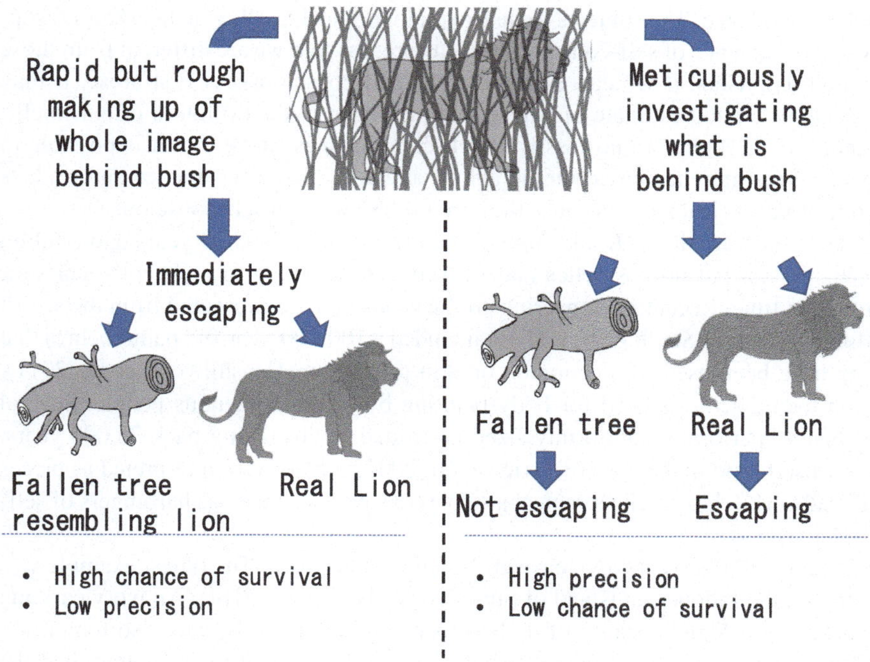

Fig. 4.4 Merit and demerit of automatically filling in the contours

4.1.11 *Poor Discussion Ability: Three Men Do Not Make a Wise Man*

Consulting others through dialogue is theoretically an effective tool for solving various problems and is expected to lead to reasonable outcomes through exchange of opinions and debate. However, whether people are good at dialogue is another question.

In the clam workshop, after the participants consulted with others in the dialogue stage, there was a modest increase in correct answers of about 15%, bringing the final level of correctness to about 65% (Table 4.1). It appears that we are not very good at dialogue, even though a popular proverb claims "Three heads are better than one."

Having already stated that the cause of oversight and false-belief lies in the evolution of our intelligence, I also believe that the cause of our poor dialogue skills can be explained in terms of the evolutionary development of our intelligence. Our dialogue ability seems to be acquired very recently and so is still immature.

Because dialogue cannot be fossilized, its origin can only be explored through indirect evidence. The subjects we use in dialogue, such as "I, you, he, she," presuppose the existence of self-consciousness (awareness that we are different from those around us). Thus, if we can estimate when self-consciousness first arose, it may provide us a good estimate of when dialogue began in the evolution of our intelligence. Again, there are no fossils of self-consciousness itself, but it is possible to make some inference based on the age of artefacts such as cosmetics and trinkets from archeological sites, because they are useless without self-consciousness.

The oldest fossils of *Homo sapiens* date from about 300,000 years ago (Hublin et al., 2017), but no cosmetics and trinkets can be found in archaeological sites from that time. Recently, pioneering artefacts began to be excavated from the site of Blombos Cave in South Africa. These included a 100,000-year-old paint (ochre) that may have been used for decoration or skin protection (Henshilwood et al., 2011). Even today, ochre is used for body painting by some indigenous people. Several artificially perforated, uniformly sized gastropod shells, dating back 75,000 years, were also found in the cave (d'Errico et al., 2005). These were interpreted as pieces of beadwork. These archaeological artefacts can be interpreted as harbingers of self-consciousness.

Around 40,000 years ago, cave art became widespread. The oldest example was found on the Indonesian island of Sulawesi (Aubert et al., 2014). An ivory carving, named "Lion Man" because of its lion face and human body, was also found in a 40,000-year-old site in Germany. These artefacts show that their creators had the ability to give and appreciate their meaning, which is as high a level of mental activity as self-consciousness. Therefore, we can make a rough estimate that self-consciousness and therefore dialogue manifested themselves around 50,000 years ago. Mithen (1996) referred to this event as "the Big Bang of human culture." The genus *Homo*, to which we belong, appeared two million years ago, while *Homo sapiens* appeared 300,000 years ago. By comparison, the beginning of dialogue

50,000 years ago is quite recent. If the interpretation of artefacts as evidence of self-consciousness is correct, and if self-consciousness is a prerequisite for the origin of language, it follows that our language is an immature human ability not older than around 50,000 years. This may explain why dialogue among participants of the clam workshop yields modest improvement in the correct answer rate.

To this point, I have described behaviors of participants in my workshops and tried to interpret them in terms of human cognitive evolution. In my view, the Biga (two-horse chariot) mode of science activity, oversight, and false-belief can all be explained as results of evolutionary adaptation of human ancestors to improve the survival rate in a harsh environment. I assume the timing of the origin of these characteristics to be around 3 million years ago, when our ancestors faced new challenges in terrestrial lifestyle. However, timing is not so relevant for my idea of emergence, because origin of these characteristics only demands a harsh environment to which our ancestors must adapt to increase their chance of survival: harsh environments have always existed during human evolution.

My view of the origin of language is in accordance with the modules or cognitive domains theory of Mithen (1996). His theory states that our minds consist of specialized modules or cognitive domains. He recognizes general, social, natural history, technical, and linguistic modules, each of which evolved with different timing and dedicated to some specific type of behavior. Around 50,000 years ago, these domains became connected into cognitive fluidity, leading to the modern human mind.

As already mentioned, around 50,000 years ago, various artefacts like the Lion Man, cave art, and other examples became abundant, indicating the beginning of cognitive fluidity in the human mind. Mithen (1996) referred to this event as the Big Bang of human culture, probably corresponding to the origin of language ability.

4.1.12 Self-reflection

To determine how the participants self-evaluated their involvement in the clam workshops, I held five additional workshops with a total of 153 participants. In these five workshops, the percentage of correct answers in the observation and dialogue stages (70.6% and 86.3%, respectively) was higher than in the previous eight workshops. The participants were all adults. At the end of each workshop, I asked them to write a short report. The feedback of the participants was analyzed by picking out participants' self-positive and self-negative words and phrases for the observation and dialogue stages.

Positive statements were judged based on words or phrases such as "I could do it," or "It was good," while negative statements were judged based on words or phrases such as "I couldn't do it," or "It was difficult." If a participant wrote more than one positive or negative comment for each stage they were counted as one; if a participant wrote both positive and negative comments for the same stage, both were counted.

During the observation stage, 39 participants among 108 participants with the correct answer and 18 among 45 participants with an incorrect answer gave comments. It is noteworthy that both the majority of participants with an incorrect answer and a majority of those with the correct answer wrote self-negative comments (Tables 4.2 and 4.3). This strongly suggests that many participants were aware that their estimates of the number of muscle scars were not always supported by sufficient observational facts and considerations.

During the dialogue stage, 72 participants among 132 participants with the correct answer and 14 among 21 participants with an incorrect answer gave comments (Table 4.4). It is significant that high percentages of both the participants with the correct answer and those with incorrect answers gave positive feedback. This suggests that a majority of participants appreciated the importance of dialogue in solving the problem.

Table 4.2 Positive and negative self-evaluative comments for the observation stage

	Number of comments (%)		
	Positive	Negative	Total
Participants giving correct answer (108)	10 (25.6%)	29 (74.4%)	39 (100%)
Participants giving incorrect answer (45)	2 (11.1%)	16 (88.9%)	18 (100%)

Table 4.3 Examples of negative self-evaluation of participants who gave correct answer for the number of muscle scars on shell inner surface at observation stage

I was not able to sketch what I saw
There is only so much that I can notice in my own observations
If I had done more "looking at," I would have missed less
I saw things in my observations but did not draw them
Not being aware of what I was looking at
I focused on a part of the inside of the shell, but did not look deeper into it

Table 4.4 Positive and negative self-evaluative comments for the dialogue stage

	Number of comments (%)		
	Positive	Negative	Total
Participants giving correct answer (132)	65 (90.3%)	7 (9.7%)	72 (100%)
Participants giving incorrect answer (21)	11 (78.6%)	3 (21.4%)	14 (100%)

Table 4.5 Explicitness levels of comments

Number of comments and percentage of total			
Level 1	Level 2	Level 3	Total
44 (29.1%)	69 (45.7%)	38 (25.2%)	151 (100.0%)

4.1.13 Depth of Participant Reflection

Participant reports were classified according to the varying explicitness of description: Level 1, statements of fact, such as "it was"; Level 2, statements with reasons for the fact, such as "it was because of"; and Level 3, statements reflecting on the fact, such as "I thought it was because of." Of the 153 participants, 151 reports were suitable for this analysis. The numbers of comments and their percentages of the total are given in Table 4.5.

Formal education puts emphasis on the reflection ability of learners. In the belief that this element can be fostered by developing the ability to express things explicitly, emphasis has been placed on, for example, recommending learners journal writing; the emphasis may be partly because of the convenience for the teacher in that explicit expressions are easier to evaluate.

In the dialogue stage of the clam workshop, both participants with correct and incorrect answers seemed to appreciate importance of dialogue. On the other hand, in the comments about the observation stage, most of the participants wrote self-negative comments, regardless of the correct or incorrect answers. Although almost 75% of the participants' comments were of low explicitness (either Level 1 or Level 2, Table 4.5), nevertheless, it can be seen that they were not satisfied with the result of having answered correctly and that they reflected on their behavior during the observation stage, including the lack of observational skills (Table 4.3). Also they appreciated importance of dialogue (Table 4.4). My results suggest that reflection functions at an implicit level and is probably an ability that humans acquired before the origin of language, an ability that may be refined through training to express it more explicitly.

4.2 Discussion and Conclusion

This chapter discusses science activity as being composed of two components: dispassionate and emotional cycles (Biga, or two-horse chariot mode). In discussing science or science education, emphasis has always been put on the analytical and rational half of science. Yet, if science works in the Biga mode, then we should pay more attention to the emotional cycle in teaching science activity. If we can stimulate the emotional part, we can more easily entice children into the fascinating world of science activity.

Oversight and false-belief are examples of pitfalls in the scientific process. Sir Francis Bacon recognized the cognitive mistakes that people will commit. I have shown that these mistakes occur as a result of the malfunctioning of abilities that our ancestors developed as means to ensure their survival. These were quite adaptive in their environments at the time, but are at odds with our present environment. By being informed of these matters, learners may become more curious about the origins of their own intelligence, and be more careful in their science activities.

Reflection is thought to contribute greatly to the improvement of learning. Often, training in written expression has been promoted to cultivate the power of reflection. However, the discrepancy between the high percentage of correct answers (70%) and high percentage of negative self-evaluation (74.4%) in the observation stage of the clam workshop suggests that many participants critically reflected on their own behavior rather than simply being pleased with the results. Secondly, because Level 3 explicitness of participants' comments was only a quarter of the whole reflection, it appears that in many cases reflection happens implicitly. This suggests that reflection is also an ability that humans had acquired before the origin of language.

This chapter describes the behaviors of participants in workshops and examined their implications. Modern education has focused on dealing with the subject on an explicit and rational horizon. However, as mentioned in this chapter, our intellectual activity is strongly linked to and, for better or worse, influenced by the implicit realm of our mental activity, including emotions, oversight, false-belief and reflection. Furthermore, language and dialogue, which are tools for rationality, are in their infancy. I conclude that the future of education should focus more on how to make use of the implicit areas of our intelligence, as well as on improving our language skills.

References

Aubert, M., Brumm, A., Ramli, M., Sutikna, T., Saptomo, E. W., Hakim, B., Morwood, M. J., van den Bergh, G. D., Kinsley, L., & Dosseto, A. (2014). Pleistocene cave art from Sulawesi Indonesia. *Nature, 514*, 223–227.

Bonnefille, R. (2010). Cenozoic vegetation, climate changes and hominid evolution in tropical Africa. *Global and Planetary Change, 72*, 390–411.

Central Council for Education (2016). Yōchien, shōgakkō, chūgakkō,kōtō gakkō oyobi tokubetsu shien gakkō no gakushū shidō yōryō tō no kaizen oyobi hitsuyō na hōsaku tō ni tsuite [Regarding improvements and necessary strategies, etc. for the government curriculum guidelines for kindergartens, elementary schools, high schools and special needs schools 2016] (Report). Ministry of Education, Culture, Sports, Science and Technology (in Japanese) (April 6, 2021). Available from URL: https://www.mext.go.jp/b_menu/shingi/chukyo/chukyo0/toushin/1380731.htm.

d'Errico, F., Henshilwood, C., Vanhaeren, M., & van Niekerk, K. (2005). *Nassarius kraussianus* shell beads from Blombos Cave: Evidence for symbolic behavior in the Middle Stone Age. *Journal of Human Evolution, 48*, 3–24.

Gani, M. R., & Gani, N. D. (2011). River-margin habitat of Ardipithecus ramidus at Aramis, Ethiopia 4.4 million years ago. Nature Commun 2(602):1–5. https://www.nature.com/articles/ncomms1610.pdf

Gibbons, A. (2009). Ardipithecus ramidus. *Science, 326*, 1598–1599.

Henshilwood, C. S., d'Errico, F., van Niekerk, K. L., Coquinot, Y., Jacobs, Z., Lauritzen, S.-E., Menu, M., & García-Moreno, R. (2011). A 100,000-year-old ochre-processing workshop at Blombos Cave South Africa. *Science, 334*, 219–222.

Hublin, J.-J., Ben-Ncer, A., Bailey, S. E., Freidline, S. E., Neubauer, S., Skinner, M. M., Bergmann, I., Cabec, A. L., Benazzi, S., Harvati, K., & Gunz, P. (2017). New fossils from Jebel Irhoud, Morocco and the pan-African origin of *Homo sapiens*. *Nature, 546*, 289–292.

Lovejoy, C. O., Latimer, B., Suwa, G., Asfaw, B., & White, T. D. (2009). Combining prehension and propulsion. *Science, 326*, 72.

Mithen, S. (1996). *The prehistory of the mind the cognitive origins of art, religion and science.* Thames and Hudson Ltd.

OECD. (2018). *The future of education and skills education 2030.* OECD, Paris (April 7, 2021). Available from URL: http://www.oecd.org/education/2030/E2030%20Position%20Paper%20(05.04.2018).pdf.

Zachos, J., Pagani, M., Sloan, L., Thomas, E., & Billups, K. (2001). Trends, rhythms and aberrations in global climate 65 Ma to Present. *Science, 292*, 686–692.

Terufumi Ohno is Professor Emeritus at Kyoto University. He has undergraduate and master's degrees from Kyoto University. He completed his Ph.D. at Bonn University. He conducted research on the ecology of fossil and living molluscan bivalves as a member of Faculty of Science as an assistant and as assistant professor. After joining the Kyoto University Museum, Prof. Ohno frequently held science workshops using his original learning material, and has continued doing so after moving to Mie Prefectural Museum and later to Takada Junior College. These workshops have given Prof. Ohno ample opportunity to observe the behaviors of participants, which he interprets in terms of the evolution of human intelligence, as discussed in this chapter.

Part II
Creative Complexity in Mathematical Sciences: The Power of Analogy in Multidisciplinary Studies

Chapter 5
Anomalous Behavior of Random Walks on Disordered Media

Takashi Kumagai

Abstract Much effort has been expended on investigation of the physical properties of disordered media (complex systems) including how the heat transfers on the media. In mathematics, these properties have been actively studied for over 30 years. In particular, probabilistic methods have been developed extensively to analyze random walks and their scaling limits on the media. This chapter provides a discussion of the behavior of random walks and diffusions on typical disordered media.

5.1 Introduction

Around the mid-1960s, mathematical physicists started investigating the anomalous behavior of heat transfer on disordered media (e.g., see Ben-Avraham & Havlin, 2000). Examples of disordered media include polymers, complex networks, and growth of mold and crystal. In this chapter, we consider "disordered media" as a subclass of " complex systems." Mathematical progress on these problems started in the late 1980s. The first systematical progress was made on fractals, which are in some sense ideal disordered media because they have exact self-similarity. Meanwhile, analytical methods and techniques were gradually developed that enabled us to analyze quantitative estimates for heat transfer on some random media that had higher complexity. Probabilistic approaches to the problems are used to investigate random walks (RWs) and diffusions. Because one cannot expect smoothness on the objects, it is not easy to construct differential operators directly. Probability theory does not require smoothness of the objects, and it works well to analyze them.

In this survey, we discuss behavior of random walks and diffusions on fractals and random media. In particular, we focus on the three random media presented in the following sections.

T. Kumagai (✉)
Research Institute for Mathematical Sciences, Kyoto University,
Kitashirakawa, Sakyouku, Kyoto 606-8502, Japan
e-mail: kumagai@kurims.kyoto-u.ac.jp

K. Nishimura et al. (eds.), *Creative Complex Systems*, Creative Economy,
https://doi.org/10.1007/978-981-16-4457-3_5

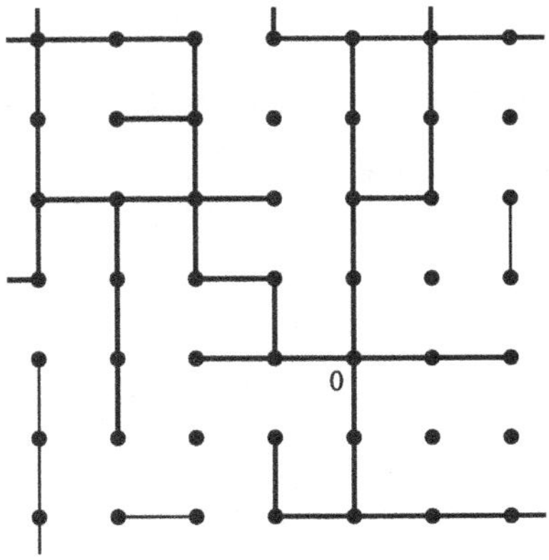

Fig. 5.1 Example of percolation cluster

5.1.1 Bond Percolation on the Lattice

The first model is the bond percolation on d-dimensional lattice $\mathbb{Z}^d$, where $\mathbb{Z}$ is the set of integers. Let $d \geq 2$. On each bond with length 1, we flip a coin and open (resp. close) the bond if it lands on heads (resp. tails). Let $p \in [0, 1]$ be the probability that the coin lands on heads (If $p \neq 1/2$, it is not a fair-coin.). We assume that flipping each coin is independent of flipping other coins. When all the coins are flipped, we have a set of open bonds; this model is called bond percolation (Fig. 5.1). Let $C(0)$ be the set of vertices in $\mathbb{Z}^d$ that is connected to the origin by open bonds and let $\theta(p)$ be the probability that the set $C(0)$ is an infinite set. Then, it is known that this model enjoys phase transition in the following sense: there exists $p_c \in (0, 1)$ such that $\theta(p) = 0$ if $p < p_c$ and $\theta(p) > 0$ if $p > p_c$. The percolation model is one of the most fundamental models in statistical physics that has phase transitions. Our interest is in how heat transfers on random media.

5.1.2 The Erdős-Rényi Random Graph

The second model is the so-called Erdős-Rényi random graph, which is a standard model in the field of disordered networks. Let $N \geq 2$ be a natural number and set $V_N := \{1, 2, \ldots, N\}$. For each pair of distinct points $i, j \in V_N$, $i \neq j$, we connect the bond $\{i, j\}$ with probability $p \in [0, 1]$ and disconnect it with probability $1 - p$. As before, whether each bond is connected is independent of the situations of other bonds. The resulting random graph is called the Erdős-Rényi random graph (Fig. 5.2). When $p = 1$, it is the complete graph with vertices V_N, so the Erdős-Rényi random graph is the bond percolation for the complete graph on V_N. Let C^N be the largest connected component of the graph. It is known that this model enjoys sharp phase transition around $p = c/N$ with $c = 1$. Namely, the following holds with high

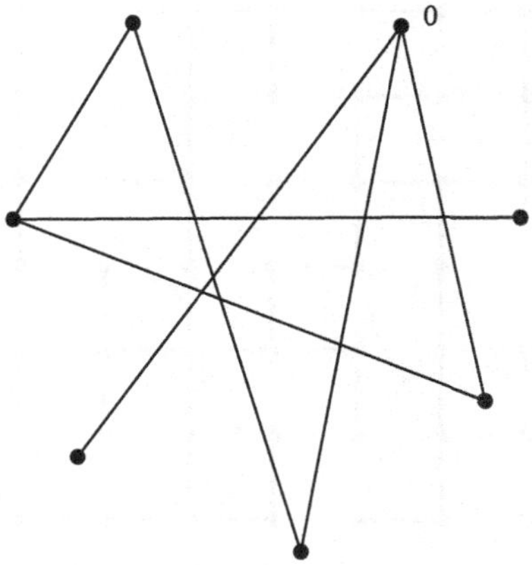

Fig. 5.2 Example of Erdős-Rényi random graph

probability:

$$c < 1 \Longrightarrow |C^N| = O(\log N),$$
$$c > 1 \Longrightarrow |C^N| \asymp N,$$
$$c = 1 \Longrightarrow |C^N| \asymp N^{\frac{2}{3}}.$$

Here $|A|$ is the number of the element in A, and we write $f(N) \asymp g(N)$ if there exist $c_1, c_2 > 0$ such that $c_1 f(N) \leq g(N) \leq c_2 f(N)$ for all N.

5.1.3 Two-Dimensional Uniform Spanning Tree

The third example of random media is the two-dimensional uniform spanning tree (2-Dim UST). Let $\Lambda_N := [-N, N]^2 \cap \mathbb{Z}^2$ and consider the graph that connects each neighboring bond with length 1. A loopless connected subgraph whose vertices consist of all the elements of Λ_N is called a spanning tree. Let $\mathcal{U}^{(N)}$ be a random graph that picks up one among all the spanning trees on Λ_N uniformly at random (Fig. 5.3). The uniform spanning tree $\mathcal{U}$ is the limit of $\mathcal{U}^{(N)}$ as $N \to \infty$. This model is extremely important in modern probability theory. Some readers may have heard of the Schramm-Loewner evolution (SLE). It is a stochastic process heavily related to the works of two Fields medalists, W. Werner and S. Smirnov. 2-Dim UST is the model that O. Schramm, who invented the SLE, studied the scaling limit in his celebrated paper in 2000 that introduced SLE for the first time.

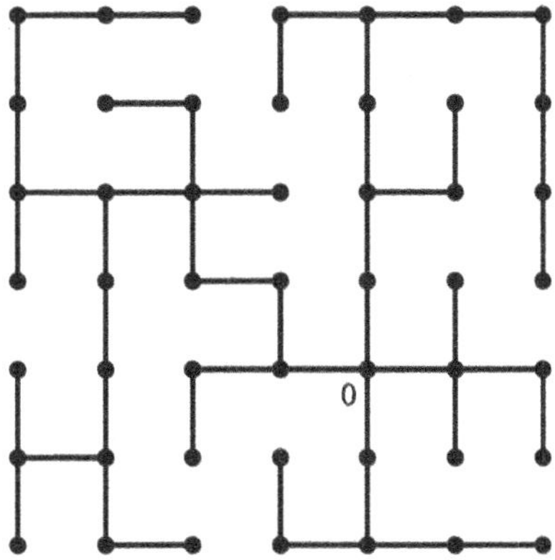

Fig. 5.3 Example of uniform spanning tree

5.2 RW on the Lattice and Brownian Motion on $\mathbb{R}^d$

Before explaining RW on random media (random graphs), let us first explain simple random walk (SRW) on the d-dimensional square lattice $\mathbb{Z}^d$ and Brownian motion on the Euclidean space $\mathbb{R}^d$ that appears as a scaling limit of the SRW. Let $Y = \{Y_n\}_{n\in\mathbb{N}}$ be the SRW on $\mathbb{Z}^d$, namely it is a random motion such that for $x, y \in \mathbb{Z}^d$ with $|x - y| = 1$,

$$P(Y_{n+1} = y|Y_n = x) = \frac{1}{2d}.$$

In other words, it is a random motion of a particle that jumps to one of the nearest neighborhoods with equal probability (Fig. 5.4).

Let us consider the scaling limit of the SRW by taking the mesh size of the lattice smaller and smaller. The geometric picture is that we take the limit $\varepsilon \to 0$ of $\varepsilon\mathbb{Z}^d$ so the spatial scaling limit is $\mathbb{R}^d$. Now let us consider the SRW εY_n on $\varepsilon\mathbb{Z}^d$. If we merely take $\varepsilon \to 0$, then the limiting process does not move at all, so we should speed up the time n depending on ε. It is known that we have the nontrivial (nondegenerate) limit process if we speed up the time by multiplying ε^{-2}, namely

$$\lim_{\varepsilon\to 0} \varepsilon Y_{[\frac{t}{\varepsilon^2}]} = B_t$$

and the limit process $\{B_t\}_{t\geq 0}$ is called Brownian motion, which is a random motion of a particle on $\mathbb{R}^d$. Brownian motion is related to the heat transfer on $\mathbb{R}^d$ because the differential operator (to be precise, the generator of the semigroup) determined by Brownian motion is

$$\frac{1}{2}\Delta := \frac{1}{2}\sum_{i=1}^{d} \frac{\partial^2}{\partial x_i^2},$$

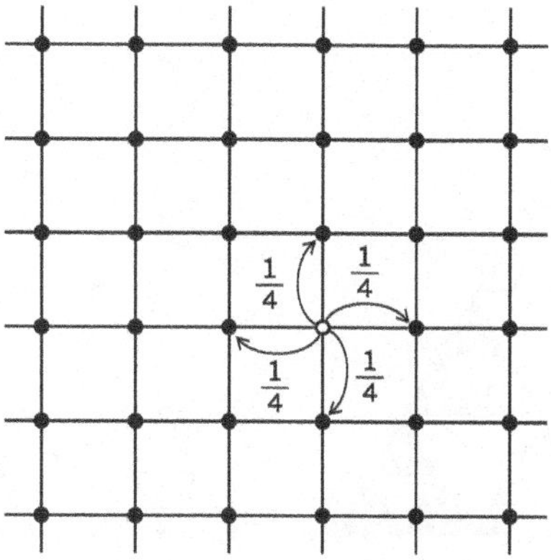

Fig. 5.4 Transition probability of two-dimensional simple random walk

that is $1/2$ times of the Laplace operator on $\mathbb{R}^d$. In fact, for a bounded continuous function f on $\mathbb{R}^d$, define $u(t, x) = E[f(B_t)|B_0 = x]$ (where E is the average with respect to Brownian motion). Then this $u(t, x)$ is the solution to the heat equation

$$\frac{\partial u}{\partial t} = \frac{1}{2}\Delta u, \qquad \lim_{t\to 0} u(t, x) = f(x).$$

The heat kernel (fundamental solution to the heat equation) is the following Gauss kernel:

$$p_t(x, y) = \frac{1}{(2\pi t)^{\frac{d}{2}}} \exp\Big(-\frac{|x-y|^2}{2t}\Big).$$

We note that the time scale factor ε^{-2} for the SRW to have the scaling limit (which is Brownian motion) is related to the fact that the Laplace operator is the second-order differential operator.

5.3 RW on Fractal Graphs and Brownian Motion on Fractals

We next consider SRWs on the fractal graphs and their scaling limits. As a typical fractal, we consider the Sierpinski gasket, which is shown on the left of Fig. 5.5. Note that a standard Sierpinski gasket is a compact one, say K. We extend it to an unbounded one by letting the left bottom vertex of the triangle as the origin and define $\hat{K} = \cup_{m=0}^{\infty} 2^m K$. Let G be the Sierpinski gasket graph as shown on the right of Fig. 5.5, where the length of each bond is 1. Now let $Y = \{Y_n\}_{n\in\mathbb{N}}$ be the SRW on G, namely the particle jumps at one of the neighboring points (which is a point that is connected by a bond) with equal probability after 1 second. Let us consider

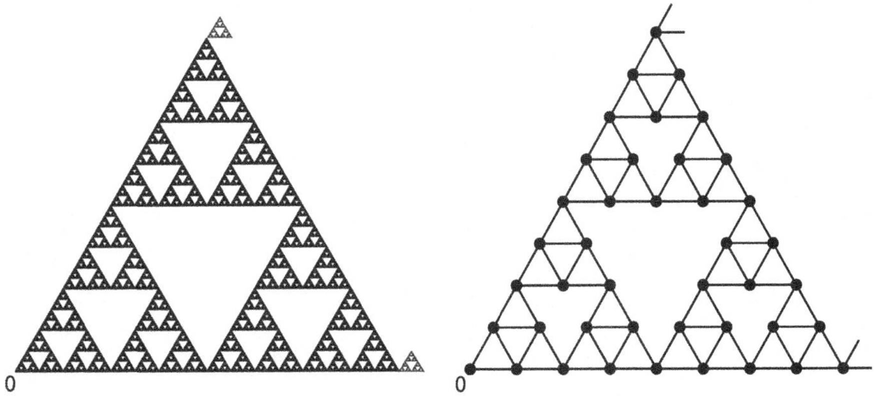

Fig. 5.5 Sierpinski gasket $\hat{K}$ and Sierpinski gasket graph G

the SRW $2^{-m}Y_n$ on $2^{-m}G$. As before, if we merely take $m \to \infty$, then the limiting process does not move at all, so we should speed up the time n depending on m. It turns out that if we speed up the time by multiplying 5^m, then we have the nontrivial (nondegenerate) limit process, namely

$$\lim_{m\to\infty} 2^{-m} Y_{[5^m t]} = B_t,$$

and the limit process $\{B_t\}_{t\geq 0}$ on the gasket is called Brownian motion, which is a random motion of a particle on $\hat{K}$. Brownian motion on the gasket was first constructed by Goldstein (1987) and Kusuoka (1987) independently.

The Laplace operator $\mathcal{L}$ that corresponds to Brownian motion was first constructed by Kigami (1989) and it can be determined as follows:

$$\mathcal{L}f(x) = \lim_{m\to\infty} 5^m \Big(\sum_{x_i : x \overset{m}{\sim} x_i} f(x_i) - 4f(x) \Big), \quad x \in \cup_{m\geq 0} 2^{-m} G \setminus \{0\}.$$

Here $x \overset{m}{\sim} y$ means that x and y are neighborhood on $2^{-m}G$. We note that the classical Laplace operator on $\mathbb{R}$ can be written as $\Delta f(x) = \lim_{m\to\infty} 2^{2m}(f(x + 2^{-m}) + f(x - 2^{-m}) - 2f(x))$ for $f \in C^2(\mathbb{R})$. Let $d_w = \log 5/\log 2$ (hence $5 = 2^{d_w}$). Naively, we can say that the Laplacian on the gasket is a " differential operator of order d_w" (One mathematical justification of this is that the domain of the so-called Dirichlet form on the gasket is a Besov space of order $d_w/2$.).

We can consider a d-dimensional gasket in a similar way in $\mathbb{R}^d$ from the family of $(d + 1)$-th contraction maps with contraction rate $1/2$ (For $d = 1$, $\hat{K} = [0, \infty)$.). The Hausdorff (fractal) dimension and the walk dimension of the d-dimensional gasket are $d_f = \log(d + 1)/\log 2$ and $d_w = \log(d + 3)/\log 2$, respectively.

Let $d(x, y)$ be the shortest distance between x and y in $\hat{K}$. It is known that there exists a heat kernel (fundamental solution of the heat equation) $p_t(\cdot, \cdot)$ and the following sub-Gaussian heat kernel estimates holds for all $t > 0, x, y \in \hat{K}$ (Barlow & Perkins, 1988):

$$c_1 t^{-\frac{d_f}{d_w}} \exp\Big(-c_2\Big(\frac{d(x,y)^{d_w}}{t}\Big)^{\frac{1}{d_w-1}}\Big) \le p_t(x,y)$$
$$\le c_3 t^{-\frac{d_f}{d_w}} \exp\Big(-c_4\Big(\frac{d(x,y)^{d_w}}{t}\Big)^{\frac{1}{d_w-1}}\Big). \quad (5.1)$$

The simple random walk on G also enjoys (5.1) for $d(x, y) \le t \in \mathbb{N}$ (Jones, 1996). Recall that for Brownian motion on $\mathbb{R}^d$, $p_t(x, y) = \frac{1}{(2\pi t)^{d/2}} \exp(-|x-y|^2/(2t))$. On fractals, we do not have explicit equality, but a generalized version of the heat kernel estimates hold, which is already good enough to deduce various analytical properties of the process. Note that when $d_f = d$ and $d_w = 2$, (5.1) boils down to the Gaussian estimates.

d_w is heavily related to properties of Brownian motion on the gasket. Indeed, by integrating (5.1), we have $c_5 t^{1/d_w} \le E^x[d(x, B(t))] \le c_6 t^{1/d_w}$; that is, d_w is the order of the average diffusion speed of particles. Given that $d_w > 2$, the behavior of the process is anomalous (for a long time, it diffuses slower than Brownian motion on $\mathbb{R}^d$, so the behavior is sub-diffusive). Set $d_s/2 = d_f/d_w$. d_s, which will appear in (5.2) again, gives the asymptotic growth of the eigenvalue counting function for Laplacian on the compact gasket K, and it is called the spectral dimension. In analysis, it is extremely important to analyze spectral properties of the Laplacian; on fractals, these properties have been extensively studied since early 1990s (Fukushima & Shima, 1992 etc.).

There are many other fractals on which natural diffusion processes are constructed and studied; for instance, on nested fractals and p.c.f. self-similar sets, and also on Sierpinski carpets. It turns out that the theory of Dirichlet forms is applicable to this area. For example, see Barlow (1998) for details.

5.4 SRW on the Percolation Cluster

In the following three sections, we discuss RWs on random media and their scaling limits. From now on, the space we consider is always random (note that there are two randomnesses, one is that of the space and the other is that of the RW). Let us write ω for the randomness of the media. That is, the random objects we denoted by C and $\mathcal{U}$ in Sect. 5.1 will be denoted by $C(\omega)$ and $\mathcal{U}(\omega)$ when we take one realization of the random graph.

In this section, we discuss the percolation cluster. Denote the SRW on the percolation cluster by $Y = \{Y_n^\omega\}_{n\in\mathbb{N}}$. Namely, Y_n^ω is located on one of the neighborhoods of Y_{n-1}^ω and it is equally distributed among all the neighborhoods. As mentioned above, ω stands for the randomness of the media; we fix $C(0) = C(0)(\omega)$ and consider SRW on it. SRW on the percolation cluster is sometimes called " the ant in the labyrinth."

5.4.1 Supercritical Case

We first consider the supercritical case, that is, when $p > p_c$. In this case, it is known that there is a unique infinite open cluster. In the following, we condition on the case $|C(0)| = \infty$. Then, the SRW on the cluster enjoys similar long-time behavior as that of the SRW on $\mathbb{Z}^d$ although there are many holes on the media. Indeed, it is known that this SRW enjoys the following Gaussian heat kernel estimates for large n almost surely with respect to the randomness of the media (Barlow, 2004):

$$c_1 n^{-\frac{d}{2}} \exp\left(-c_2 \frac{|x-y|^2}{n}\right) \le p_n^{\omega}(x,y) + p_{n+1}^{\omega}(x,y) \le c_3 n^{-\frac{d}{2}} \exp\left(-c_4 \frac{|x-y|^2}{n}\right).$$

Furthermore, the scaling limit of the SRW is similar to that of the SRW on $\mathbb{Z}^d$. That is, there exists a (nonrandom) constant $\sigma > 0$ such that the following holds almost surely with respect to the randomness of the media (Sidoravicius & Sznitman, 2004; Berger & Bivskup, 2007; Mathieu & Piatnitski, 2007):

$$\lim_{\varepsilon \to 0} \varepsilon Y^{\omega}_{[\frac{t}{\varepsilon^2}]} = \sigma B_t.$$

As we see, for the supercritical case, although there are many holes in the media, the long-time behavior of the SRW is similar to that of the SRW without holes. [In fact, if we search more detailed properties of the SRW, we can find differences between the two SRWs. We omit details and refer to Biskup (2011) and Kumagai (2014).]

5.4.2 Critical Case

Alexander and Orbach (1982) conjectured that the behavior of SRW on the critical percolation is completely different from that of SRW on $\mathbb{Z}^d$. As before, let $p_n^{\omega}(x, y)$ the heat kernel for the SRW. We call the following quantity (if the limit exists) spectral dimension:

$$d_s := -2 \lim_{n \to \infty} \frac{\log p_{2n}^{\omega}(x,x)}{\log n}. \tag{5.2}$$

One mathematical formulation of the Alexander-Orbach conjecture is that the spectral dimension for the SRW on the critical percolation is $4/3$ regardless of the dimension $d \ge 2$. For SRW on $\mathbb{Z}^d$, it holds that $d_s = d$, so this conjecture says the SRW on the critical percolation is anomalous like those on fractals.

To tackle this conjecture, the first problem is that there is no infinite cluster at $p = p_c$ (that is $\theta(p_c) = 0$) at least for $d = 2$ and $d \ge 11$. [In fact, it is a major open problem in this area whether $\theta(p_c) = 0$ for all $d \ge 2$ or not. It is believed that it is the case.] It is known that RWs on finite graphs converge to the stationary state under very mild conditions, so the limit in (5.2) will be 0, which is not what we want. So, we consider the so-called incipient infinite cluster (IIC), which is defined as follows. Consider the conditional probability that $C(0)$ intersects with the boundaries of the box of length N centered at 0 and then take $N \to \infty$; IIC is the (unique) infinite cluster on the probability space. It is known that at $p = p_c$, with high probability there is an open cluster with length of order n in the box of size n. So one can naturally believe that the mesoscopic behavior for the RW on the large finite cluster is similar to the long-time behavior of the RW on the IIC.

In general, analysis at critical probability is very difficult. So far, the IIC is rigorously constructed only for $d = 2$ and $d \geq 19$ (The former uses planar properties of $d = 2$ and the latter uses the renormalization technique at criticality called the lace expansion.). Let $Y = \{Y_n^{\omega}\}_{n \in \mathbb{N}}$ be the SRW on IIC and $p_n^{\omega}(x, y)$ be its heat kernel. Recently (5.2) has been proved with $d_s = 4/3$ almost surely with respect to the randomness of the media when the dimension is very high, namely the Alexander-Orbach conjecture is proved affirmatively in this case (Kozma & Nachmias, 2009; cf. Barlow et al., 2008). It was also revealed that the number 4/3 comes from the Hausdorff dimension of IIC $d_f = 2$ and the walk dimension $d_w = 3$ via a formula $d_s = 2d_f/d_w$. It is conjectured that the Alexander-Orbach conjecture does not hold for $d \leq 5$, but there is no mathematically rigorous proof. Of note, disproving the conjecture for $d = 2$ is one of the challenging open problems in this area.

5.5 SRW on the Erdős-Rényi Random Graph

As observed in Sect. 5.1.2, the Erdős-Rényi random graph enjoys phase transition around $p = 1/N$. Here we fix $\lambda \in \mathbb{R}$, choose $p = N^{-1} + \lambda N^{-4/3}$, and study the spatial scaling limit at the critical window. When p is in this critical window, it is known that $|C^N| \asymp N^{2/3}$ (Aldous, 1997).

Let us first explain the geometrical scaling limit of the random graph. We regard C^N as a metric space with origin. Then it is proved that when $N \to \infty$, there exists a random compact set $\mathcal{M} = \mathcal{M}_\lambda$ such that the following holds (Addario-Berry et al. 2012),

$$N^{-\frac{1}{3}} C^N \longrightarrow \mathcal{M}.$$

Here the convergence is in the sense of Gromov-Hausdorff, but we omit details. Let $\{Y_n^{C^N}\}_{n \geq 0}$ be the SRW on C^N. Then the following holds (Croydon 2012),

$$\lim_{N \to \infty} N^{-\frac{1}{3}} Y_{[Nt]}^{C^N} = B_t^{\mathcal{M}}.$$

Here $\{B_t^{\mathcal{M}}\}_{t \geq 0}$ is the Brownian motion on $\mathcal{M}$. Furthermore, there exists the heat kernel $p_t^{\mathcal{M}}(\cdot, \cdot)$ for Brownian motion such that the following estimates hold for all $x, y \in \mathcal{M}$ and $t \leq 1$ (Croydon, 2012),

$$p_t^{\mathcal{M}}(x,y) \le c_1 t^{-\frac{d_f}{d_w}} \ell(t^{-1})^{\theta} \exp\left\{-c_2 \left(\frac{d(x,y)^{d_w}}{t}\right)^{\frac{1}{d_w-1}} \ell\left(\frac{d(x,y)}{t}\right)^{-\theta}\right\}, \tag{5.3}$$

$$p_t^{\mathcal{M}}(x,y) \ge c_3 t^{-\frac{d_f}{d_w}} \ell(t^{-1})^{-\theta} \exp\left\{-c_4 \left(\frac{d(x,y)^{d_w}}{t}\right)^{\frac{1}{d_w-1}} \ell\left(\frac{d(x,y)}{t}\right)^{\theta}\right\}. \tag{5.4}$$

Here $\theta > 0$, $\ell(x) := 1 \vee \log x$, $d_f = 2, d_w = d_f + 1 = 3$ and $d(\cdot,\cdot)$ is the metric naturally defined on $\mathcal{M}$. As discussed above, Brownian motion on fractals enjoy similar sub-Gaussian heat kernel estimates with $\ell(x) = 1$. SRWs around critical probability and their scaling limits enjoy heat kernel estimates similar to the sub-Gaussian estimates, and because of the randomness of the media, there is oscillation of the logarithmic order.

5.6 SRW on the 2-Dim UST

Let us first explain the geometrical scaling limit of the 2-Dim UST. In the paper of Schramm (2000), topological properties of a candidate of a scaling limit of the UST were analyzed. The space is $\mathbb{R}^2$ as a set, but the topological structure of the space is given by the embedding of some tree into $\mathbb{R}^2$, and it is very different from the one we usually consider using the Euclidean metric. Later, Lawler-Schramm-Werner proved that the scaling limit exists uniquely. UST can be constructed as a collection of some random paths called the loop-erased RW, and it is known that the scaling limit of the loop-erased RW is SLE_2, which is in the class of the Schramm-Loewner evolutions. The scaling limit of the 2-Dim UST is thus heavily related to the theory of SLE.

We next discuss SRW on the 2-Dim UST. As before, we regard $\mathcal{U}$ as a metric space with origin and let $X^{\mathcal{U}}$ be the SRW on $\mathcal{U}$ starting at 0. Then the following holds (Barlow et al., 2017; Holden & Sun, 2018):

$$\lim_{\varepsilon\to 0} \varepsilon X^{\mathcal{U}}_{\varepsilon^{-\frac{13}{4}}t} = Y_t. \tag{5.5}$$

Here $\{Y_t\}_{t\ge 0}$ is a stochastic process on $\mathbb{R}^2$, but it is completely different from Brownian motion on $\mathbb{R}^2$. Indeed, there exists the heat kernel of $\{Y_t\}_{t\ge 0}$ such that (5.3) and (5.4) hold with $\ell(x) := 1 \vee \log x$ and $d_f = 8/5, d_w = d_f + 1 = 13/5$. Here $d(\cdot,\cdot)$ is the metric on the tree that embeds into $\mathbb{R}^2$, and it is completely different from the Euclidean metric. $d_f = 8/5$ is the Hausdorff dimension of $\mathbb{R}^2$ with respect to the metric $d(\cdot,\cdot)$ (note that if we use the Euclidean metric, then the dimension of $\mathbb{R}^2$ is clearly 2), hence if we observe the exponents with respect to the Euclidean metric, then it should be multiplied by $5/4$. Indeed, the exponent $13/4 = (5/4)\cdot d_w$ appearing in (5.5) is the walk dimension of the process with respect to the Euclidean metric.

5.7 Conclusions

As observed in several concrete examples, SRWs on disordered random media and their scaling limits enjoy similar properties as those on fractals, which are anomalous and quite different from those on $\mathbb{Z}^d$ or on $\mathbb{R}^d$. These examples may have various applications such as dynamics on the Internet (for instance, how the viruses spread out on the Internet), and various other open problems. The interested reader may refer to Kumagai (2014) and references therein.

Acknowledgements This research was partly supported by JSPS KAKENHI Grant Number JP17H01093.

References

Addario-Berry, L., Broutin, N., & Goldschmidt, C. (2012). The continuum limit of critical random graphs. *Probab. Theory Relat. Fields*, *152*, 367–406.
Aldous, D. (1997). Brownian excursions, critical random graphs and the multiplicative coalescent. *Ann. Probab.*, *25*, 812–854.
Alexander, S., & Orbach, R. (1982). Density of states on fractals: "fractons". *Journal de Physique Lettres* **43** (1982), L625–L631.
Barlow, M. T. (1998). Diffusions on fractals. In *Lectures on Probability Theory and Statistics* (Vol. 1690). Springer.
Barlow, M. T. (2004). Random walks on supercritical percolation clusters. *Ann. Probab.*, *32*, 3024–3084.
Barlow, M. T., Croydon, D. A., & Kumagai, T. (2017). Subsequential scaling limits of simple random walk on the two-dimensional uniform spanning tree. *Ann. Probab.*, *45*, 4–55.
Barlow, M. T., Járai, A. A., Kumagai, T., & Slade, G. (2008). Random walk on the incipient infinite cluster for oriented percolation in high dimensions. *Comm. Math. Phys.*, *278*, 385–431.
Barlow, M. T., & Perkins, E. A. (1988). Brownian Motion on the Sierpiński gasket. *Probab. Theory Relat. Fields*, *79*, 543–623.
Ben-Avraham, D., & Havlin, S. (2000). *Diffusion and reactions in fractals and disordered systems*. Cambridge University Press.
Berger, N., & Biskup, M. (2007). Quenched invariance principle for simple random walk on percolation clusters. *Probab. Theory Relat. Fields*, *137*, 83–120.
Biskup, M. (2011). Recent progress on the random conductance model. Probability Surveys, 8, 294–373.
Croydon, D. A. (2012). Scaling limit for the random walk on the largest connected component of the critical random graph. *Publ. Res. Inst. Math. Sci.*, *48*, 279–338.
Fukushima, M., & Shima, T. (1992). On a spectral analysis for the Sierpiński gasket. *Potential Analaysis*, *1*, 1–35.
Goldstein, S. (1987). Random walks and diffusions on fractals. In *Percolation theory and ergodic theory of infinite particle systems* (pp. 121–129). Minneapolis, Minnesota, 1984–1985. IMA Volumes in Mathematics and Applications (Vol. 8). Springer.
Holden, N., & Sun, X. (2018). SLE as a mating of trees in Euclidean geometry. *Comm. Math. Phys.*, *364*, 171–201.
Jones, O. D. (1996). Transition probabilities for the simple random walk on the Sierpiński graph. *Stoch. Proc. Their Appl.*, *61*, 45–69.
Kigami, J. (1989). A harmonic calculus on the Sierpiński space. *Japan J. Appl. Math.*, *6*, 259–290.

Kozma, G., & Nachmias, A. (2009). The Alexander-Orbach conjecture holds in high dimensions. *Invent. Math.*, *178*, 635–654.
Kumagai, T. (2014). Random walks on disordered media and their scaling limits. In *Lecture Notes in Mathematics* (Vol. 2101). Springer.
Kusuoka, S. (1987). A diffusion process on a fractal. Probabilistic methods in mathematical physics (pp. 251–274) (Katata, Kyoto). Academic Press, 1987.
Mathieu, P., & Piatnitski, A. (2007). Quenched invariance principles for random walks on percolation clusters. *Proc. Roy. Soc. A*, *463*, 2287–2307.
Schramm, O. (2000). Scaling limits of loop-erased random walks and uniform spanning trees. *Israel J. Math.*, *118*, 221–288.
Sidoravicius, V., & Sznitman, A.-S. (2004). Quenched invariance principles for walks on clusters of percolation or among random conductances. *Probab. Theory Relat. Fields*, *129*, 219–244.

Takashi Kumagai is a professor of mathematics at the Research Institute for Mathematical Sciences (RIMS) at Kyoto University in Japan. His research focuses on anomalous diffusions on disordered media, such as fractals and random media. Kumagai completed his Ph.D. at Kyoto University in 1994, and after working at Osaka University and Nagoya University, he accepted a position at Kyoto University in 1998. Kumagai was an invited speaker at the 2014 ICM in Seoul, and gave a Medallion Lecture at the Conference on Stochastic Processes and their Applications in Moscow in 2017. His awards include the Spring Prize of the Mathematical Society of Japan (2004), JSPS Prize (2012), Inoue Prize for Science (2017), Osaka Science Prize (2017) and Humboldt Research Award (2017).

Chapter 6
Pollution, Human Capital, and Growth Cycles

Takuma Kunieda and Kazuo Nishimura

Abstract To investigate the growth effect of pollution, we apply an optimal growth framework in which human and physical capital accumulation are two growth engines. Pollution is emitted from the stock of physical capital and has a negative impact on the formation of human capital. In this simple growth model, sustainable endogenous growth never occurs and a unique steady state emerges because of the negative impact of pollution. The model shows that (i) if the extent of the external effect of pollution is relatively small, the steady state is stable and the economy starting in the neighborhood of the steady state converges to it; (ii) if the extent of the external effect is relatively large, the steady state is unstable and the economy diverges away from it; and (iii) a Hopf bifurcation occurs at a certain intermediate extent of the external effect. The numerical analysis illustrates the global dynamic behavior in which the economy exhibits a closed orbit as sufficient time passes if the steady state is unstable.

Keywords Pollution · Human capital · Hopf bifurcation · Limit cycle · Endogenous business cycles

JEL Classifications O41 · O44 · E32

6.1 Introduction

Many researchers have empirically demonstrated that pollution negatively impacts human health (e.g., Bell et al., 2004; Chay & Greenstone, 2003; Currie & Neidell, 2005; Graff Zivin & Neidell, 2012, 2013; Pope et al., 2002). If pollution harms

T. Kunieda (✉)
School of Economics, Kwansei Gakuin University, 1-155 Uegahara Ichiban-cho, Nishinomiya 662-8501, Hyogo, Japan
e-mail: tkunieda@kwansei.ac.jp

K. Nishimura
Research Institute for Economics and Business Administration, Kobe University, 2-1 Rokkoudaicho, Nadaku, Kobe 657-8501, Hyogo, Japan
e-mail: nishimura@rieb.kobe-u.ac.jp

K. Nishimura et al. (eds.), *Creative Complex Systems*, Creative Economy,
https://doi.org/10.1007/978-981-16-4457-3_6

human health, it probably impedes the formation of human capital, thereby reducing economic growth, because illness has direct and indirect negative impacts on mental and physical functioning that interfere with learning and working performance. In this chapter, we investigate how pollution that impedes the formation of human capital affects the dynamic behavior of an economy by applying an optimal growth framework.

In our model, there are two growth engines: human capital accumulation and physical capital accumulation, as in the model of Lucas (1988). Although without pollution, the economy would experience sustainable endogenous growth, the economy never exhibits endogenous growth because of the negative external effect of pollution. In our model, the stock of physical capital emits pollution if sufficient physical capital accumulates, and this negatively affects the formation of human capital. More concretely, whereas one unit of investment in physical capital production produces one unit of physical capital, one unit of investment in human capital production produces less than one unit of human capital, being impeded by pollution. As physical capital accumulates further, the negative external effect of pollution on the formation of human capital is strengthened. Therefore, the economy is prevented from experiencing endogenous growth.

The outcome that the economy cannot experience endogenous growth is intuitive. The representative agent does not intend to invest much in human capital production because pollution disturbs the formation of human capital. Instead, the agent invests more in physical capital production, thereby increasing the supply of physical capital. As a result, the shadow price of physical capital (and general goods) declines in each period relative to the case without pollution. Our model proves that the value of consumption (i.e., the shadow price of general goods multiplied by consumption) is constant in each period. Then, if the shadow price becomes low, the current consumption becomes large. As such, the allocative inefficiency coming from the investment decision causes overconsumption in each period relative to the case without pollution, and thus, the amount of general goods produced decreases. Therefore, both human capital and physical capital accumulate to a lesser extent, and endogenous growth never occurs.

The investigation of the local dynamics shows that (i) if the extent of the external effect of pollution is relatively small, the steady state is stable and the economy starting in the neighborhood of the steady state converges to it; (ii) if the extent of the external effect is relatively large, the steady state is unstable and the economy diverges away from it; and (iii) a Hopf bifurcation occurs at a certain intermediate extent of the external effect and a limit cycle emerges around the steady state. The Hopf bifurcation in the optimal growth model was first obtained by Benhabib and Nishimura (1979); we apply it to the current growth model. Additionally, the numerical analysis illustrates the global behavior of the dynamical system in which the economy exhibits a closed orbit as sufficient time passes if the steady state is unstable.

Our study belongs to the literature on growth and the environment. Many researchers have investigated growth and the environment over the past 40 years by applying an optimal growth framework with infinitely lived agents.[1] Among others, Forster (1973), Tahvonen and Kuuluvainen (1993), and van der Ploeg and Withagen (1991) studied the growth effect of pollution with Ramsey-type growth models when pollution affects an instantaneous utility function or a neoclassical production function. Bovenberg and Smulders (1995) and Xepapadeas (1997) also investigated the growth effect of pollution by employing endogenous growth models, á la Romer (1986) and á la Lucas (1988), respectively. Unlike our study, these studies did not obtain endogenous business cycles in equilibrium. In contrast, Wirl (2004) and Bosi and Desmarchelier (2018a) showed that a limit cycle emerges in equilibrium by extending the model of Ayong Le Kama (2001), in which pollution negatively affects an environmental resource that has a positive effect on an instantaneous utility function. Furthermore, Bosi and Desmarchelier (2018b, c) developed Ramsey-type growth models in which pollution affected the disease transmission mechanism (which indirectly impacts the aggregate labor supply) and/or consumption demand, and derived a limit cycle. In our model, a limit cycle is also derived, but unlike these models, pollution has a negative effect on the formation of human capital.

The remainder of the chapter is organized as follows. The next section presents the model, the growth engine of which is human and physical capital. In Sect. 6.3, we derive an equilibrium in which we obtain differential equations with respect to human and physical capital and investigate the dynamical system locally. In Sect. 6.4, we conduct a numerical exercise to observe the global behavior of the dynamical system and Sect. 6.5 concludes our study.

6.2 Model

An economy goes from time $t = 0$ to $t = +\infty$ in continuous time and is inhabited by identical infinitely lived agents. Their population is constant and normalized to one. A representative agent produces general goods with a Cobb-Douglas production technology $y = Ah^{\alpha}k^{1-\alpha}$ $(0 < \alpha < 1)$, the inputs of which are human capital h and physical capital k, where A is the technology level. Because the general goods are used for investment or consumption in each period, the flow budget constraint is given by

$$Ah^{\alpha}k^{1-\alpha} = c + i^{h} + i^{k}, \tag{6.1}$$

where c is consumption and i^h and i^k are investments in the formation of human and physical capital, respectively. The accumulation of human and physical capital follows the equations below:

$$\dot{h} = \bar{k}^{-\sigma} i^{h} - \delta h \tag{6.2}$$

[1] See Xepapadeas (2005) for a comprehensive survey of the literature.

$$\dot{k} = i^k - \delta k, \tag{6.3}$$

where the depreciation rates of human and physical capital are the same, which is given by δ. In Eq. (6.2), $\bar{k}^{-\sigma}$ ($\sigma > 0$) is an external effect that the accumulation of physical capital has on the formation of human capital. If k is less than one, investment in human capital production is enhanced by the external effect, and if k is greater than one, it is disturbed by the effect. One may imagine that if sufficient physical capital accumulates, it begins to emit pollution that negatively affects the production of human capital. In what follows, we simply call σ the extent of the external effect. Because our interest is in the situation in which physical capital accumulates sufficiently that it emits pollution, we focus our following analysis on such a case unless otherwise stated. As σ increases, the extent of the negative external effect becomes large.

6.2.1 Utility Maximization

The representative agent maximizes lifetime utility $\int_0^\infty e^{-\rho t} \ln(c)dt$ subject to Eqs. (6.1)–(6.3); namely, the agent solves the following maximization problem:

$$\max \int_0^\infty e^{-\rho t} \ln(c)dt$$

subject to:

$$\begin{aligned} y &= Ah^\alpha k^{1-\alpha} = c + i^h + i^k \\ \dot{h} &= \bar{k}^{-\sigma} i^h - \delta h \\ \dot{k} &= i^k - \delta k, \end{aligned}$$

where $\rho > 0$ is the subjective discount rate. The current-value Hamiltonian is set as follows:

$$H := \ln(c) + \lambda(Ah^\alpha k^{1-\alpha} - c - i^h - i^k) + \lambda^h(\bar{k}^{-\sigma} i^h - \delta h) + \lambda^k(i^k - \delta k),$$

where λ, λ^h, and λ^k are the shadow prices of general goods, human capital, and physical capital, respectively. The first-order conditions are given by

$$\frac{\partial H}{\partial c} = \frac{1}{c} - \lambda = 0 \tag{6.4}$$

$$\frac{\partial H}{\partial i^h} = -\lambda + \lambda^h \bar{k}^{-\sigma} = 0 \tag{6.5}$$

$$\frac{\partial H}{\partial i^k} = -\lambda + \lambda^k = 0 \tag{6.6}$$

$$\frac{\partial H}{\partial h} = -\dot{\lambda}^h + \rho\lambda^h \Leftrightarrow \lambda\alpha\frac{y}{h} - \delta\lambda^h = -\dot{\lambda}^h + \rho\lambda^h \tag{6.7}$$

$$\frac{\partial H}{\partial k} = -\dot{\lambda}^k + \rho\lambda^k \Leftrightarrow \lambda(1-\alpha)\frac{y}{k} - \delta\lambda^k = -\dot{\lambda}^k + \rho\lambda^k, \tag{6.8}$$

where one should note that in equilibrium, it holds that $\bar{k} = k$. The transversality conditions are given by

$$\lim_{t\to\infty} e^{-\rho t}\lambda^h h = \lim_{t\to\infty} e^{-\rho t}\lambda^k k = 0. \tag{6.9}$$

6.3 Equilibrium

In this section, we define a dynamic competitive equilibrium and characterize the dynamical system.

Proposition 1 The shadow prices of human and physical capital satisfy the following differential equations in equilibrium:

$$\dot{\lambda}^h = \left(\rho + \delta - \alpha\frac{k^{-\sigma}y}{h}\right)\lambda^h \tag{6.10}$$

and

$$\dot{\lambda}^k = \left(\rho + \delta - (1-\alpha)\frac{y}{k}\right)\lambda^k. \tag{6.11}$$

Proof From Eq. (6.5), we have $\lambda = \lambda^h k^{-\sigma}$. Substituting this into Eq. (6.7) yields Eq. (6.10). Similarly, from Eq. (6.6), we have $\lambda = \lambda^k$. Substituting this into Eq. (6.8) yields Eq. (6.11). □

Proposition 2 The laws of motion of human and physical capital satisfy the following differential equations in equilibrium:

$$\dot{h} = \left((\sigma + \alpha - 1)h + \alpha k^{1-\sigma}\right)\left(\frac{Ah^{\alpha-1}k^{1-\alpha-\sigma}}{\sigma}\right) - \delta(h + k^{1-\sigma}) - \frac{1}{\lambda^k k^{\sigma}} \tag{6.12}$$

and

$$\dot{k} = \left((1-\alpha)h - \alpha k^{1-\sigma}\right)\left(\frac{Ah^{\alpha-1}k^{1-\alpha}}{\sigma}\right). \tag{6.13}$$

Proof See the Appendix.

A dynamic competitive equilibrium is defined as a sequence of the shadow prices $\{\lambda^h, \lambda^k\}$ and human and physical capital stocks $\{h, k\}$ for $t \geq 0$ that satisfy Eqs. (6.10)–(6.13) and the transversality conditions (6.9), given h_0 and k_0. In the equilibrium dynamics, λ^h and λ^k are non-predetermined variables that can jump and h and k are state variables that cannot jump.

6.3.1 Dynamical System

By using the transversality conditions (6.9), we can reduce the four-dimensional dynamical system consisting of Eqs. (6.10)–(6.13) to a two-dimensional dynamical system with respect to h and k. To do so, we prepare a useful lemma below.

Lemma 1 It holds that

$$\lambda^h h + \lambda^k k = \frac{1}{\rho} \tag{6.14}$$

for all $t \geq 0$ in equilibrium.

Proof See the Appendix.

Equation (6.14) implies that the sum of the current values of human and physical capital is equal to the sum of the present values of consumption flows.[2] This equation holds regardless of the presence of the negative external effect of pollution. Before arranging the dynamical system by using Eq. (6.14), we observe the relationship among human capital, physical capital, and consumption in Remark 1.

Remark 1 It holds that

$$c = \rho(k^\sigma h + k) \tag{6.15}$$

for all $t \geq 0$ in equilibrium.

Proof From Eqs. (6.4), (6.5), and (6.6), it follows that $\lambda c = 1$ and $\lambda = \lambda^k = \lambda^h k^{-\sigma}$ in equilibrium. Substituting these equations into Eq. (6.14) yields Eq. (6.15). □

---

[2] Because $\lambda c = 1$, the sum of the present values of consumption flows is computed as $\int_0^\infty e^{-\rho t} \cdot 1 dt = 1/\rho$, which is the right-hand side of Eq. (6.14).

Note from Remark 1 that overconsumption occurs in each period when the accumulation of physical capital emits pollution. To consider this, suppose that $k > 1$. If there were no negative external effects of pollution, consumption would be determined by $c = \rho(h + k)$ given human and physical capital, which implies that a certain proportion of the sum of human and physical capital is optimally consumed in each period. However, now that there is a negative external effect of pollution, the representative agent consumes more than the optimal level because $k^\sigma > 1$.

From Eqs. (6.5), (6.6), and (6.14), we have $1/\lambda^k = \rho(k^\sigma h + k)$. Substituting this equation into Eq. (6.12) yields the following equation:

$$\dot{h} = \left((\sigma + \alpha - 1)h + \alpha k^{1-\sigma}\right)\left(\frac{Ah^{\alpha-1}k^{1-\alpha-\sigma}}{\sigma}\right) - (\rho + \delta)(h + k^{1-\sigma}). \tag{6.16}$$

Given h_0 and k_0, the equilibrium path with respect to h and k is given by Eqs. (6.13) and (6.16). Because of the transversality conditions, the shadow prices, λ^h and λ^k, jump such that the economy reaches the manifold that includes the set of general solutions of Eqs. (6.13) and (6.16). In what follows, our analysis focuses on the dynamical system with respect to h and k that is given by Eqs. (6.13) and (6.16), or equivalently,

$$\begin{cases} \dot{h} = \left((\sigma + \alpha - 1)h + \alpha k^{1-\sigma}\right)\left(\frac{Ah^{\alpha-1}k^{1-\alpha-\sigma}}{\sigma}\right) - (\rho + \delta)(h + k^{1-\sigma}) \\ \dot{k} = \left((1-\alpha)h - \alpha k^{1-\sigma}\right)\left(\frac{Ah^{\alpha-1}k^{1-\alpha}}{\sigma}\right). \end{cases} \tag{6.17}$$

6.3.2 Steady State

From Eq. (6.17) , we obtain the steady-state values of human and physical capital as follows:

$$h^* := \left[\left(\frac{(1-\alpha)A}{\rho+\delta}\right)^{\frac{1-\sigma}{\alpha}}\left(\frac{\alpha}{1-\alpha}\right)\right]^{\frac{1}{\sigma}} \tag{6.18}$$

and

$$k^* := \left[\left(\frac{(1-\alpha)A}{\rho+\delta}\right)^{\frac{1}{\alpha}}\left(\frac{\alpha}{1-\alpha}\right)\right]^{\frac{1}{\sigma}}. \tag{6.19}$$

It is straightforward to show that both h^* and k^* increase as σ decreases if $A\alpha^\alpha(1-\alpha)^{1-\alpha} > \rho + \delta$. In particular, Remark 2 clarifies the case in which σ is close to zero.

Remark 2 Suppose that $A\alpha^{\alpha}(1-\alpha)^{1-\alpha} > \rho + \delta$. Then, it follows that $\lim_{\sigma \to 0} h^* = \infty$ and $\lim_{\sigma \to 0} k^* = \infty$.

Proof The claim immediately follows from Eqs. (6.18) and (6.19). □

Remark 2 implies that without the negative external effect of pollution, the economy would exhibit endogenous growth, the engines of which are the accumulation of human and physical capital. If there were no external effects of pollution, that is, if $\sigma = 0$, the growth rate on the balanced growth path in equilibrium would be equal to $A\alpha^{\alpha}(1-\alpha)^{1-\alpha} - (\rho + \delta)$. However, the negative external effect that pollution has on human capital formation produces finite steady-state values of human and physical capital. Note that although the extent of the negative external effect is infinitesimal but not zero, the steady state appears. The intuition behind the outcome that the economy does not exhibit endogenous growth is as follows. Suppose that sufficient physical capital accumulates such that k is greater than one; thus, physical capital emits pollution at a certain time. Because the production of human capital is disturbed by pollution, the representative agent refrains from investing in human capital production. Rather, the agent invests more in physical capital production. Then, the supply of physical capital increases, and the shadow price of physical capital (and general goods) decreases in each period relative to the case without pollution. Because the value of current consumption is one (i.e., $\lambda c = 1$), if the shadow price of general goods becomes low, current consumption becomes large. The allocative inefficiency regarding the investment decision causes overconsumption, as seen in Remark 1, and the lower amount of general goods production in each period relative to the case without pollution. Eventually, both human capital and physical capital accumulate less, and thus, endogenous growth never occurs.

6.3.3 Dynamics of *h* and *k*

Linearizing the dynamical system (6.17) around the steady state, we obtain the local system as follows:

$$\begin{pmatrix} \dot{h} \\ \dot{k} \end{pmatrix} = J \begin{pmatrix} h - h^* \\ k - k^* \end{pmatrix}, \tag{6.20}$$

where J is the Jacobian of the system, which is given by

$$J = \begin{pmatrix} -\frac{(1-\alpha)(\rho+\delta)}{\alpha\sigma} & \left(\frac{(1-\alpha)(1-\sigma)-\alpha\sigma^2}{\alpha\sigma}\right)(\rho+\delta)^{\frac{1+\alpha}{\alpha}}(A(1-\alpha))^{-\frac{1}{\alpha}} \\ \frac{1}{\sigma}(\rho+\delta)^{\frac{\alpha-1}{\alpha}}(A(1-\alpha))^{\frac{1}{\alpha}} & -\left(\frac{1-\sigma}{\sigma}\right)(\rho+\delta) \end{pmatrix}.$$

Let $\mathrm{Tr}(J)$ and $\mathrm{Det}(J)$ denote the trace and determinant of J, respectively, which are computed as

$$\mathrm{Tr}(J) = (\rho + \delta)\left(1 - \frac{1}{\alpha\sigma}\right) \tag{6.22}$$

and

$$\mathrm{Det}(J) = (\rho + \delta)^2. \tag{6.23}$$

Because $\mathrm{Det}(J) > 0$, the real parts of the eigenvalues have the same sign and whether the sign is positive or negative is determined by the trace. From Eq. (6.22), if $\alpha\sigma < 1$ (> 1), the sign of the real parts is negative (positive). The following proposition characterizes the steady state and the local dynamic behavior of the economy in terms of the range of parameter values of σ.

Proposition 3 The following hold.

- If $0 < \sigma < 1/\alpha$, the steady state is stable, and (h, k) starting with any initial values of (h_0, k_0) in the neighborhood of the steady state converges to the steady state.
- If $1/\alpha < \sigma$, the steady state is unstable, and (h, k) starting with any initial values of (h_0, k_0) in the neighborhood of the steady state diverges away from the steady state.

Proof See the Appendix.

Furthermore, if σ increases from 0, a Hopf bifurcation occurs as summarized in the Remark 3.

Remark 3 A Hopf bifurcation occurs at $\sigma = 1/\alpha$, and a limit cycle appears at a certain value of $\sigma \in (1/\alpha - \epsilon, 1/\alpha + \epsilon)$ where $\epsilon > 0$.

Proof At $\sigma = 1/\alpha$, $\mathrm{Tr}(J)$ becomes zero with $\mathrm{Det}(J)$ remaining positive. Additionally, the derivative of the real parts of the eigenvalues with respect to σ is positive. Then, a Hopf bifurcation occurs at $\sigma = 1/\alpha$. □

For the criterion to determine whether the Hopf bifurcation is subcritical or supercritical, see Nishimura and Shigoka (2019).

6.4 Numerical Analysis

In the previous section, we investigated the local dynamics of the economy. To observe the global dynamic behavior, we perform a numerical analysis in this section.[3] In particular, we shall see that the economy can exhibit cyclical behavior when the steady state is unstable as in the second case of Proposition 3. Throughout

[3] To perform the numerical analysis, we use MATLAB R2020a with ode45.

the numerical analysis in this section, we set the parameter values as $\alpha = 0.33$, $\rho = 0.05$, $\delta = 0.10$, and $A = 1$. Regarding σ, we examine three cases: $\sigma = 1$, $\sigma = 1/0.3$, and $\sigma = 1/0.33$. We iterate 200,000 times for the simulation. From Proposition 3, $\sigma = 1$ yields a stable steady state and $\sigma = 1/0.3$ yields an unstable steady state. $\sigma = 1/0.33$ is the value of σ that produces a Hopf bifurcation.

Case 1: $\sigma = 1$

Figure 6.1 shows the case of $\sigma = 1$. In this case, the steady-state values of human and physical capital are given by $h^* = 0.491$ and $k^* = 45.92$, respectively. The initial values of human and physical capital are $h_0 = 0.01$ and $k_0 = 2.4$, respectively. Starting from the initial values, human capital increases first and overshoots, while physical capital remains relatively unchanged. However, because of the negative external effect of pollution, human capital starts to decline, and physical capital significantly increases. Then, the economy converges to the stable steady state. We examined various initial values, but the convergence outcomes were all the same, which implies that the economy is globally stable if the condition of the first part of Proposition 3 holds.

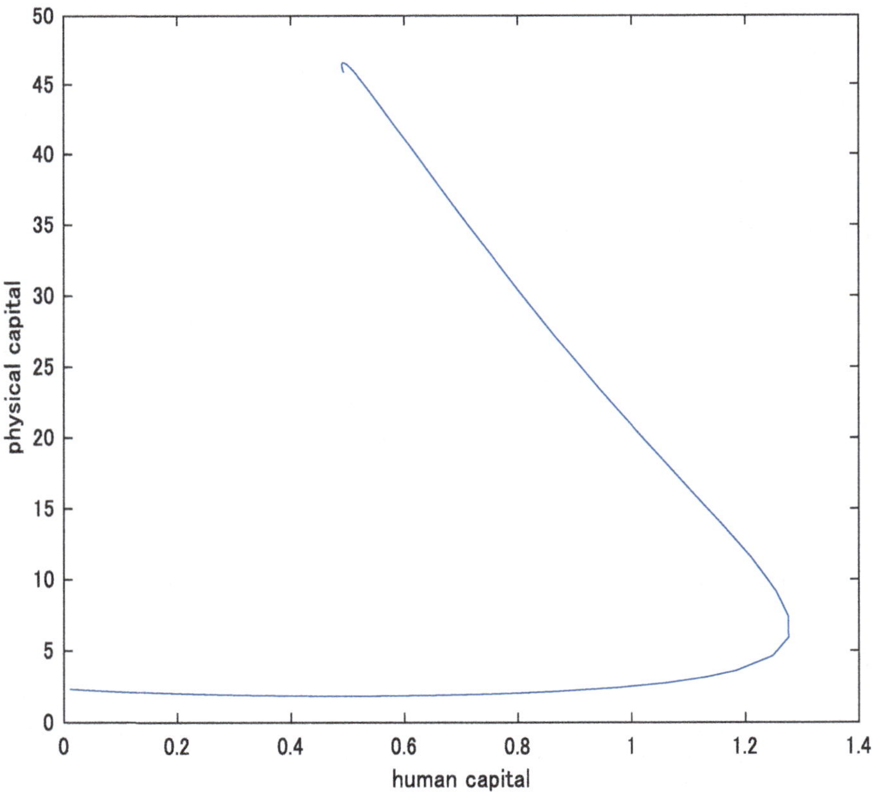

Fig. 6.1 Global dynamics $\sigma = 1$, $h_0 = 0.01$, $k_0 = 2.4$, $h^* = 0.491$, $k^* = 45.92$

Case 2: $\sigma = 1/0.3$

The outcome of this case is presented in Fig. 6.2. In this case, the steady-state values are $h^* = 0.0338$ and $k^* = 3.152$, both of which are smaller than those in Case 1. Of course, this is because the extent of the negative external effect of pollution is more severe than in Case 1. Panel **a** in Fig. 6.2 examines the case in which the economy starts away from the steady state. More concretely, the initial values are $h_0 = 0.01$ and $k_0 = 2.4$, which are the same as those in Case 1. Although the dynamic courses of both human and physical capital on the initial dates are similar to those of Case 1, the economy does not converge to the steady state but exhibits a closed orbit that surrounds the steady state, implying that endogenous business cycles occur. In panel **b** of Fig. 6.2, the economy starts from initial values that are closer to the steady state, which are $h_0 = 0.033$ and $k_0 = 3.15$. Because the steady state is unstable and the eigenvalues of J are imaginary numbers with positive real parts, the economy exhibits oscillation with the amplitude widening over time. Eventually, we obtain the same closed orbit as that in the case of panel **a**.

Case 3: $\sigma = 1/0.33$

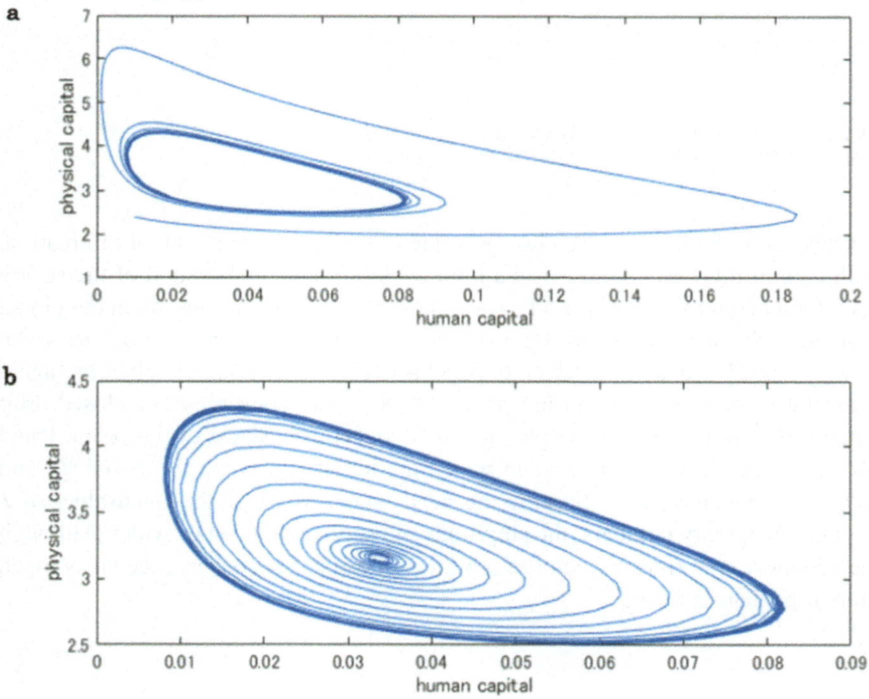

Fig. 6.2 Global dynamics $\sigma = 1/0.3$, **a** $h_0 = 0.01$, $k_0 = 2.4$, $h^* = 0.0338$, $k^* = 3.152$, **b** $h_0 = 0.033$, $k_0 = 3.15$, $h^* = 0.0338$, $k^* = 3.152$

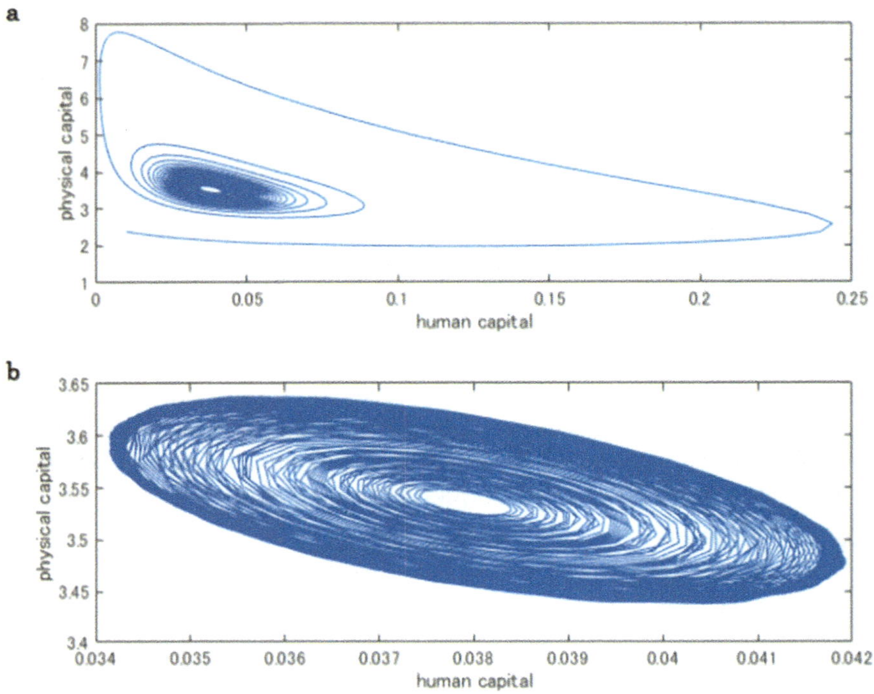

Fig. 6.3 Global dynamics $\sigma = 1/0.33$, **a** $h_0 = 0.01$, $k_0 = 2.4$, $h^* = 0.03779$, $k^* = 3.535$, **b** $h_0 = 0.0377$, $k_0 = 3.53$, $h^* = 0.03779$, $k^* = 3.535$

Because $\alpha = 0.33$, $\sigma = 1/0.33$ is the value of σ that causes the Hopf bifurcation. Although in this case, there exists a limit cycle in the neighborhood of the steady state for a certain value of $\sigma \in (1/\alpha - \epsilon, 1/\alpha + \epsilon)$, we are curious about the global dynamic behavior. The steady-state values of human and physical capital are given by $h^* = 0.03379$ and $k^* = 3.535$. In panel **a** in Fig. 6.3, the initial values are again $h_0 = 0.01$ and $k_0 = 2.4$. As in Case 2, the economy converges to a closed orbit when sufficient time passes, implying that endogenous business cycles occur. Panel **b** of Fig. 6.3 illustrates the case in which the initial values are $h_0 = 0.0377$ and $k_0 = 3.53$, which are even closer to the steady state. Although the eigenvalues of J are pure imaginary numbers, the amplitude of oscillation becomes wider. Although the economy converges to a closed orbit as in Case 2, the convergence takes much more time than in Case 2.

6.5 Conclusion

We have investigated the growth effect of pollution that is emitted from the stock of physical capital and has a negative impact on the formation of human capital by applying a simple optimal growth framework in which there are two growth engines: human and physical capital accumulation. Because of the negative impact, sustainable endogenous growth never occurs, and a unique steady state emerges despite that the extent of the external effect of pollution is infinitesimal. The analysis shows that if the extent of the external effect is relatively small, the steady state is stable, and the economy starting in the neighborhood of the steady state converges to it. If the extent of the external effect is relatively large, the steady state is unstable, and the economy starting in the neighborhood of the steady state diverges away from it. Furthermore, a Hopf bifurcation occurs at a certain intermediate extent of the external effect. We have also performed a numerical analysis to observe the global dynamic behavior, in which the economy exhibits a closed orbit once sufficient time passes if the steady state is unstable.

One can extend our model such that pollution abatement technologies are introduced and pollution impacts not only the formation of human capital but also consumption demand. Investigating what would happen to the dynamic behavior under these extensions is left for future research.

Acknowledgements This work was financially supported by the Japan Society for the Promotion of Science: Grants-in-Aid for Scientific Research (Nos. 16H03598, 16K03685, 20H05633, and 20K01647).

Appendix

Proof of Proposition 2 From Eqs. (6.5) and (6.6), it follows that

$$\frac{\dot{\lambda}^k}{\lambda^k} = \frac{\dot{\lambda}^h}{\lambda^h} - \sigma \frac{\dot{k}}{k}. \tag{6.24}$$

From Eqs. (6.10), (6.11), and (6.24), we obtain Eq. (6.13). Equations (6.1)–(6.4) and (6.6) eliminate c, i^h, and i^k and we obtain

$$\frac{1}{\lambda^k} + \dot{k} + \delta k + k^\sigma (\dot{h} + \delta h) = y. \tag{6.25}$$

Substituting Eq. (6.13) into Eq. (6.25) yields Eq. (6.12). □

Proof of Lemma 1 Define $x = \lambda^h h + \lambda^k k$. Then, it follows from Eqs. (6.2), (6.3), (6.5), (6.10), and (6.11) that

$$\begin{aligned}\dot{x} &= \dot{\lambda}^h h + \lambda^h \dot{h} + \dot{\lambda}^k k + \lambda^k \dot{k} \\ &= \lambda^h (k^{-\sigma} i^h - \delta h) + \lambda^k (i^k - \delta k) + (\rho + \delta)(\lambda^h h + \lambda^k k) \\ &\quad - \lambda^h k^{-\sigma} \alpha y - \lambda^k (1-\alpha) y = \lambda^k (i^h + i^k) + \rho x - \lambda^k y. \end{aligned} \tag{6.26}$$

Equations (6.1) and (6.26) eliminate i^h and i^k and we obtain

$$\dot{x} = \rho x - \lambda^k c. \tag{6.27}$$

Applying Eqs. (6.4) and (6.6) to Eq. (6.27) yields

$$\dot{x} = \rho x - 1. \tag{6.28}$$

Solving Eq. (6.28), we have

$$e^{-\rho t} x = x_0 + \frac{1}{\rho}(e^{-\rho t} - 1). \tag{6.29}$$

From the transversality conditions, it follows that $\lim_{t \to \infty} e^{-\rho t} x = e^{-\rho t}(\lambda^h h + \lambda^k k) = 0$. Therefore, from Eq. (6.29) we obtain $x_0 = 1/\rho$ and thus $x = 1/\rho$ for all $t \geq 0$. □

Proof of Proposition 3 Because $\text{Det}(J) > 0$, the real parts of the eigenvalues have the same sign. If $0 < \sigma < 1/\alpha$, it follows that $\text{Tr}(J) < 0$, and thus, the real parts of the eigenvalues are negative. Therefore, the steady state is stable. If $1/\alpha < \sigma$, it follows that $\text{Tr}(J) > 0$, and thus, the real parts of the eigenvalues are positive. Therefore, the steady state is unstable. □

References

Ayong Le Kama, A. D. (2001). Sustainable growth, renewable resources and pollution. *Journal of Economic Dynamics and Control, 25*(12), 1911–1918.

Bell, M. L., Samet, J. M., & Dominici, F. (2004). Time-series studies of particulate matter. *Annual Review of Public Health, 25*, 247–280.

Benhabib, J., & Nishimura, K. (1979). Hopf bifurcation and the existence and stability of closed orbits in multi-sector models of economic growth. *Journal of Economic Theory, 21*(3), 421–444.

Bosi, S., & Desmarchelier, D. (2018). Natural cycles and pollution. *Mathematical Social Sciences, 96*, 10–20.

Bosi, S., & Desmarchelier, D. (2018). Limit cycles under a negative effect of pollution on consumption demand: The role of an environmental Kuznets curve. *Environmental and Resource Economics, 69*, 343–363.

Bosi, S., & Desmarchelier, D. (2018). Pollution and infectious diseases. *International Journal of Economic Theory, 14*(4), 351–372.

Bosi, S., & Desmarchelier, D. (2019). Pollution effects on disease transmission and economic stability. *International Journal of Economic Theory, 17*(2), 169–189.

Bovenberg, A. L., & Smulders, S. A. (1995). Environmental quality and pollution-augmenting technological change in a two-sector endogenous growth model. *Journal of Public Economics, 57*(3), 369–391.

Chay, K. Y., & Greenstone, M. (2003). The impact of air pollution on infant mortality: Evidence from geographic variation in pollution shocks induced by a recession. *Quarterly Journal of Economics, 118*(3), 1121–1167.

Currie, J., & Neidell, M. (2005). Air pollution and infant health: What can we learn from California's recent experience? *Quarterly Journal of Economics, 120*(3), 1003–1030.

Forster, B. A. (1973). Optimal capital accumulation in a polluted environment. *Southern Economic Journal, 39*(4), 544–557.

Graff Zivin, J., & Neidell, M. (2012). The impact of pollution on worker productivity. *American Economic Review, 102*(7), 3652–3673.

Graff Zivin, J., & Neidell, M. (2013). Environment, health, and human capital. *Journal of Economic Literature, 51*(3), 689–730.

Lucas, R. E. (1988). On the mechanics of economic development. *Journal of Monetary Economics, 22*(1), 3–42.

Nishimura, K., & Shigoka, T. (2019). Hopf bifurcation and the existence and stability of closed orbits in three-sector models of optimal endogenous growth. *Studies in Nonlinear Dynamics and Econometrics, 23*(4). https://doi.org/10.1515/snde-2019-0017

Pope, C. A., Burnett, R. T., Thun, M. J., Calle, E. E., Krewski, D., Ito, K., & Thurston, G. D. (2002). Lung cancer, cardiopulmonary mortality, and long-term exposure to fine particulate air pollution. *Journal of the American Medical Association, 287*(9), 1132–1141.

Romer, P. M. (1986). Increasing returns and long-run growth. *Journal of Political Economy, 94*(5), 1002–1037.

Tahvonen, O., & Kuuluvainen, J. (1993). Economic growth, pollution, and renewable resources. *Journal of Environmental Economics and Management, 24*(2), 101–118.

van der Ploeg, F., & Withagen, C. (1991). Pollution control and the Ramsey problem. *Environmental and Resource Economics, 1*, 215–236.

Wirl, F. (2004). Sustainable growth, renewable resources and pollution: Thresholds and cycles. *Journal of Economic Dynamics and Control, 28*(6), 1149–1157.

Xepapadeas, A. (1997). Economic development and environmental pollution: Traps and growth. *Structural Change and Economic Dynamics, 8*(3), 327–350.

Xepapadeas, A. (2005). Economic growth and the environment. In K.-G. Mäler & J. R. Vincent (Eds.), *Handbook of environmental economics* (Vol. 3, pp. 1219–1271). Elsevier.

Takuma Kunieda Ph.D., is a Professor at Kwansei Gakuin University. After obtaining a bachelor's degree in agriculture and a master's degree in economics from Kyoto University, Kunieda received a Ph.D. in economics from Brown University. Before joining Kwansei Gakuin University, Kunieda worked at the Department of Economics and Finance at City University of Hong Kong. His research interests include macroeconomic theory and empirical macroeconomics, covering economic growth, endogenous business cycles, inequality, and financial market imperfections. He has published research articles in high-quality academic journals such as *Journal of Monetary Economics, Economic Inquiry, Journal of International Money and Finance*, and *Journal of Mathematical Economics.*

Kazuo Nishimura Ph.D., is a Specially Appointed Professor of the Research Institute for Economics and Business Administration at Kobe University in Japan. He is also Professor Emeritus of Kyoto University and a member of the Japan Academy. He received his Ph.D. from the University of Rochester in 1977. Nishimura served as President of The Japanese Economic Association in 2000–2001 and has been a fellow of the Econometric Society since 1992. He is known for contributions in complexity economics and has served as an external professor of the Santa Fe Institute, 2008–2017.

Chapter 7
Productive Consumption in a Two-Sector Model of Economic Development

Ichiroh Daitoh and Kazuo Nishimura

Abstract In low-income countries, labor productivity crucially depends on per capita consumption that contributes to good nutrition, health, and education. A higher level of per capita consumption improves each worker's labor productivity. This concept of productive consumption was first formulated in the aggregate growth model by Steger (J Econ Dyn Control (2002) 26:1053–1068) and analyzed using numerical simulations. In this chapter, we investigate the stability of the labor-efficiency model of productive consumption by extending Steger's model to a two-sector model. We find that a steady state uniquely exists and is saddle-point stable, providing theoretical support for Steger's simulation results.

Keywords Developing country · Aggregate economic growth · Productive consumption externality · Stability · Two-sector model

JEL Classification Codes O1 · O4 · E1

7.1 Introduction

In low-income countries, labor productivity crucially depends on a per capita consumption level that contributes to good nutrition (i.e., intake of calories, vitamins and minerals, etc.), health (including access to medical services), and (basic) education.[1] Thus, consumption not only satisfies current needs but also increases

Declarations of Interest: none.

I. Daitoh (✉)
Faculty of Business and Commerce, Keio University, 2-15-45, Minato-ku, Tokyo 108-8345, Japan
e-mail: idaitoh@fbc.keio.ac.jp

K. Nishimura
RIEB, Kobe University, 2-1 Rokkoudaicho, Nadaku, Kobe 657-8501, Japan
e-mail: nishimura@rieb.kobe-u.ac.jp

[1] A higher per capita consumption improves education because it leads to better nutrition and health, contributing to the ability to learn. In addition, it enables greater availability of and access to learning materials (e.g., notebooks and pencils) and to educational services provided by teachers.

K. Nishimura et al. (eds.), *Creative Complex Systems*, Creative Economy,
https://doi.org/10.1007/978-981-16-4457-3_7

the productivity of labor. This concept of "productive consumption," which has also been well known in the context of efficiency wage theory since Leibenstein's (1957a, b) seminal studies, remains relevant today in low-income economies.

Productive consumption had been studied in the labor, development, and agricultural economics literature from a microeconomic perspective. Lazear (1977), in investigating why people attend school, treated education as a joint product, simultaneously producing potential wage gains and utility. Bliss and Stern (1978) emphasized the link between productivity and consumption in the determination of wages. This link was shown to have solid empirical support in terms of the nutrition–productivity relation. Strauss (1986) showed a highly significant positive effect of caloric intake on family farm labor productivity using household-level data from Sierra Leone. On the basis of this study, Deolalikar (1988) found a significant effect of nutritional status, proxied by weight-for-height, in determining labor productivity in rural South India. Suen and Mo (1994) developed a microeconomic theory of productive consumption goods, pointing out that their true cost is the money price less their marginal value product. They demonstrated that the demand for productive goods tends to be relatively unresponsive to exogenous changes in prices and income. All the studies above conducted static analyses.

Dynamic analysis of productive consumption can be found in the literature on the efficiency wage hypothesis. This hypothesis was analyzed in static models in the 1970s and 1980s, focusing on rural labor markets in developing economies (Stiglitz, 1976, Bliss & Stern, 1978; Gersovitz, 1983; Dasgupta & Ray, 1986). However, in the 1990s, it was often used as a theory that could explain wage rigidity and involuntary unemployment in Keynesian macroeconomics. In this strand of research, the dynamic implications of productive consumption have been extensively studied by several authors (Dasgupta, 1993; Ray & Streufert, 1993; Banerji & Gupta, 1997; Jellala & Zenoub, 2000). However, these studies focused on the labor market and did not always investigate its macroeconomic consequences.[2]

In the early 2000s, Steger (2000a, 2002) proposed two models to analyze the growth process of an entire economy with productive consumption effects.[3] The labor-efficiency model in his 2002 study[4] provided broad explanations for the main stylized facts on aggregate economic growth in low-income developing countries. Steger (2000b) summarized them in four points: (1) a big diversity in the growth rates

[2] On a macroeconomic level, Wheeler (1980) empirically examined the relation between the growth rate of output and nutritional and health status and education using data on 54 countries in Africa, Asia, and South America.

[3] Whereas an increase in per capita consumption accelerates (disembodied) human capital accumulation in Steger (2000a), it increases the efficiency of labor at the same point in time in Steger (2002) (see also Gupta 2003). Both studies considered the optimal path along which the representative consumer takes this effect into account and controls it. By contrast, we regard it as an externality in this chapter.

[4] Dinda (2008) investigated the growth process in a one-sector AK-type model by incorporating social capital, which is formed by human capital accumulation as a result of productive consumption. Daitoh (2010) extended Steger's human-capital-accumulation formulation to the endogenous growth model under the "productive consumption hypothesis" and provided the conditions for a unique saddle-point stable steady state or multiple steady states.

of per capita income, (2) a positive correlation between the savings rate and per capita income, (3) β-divergence (a positive correlation between the growth rate and the level of per capita income) and, more generally, (4) a hump-shaped pattern of growth (β-divergence for the lower range of per capita income and β-convergence for the upper range of per capita income). Although Steger (2000b) reproduced stylized facts 2 and 3 in his linear (AK) growth model with subsistence consumption (Stone–Geary preferences), this basic model could not explain stylized fact 4, which is a well-documented empirical regularity (e.g., Cho, 1994).[5] In contrast, the labor-efficiency models with exogenous and endogenous labor supply in Steger (2002) illustrated stylized fact 4, as well as stylized facts 2 and 3, focusing on the transition path to an asymptotic balanced-growth equilibrium derived by numerical simulations.[6]

In this chapter, we extend Steger's labor-efficiency model of productive consumption to a two-sector model and investigate the existence, uniqueness, and stability of a balanced-growth (steady-state) equilibrium. Using a two-sector optimal growth model studied by Srinivasan (1964) and Uzawa (1964), we introduce a productive consumption externality into the production functions.[7] Both technologies are assumed to exhibit *constant* returns to scale in capital and labor *from the social perspective.*

We find that a steady state uniquely exists and is locally saddle-point stable, giving theoretical support to the simulation results in Steger (2002). It means that a low-income economy in which an increase in consumption improves labor efficiency will follow the transition path converging to a unique steady state given the initial level of capital. Our results show that an equilibrium path is uniquely determined and converges to a steady state even if the production consumption externality is introduced. This is consistent with earlier results, including those of Srinivasan (1964) and Uzawa (1964).

The remainder of the chapter is organized as follows. We discuss the model in Sect. 7.2 and investigate the local dynamics in Sect. 7.3. Section 7.4 concludes.

[5] Steger (2000b) explained stylized facts 1 and 4 with an extended model incorporating distortive government policy and the Jones–Manuelli type production function.

[6] In Steger's (2002) simulation, the speed of convergence to the asymptotic balanced-growth equilibrium was remarkably slow, i.e., half-life times were close to 200 years.

[7] Wichmann (1997) assumed that the nutrition–productivity relationship (a rise in agricultural and industrial labor productivity due to higher consumption of agricultural goods) was an externality.

7.2 The Model

Consider a perfectly competitive closed economy where a pure consumption good c (numeraire) and a pure investment good x are produced with capital $k_i(i = c, x)$ and labor $n_i(i = c, x)$. The total number of workers in the entire economy is normalized to one. In sector i $(i = c, x)$, the labor input in efficiency units is $h_i n_i$, where n_i is physical labor (each worker's number of working hours) and h_i is the level of human capital (labor productivity) embodied by each worker.

We introduce the "productive consumption hypothesis" in the form that each worker's labor productivity depends positively on the average level of consumption $\overline{c}$ in the society; that is, $h_i = h_i(\overline{c})$ with $h_i'(\overline{c}) > 0$. This effect works as an externality.[8] To make the analysis clear, we use the specification $h_i(\overline{c}) = \overline{c}^{\theta_i}(0 < \theta_i < 1)$. The production functions of the consumption and investment goods are, respectively:

$$c = k_c^{\alpha_1}(\overline{c}^{\theta_c} n_c)^{\alpha_2}, \alpha_1 > 0, \alpha_2 > 0, \alpha_1 + (1 + \theta_c)\alpha_2 = 1 \tag{7.1}$$

$$x = k_x^{\beta_1}(\overline{c}^{\theta_x} n_x)^{\beta_2}, \beta_1 > 0, \beta_2 > 0, \beta_1 + (1 + \theta_x)\beta_2 = 1, \tag{7.2}$$

Equations (7.1) and (7.2) imply that production technologies are assumed to exhibit *constant* returns to scale in capital $k_i(i = c, x)$ and labor $n_i(i = c, x)$ *from the social perspective.*

In what follows, we confine ourselves to the range of parameter values for incomplete specialization (both goods c and x are produced), and investigate the dynamic properties of the model. The capital accumulation function at any point in time t is:

$$\dot{k}(t) = k_x(t)^{\beta_1}(\overline{c}^{\theta_x}(t) n_x(t))^{\beta_2} - \delta k(t), \tag{7.3}$$

where $\delta > 0$ is the depreciation rate of capital.

The representative consumer's instantaneous utility function is $u(c) = c$, where c is per capita consumption. Given the expected time paths of the consumption externalities $\{\overline{c}(t)^{\theta_i}\}_{t=0}^{\infty}(i = c, x)$, the representative consumer chooses the time path of $\{n_c(t)\}_{t=0}^{\infty}$, $\{k_c(t)\}_{t=0}^{\infty}$, $\{n_x(t)\}_{t=0}^{\infty}$ and $\{k_x(t)\}_{t=0}^{\infty}$ to

$$\text{maximize} \int_0^{\infty} k_c(t)^{\alpha_1}(\overline{c}^{\theta_c}(t) n_c(t))^{\alpha_2} e^{-\rho t} dt$$

$$\text{subject to } \dot{k}(t) = k_x(t)^{\beta_1}(\overline{c}^{\theta_x}(t) n_x(t))^{\beta_2} - \delta k(t),$$

---

[8] We assume away factor-input externalities to clarify the role of the production consumption externality. Eliminating capital-input externalities fits with reality because capital accumulation tends to be scarce in the low-income economies in which the productive consumption hypothesis appears to hold.

$$n_c(t) + n_x(t) = n, k_c(t) + k_x(t) = k(t) \tag{7.4}$$

where $\rho > 0$ is a constant time discount rate. The last two equalities are, respectively, the labor and capital constraints at a point in time. In what follows, we normalize the total number of working hours endowed to each worker to one ($n = 1$).

Let us define the Lagrangian function as:

$$\begin{aligned} L = {} & k_c(t)^{\alpha_1}(\overline{c}(t)^{\theta_c} n_c(t))^{\alpha_2} + p(t)\{k_x(t)^{\beta_1}(\overline{c}^{\theta_x} n_x(t))^{\beta_2} - \delta k(t)\} \\ & + w(t)\{n - n_c(t) - n_x(t)\} + r(t)\{k(t) - k_c(t) - k_x(t)\}, \end{aligned} \tag{7.5}$$

where $p(t)$ is a costate variable (i.e., the imputed price of the investment good), and $w(t)$ and $r(t)$ are Lagrangian multipliers. From the first-order conditions, the input coefficients $a_{ij}(i = k, n;\ j = c, x)$, e.g., $a_{nc} = \frac{n_c}{c}$, may be obtained:

(i)

$$\frac{\alpha_2 c}{n_c} = w, \text{ or } a_{nc} = \frac{\alpha_2}{w}, \tag{7.6}$$

$$\frac{\alpha_1 c}{k_c} = r, \text{ or } a_{kc} = \frac{\alpha_1}{r}, \tag{7.7}$$

$$p\beta_2 \frac{x}{n_x} = w, \text{ or } a_{nx} = \frac{p\beta_2}{w}, \tag{7.8}$$

$$p\beta_1 \frac{x}{k_x} = r, \text{ or } a_{kx} = \frac{p\beta_1}{r}, \tag{7.9}$$

(ii)

$$\dot{p}(t) = \rho p(t) - \frac{\partial L}{\partial k(t)} = (\rho + \delta)p(t) - r(k(t), p(t)) \tag{7.10}$$

$$\dot{k}(t) = k_x(t)^{\beta_1}(\overline{c}^{\theta_x}(t) n_x(t))^{\beta_2} - \delta k(t).$$

Because the Lagrangian function (7.5) is concave in the control variables (n_c, k_c, n_x, k_x) and the state variable (k), the time paths that satisfy the first-order conditions and the transversality condition ($\lim_{t\to\infty} p(t)k(t)e^{-\rho t} = 0$) are the solution path of Problem (7.4).

We consider the market equilibrium path on which the derived consumption path coincides with the expected path of the consumption externality. Setting $c = \overline{c}$ in

Eq. (7.1) and solving the resulting equation, we obtain:

$$c = k_c^{\frac{\alpha_1}{\tau_c}} n_c^{\frac{\alpha_2}{\tau_c}}, \tag{7.11}$$

where $\tau_c \equiv 1 - \theta_c \alpha_2 = \alpha_1 + \alpha_2$ is the sum of the indices of the inputs, considering the productive consumption externality in the consumption good sector. The market equilibrium path is characterized by the autonomous dynamical system of capital k and its imputed price p:

$$\dot{k}(t) = x(k(t), p(t)) - \delta k(t), \tag{7.12}$$

$$\dot{p}(t) = (\rho + \delta)p(t) - r(k(t), p(t)). \tag{7.13}$$

We define a steady state (k^*, p^*) by $\dot{k}(t) = \dot{p}(t) = 0$, which exists uniquely.

Proposition 1 *Consider the two-sector growth model with a productive consumption externality in both the consumption and the investment goods sectors. There uniquely exists a steady state.*

Proof See Appendix.

7.3 Local Dynamics

We will show that the steady state is locally saddle-point stable in the present model. An equilibrium path converging to the steady state is the solution of Problem (7.4) because it satisfies the transversality condition. Let us first derive the Jacobian matrix for the linearized system of Eqs. (7.12) and (7.13) evaluated at the steady state:

$$J^* = \begin{bmatrix} \frac{\partial x^*}{\partial k} - \delta & \frac{\partial x^*}{\partial p} \\ -\frac{\partial r^*}{\partial k} & \rho + \delta - \frac{\partial r^*}{\partial p} \end{bmatrix}. \tag{7.14}$$

Full employment conditions for capital and labor hold on the market equilibrium path:

$$a_{kc}c + a_{kx}x = k, \tag{7.15}$$

$$a_{nc}c + a_{nx}x = n = 1. \tag{7.16}$$

In our two-sector model with incomplete specialization, the capital–labor endowment ratio k lies in the cone of diversification. Then, an infinitesimal increase in k

does not affect the factor prices (r, w) by the factor price equalization theorem (see below), and, thus, the input coefficients $a_{ij}(i = k, n, j = c, x)$. Totally differentiating Eqs. (7.15) and (7.16), we obtain $a_{kc}dc + a_{kx}dx = dk$ and $a_{nc}dc + a_{nx}dx = 0$. Eliminating dc and rearranging the terms yields:

$$\frac{\partial x^*}{\partial k} = \frac{a_{nc}}{a_{kx}a_{nc} - a_{kc}a_{nx}}. \tag{7.17}$$

Substituting Eqs. (7.6) through (7.9) into Eq. (7.17) and using $p^*(\rho + \delta) = r(k^*, p^*)$ lead to:

$$\frac{\partial x^*}{\partial k} = \frac{\alpha_2/w}{(p\beta_1/r)(\alpha_2/w) - (\alpha_1/r)(p\beta_2/w)} = \frac{\alpha_2(\rho + \delta)}{\alpha_2\beta_1 - \alpha_1\beta_2} \tag{7.18}$$

Thus, we obtain:

$$\frac{\partial x^*}{\partial k} - \delta = \frac{\alpha_1\rho + \delta\{\alpha_1\beta_2 + (1 + \theta_x)\beta_2\alpha_2\}}{\alpha_2\beta_1 - \alpha_1\beta_2}. \tag{7.19}$$

For price relationships, let us define $\hat{a}_{ij}(i = k, n;\ j = c, x)$ in the following:

$$\hat{a}_{nc} = (1 + \theta_c)a_{nc} = \frac{(1 + \theta_c)\alpha_2}{w}, \tag{7.20}$$

$$\hat{a}_{kc} = a_{kc} = \frac{\alpha_1}{r}, \tag{7.21}$$

$$\hat{a}_{nx} = (1 + \theta_x)a_{nx} = \frac{(1 + \theta_x)p\beta_2}{w}, \tag{7.22}$$

$$\hat{a}_{kx} = a_{kx} = \frac{p\beta_1}{r}. \tag{7.23}$$

Using these definitions, simple calculations yield:

$$\hat{a}_{kc}r + \hat{a}_{nc}w = 1, \tag{7.24}$$

$$\hat{a}_{kx}r + \hat{a}_{nx}w = p. \tag{7.25}$$

By the factor price equalization theorem, factor prices (r, w) are uniquely determined by output prices $(1, p)$, independently of factor endowments. Then, we obtain:

$$\frac{\partial r^*}{\partial k} = 0. \tag{7.26}$$

Therefore, the roots of J^* are $(\partial x^*/\partial k) - \delta$ and $\rho + \delta - (\partial r^*/\partial p)$.

We have $1 = (a_{kc})^{\frac{\alpha_1}{\tau_c}} (a_{nc})^{\frac{\alpha_2}{\tau_c}}$ from Eq. (7.11). Substituting Eqs. (7.6) and (7.7) into this and rearranging the terms, we obtain $(r/\alpha_1)^{\frac{\alpha_1}{\tau_c}} (w/\alpha_2)^{\frac{\alpha_2}{\tau_c}} = 1$. Total differentiation yields:

$$\hat{a}_{kc} dr + \frac{\hat{a}_{nc}}{1 + \theta_c} dw = 0. \tag{7.27}$$

Similarly, we have $(a_{kx})^{\beta_1} (a_{nx})^{\beta_2} = (x/c)^{1-\beta_1-\beta_2}$ from Eq. (7.2). Substituting Eqs. (7.8) and (7.9) into this and rearranging the terms, we obtain $p^{\beta_1+\beta_2} = (r/\beta_1)^{\beta_1} (w/\beta_2)^{\beta_2} (x/c)^{1-\beta_1-\beta_2}$. Total differentiation with (x/c) fixed yields:

$$\hat{a}_{kx} dr + \frac{\hat{a}_{nx}}{1 + \theta_x} dw = (\beta_1 + \beta_2) dp. \tag{7.28}$$

Eliminating dw from Eqs. (7.27) and (7.28) and rearranging the terms, we obtain the change in the rental rate of capital corresponding to a change in the price of the investment good:

$$\begin{aligned} \frac{\partial r^*}{\partial p} &= -(\beta_1 + \beta_2) \frac{(1 + \theta_x)\hat{a}_{nc}}{(1 + \theta_c)\hat{a}_{kc}\hat{a}_{nx} - (1 + \theta_x)\hat{a}_{kx}\hat{a}_{nc}} \\ &= -(\beta_1 + \beta_2)\left(\frac{r}{p}\right) \frac{\alpha_2}{\alpha_1\beta_2 - \alpha_2\beta_1} \end{aligned} \tag{7.29}$$

Because $p^*(\rho + \delta) = r(k^*, p^*)$ holds at the steady state, we obtain:

$$\rho + \delta - \frac{\partial r^*}{\partial p} = (\rho + \delta)\left(1 - \frac{p^*}{r^*}\frac{\partial r^*}{\partial p}\right) = \frac{(\rho + \delta)(\alpha_1 + \alpha_2)\beta_2}{\alpha_1\beta_2 - \alpha_2\beta_1}. \tag{7.30}$$

This establishes the next proposition.

Proposition 2 *Consider the two-sector growth model with a productive consumption externalityin both the consumption and the investment goods sectors. Whenever the capital intensity in the investment good sector differs from that in the consumption good sector from the private perspective ($\alpha_1/\alpha_2 \neq \beta_1/\beta_2$), the steady state is locally saddle-point stable.*

Proof Under $\alpha_1/\alpha_2 \neq \beta_1/\beta_2$, the two roots (Eqs. (7.19) and (7.30)) of J^* always have the opposite sign, implying the local saddle-point stability of the steady state. (Q.E.D.)

The inclusion of this factor intensity condition ($\alpha_1/\alpha_2 \neq \beta_1/\beta_2$) differentiates our two-sector model from an aggregative (one-sector) model. This proposition implies that the equilibrium path converges to the steady state when choosing the initial price level on the stable manifold.

7.4 Concluding Remarks

In this chapter, we investigated the general property of the labor-efficiency model of productive consumption, that is, the existence, uniqueness, and stability of a steady-state equilibrium by extending the labor-efficiency model of Steger (2002) to a two-sector constant-return-to-scale dynamic model with a productive consumption externality. We have proved that a steady state uniquely exists and is saddle-point stable, providing theoretical support for the simulation results of Steger (2002).

Productive consumption plays an important role in low-income developing economies even today. In future research, we aim to study the diversity of growth rates of developing countries in a model with productive consumption.

Acknowledgements We thank Takumi Naito (Waseda University), Masatoshi Tsumagari, Takako Fujiwara-Greve, and Fukunari Kimura (Keio University) for useful discussions and comments on the earlier version of this chapter. We acknowledge the JSPS Grant-in-Aid for Scientific Research Nos. 16H03598, 16K03654, 20H05633 and 24530241 for supporting this project. We thank Edanz Group (https://jp.edanz.com/ac) for editing a draft of this manuscript.

Appendix: Existence and Uniqueness of a Steady State

In this appendix, we prove that a steady state uniquely exists. First, by substituting the first-order conditions Eqs. (7.6) and (7.7) into the consumption good production function $c = k_c^{\alpha_1} n_c^{\alpha_2} \overline{c}^{\theta_c \alpha_2}$ and using $\alpha_1 + (1 + \theta_c)\alpha_2 = 1$, we obtain:

$$r^{\alpha_1} w^{\alpha_2} = \alpha_1^{\alpha_1} \alpha_2^{\alpha_2}. \tag{7.31}$$

Next, by substituting Eqs. (7.6), (7.7), (7.8), (7.9), and (7.11) into the investment good production function $x = k_x^{\beta_1} (\overline{c}^{\theta_x} n_x)^{\beta_2}$ and using $\beta_1 + (1 + \theta_x)\beta_2 = 1$ and the steady-state condition $p = r/(\rho + \delta)$, we obtain:

$$x = A\phi(r, w)c, \tag{7.32}$$

where

$$A = \left(\frac{1}{\rho + \delta}\right)^{\frac{\beta_1 + \beta_2}{1-\beta_1-\beta_2}} \left(\beta_1^{\beta_1} \beta_2^{\beta_2}\right)^{\frac{1}{1-\beta_1-\beta_2}} \left(\alpha_1^{\alpha_1} \alpha_2^{\alpha_2}\right)^{\frac{1}{\alpha_1+\alpha_2}} \text{ and}$$

$$\phi(r, w) = r^{\frac{\beta_2}{1-\beta_1-\beta_2} - \frac{\alpha_1}{\alpha_1+\alpha_2}} w^{-\left(\frac{\beta_2}{1-\beta_1-\beta_2} + \frac{\alpha_2}{\alpha_1+\alpha_2}\right)}.$$

The full employment condition for capital and the first-order conditions of Eqs. (7.7) and (7.9) imply that:

$$k = k_x + k_c = \left(\frac{p\beta_1}{r}\right)x + \left(\frac{\alpha_1}{r}\right)c \text{ i.e., } rk = p\beta_1 x + \alpha_1 c, \tag{7.33}$$

Eliminating c by substituting Eq. (7.32) into Eq. (7.33) and using the steady-state condition $x = \delta k$, we obtain:

$$rk = p\beta_1 x + \alpha_1 \frac{x}{A\phi(r, w)} = \left\{\left(\frac{\beta_1}{\rho + \delta}\right)r + \frac{\alpha_1}{A\phi(r, w)}\right\}\delta k.$$

Dividing both sides by k, collecting the terms with r on the left-hand side, and rearranging the terms, we obtain:

$$r\phi(r, w) = \frac{\alpha_1 \delta(\rho + \delta)}{A\{\rho + (1 - \beta_1)\delta\}}. \tag{7.34}$$

Equation (7.34) determines the steady-state rental rate r^* of capital because Eq. (7.31) implies that w is a decreasing function of r. By substituting $w = (\alpha_1/r)^{\alpha_1/\alpha_2}\alpha_2$ into Eq. (7.34), the left-hand side can be written as:

$$r\phi(r, w(r)) = r^{1 + \frac{\beta_2}{1-\beta_1-\beta_2}\left(1 + \frac{\alpha_1}{\alpha_2}\right)}\left[\alpha_1^{\alpha_1/\alpha_2}\alpha_2\right]^{-\left(\frac{\beta_2}{1-\beta_1-\beta_2} + \frac{\alpha_2}{\alpha_1+\alpha_2}\right)}. \tag{7.35}$$

Because the $r\phi(r, w(r))$ curve starting from the origin is monotonically increasing in r and goes to infinity as r increases, it crosses the horizontal line representing the right-hand side of Eq. (7.34) at only one point. Therefore, the steady state value r^* that satisfies Eq. (7.34) exists uniquely.

References

Banerji, S., & Gupta, M. R. (1997). The efficiency wage given long-run employment and concave labor constraint. *Journal of Development Economics, 53*, 185–195.

Bliss, C., & Stern, N. (1978). Productivity, wages and nutrition part I: Theory. *Journal of Development Economics, 5*, 331–362.

Cho, D. (1994). Industrialization, convergence, and patterns of growth. *Southern Economic Journal, 61*, 398–414.

Daitoh, I. (2010). Productive consumption and population dynamics in an endogenous growth model: Demographic trends and human development aid in developing economies. *Journal of Economic Dynamics and Control, 34*, 696–709.

Dasgupta, P. (1993). *An inquiry into well-being and destitution*. Oxford University Press.

Dasgupta, P., & Ray, D. (1986). Inequality as a determinant of malnutrition and underemployment: Theory. *Economic Journal, 96*, 1011–1034.

Deolalikar, A. B. (1988). Nutrition and labor productivity in agriculture: Estimates for rural south India. *Review of Economics and Statistics, 70*, 406–413.

Dinda, S. (2008). Social capital in the creation of human capital and economic growth: A productive consumption approach. *Journal of Socio-Economics, 37*, 2020–2033.

Gersovitz, M. (1983). Savings and nutrition at low income. *Journal of Political Economy, 91*, 841–855.
Gupta, M. R. (2003). Productive consumption and endogenous growth: A theoretical analysis. *Keio Economic Studies, 40*, 45–57.
Jellala, M., & Zenoub, Y. (2000). A dynamic efficiency wage model with learning by doing. *Economics Letters, 66*, 99–105.
Lazear, E. (1977). Education: Consumption or production? *Journal of Political Economy, 85*, 569–597.
Leibenstein, H. (1957a). *Economic backwardness and economic growth.* Wiley.
Leibenstein, H. (1957b). The theory of underemployment in backward economies. *Journal of Political Economy, 65*, 91–103.
Ray, D., & Streufert, P. (1993). Dynamic equilibria with unemployment due to undernourishment. *Economic Theory, 3*, 61–85.
Srinivasan, T. N. (1964). Optimal savings in a two-sector model of growth. *Econometrica, 32*(3), 358–373.
Steger, T. M. (2000a). Productive consumption and growth in developing countries. *Review of Development Economics, 4*, 365–375.
Steger, T. M. (2000b). Economic growth with subsistence consumption. *Journal of Development Economics, 62*, 343–361.
Steger, T. M. (2002). Productive consumption, the intertemporal consumption trade-off and growth. *Journal of Economic Dynamics and Control, 26*, 1053–1068.
Stiglitz, J. (1976). The efficiency wage hypothesis, surplus of labor and distribution of income in the LDCs. *Oxford Economic Papers, 28*, 185–207.
Strauss, J. (1986). Does better nutrition raise farm productivity? *Journal of Political Economy, 94*, 297–320.
Suen, W., & Mo, P. H. (1994). Simple analytics of productive consumption. *Journal of Political Economy, 102*, 372–383.
Uzawa, H. (1964). Optimal growth in a two-sector model of capital accumulation. *Review of Economic Studies, 31*, 1–24.
Wheeler, D. (1980). Basic needs fulfillment and economic growth: A simultaneous model. *Journal of Development Economics, 7*, 435–451.
Wichmann, T. (1997). *Agricultural technical progress and the development of a dual economy.* Physica-Verlag.

Ichiroh Daitoh Ph.D. is a Professor of the Faculty of Business and Commerce at Keio University in Japan. He received his Ph.D. from Keio University in 1999. He is interested in economic growth and development, international trade, and the environment. Daitoh has published journal papers not only on endogenous growth and population dynamics but also on the compatibility between poverty reduction and environmental protection in developing countries.

Kazuo Nishimura Ph.D. is a Specially Appointed Professor of the Research Institute for Economics and Business Administration at Kobe University in Japan. He is also Professor Emeritus of Kyoto University and a member of the Japan Academy. He received his Ph.D. from the University of Rochester in 1977. Nishimura served as President of The Japanese Economic Association in 2000–2001 and has been a fellow of the Econometric Society since 1992. He is known for contributions to the complexity economics literature and served as an external professor of the Santa Fe Institute during 2008–2017.

Chapter 8
Time and Mnemonic Morphism

Goro C. Kato and Kazuo Nishimura

Abstract This chapter provides a temporal topos theoretic formulation for the relationship of the conscious (cognitive) states of an entity for the past, present, and future in terms of nonfunctorially induced mnemonic morphisms. The past is recalled by memories, and the future consists of possible states with a probability space of possibilities. When our present consciousness and behavior change, through our memory our interpretation of the past changes, and simultaneously the range of probability distribution of future states will be affected. We describe the internal relationship among the conscious states in terms of the temporal topos (t-topos) theory. A conscious entity can be expressed as a pair (p, δ^p), where p is an object of the t-topos $S^\wedge$ over t-site S, and δ^p is a uniquely determined morphism from an initial object α of t-topos to p. For three objects U, V, and W in the t-site S (which correspond to the past, present, and future), we introduce noncanonical (i.e., nonfunctorial) mnemonic morphisms among the three states of p defined over U, V, and W, respectively. Then we study the interplay among conscious states δ^p_U, δ^p_V and δ^p_W of p corresponding to the past, present, and future.

Keywords Category · Sheaf · Topos · Mnemonic morphism

G. C. Kato
Mathematics Department, California Polytechnic State University, San Luis Obispo, CA 93407, USA
e-mail: gkato@calpoly.edu

K. Nishimura (✉)
Research Institute for Economics and Business Administration, Kobe University,
2-1 Rokkoudaicho, Nadaku, Kobe 657-8501, Hyogo, Japan
e-mail: nishimura@rieb.kobe-u.ac.jp

K. Nishimura et al. (eds.), *Creative Complex Systems*, Creative Economy,
https://doi.org/10.1007/978-981-16-4457-3_8

8.1 Introduction

This chapter provides a categorical formulation for a relationship among conscious states of the past, present, and future. We look back at the past and ponder the future. It is in the present state to view the past, and it is also the present that envisions the future. Because we cannot view our past without its memory, and we cannot imagine a future that we cannot conceive, the present state $p(V)$ of a conscious entity p during V exists in memories, and the future consists of possible events and a probability space of possibilities. If our present consciousness changes, through our memory our interpretation of the awareness of the past changes. The present cognitive state influences the range of probability distribution of future events. If the consciousness δ^p_V of the present is influenced by memory, then the interpretation of the past is affected, and likewise, if the consciousness of the present is changed, then the probability distribution of the future will be affected. In this chapter, we provide a sheaf theoretic formulation for such a relationship among the conscious states of the past, present, and future.

As an introduction, we recall a brief history of categories and sheaves. The first paper on a category was published in Eilenberg and MacLane (1945). Since then, algebraic geometry, the theory of holomorphic functions in several complex variables (i.e., complex analytic geometry), D-module theory, algebraic topology, and other fields have been developed in the language of categories and sheaves. The concept of a sheaf was discovered during World War II by Leray (1950) and Oka (1950) in algebraic topology and several complex variables, respectively. Grauert and Remmert (1984) provide some background on the history and development of sheaves.

The notion of a topos, which is the category of presheaves, has been applied to quantum physics and quantum gravity (e.g., Butterfield & Isham, 1999, 2001; Döring & Isham, 2011; Kato, 2004). Later it was applied to cognitive sciences as in Kato and Nishimura (2013). One of our fundamental axioms in the temporal topos (abbreviated as t-topos) theory is that for any entity (e.g., a particle), there exists a presheaf p corresponding to the entity. The theory of t-topos has been developed to explain physical and cognitive phenomena.

In Sect. 8.2, we give a short review of some of the basic concepts in categories and sheaves. In Sect. 8.3, we introduce the notion of a mnemonic morphism as a noncanonical (i.e., nonfunctorial) morphism that is comparable with cognitive states. Section 8.4 is the conclusion.

8.2 Functor and Sheaf

Here we provide basic notions regarding categories and sheaves. For general references related to categories, we recommend the works of Mitchell (1965), Schubert (1972), Gelfand and Manin (1996), Kato (2006).

A category S consists of objects and morphisms. For two objects U and V in category S, the set of morphisms from U to V is denoted as $Hom_S(U, V)$. For two morphisms f: U to V and g: V to W, the composition of morphism f followed by morphism g, written as $g \cdot f$, is defined as a morphism from U to W. That is, we have the following set theoretic mapping induced by the composition of morphisms; for (f, g) in $Hom_S(U, V)$x$Hom_S(V, W)$, there corresponds $g \cdot f$ in $Hom_S(U, W)$. Namely, we have the following mapping[1]:

$$Hom_S(U, V)\text{x}Hom_S(V, W) \rightarrow Hom_S(U, W).$$

A functor F from category C to category D is an assignment as follows. F takes an object V in C to an object $F(V)$ in D. Furthermore, for a morphism f: $U \rightarrow V$ in C, there is canonically (i.e., functorially) induced a morphism $F(f)$: $F(U) \rightarrow F(V)$ in D if F is a covariant functor. On the other hand, if F is a contravariant functor, we get a morphism $F(f)$: $F(V) \rightarrow F(U)$ in D. Because all the functors appearing in this chapter are contravariant functors, we only consider a contravariant functor in what follows.

Furthermore, for F to be a functor, it must satisfy the following properties. For the composed morphism $g \cdot f$: $U \rightarrow W$ of morphisms f: $U \rightarrow V$ and g $V \rightarrow W$ in C, a contravariant functor F takes the composed morphism $g \cdot f$ to $F(g \cdot f)$: $F(W) \rightarrow F(U)$ in D. On the other hand, as the composition of $F(g)$: $F(W) \rightarrow F(V)$ and $F(f)$: $F(V) \rightarrow F(U)$ in D, we have $F(f) \cdot F(g)$: $F(W) \rightarrow F(U)$, and then the equality $F(f) \cdot F(g) = F(g \cdot f)$ must hold. The last property of F being a functor is $F(1_U) = 1_{F(U)}$, where 1_U: $U \rightarrow U$ is the identity morphism on U, and $1_{F(U)}$: $F(U) \rightarrow F(U)$ is the identity morphism on $F(U)$.

A site S is a category with a Grothendieck topology. The concept of a site is a generalization of a topological space so that one can build a categorical theory of sheaves on a site. The works of Artin (1962) and Kashiwara and Schapira (2006) provide definitions of a site and Grothendieck topology.

By definition, a presheaf is a contravariant functor F from a site to any category. Then a presheaf F is a sheaf if the following sheaf axiom is satisfied, which we describe through a more classical set-theoretic way as done in Gelfand and Manin (1996) and Kato (2006).

Sheaf Axiom: Let $\{U_i\}$ be a covering of U in S. For sections s_i in $F(U_i)$ and s_j in $F(U_j)$, if the functorially induced "restriction" morphisms by F of s_i and s_j to $F(U_i \text{ x}_U U_j)$ coincide over the product $U_i \text{ x}_U U_j$ of U_i and U_j, then there exists a unique section s in $F(U)$ so that s restricted to each U_i coincides with s_i.

Notice that in the case of a topological space, the product $U_i \text{ x}_U U_j$ is the set-theoretic "intersection" of U_i and U_j. The works of Artin (1962), Gelfand and Manin (1996), Schubert (1972), and Makkai and Reyes (1977) provide a more precise categorical definition of a covering in terms of a site.

---

[1] Assignments $Hom_S(U, -)$ and $Hom_S(-, V)$ are examples of covariant and contravariant functors from category S to the category of sets, respectively.

8.3 Past-Present-Future as Objects of a Temporal Site

We review the canonically (i.e., functorially) induced morphism. Then we introduce the notion of a noncanonical mnemonic morphism. There is a difference between those mnemonic morphisms from the past to the present and from the present to the future. A mnemonic morphism from the past to the present is memory-induced and a mnemonic morphism from the present to the future is consciousness-induced. Note also that because t-topos theory is nondeterministic (as foundations for quantum physics), there can be more than one choice of an object in t-site corresponding to a future state.

By the definitions of initial and terminal objects of a category, for any object p in $S^\wedge$, we have uniquely determined morphisms δ^p from α to p and σ_p from p to ω, respectively. The following commutative diagram is most fundamental for the t-topos theoretic methods.

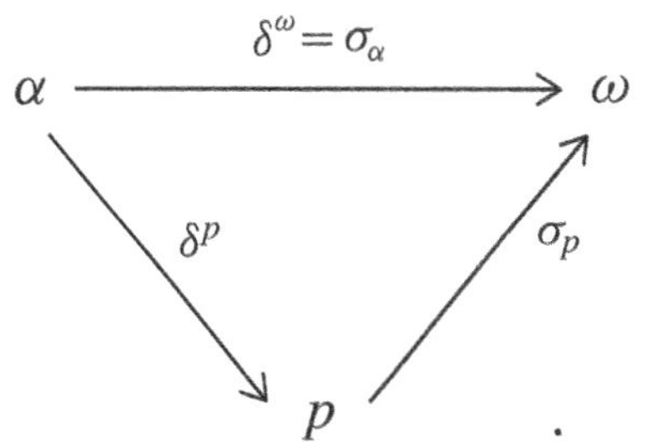

(1)

It is our ontological view that an entity p is associated with a pair of morphisms (δ^p, σ_p) satisfying the commutativity in Diagram (1).

Let p be an object of t-topos $S^\wedge$, corresponding to a conscious entity like a human. Let U, V, and W be objects of t-site such that both morphisms f from U to V, and g from V to W are t-linear morphisms, respectively, in the sense of t-topos theory. Note that $p(U)$, $p(V)$, and $p(W)$ are mutually in the same t-light cone. Definitions of a t-light cone and a t-linear morphism are provided by Kato (2004, 2005, 2013, 2017). For such t-linear morphisms f: $U \to V$ and g: $V \to W$ in the t-site S, we let U, V, and W correspond to the past, present, and future, respectively. We will study the interplay among the cognitive states δ^p_U, δ^p_V, and δ^p_W of the conscious entity p, corresponding to U, V, and W, respectively.

For a t-linear morphism f: $U \to V$, and for an object p in $S^\wedge$, we have the canonically induced morphism $p(V)$ to $p(U)$, by the definition of F being a contravariant functor, so that the following diagram.

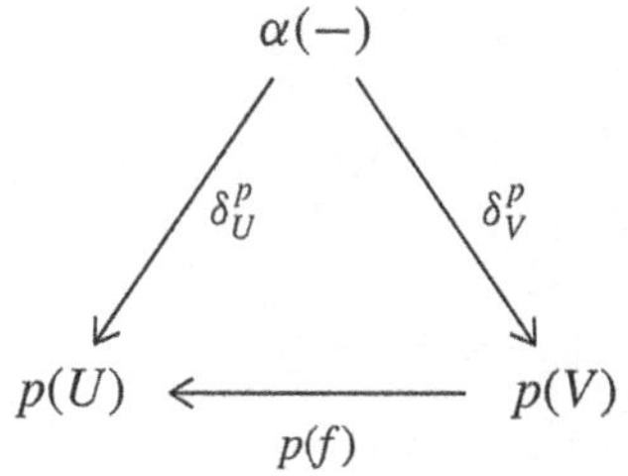

(2)

becomes commutative. Note that we use $\alpha(-)$ in commutative Diagram (2) for both U and V because an initial object α is a constant functor; that is, $\alpha(U) = \alpha(V)$ for any objects U and V in t-site S. Next, we want to define a noncanonical morphism from $p(U)$ to $p(V)$.

For a conscious entity p, we define a noncanonical morphism, called a mnemonic morphism

$$p_V^U : p(U) \to p(V),$$

so that the following diagram.

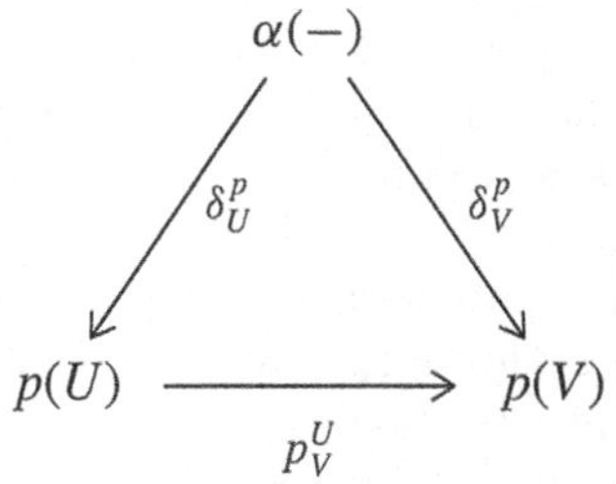

(3)

becomes commutative; that is, $\delta_V^p = p_V^U \cdot \delta_U^p$. Note that the mnemonic morphism is in the opposite direction to the canonically induced morphism $p(f)$: $p(V) \to p(U)$ in Diagram (2).

Next, we consider a mnemonic morphism p_W^V for V and W corresponding to the present and the future:

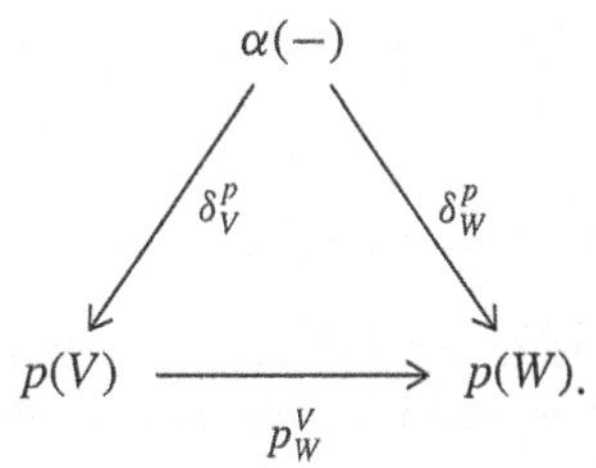

(3')

The combined commutative diagram of Diagrams (3) and (3') is given as follows.

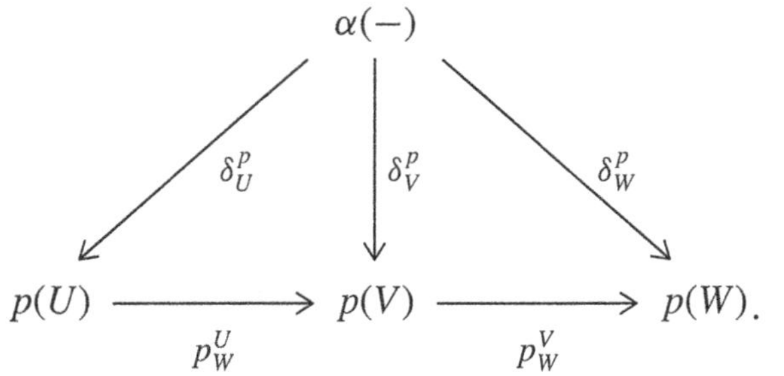

(4)

Commutative Diagram (4) should be read as follows. The present interpretation δ^p_V of the past cognitive state δ^p_U of p depends upon the mnemonic morphism p^U_V. That is, the present cognitive state δ^p_V is the composition of the past cognitive state δ^p_U and the mnemonic morphism p^U_V, which is expressed in the commutativity of Diagram (3); that is

$$\delta^p_V = p^U_V \cdot \delta^p_U. \tag{8.1}$$

Equation (8.1) implies that the mnemonic morphism p^U_V connects the past cognitive state δ^p_U to the present cognitive state δ^p_V; that is, memory of the past. This means that the interpretation of the past cognitive state δ^p_U from the point of view of the present state δ^p_V over V depends upon the mnemonic morphism p^U_V. Namely, the present view of the past conscious state can be controlled by a memory. Consequently, the past consciousness state can be varied through the memory, which is viewed as the present interpretation.

For a nondeterministic theory of t-topos, there can be several choices of t-linear morphisms with the domain object V toward the future t-site objects. See Kato (2004, 2005, 2013, 2017) for nondeterministic aspect and an ur-wave state of an entity in terms of t-topos theory relating to the foundations of quantum physics and quantum gravity. Then a mnemonic morphism (which is affected by the current awareness) among several choices is determined from $p(V)$ toward the future state $p(W)$ so that the right-hand side triangle in Diagram (4) becomes commutative. Namely, the present awareness can affect the range of the nondeterministic arbitrariness among the choices for the future state. Then morphism δ^p_W, the renewed cognitive state, is different from the (immediate) past δ^p_V by the mnemonic morphism p^V_W, which is expressed as the composition $\delta^p_W = p^V_W \cdot \delta^p_V$.

8.4 Conclusion

We have provided a t-topos theoretic formulation for cognitive states of the past, present, and future. If our present consciousness changes, our awareness of the past through our memory changes, and the range of probability distribution of future events is affected. For sheaf theoretic formulations for cognitive functions, see Kato and Nishimura (2013, 2017). Any entity is associated with a presheaf, which is the t-topos' ontological hypothesis toward the physical and cognitive interpretation of existence. Hence, any conscious entity p is defined as not only a presheaf (prestack) but also a sheaf (stack). Consequently, we can give a definition for being animate (or inanimate) as follows. An entity p is animate only if, for t-linear cognitive states δ_U^p and δ_V^p, there exists a mnemonic morphism p_V^U from the preceding state δ_U^p to δ_V^p satisfying $\delta_V^p = p_V^U \cdot \delta_U^p$.

Acknowledgements This project was supported by the Japan Society for the Promotion of Science (JSPS) Grant-in-Aid for Scientific Research (S) No. 20H05633 and JSPS Grant-in-Aid for Scientific Research (B) No. 16H03598.

References

Artin, M. (1962). *Grothendieck topologies*. Notes on a Seminar, Harvard University.
Butterfield, J., & Isham, C. J. (1999). On the emergence of time in quantum gravity. In J. Butterfield (Ed.), *The arguments of time*. Oxford University Press.
Butterfield, J., & Isham, C. J. (2001). Spacetime and the philosophical challenge of quantum gravity. In C. Callender, & N. Huggett (Eds.), *Physics meets philosophy at the Planck scale: Contemporary theories in quantum gravity*. Cambridge University Press.
Döring, A., & Isham, C. (2011) What is a thing?: Topos theory in the foundations of physics. In B. Coecke (Ed.), *New Structures for physics* (pp. 753–940). Lecture Notes in Physics, 813. Springer.
Eilenberg, S., & MacLane, S. (1945). General theory of natural equivalences. *Transactions of the American Mathematical Society, 58*, 231–294.
Gelfand, S. I., & Manin, Y. O. (1996). *Methods of Homological Algebra*. Springer.
Grauert, H., & Remmert, R. (1984). Coherent analytic sheaves. *Grundlehren der Mathematischen Wissenschaften, 265*. Springer.
Kashiwara, M., & Schapira, P. (2006). Categories and sheaves. *Grundlehren der Mathematischen Wissenschaften, 332*. Springer.
Kato, G. (2004). Elemental principles of t-topos. *Europhysics Letters, 64*(4), 467–472.
Kato, G. (2005). Elemental t.g. principles of relativistic t-topos (Presheafification of matter, space, and time). *Europhysics Letters, 71*(2), 172–178.
Kato, G. (2006). *The heart of cohomology*. Springer.
Kato, G., & Nishimura, K. (2013). Grasping a concept as an image or as a word—A categorical formulation of visual and verbal thinking processes. *Journal of Scientific Research and Reports, 2*(2), 682–691.
Kato, G., & Nishimura, K. (2017). An integrated brain function -sheaf theoretic approach to brain as a conscious entity. *Annals of Cognitive Science1*, 2, 39–43, The Scholarly Pages.
Kato, G. (2013). *Elements of temporal topos*. http://www.Theschoolbook.com
Kato, G. (2017). Topos theoretic approach to space and time. In S. Wuppuluri, & G. Ghirardi (Eds.), *Space, time and the limits of human understanding* (pp. 313–326). Springer.
Leray, L. (1950). L'anneau spectral et l'anneau filtré d'homologie d'un espace localement compact et d'un application continue. *Journal des Mathematiques Pures Appliquees, 29*.
Makkai, M., & Reyes, G. (1977). *First order categorical logic: Model-theoretical methods in the theory of topoi and related categories*. Springer.
Mitchell, B. (1965). *Theory of categories*, Academic Press.
Oka, K. (1950). Sur quelques notions arithmétiques. *Bulletin de la Société Mathématique de France, 78*.
Schubert, H. (1972). *Categories*. Springer.

Goro C. Kato Ph.D. is Professor Emeritus of mathematics and an advisor for the Alexander von Humboldt Institute at California Polytechnic State University, San Luis Obispo, CA, USA. He received his Ph. D. in mathematics under Professor Saul Lubkin from the University of Rochester, NY, in 1979. His fields of interest are zeta invariants associated with p-adic cohomology, homological algebra, and quantum gravity in terms of temporal topos. Kato is a member of the Association of the Institute for Advanced Study, Princeton, NJ, USA.

Kazuo Nishimura Ph.D. is a Specially Appointed Professor of the Research Institute for Economics and Business Administration at Kobe University in Japan. He is also Professor Emeritus of Kyoto University and a member of the Japan Acadsemy. He received his Ph.D. from the University of Rochester in 1977. Nishimura served as President of The Japanese Economic Association in 2000–2001 and has been a fellow of the Econometric Society since 1992. He is known for contributions in complexity economics and has served as an external Professor of the Santa Fe Institute, 2008–2017.

Chapter 9
Universality and the Role of Limitations Influencing Interdisciplinary Scientific and Cultural Advances

Paul G. Mezey and Masatoshi Murase

Abstract Universality and limited validity are challenging concepts in most human efforts to systematize our current understanding of science and culture, or, in a broader context, to better understand nature and human society. In this chapter, we discuss some of the advantages of considering these two types of complementary principles as manifestations of "abstract forces," acting on any evolving process of describing phenomena, in fact, forming some of the most relevant, opposing, but co-existing features of a common framework of "understanding."

Keywords Universality · Nonuniversality · Duality · Mathematics

9.1 Introduction

Universality is a simplifying concept, a seemingly desirable feature of an abstract idea, assumed to be generally valid and applicable in all instances. Ideally, it can provide "ready answers" without the need to ever check if no more detailed analysis is required, which may question the basis on which this "universal" idea was originally proposed.

There are many examples of evolution in nature and of evolution in human societies and culture, where some type of universality is often assumed. These may

P. G. Mezey · M. Murase (✉)
International Research Unit of Advanced Future Studies, Yukawa Institute for Theoretical Physics, Kyoto University, Kitashirakawa, Kyoto 606-8502, Japan
e-mail: murase@yukawa.kyoto-u.ac.jp
URL: http://www2.yukawa.kyoto-u.ac.jp/~future/?lang=en

P. G. Mezey
Scientific Modelling and Simulation, Department of Chemistry and Department of Physics, Memorial University of Newfoundland, St. John's, NL, Canada

Institute of Chemistry, Eötvös University of Budapest, Budapest, Hungary

Faculty of Chemistry and Chemical Engineering, Babes-Bolyai University, 11 Arany Janos st., 400028 Cluj-Napoca, Romania

Szabolcska M. Street 7, 2.em 2.sz, 1114 Budapest, Hungary

K. Nishimura et al. (eds.), *Creative Complex Systems*, Creative Economy,
https://doi.org/10.1007/978-981-16-4457-3_9

be manifested as cross-species universality in nature, cross-cultural universality in society, or cross-discipline universality in scientific studies. However, excessive reliance on such universalities and ignoring limits to validity have often been the sources of conflict and negative events. As society becomes more interconnected and interdisciplinary studies become more widely practiced in science, recognition of the concept proposed here, the "Universality of Nonuniversality" is expected to provide further motivation for exploring the levels of commonalities in diverse fields. The important roles of both Eastern and Western philosophical histories in current examples of studies provide a background for this proposal (Murase, 2008a, b, 2011, 2018; Murase & Murase, 2020).

Here we pay special attention to three areas:

A. The use of specific types of properties for organizing objects, activities, information, etc., exploiting the perceived universality of these specific properties for all items considered, yet paying attention to possible mistakes in the evaluation, or even to questioning the actual existence of these properties,
B. The assumed universality of some favored reference when comparisons are made, where it is often preferred to assume the existence of some universally acceptable reference, for example, the average income within a given subfield of contemporary occupations,
C. The currently accepted fundamental laws of some scientific fields, or the axioms of mathematics, where the exceptions, or the direct violations of those laws, or the antinomies of mathematics are often the very source of further progress.

9.2 Universality and Limitations of a Selected Property Assumed to be Present for a Family of Entities

In many studies, including interdisciplinary investigations, an assumption of universality and an associated "promotion" of one feature to "universal" status can be used to generate simplified, reduced, abstract models of reality, where many of the special peculiarities of individual cases are no longer relevant. One example discussed here is the influence on scientific research approaches generated by some useful, but artificial choice of data management, which, by the simple fact of being used so often, assumes some "universality" status that may actually hinder further scientific developments.

Pharmaceutical drug companies and even environmental researchers involved in the computational aspects of toxicology (Mezey, 1999) use huge databanks on millions of different molecules and their various properties. These data are crucial in finding relations, combinations of properties, and even synthetic pathways for the generation of similar, but new and possibly more advantageous molecules with fewer and lesser side effects. However, if the databank has specific emphasis on certain molecular properties, and the search methods of the databank are designed specifically for those, then alternative searches based on some other combination of preferences may not be efficient. This fact may influence actual research approaches, and the potential for success, if the scientists of the pharmaceutical company are constrained by the presumed "universality" of the established databank structure, which may be quite different in another pharmaceutical company.

9.3 An Abstract Nonexistent Entity as the Universal Reference

If the goal is to express some fundamental aspect of a group of objects, or ideas, or people, then it is often desirable to avoid references to specific examples. Individual examples may be influenced, perhaps dominantly, by aspects other than the one that is the main subject. In such cases, it is often more reliable to consider many examples collectively; in fact, it appears that as more examples are considered, the conclusions are more reliable. If the aspect studied can be characterized by numbers, then the collective view of the specific aspect can be characterized by the average of the numbers that individually characterize each entity. However, such a "universal" approach has pitfalls. For example, it is possible, that this average is never realized by any of the entities studied, and it may be an impossible outcome for any one case. One such example is the usual custom to refer to the benefits of some new health policy for an average citizen. However, such an average citizen is never realized, because an abstract average citizen can be neither a woman, nor man, not even any transgender person: the average citizen does not and cannot exist.

The pitfalls of averages, even for entities that are closely comparable, are many. However, the use of averages, as an example of universality, is an extremely useful concept. Universality is a simplifying idea, yet it relies on an abstraction that by itself can generate problems if, after getting used to this abstraction, it is almost promoted to "reality" status.

As soon as a property is considered by itself, it is removed from the actual object or phenomenon to which the property originally belongs, and the property becomes an abstract concept, and without other properties it is nonexistent.

9.4 Mathematics Taken as the Universal Language of Science

In many instances, universality can be used to generate reduced, abstract models of reality, where many of the special peculiarities of individual cases are no longer relevant. In most such instances, the abstract models are mathematical, which means that they belong to the most abstract field of all the sciences. Interestingly, for these very same reasons, some people do not consider mathematics as a formal science, but as a special aspect of philosophy.

The recognition that mathematics can offer a sound basis for universality in studying nature is one that is well manifested by the quote of Galileo: "The laws of nature are written in the language of mathematics … the symbols are triangles, circles and other geometrical figures, without whose help it is impossible to comprehend a single word," as it was quoted in *Mathematical Thought from Ancient to Modern Times* (Kline, 1990).

However, one should not miss the fact that even mathematically well-defined approaches may be affected by some of the unavoidable internal conflicts of mathematics itself, called antinomies, such as the rather famous antinomy of Russell, and other antinomes which, in rare cases, may degrade our trust in a mathematical analysis. Some of these antinomies are described and illustrated in the Appendix. Yet, antinomies in general do not reduce the value of most mathematical approaches. Overall, a mathematical approach is extremely useful for extracting the essence from complex issues, using both precise definitions and statements. It also has the benefit of unquestionable truthfulness in many cases thanks to the use of formal proofs, a luxury that is rarely if ever attainable in many of the more applied areas of culture. In addition, mathematics can handle fuzziness, and for transferring ideas, mathematically treated analogies can provide a useful foundation for exploiting the real benefits of knowledge acquired earlier, and using them in various interdisciplinary advances.

9.5 Some Universality of Inertia of Organizational Traditions

Large organizations in human societies have typically evolved to have rather complex internal structures, and complex relations with the rest of the society. As a consequence, if circumstances change, it is not a simple task to rationally modify these relations. It is often assumed that smaller organizations are more flexible and better able to react to changing circumstances, but there are many counter-examples where this has turned out to be unfulfilled.

Large organizations, by their size, often have features that imitate universality, and those features are prone to be the more difficult to change when change is needed or when change is desired (and these two conditions do not always coincide). One example is the historically slow adoption of some new technologies even in those segments of societies where rapid exploitation of new technologies could provide important, even strategic advantages such as in the military.

9.5.1 *Slow Adoption of the Steam Engine by the Royal Navy*

The example here is the slow adoption of the steam engine, the first truly powerful technological invention to replace the exploitation of human physical force, or the strength of animals (the "beasts of burden"), or some lucky uses of the wind and sun. Whereas the first steam engine in Britain was produced by Thomas Newcomen in 1712, this invention had several disadvantages, and the inventor of a far more useful version of the steam engine was James Watt, somewhat later in the same century. However, in Britain, the steam engine was first introduced for boats much later in the next century, in 1802, by William Symington, although mostly for use on riverboats rather than at sea. In a later development, the first Royal Navy steamboat was introduced in 1820.

Not surprisingly, there were technical reasons, considerations of size and weight, and other complex issues with the steam engines themselves that delayed the implementation, but there was also the universality of slow adaptation of new possibilities by large organizations. It is noteworthy that in a field as important as military power, such a major invention took so long to be adopted, where reliable energy-producing solutions, such as energy sources that can be made available on the open sea, have been so rare before.

9.5.2 *Resistance to Change at Traditional Universities*

Universities have had quite considerable autonomy in many countries, and this has led to an enhanced chance for the development of "mini-universes," in some sense reflected in the very names of such organizations. It is not surprising then that the concept of universality has always had special relevance in educational institutions that have often assumed the role of moral and intellectual "upper chamber" over the "commons" of society in general.

Of course, it is often revealing to consider an insider's view of this mini-universe. The rather sarcastic book by Cornford (1908) titled *Microcosmographia Academica*, with insights that are over 100 years old, still provides a rather universal but university-specific description of political manipulations in large, and by their own standards, important organizations.

9.6 Summary

The duality and twin-relations between the principles of universality and limited validity are discussed in terms of examples taken from history, society, science, mathematics, and education, pointing out the useful role of the formal conflicts between these two principles. Given that nature and life are characterized by so many contradictions and infinite possibilities, universality and limited validity play very important roles for our future evolutionary dynamics.

Appendix

Some Antinomies of Mathematics

The antinomies of mathematics are actual contradictions that cannot be resolved within mathematics. They are actual mathematical patterns where the logical process itself fails.

One of the most famous examples of such antinomies is Russell's Antinomy, which can be described as follows:

Let us define R as the set of all sets that are not members of themselves (not elements of themselves). Here we deal with only two considerations: (1) is R not a member of itself, or (2) is R a member of itself?

(1) If R is not a member of itself, then by its definition, it must be a member of itself, a contradiction.
(2) If R is a member of itself, then by its definition, it must be a set that is not a member of itself, again a contradiction.

That is, an unavoidable contradiction, an antinomy, where mathematics fails. This is called Russell's Antinomy, although some other mathematicians, including Zermelo, had knowledge of this problem earlier.

Another simple antinomy can be illustrated as follows:

Consider the English language, all the letters used, blank space, and the integer numbers 0 to 9, and possibly some additional symbols such as + and −, in total, say, a list A of 50 symbols.

Now consider all possible strings of length 200 made up in all possible ways using some or all of these symbols from list A. Clearly, there are only a finite number (50^{200}) of such strings. Consider now all those of these strings which express an integer number, for example, by numerical symbols, + , −, or by descriptions using words, or a combination of these (note that infinity is not a number). Let us refer to this collection of numbers as Small Numbers. Clearly, there are only a finite number of integers in this Small Numbers collection. Consequently, there must be one, a greatest number among them, in the collection of Small Numbers.

Consider now the following sentence:

1+ the greatest number expressible by a string of 200 length using all possible ways the 50 allowed symbols taken from (list A).

where in this expression the parenthetical (list A) part actually means a listing of the 50 symbols.

Thereby, a string of actually less than 200 symbols long is created, which can be completed by up to 200 by adding sufficient number of the symbol of blank space.

Consequently, we have a string of symbols expressing an integer number.

This very string of 200 symbols describes a number, that is:

1. One greater than the greatest integer included among the family of Small Numbers, hence it is not included in that family of Small Numbers,
2. However, it is described by 200 symbols, hence, it is included in the family of Small Numbers.

That is, this number, by definition, is not included, yet, also by the same definition, it is included in the family of Small Numbers.

This is another annoying antinomy of mathematics.

Fortunately, such antinomies seldom affect the efficient use of logic and mathematics for the purposes of scientific inquiry, including efforts to generate, describe, rationalize, and use the tools of mathematics in interdisciplinary studies.

References

Cornford, F. M. (1908). *Microcosmgraphia Academica (being a guide for the young academic politician).* Bowes & Bowes Publishers Ltd.

Kline, M. (1990). *Mathematical thought from ancient to modern times.* Oxford University Press.

Mezey, P. G. (1999). Holographic electron density shape theorem and its role in drug design and toxicological risk assessment. *Journal of chemical information and computer sciences, 39*, 224–230.

Murase, M. (2008a). Endo-exo circulation as a paradigm of life: Towards a new synthesis of Eastern Philosophy and Western Science. *Progress of Theoretical Physics Supplement No., 173*, 1–10.

Murase, M. (2008b). Environmental pollution and health: An interdisciplinary study of the bioeffects of electromagnetic fields. *SNSAI, An Environmental Journal for the Global Community, 3*, 1–35.

Murase, M. (2011). The origin and evolution of life by means of endo-exo (or self-nonself) circulation. *Viva Origino, 39*(1), 7–10.

Murase, M. (2018). A self-similar dynamic systems perspective of "Living" nature: The self-nonself circulation principle beyond complexity. *The Kyoto Manifesto for Global Economics.* In: S. Yamash'ta, T. Yagi, & S. Hill (Eds.),*The Platform of Community, Humanity, and Spirituality* (pp. 257–283). Springer.

Murase, M., & Murase, T. (2020). *A grand unified life theory: An extension of the self-nonself circulation theory.*

Paul G. Mezey is a Hungarian-Canadian mathematical chemist. He was the Canada Research Chair in Scientific Modeling and Simulation in the Department of Chemistry at the Memorial University of Newfoundland. Mezey received a Master's degree in chemistry, a Ph.D. in Chemistry, and a Master's degree in mathematics, all from Eötvös Lorånd University in Budapest, in the years 1967, 1970, and 1972, respectively. From 1982 to 2003, he was a professor of chemistry and mathematics at the University of Saskatchewan, where he received a D.Sc. in 1985 in mathematical chemistry. He was a faculty member at Memorial University from 2003 to 2018. Mezey is Editor-in-Chief of the *Journal of Mathematical Chemistry.*

Masatoshi Murase received his Ph.D. degree from The University of Tokyo in 1987. Since 1992, he has been an associate professor at the Yukawa Institute for Theoretical Physics, Kyoto University. In 1987 and 1988 he was a visiting scientist at the Duke University Medical Center, Durham, NC, USA. Since 2010, he has been a member of the Cooperation Promotion Committee of the International Research Unit of Integrated Complex System Science, Kyoto University. Between 2015 and 2020 he was a director of the Research Promotion Strategy Office of the International Research Unit of Advanced Future Studies, Kyoto University.

Chapter 10
Some Conceptual Principles with Mathematical Background for Interdisciplinary Developments in the Sciences and Beyond

Paul G. Mezey and Masatoshi Murase

Abstract This chapter discusses some of the principles of interdisciplinarity and emphasizes those principles that may provide some relevant components to a potentially unified background to many areas of natural sciences. This extension even reaches into fields that are typically regarded as humanities. The discussion focuses on the three concepts of "paradox," "analogy," and "fractals," which are prevalent in the basis of our understanding of nature and related principles. Although these concepts are well defined in mathematical terms, the current description relies on limited hints towards the mathematical background and avoids detailed mathematical treatment. Emphasis is on the conceptual aspects, while some parts of the mathematical basis, relying on fuzzy sets, fuzzy logic, and a treatment of analogies using the functor model of category theory, are presented in the Appendices 1 and 2.

Keywords Principle · Paradox · Analogy · Fractal · Category theory · Mathematical background

---

P. G. Mezey · M. Murase (✉)
International Research Unit of Advanced Future Studies, Yukawa Institute for Theoretical Physics, Kyoto University, Kitashirakawa, Kyoto 606-8502, Japan
e-mail: murase@yukawa.kyoto-u.ac.jp
URL: http://www2.yukawa.kyoto-u.ac.jp/~future/?lang=en

P. G. Mezey
Scientific Modelling and Simulation, Department of Chemistry and Department of Physics, Memorial University of Newfoundland, St. John's, NL, Canada

Institute of Chemistry, Eötvös University of Budapest, Budapest, Hungary

Faculty of Chemistry and Chemical Engineering, Babes-Bolyai University, 11 Arany Janos st., 400028 Cluj-Napoca, Romania

K. Nishimura et al. (eds.), *Creative Complex Systems*, Creative Economy,
https://doi.org/10.1007/978-981-16-4457-3_10

10.1 Introduction

Interdisciplinary research often involves transfers of methodologies and ideas from one field to another, and these transfers often define the very range of the actual level of interdisciplinarity. Such transfers usually show some characteristics that are approximate, because the different fields playing roles in the actual interdisciplinary topics are, indeed, different, and instead of strict relations, often "softer" rules may apply. In such cases, the degree of approximation can be treated on a mathematical level using the concepts and methodologies of Fuzzy Sets and Fuzzy Logic, as originally introduced by Zadeh (1965). Another useful tool is the extension of the concept of similarity to a broader and multi-level version, which in the mathematical sense can be regarded as an analogy. Again, the ideas of analogies are treated on a mostly conceptual level, without much detail of the relevant mathematics. While avoiding a more involved, formal mathematical treatment, only a brief introduction is given. In the context of analogies and families of similarities, this may be based on the functor model of category theory, which can transform families of sets into families of sets and the relations among the first family of sets into relations among the second family of sets (Fig. 10.1). Some elements of the mathematical background of functors are presented in the Appendices 1 and 2.

The motivation for exploring such approaches is justified by the perceived need for making better use of already available methodologies and treatments within a broader range of natural sciences and beyond.

Research on interdisciplinarity is a rapidly growing component of the general trend of encouraging research into approaches needed to take advantage of current and expected near-future advances of science. Such studies can provide ideas and

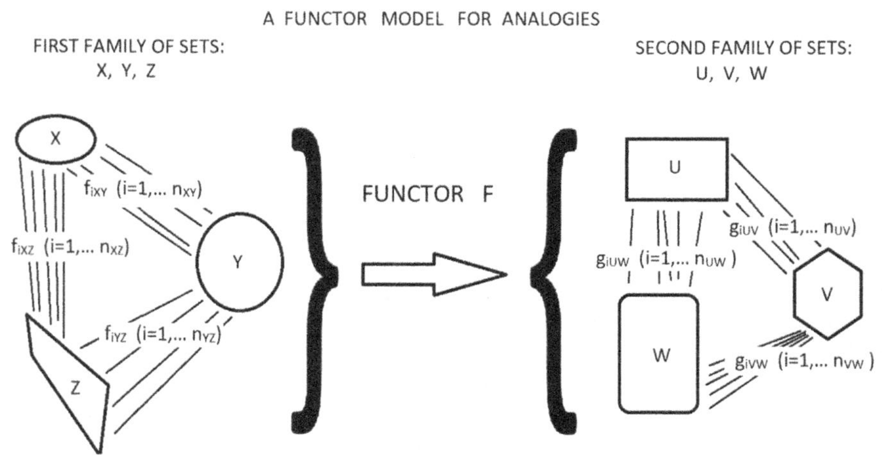

Fig. 10.1 A sketch of a Functor F, converting families of sets into families of sets and converting the relations among the first family of sets into the relations among the second family of sets. For more detail, see Appendix 2

suggestions that may contribute to the initiatives of research establishments, such as the International Research Unit of Advanced Future Studies (Murase, 1996, 2000, 2008a, 2008b, 2011, 2018).

The International Research Unit of Advanced Future Studies has developed an approach that focuses on three key concepts to enhance interdisciplinary research:

Concept A: Paradox. In this context, opposing positions should not be considered as mutually exclusive, but instead, as complementary with each other.
Concept B: Analogy. This concept considers the multiple-level similarities between two different phenomena.
Concept C: Fractal. The fractal approach focuses on self-similarity or nested hierarchy.

Interestingly, these three fundamental concepts, manifested as characteristics, are not only the properties typical of living things, but also of many of the other features of nature and even the essence of thinking. Based on these characteristics, it may be also possible to think about the question of evolution. Murase (1996, 2018) introduced a new idea about aging: aging is one of the evolutionary processes. This idea is indeed relating to Concept A.

In a different context, it is also impressive to consider effective learning on the basis of an evolutionary perspective, because learning can be analogous to evolution. Here, Concept B plays an important role in understanding subjective processes of learning through the objective dynamics of evolution. Murase and Murase (2021a, 2021b) suggested the idea that learning is conducted based on evolutionary trends of human beings. Furthermore, macroscopic dynamics could demonstrate the analogous dynamical phenomena typical of microscopic levels, which is characterized by "fractal" features (Murase, 2018). It is clear that these three main concepts are not completely independent of one another; rather, they are closely related. This is the main reason that interdisciplinary studies should be conducted to investigate the fundamental principles behind such diversities.

Interdisciplinarity typically involves transfers of ideas between fields, usually involving some novelty being applied to some new problems. Our understanding of new problems and finding new solutions usually follows a well-tested, time-honored path: we try to use similar logical steps that have already been used with success in the case of a problem that we judge to be similar. It is no surprise then that similarities, in fact, entire families of similarities play very important roles in most of our learning processes, and in our never fully routine everyday activities. One may find that exploiting similarities is even more important in scientific research and in understanding cultural and religious traditions and connections, artistic activities, and economic processes. Such families of similarities, on a level higher than individual similarities, are often regarded as analogies.

Difficulties in recognizing analogies often arise because most comparisons of different fields of science involve some imprecise concepts that are not fully equivalent, where only some approximate correspondence can be established, and even the levels of those approximations are not precise, and often not fully understood. Clearly, some uncertainty is inevitable for most of these comparisons, and in such cases, chance, probability, and general imprecision always complicate the processes one may follow. In other words, the comparisons have some fuzzy aspects, and therefore, it is natural to rely on the conceptual tools, and in a more formal way, the mathematical tools available for fuzzy problems. In this context, a highly developed branch of contemporary mathematics, the theory of fuzzy sets and fuzzy logic, is the natural choice for treating such problems (Maggiora & Mezey, 1999; Mezey, 1993, 1998, 2000, 2016; Szekeres et al., 2005; Zadeh & Yager, 1987).

In a mathematical sense, analogies may be treated in a rather comprehensive way using the Functor model of Category Theory. This approach appears especially suitable for the study of interdisciplinarity, where the very connections between different fields is the subject of the study.

Recognition of analogies bridging the gaps between often rather different fields has been very important in the development of most aspects of human activities. Transfers of analogous solutions between various fields of science have been perhaps the most easily demonstrable cases. For example, analogous structures are observed in the equations describing Coulomb's Law in electrodynamics and Newton's Law of gravity, even though the participating quantities are fundamentally different: electric charge in the first case, and mass in the second.

Similarly, analogous value systems of human cultures, originally developed almost independently, show strong resemblances, even if they may appear rather different. Such analogies have always had special roles in finding harmonies between different cultures during the development of humanity. In motivating and expanding various art forms, the analogies discovered between the inspirational aspects of the initial stages of their creation, and the analogies in the actual intellectual processes during their actual creation, all have had major influences during the history of art. Such analogies have often played their roles without acknowledgement, sometimes on a subconscious level; for example, new trends in composing music have also motivated new styles in poetry, and even in the visual arts, such as painting. In addition, analogies have always played an important role in the interpretation of seemingly different art forms. On a somewhat less spiritual level, one can also note the transfer of analogous financial and management approaches between rather different types of economic and industrial activities.

In much of human activity, both on the individual level and in groups, even in the large group we call society, many of the main principles are related to one another on some abstract level, and often on some practical levels. In this chapter, we discuss some of these principles with the aim of pointing out some patterns and regularities that allow for a formal, mathematical treatment.

10.2 Recognition of Inherent Fuzziness and a Fuzzy Set and Fuzzy Logic Approach to Interdisciplinarity

Even within an established and clearly defined scientific field, like classical thermodynamics, the concepts themselves, and the actual approaches to the solutions of various problems, inevitably involve some levels of uncertainty, which, traditionally, could be addressed by the applications of probability theory. Yet, this traditional approach has not always been conceptually well justified. This is especially the case if the very concepts used are "too pedantic," do not fit well with the actual complexity, and where the concepts do not account for often unknown influences affecting the problem. Consequently, when the concepts used do not fit the characteristics of the problem, then a possible alternative to the use of probability theory is apparently more suitable: the use of fuzzy sets and fuzzy logic methods, as introduced originally by Zadeh and Yager (1987). Such considerations are especially relevant if the topic of study is interdisciplinary, where the disciplines involved often may have somewhat conflicting choices of preferred definitions and not well-matched approaches to solutions. Recognition of the inherent fuzziness and the actual use of fuzzy set and fuzzy logic approaches in interdisciplinary areas are gaining momentum as modern science evolves; some examples in the field of chemistry are listed in the literature (Maggiora and Mezey 1999; Mezey, 1993, 1998, 2000, 2011, 2012, 2014, 2016; Szekeres et al., 2005).

However, if we acknowledge the limitations of our knowledge, then it becomes clear that fuzziness and fuzzy set models are relevant to nearly all branches of science. They are also relevant to our cultural heritage of intellectual processes, to the way we view human interactions, to society, and to life processes themselves. Evolution involves inherent fuzziness because it is governed by a multitude of environmental factors, often with conflicting influences. Hence, some level of randomness and fuzzy processes are at the very core of the principles of evolution. This naturally applies to the evolution of life-forms on Earth, but also to the evolution of human intellect, language, artistic ideas, science, and culture in general.

Fuzzy set models can be regarded as a softening of criteria for membership of a given set. For example, a given asymmetric object may be regarded as "almost symmetric" with respect to a mirror plane because a fuzzy, imperfect photo of the object may suggest that the object has mirror symmetry. However, a sharp photo will show that the object is not symmetric. This type of "fuzzy symmetry" may be treated mathematically, and one may say that the object belongs to the family of mirror-symmetry objects not with a full, 100% membership, but only, say, with a 70% membership. Then, this family of fuzzy mirror-symmetry objects can be regarded as a fuzzy set, with the generous chance that even nonsymmetric, but "almost symmetric" objects can be regarded as "approximate members" of this family.

Such an approach can be beneficial if interdisciplinary studies are involved. This is because when a multitude of diverse concepts are in play, as is common for interdisciplinary studies, then it is more likely that the precise definitions useful in one field are not 100% suitable for other fields. Thus, fuzziness is a natural occurrence in interdisciplinary combination of scientific fields.

Some of the formal mathematical aspects of fuzzy sets and fuzzy approaches are briefly reviewed in Appendix 1.

10.3 Recognizing Analogy as a System of Similarities, and Approaches for Analyzing Analogies as Motivated by the Functor Model of Category Theory

Natural sciences are rich in analogies that are far more than just simple similarities. This has been very useful in many instances, although there are many potential analogies that have not been fully exploited. In addition, it is highly possible that there are many more analogies that have not been recognized.

An example of strong analogy that has been referred to on several occasions (Mezey, 2019a, 2019b, 2019c) relates to a pair of relations that follow the very same pattern: the law of gravity describes how masses interact and the main law for electricity describes how electric charges interact. It is tempting to think of a single abstract law of physics, that can employ two types of variables: masses or electric charges. This abstract law may be called the Inverse Square Law of physics.

This Inverse Square Law can be stated as follows: the force denoted by F depends on the distance between two objects A and B affected by this force in a simple way:

$$F = c \times p_{\mathrm{A}} \times p_{\mathrm{B}} \times 1/r^2, \tag{10.1}$$

where r is the distance between the two objects A and B, p_{A} and p_{B} are the given type of properties of A and B, and c is a constant for the type of interaction.

According to the Inverse Square Law, the force F between two objects A and B is proportional to the inverse square of the distance r between them.

What are the properties p_{A} and p_{B}? Here we can observe an interesting analogy.

In one case, when p_{A} and p_{B} are the masses m_{A} and m_{B}, respectively, we are dealing with the law describing gravity:

$$F_{\text{gravity}} = c_{\mathrm{g}}(m_{\mathrm{A}} m_{\mathrm{B}})/r^2, \tag{10.2}$$

where c_{g} is the gravitational constant, and the above equation is the well-known Newton's Law of gravity (Newton, 1687).

In another case, p_{A} and p_{B} are the electric charges q_{A} and q_{B}, respectively, in the law of electric attraction or repulsion:

$$F_{\text{electric repulsion}} = c_{\mathrm{e}}(q_{\mathrm{A}} q_{\mathrm{B}})/r^2, \tag{10.3}$$

where c_e is the Coulomb Law constant, and the above equation is Coulomb's Law of electric repulsion (Roller & Roller, 1954).

It is evident, that the structures of these two laws as described by Eqs. (10.2) and (10.3) are fundamentally identical, and there is a very strong analogy between these two, but seemingly for very different physical problems.

In these two laws, there are several similarities, beyond the inverse square dependence on distance: both of the relevant quantities occur on the first power, and they both appear as multiplying factors to the overall result. Whenever sets of similarities collectively appear, one may regard such a case as more than a mere similarity: it is a set of similarities, and one may consider it as an analogy.

The mathematical treatment of sets of similarities that form analogies can be well treated using some of the tools of category theory of mathematics, specifically, the tools of functors (Mezey, 2019a, 2019b, 2019c). A functor is a transformation that transforms not only a family of sets into another family of sets, but also transforms the relations among the sets within the first family into the relations among the sets within the second family. In this way, a functor can treat an analogy in a complete way, interrelating the two families of subjects involved in the analogy, as well as the relations within the two families. Here, the relations may be regarded as individual similarities, and a functor is able to deal with the whole complexity of entire families of similarities. In Appendix 2, some of the basic mathematical properties of functors are illustrated, with an emphasis on analogies.

One set of analogies, partially motivated by chemistry, is related to the logical processing involved in both scientific and cultural analyses. As mentioned earlier in the Introduction as Concept A, a paradox is a fundamental feature of most aspects of the intellectual processing of information. However, most thought processes are actually multi-dimensional, and there are mental activities in the human brain beyond the conscious level, and various background thoughts are also occurring simultaneously. As a rather simple chemical analogy, this multi-dimensionality may remind us of the simple two-dimensional geometrical arrangements of substituents on a benzene ring of many organic molecules, where relative to a chosen location on the ring one may have *para-*, *meta-*, and *ortho-* substituents. On the benzene ring, the *para-* position is directly opposite the original position, the *ortho-* position sits adjacent to the original position, and the *meta-* position is between the *para-* and *ortho-* positions. Using this analogy, a higher-dimensional extension of the traditional description of a paradoxical conflict can be made. One may consider a lesser level of paradoxical conflict as a "metadox conflict," which is a less direct conflict than a paradox, and one may also consider an "orthodox conflict," with a much weaker paradoxical feature, and with more features of approximate agreement than disagreement (an orthodox conflict may be relevant, but should not be confused with philosophical and religious orthodoxy). Such an analogy, borrowed from chemistry, is expected to provide some versatility that may be often required when dealing with relations involving some not well-defined levels of conflicts, which are potential paradoxes only in extreme cases.

Other analogies involving more diverse aspects of human activity, such as mathematical descriptions of chemical reactions using the concepts of reaction paths, paths for tourists, or cultural paths of understanding, might be found in many aspects of human activities. One example is quoted in connection to the formal Intrinsic Reaction Coordinate concept of Fukui, which was designed for the multidimensional description of molecular transformations. This scenario is also referred to reaction paths in the high-dimensional abstract space spanned by the typically very large number of atomic nuclear coordinates for many-atom molecules, and this concept may apply to alternative paths in touristic adventures and paths in understanding the process of abstract, cultural experiences (Mezey, 2018).

10.4 Some General Conclusions

The concept of paradox is one of the central features in many evolutionary processes, in understanding, and in the process of extracting the general principles applicable in the sciences, in cultural developments, and in society. In this context, it is relevant to consider suitable mathematical models that can connect a formal treatment with the needed flexibility, one that is often manifested as fuzziness of applying such principles to actual problems. While a fuzzy set approach has been proven to be very useful in many scientific applications, and here we have various examples quoted from the field of chemistry, the generalizations of many of these approaches are further enhanced if the ideas of analogies are also used in a mathematically well-defined fashion. Analogies, as sets of interrelated similarities, can be described mathematically using the functor model of category theory, as illustrated in this contribution, in part by some chemical examples in the literature.

Appendix 1

Main Concepts and Definitions of Fuzzy Set Theory

Following Zadeh's original exposition (Zadeh, 1965), the fundamentals of fuzzy sets can be described as shown below:

Take X as a set of elements denoted by x, thus

$$X = \{x\}. \tag{10.4}$$

A fuzzy set A in X is characterized by a membership function $f_A(x)$ assigning to each element x of X a real number from the interval [0, 1] that represents the "grade of membership" of x in A.

For a fuzzy set A, if this grade of membership $f_A(x)$ of element x is close to 1, then x is a "strong member" of fuzzy set A. However, if this grade of membership $f_A(x)$ of element x is close to 0, then x is a "weak member" of fuzzy set A. This fuzzy set approach also includes the traditional set approach as well, because if for a formally fuzzy set A the membership function $f_A(x)$ can take only one of the values of 0 or 1, then we find that set A is actually a traditional ordinary set and is not fuzzy. In this case, the element x of X is either a "full member" of set A, or not a member at all of set A.

For the complement A′ of a fuzzy set A, the membership function $f_{A'}(x)$ is defined as

$$f_{A'}(x) = 1 - f_A(x). \tag{10.5}$$

Unions and intersections of fuzzy sets are also defined in terms of the respective membership functions:

For the union C of two fuzzy sets A and B,

$$F_C(x) = \mathrm{Max}\{f_A(x), f_B(x)\}, \tag{10.6}$$

whereas for the intersection D of two fuzzy sets A and B,

$$f_D(x) = \mathrm{Min}\{f_A(x), f_B(x)\}. \tag{10.7}$$

Appendix 2

A Functorial Approach to Analogies

The concept of analogy is commonly perceived as being more than similarity. If one finds several types of similarities between two phenomena, then a comparison between the two often suggests a deeper level of similarity: an analogy. In a formal sense, analogies can be considered as systems of similarities. If such systems of similarities, taken as families of "interconnected" similarities, can be characterized and described by mathematical means, then one should be able to take advantage of the formalism so created.

For a systematic study of similarities, one may use simple, common sense approaches without sophisticated mathematical tools (Mezey, 2019a, 2019b, 2019c). However, some of the more formal mathematical treatments also offer advantages, and in some recent developments the application of the functor model of category theory has been suggested for the analysis of analogies (Mezey, 2019a, 2019b, 2019c). Although functorial models are powerful and adaptable tools, and connect to some deeper chapters of category theory, functors and their use in describing analogies can be presented in a rather simple way, without the need for excessive detail on the category theory background.

Here, we present a somewhat simplified, but conceptually sufficient description of a special but relevant functor model. We may regard a functor as a transformation that operates on two levels, transforming two types of mathematical entities. First, a functor establishes a connection between two families of sets: a "sets to sets" connection. Second, the functor also establishes a connection between some mappings and transformations present among the first family of sets, connecting them to the mappings and transformations present among the second family of sets: a "mappings to mappings" connection. That is, a functor transforms a family of sets to another family of sets, and at the same time, it transforms the relations among members of the first family of sets to the relations among members of the second family of sets.

Given that our primary interest is analogies used in interdisciplinary areas, one may consider two scientific fields, with sets of concepts, problems, and approaches in both, as well as connections, relations, and comparisons in both. An analogy between these two fields of science may be modelled by a functor, where the concepts, problems, and approaches of one field are related to those of the other scientific field, and beyond all these, the individual mappings among the sets within one family of problems in the first scientific field are also connected to those of the second scientific field. Whereas individual mappings can be regarded as expressions of similarities, the complete functor itself can then be taken as an expression of a higher level of similarity: an analogy. In other words, this entire comparison can then be taken as a description of the similarity of similarities; that is, an expression of an analogy.

A simplified example may provide some insight of the power of a functorial approach. In this case, we use the example of only three sets in the first family of sets, X, Y, and Z, and another three sets, U, V, and W, in the second family of sets. We assume that among the sets of each family of sets, there are several mappings, such as

$$f_{iXY}(i = 1, \ldots n_{XY}), \tag{10.8}$$

$$f_{iXZ}(i = 1, \ldots n_{XZ}), \tag{10.9}$$

$$f_{iYZ}(i = 1, \ldots n_{YZ}), \tag{10.10}$$

and

$$g_{iUV}(i = 1, \ldots n_{UV}), \tag{10.11}$$

$$g_{iUW}(i = 1, \ldots n_{UW}), \tag{10.12}$$

$$g_{iVW}(i = 1, \ldots n_{VW}). \tag{10.13}$$

Here we also assume that for each index i, and for all choices of sets A and B, the mapping f_{iAB} maps set A to set B.

A functor F can then be defined for a functorial relation between the two families of sets and for the corresponding mappings, as follows:

Functor F maps the first three sets, X, Y, and Z to sets U, V, and W, respectively. It also converts for each index i the mapping f_{iXY} into g_{iUV}, the mapping f_{iXZ} into g_{iUW}, and the mapping f_{iYZ} into g_{iVW}.

Indeed, the above construction illustrates the most fundamental properties of functors, especially in the context of modelling analogies. In the above example, one may also regard the mappings f_{iST} as expressions of similarities. Consequently, the above functor can be regarded as an expression of a higher level of similarity between the properties of the family of the first three sets, X, Y, and Z, and their interrelations, on the one hand, and the family of the other three sets, U, V, and W, and their interrelations, on the other hand, where the actual similarities within each family of sets are described by the individual mappings f_{iST}, as presented in Eqs. (10.8)–(10.13).

Such functor models can be used as mathematical tools to model various complex problems of comparisons and analogies, which may occur on several levels of interdisciplinary studies. Analogies often may appear to be exploitable in a new field if various levels of comparisons, and possible re-uses of earlier approaches and methodologies, which were originally introduced for a different set of problems, are possible. Functors are useful to formalize such analogies, and can be beneficial, especially if interdisciplinary considerations apply.

References

Maggiora, G. M., & Mezey, P. G. (1999). A fuzzy set approach to functional group comparisons based on an asymmetric similarity measure. *International Journal of Quantum Chemistry, 74*, 503–514.

Mezey, P. G. (1993). *Shape in chemistry: An introduction to molecular shape and topology*. Wiley-VCH Publishers.

Mezey, P. G. (1998). The proof of the metric properties of a fuzzy chirality measure of molecular electron density clouds. *Journal of Molecular Structure: THEOCHEM, 455*, 183–190.

Mezey, P. G. (2014). Fuzzy electron density fragments in macromolecular quantum chemistry, combinatorial quantum chemistry, functional group analysis, and shape-activity relations. *Accounts of Chemical Research, 47*, 2821–2827.

Mezey, P. G. (2016). A Trigonometrically scaled multiple tiling approach for error reduction of models built from fuzzy fragments. *Journal of Computational Methods in Sciences and Engineering, 16*, 729–732.

Mezey, P. G. (2000). Transferability, adjustability, and additivity of fuzzy electron density fragments. In P. G. Mezey, & B. Robertson (Eds.), *Electron, spin and momentum densities and chemical reactivity*. Springer.

Mezey, P. G. (2011). Linear scaling methods using additive fuzzy density fragmentation. In R. Zalesny, M. Papadopoulos, P. G. Mezey, & J. Leszczynski (Eds.), *Linear-scaling techniques in computational chemistry and physics, methods and applications*. Springer.

Mezey, P. G. (2012). Fuzzy electron density fragments as building blocks in crystal engineering design. In E. Tiekink, & J. Zukerman-Schpector (Eds.), *The importance of pi-interactions in crystal engineering: Frontiers in crystal engineering*. Wiley.
Mezey, P. G. (2018). The intrinsic reaction coordinate and a path of ascent to Mount Hiei: In memory of professor kenichi fukui, a nobel prize winning chemist, and a pioneer of many new paths in mathematical chemistry. In G. Fang, M. Amini, H. Chen, N. Fukuda, H. Hosoya, M. Kawai, J. E. LeBlanc, P. G. Mezey, I. Naruki, T. Okada, E. Rambo, M. Spivakovsky, S. Takeuchi, K. F. Taylor, H. Wong, S. Yamanaka, M. Yokotani, P. Zizler, & S. Arimoto (Eds.), *Mathematics and chemistry interdisciplinary joint research and the fukui project XXV* (Vol. 60, pp. 37–42), Bulletin of National Institute of Technology, Tsuyama College.
Mezey, P. G. (2019a). A Functorial approach to analogous molecular systems. In *Proceedings of the international conference on computational methods in science and engineering 2019 (ICCMSE 2019), AIP Conference Proceedings 2186, 020001*. https://doi.org/10.1063/1.5137911.
Mezey, P. G. (2019b). The role of analogies and data structures in cultural, environmental, and scientific developments. In Z. Androvičová, & E. Belaňová (Eds.), *Selected aspects of integrated environmental management: Culture and environment*. Technical University of Zvolen.
Mezey, P. G. (2019c). Analogues of IRC, catchment regions, and symmetry relations on the upside-down potential energy surfaces, UDPES. In G. Fang, Y. Aoki, H. Chen, N. Fukuda, H. Hosoya, A. Imamura, M. Kawai, J. E. LeBlanc, P. G. Mezey, I. Naruki, T. Okada, M. Spivakovsky, K. F. Taylor, H. Wong, S. Yamanaka, M. Yokotani, P. Zizler, & S. Arimoto (Eds.), *Mathematics and chemistry interdisciplinary joint research and the fukui project XXXIV*, Bulletin of National Institute of Technology, Tsuyama College 61 (in press).
Murase, M. (1996). Alzheimer's disease as subcellular 'Cancer'—The scale-invariant principles underlying the mechanisms of aging. *Progress of Theoretical Physics, 95*(1), 1–36.
Murase, M. (2000). *Life as History: The Construction of Self-Nonself Circulation Theory*. Kyoto University Press.
Murase, M. (2008a). Endo-exo circulation as a paradigm of life: Towards a new synthesis of Eastern Philosophy and Western Science. *Progress of Theoretical Physics Supplement, 173*, 1–10.
Murase, M. (2008b). Environmental pollution and health: An interdisciplinary study of the bioeffects of electromagnetic fields. *SNSAI, an Environmental Journal for the Global Community, 3*, 1–35.
Murase, M. (2011). The origin and evolution of life by means of endo-exo (or self-nonself) circulation. *Viva Origino, 39*(1), 7–10.
Murase, M., & Mursae, T. (2021a). Transdisciplinary study of how to integrate shattered world: The self-nonself circulation principle of "living" wholeness, In S. Hill, Y. Tadashi, & Y. Stomu (Eds.), *The Kyoto Manifesto II*, Springer.
Murase, M., & Mursae, T. (2021b). The self-nonself circulation principle of "living" nature: how to survive shattered world. In S. Hill, Y. Tadashi, & Y. Stomu (Eds.), *The Kyoto Manifesto II*. Springer.
Murase, M. (2018). A self-similar dynamic systems perspective of "living" nature: The self-nonself circulation principle beyond complexity. In S. Yamash'ta, T. Yagi, & S. Hill (Eds.), *The Kyoto manifesto for global economics: The platform of community humanity, and spirituality*. Springer.
Newton, I. (1687). *Philosophiæ naturalis principia mathematica (the Principia)*. Royal Society.
Roller, D., & Roller, D. H. D. (1954). *The development of the concept of electric charge: Electricity from the greeks to Coulomb*. Harvard University Press.
Szekeres, Z., Exner, T., & Mezey, P. G. (2005). Fuzzy fragment selection strategies, basis set dependence, and HF-DFT comparisons in the applications of the ADMA method of macromolecular quantum chemistry. *International Journal of Quantum Chemistry, 104*, 847–860.
Zadeh, L. A. (1965). Fuzzy sets. *Information and Control, 8*, 338–353.
Zadeh, L. A., & Yager, R. R. (1987). *Fuzzy sets and applications: Selected papers*. John Wiley.

Paul G. Mezey is a Hungarian-Canadian mathematical chemist. He was the Canada Research Chair in Scientific Modeling and Simulation in the Department of Chemistry at the Memorial University of Newfoundland. Mezey received a Master's degree in chemistry, a Ph.D. in Chemistry, and a Master's degree in mathematics, all from Eötvös Loránd University in Budapest, in the years 1967, 1970, and 1972 respectively. From 1982 to 2003, he was a professor of chemistry and mathematics at the University of Saskatchewan, where he received a D.Sc. in 1985 in mathematical chemistry. He was a faculty member at Memorial University from 2003 to 2018. Mezey is Editor-in-Chief of the *Journal of Mathematical Chemistry*.

Masatoshi Murase received his Ph.D. degree from The University of Tokyo in 1987. Since 1992, he has been an associate professor at the Yukawa Institute for Theoretical Physics, Kyoto University. In 1987 and 1988 he was a visiting scientist at the Duke University Medical Center, Durham, NC, USA. Since 2010, he has been a member of the Cooperation Promotion Committee of the International Research Unit of Integrated Complex System Science, Kyoto University. Between 2015 and 2020 he was a director of the Research Promotion Strategy Office of the International Research Unit of Advanced Future Studies, Kyoto University.

Chapter 11
The Role of Paradox in the Development of Interdisciplinary Scientific and Cultural Advances

Masatoshi Murase and Paul G. Mezey

Abstract Paradoxes are often regarded as relatively rare events in scientific and cultural developments. Yet this perception may itself be somewhat paradoxical, since many of the most significant developments in science and culture have been originally motivated, in some cases, even triggered by paradoxes. Many paradoxes show common features beyond their fundamental contradictory aspects. In addition to the recognition of these main characteristics, the possibilities for the advantageous, highly beneficial use of these additional features may provide novel ideas and new efficient approaches in the advancement of future studies, especially in the interdisciplinary areas.

Keywords Evolution · Paradox · Development · Discontinuity · Life · Earth · Nature · History

11.1 Introduction

A paradox is often viewed as a self-contradictory statement based on seemingly well-established facts and arrived at by apparently flawless logical reasoning and well-justified thought processes. Yet, such paradoxes do occur, and they are often the source of new trends in our understanding of nature, science, culture, and

M. Murase (✉) · P. G. Mezey
International Research Unit of Advanced Future Studies, Yukawa Institute for Theoretical Physics, Kyoto University, Kitashirakawa, Kyoto 606-8502, Japan
e-mail: murase@yukawa.kyoto-u.ac.jp
URL: http://www2.yukawa.kyoto-u.ac.jp/~future/?lang=en

P. G. Mezey
Scientific Modelling and Simulation, Department of Chemistry and Department of Physics, Memorial University of Newfoundland, St. John's, NL, Canada

Institute of Chemistry, Eötvös University of Budapest, Budapest, Hungary

Faculty of Chemistry and Chemical Engineering, Babes-Bolyai University, 11 Arany Janos st., 400028 Cluj-Napoca, Romania

K. Nishimura et al. (eds.), *Creative Complex Systems*, Creative Economy,
https://doi.org/10.1007/978-981-16-4457-3_11

society. Paradoxes have often provided motivation for new advances and inventions throughout human history.

Paradoxes are likely to have a far stronger influence on many interdisciplinary developments than is commonly perceived. Using analogies, the approaches, methodologies, and results successfully employed in one field are often taken as starting points for new advances in another field; in fact, exploiting the very aspects of interdisciplinarity. In such cases, some approaches with a perfectly logical justification in one field might provide some seemingly illogical surprises in another field. Thus, interdisciplinary areas are perhaps more prone to apparent paradoxes than the more traditional individual areas of science or culture.

Yet, paradoxes do appear even within narrow fields and narrow ranges of human activities, or in fields that the general public would regard as narrow. For example, the introduction of quantum mechanics in theoretical physics, by itself not usually regarded as a highly interdisciplinary field by the general public, has revealed many paradoxes, with some of them still with us today. Several types of paradoxes are quite prevalent in a more general context as well. It appears that some interdisciplinarity can be applied even to paradoxes themselves. Analogous paradoxes appear in many closely related fields, and also in seemingly rather distant fields, providing a new dimension to interdisciplinarity and the search for analogies. In this chapter, some of these aspects of paradoxes are reviewed.

The following discussion examines several general paradoxes, but also two types of Special Paradox that can be exploited often, sometimes in unexpected ways, within various interdisciplinary contexts.

Special Paradox of Type A: Using a false premise, being aware and fully acknowledging its falseness, but using it in a controlled, consistent, and well-defined way. In many cases, such a process may allow one, paradoxically, to arrive at a true and useful conclusion.

Special Paradox of Type B: Using a properly defined but otherwise unspecified, undetermined component in a tentative way to reach a conclusion, where the logical process needs such a component, or it is unclear how the logical process can be completed without such a component. Paradoxically, the conclusion of the logical process does not require a precise description of this component, but has merely a "catalytic" role, using a term from a chemical analogy.

11.2 The Continuity–Discontinuity Paradox

Actual changes in nature, changes in our understanding of nature using science, and changes in technology, history, or culture may occur in various ways, but there is a common feature one can identify in most cases: the fact that the seemingly continuous changes occurring over relatively long time periods are eventually disrupted by sudden, discontinuous changes. Distinctions are not always indisputably obvious and recognizable between these two fundamentally different but still related types of changes, and transitions between them usually do not occur in regular, predictable patterns. Nevertheless, the Continuity–Discontinuity Paradox is one type of paradox that is with us in many levels of the human experience.

11.2.1 Some of the Paradoxes in the History of the Development of Life on Earth

The terminology we often employ in the study of paradoxes related to the evolution of life usually has some connection to our everyday life, and to the history of societies where, for example, actual revolutions, with all their paradoxical aspects are described. Such terminologies are also employed in more abstract fields. For many natural processes, the Continuity–Discontinuity Paradox is apparent in the evolution–revolution alternatives, yet our focus is often on the developments of human societies where historians pay particular attention to sudden revolutions. Even more fundamentally, such continuity–discontinuity alternatives are also at the foundations of the development of various life-forms on Earth, and this is the very subject to which the word "evolution" brings the most associations in everyday conversations (Darwin, 1859; Murase, 1992, 1996, 2000, 2008, 2018; Murase & Murase, 2021a; Maynard-Smith & Szathmary, 1995). In a broader context, the Continuity–Discontinuity Paradox (also known as the Evolution–Revolution Paradox) can relate to any system capable of development (Bak, 1996; Kauffman, 1993; Murase, 1992, 1996; Murase & Murase, 2021a). In the history of these life-forms, some sequences of minor mutations have often provided some seemingly "continuous" changes for "slightly" changing generations of organisms. However, some major mutations, often triggered by sudden environmental changes, have led to sudden, practically "discontinuous" or "revolutionary" changes in the history of various species.

Hohenegger (in Chap. 17 in this book) suggested the interesting concept of evolution as a plausible motor of the changing biosphere. It is also true that the so-called phase transition (such as the sudden appearance or disappearance of order) also relates to the revolution of life history (Kauffman, 1993; Murase, 1996) as well as physical phenomena (Bak, 1996; Nicolis & Prigogine, 1977). There is always underlying gradual processes before the sudden change of a new phase, and interestingly, some complex systems show the self-organized phenomena for such phase transitions. Bak (1996) named such situations as self-organized critical phenomena. On this basis, it is possible to explain some of the essential features of how earthquakes take place, how mass extinctions occur during the history of life, how economic crisis happens without any obvious symptoms, and how we can realize the new ideas among the ideas of common knowledge. There must be some underlying common features or some universal principles behind such diverse phenomena.

There are many illustrative examples in our contemporary interpretation of many scientific facts, where the Continuity–Discontinuity Paradox plays a fundamental role. Although the list of such paradoxes is very long, the types of paradoxes may be dependent on the specific field of science.

However, in some cases, a commonly occurring feature may be found that provides a suitable mathematical background. As such, a less than universal but still rather widely applicable mathematical model has been described earlier, where, also somewhat paradoxically, a continuous transformation is suggested that is able to interconnect the original, seemingly paradoxical discrete and continuous representations in a consistent manner (Mezey, 2000a, 2000b).

11.3 The De-quantization–Re-quantization Paradox

The De-quantization–Re-quantization Paradox may be regarded as an inverted form of the Continuity–Discontinuity Paradox that is applicable to many processes. In the Continuity–Discontinuity Paradox, the usual change is an expected slow process, and the process usually develops in an apparently continuous way. When the usual process is disrupted, it may be unexpected. In such a paradox, the discontinuity or "revolutionary event" is the apparent exception, and is regarded as the "paradox-generating" component. However, in the case of the De-quantization–Re-quantization Paradox, the more "expected" cases are the discontinuous, quantized ones, and continuity appears as seemingly artificial or inconsistent. Many such De-quantization–Re-quantization Paradoxes typically involve the usual characterizations by integer numbers. However, the act of removing this restriction and allowing the possibility of continuous changes is regarded as the actual "de-quantization," and represents the unusual "paradox-generating" component.

The very expressions of "de-quantization" and "re-quantization" immediately bring to mind the field of quantum mechanics, as applied to quantum physics and quantum chemistry, and, indeed, there are many interesting examples of this paradox in natural science. Here we briefly describe some of these, although we do not use examples that require special physical or mathematical backgrounds.

The word "quantization", became popular after the introduction of Quantum Theory, initiated by Planck (1901a, 1901b), who, reluctantly, proposed the revolutionary idea that energy leaving a hot body in the form of light exists as well-specified packages, or "quanta", and that such a process is not continuous in terms of energy. This idea provided a solution for a long outstanding problem of theoretical physics, but itself appeared paradoxical at that time, because continuity in such processes had been regarded as natural and self-evident.

However, in the case of the De-quantization–Re-quantization Paradox, contrast is provided by artificially modifying an intrinsically "quantized", integer-related, or discrete model, originally incapable of continuous changes, to change into a more "restriction-free", artificial model, to become more flexible, and capable of continuous change. This step of making the model more flexible, represents a de-quantization. However, after taking advantage of this step, discrete or integer constraints can be re-introduced; a paradoxical step in the opposite direction. Then, restricting the model again, and allowing only discrete, specific, arrangements again, without the possibilities of continuous transitions among them; this step represents a re-quantization. Such a tentative lifting of the quantization restriction, or, in a more general sense, lifting the discrete model restriction, does allow one to use the freedom provided by concepts and mathematical tools of continuous functions, which are otherwise not available to the original quantized, or discrete model. However, all these steps are performed with a clear understanding of the artificial aspects of these "introduced" continuous features. Nevertheless, with the recognition of this artificiality, it is still possible to arrive at valid conclusions, which, after returning to the "reality" of the discrete, quantized nature of the real systems studied, the conclusions can be verifiably established.

There are many examples where discrete descriptions can be replaced by artificially extended continuous descriptions. After exploiting the possibilities this creates, eventually the description is returned to the discrete model, and some new insight or new knowledge can be obtained. One example may involve formal, continuous transformations between individual molecules, which, surprisingly, can be the source of rigorously established new knowledge.

Often, the model may be regarded as one using transformations in some abstract space and returning to reality after the mathematical freedom of this abstract space is utilized. In the following paragraphs, some of these examples are briefly described.

11.3.1 The Chemical Space, the Z Space (Nuclear Charge Space), and the Universal Molecule

In many areas of natural science, it is beneficial to think of the various objects studied there to form an abstract collection. These objects may be thought of as inhabiting some abstract space, and the relations among them can then be modeled by studying some of the structural properties of the abstract space. Often, this abstract space can be defined by some abstract coordinates, and the relations among the objects of the space can be modeled according to changes of the coordinates. By analogy with Cartesian coordinates used in three-dimensional space, the changes of these abstract coordinates in the abstract space can be regarded as continuous, or, possibly, as discontinuous if the property they describe can be altered in a sudden, discontinuous way, just like a change of integer numbers, such as a change from 6 to 7.

One such example is the so-called chemical space of molecules, which has been used in various forms and with various foundations (Fink et al., 2005; Kirkpatrick & Ellis, 2004; Smith et al., 2018; von Lilienfeld, 2013). In some of the early approaches regarding molecules collectively, such as systematic collation of experimental molecular data by pharmaceutical companies, some of the formal coordinates of an abstract Chemical Space have been taken as some experimentally measured molecular property values, in addition to composition and structural information presented numerically.

However, fundamental, quantum-chemistry-based approaches have also appeared, using some formal, continuous transformations between different molecules, indeed, as if those molecules were represented by subsets of some abstract space. This abstract space may be defined in terms of fictitious, continuously changing nuclear charges of the atoms forming the molecules. The idea for such "continuous" nuclear charges, which in protonic charge units are in reality integers and are usually denoted by the letter *Z*, has led to a formal "Z-space model" being proposed on a quantum-chemical basis (Mezey, 1981a, 1981b, 1983, 1985), independent of experimentally based pharmaceutical applications. This Nuclear Charge Space *Z*, when combined with the space of nuclear coordinates of all the atoms of the molecule (a metric space, usually denoted by *M*), forms an abstract, multidimensional "Product Space", $M*Z$, a quantum-chemically based Chemical Space. This provides an approach to represent all molecules and all of their conformations and shape changes within a single, abstract Chemical Space. This approach provides a theoretical, mathematical foundation for many of the experimentally motivated and organized "data-bank-type" approaches to the idea of Chemical Space.

The foundation of this approach is paradoxical, involving a contrast between the actual integer numbers representing the number of protons in atomic nuclei, on the one hand, and their continuous treatment in the Chemical Space, on the other hand. Nevertheless, the above theoretical basis for the Chemical Space (Mezey, 1981a, 1981b, 1983, 1985) is justified by some simple facts of chemistry, also leading in surprisingly simple ways to various new scientific results.

To appreciate this, it is worthwhile recalling some simple facts. Molecules contain nothing but atomic nuclei, characterized by the positive charge of the nucleus, which depends on the number of protons in those nuclei. Thus, the nuclear charge cannot have any value, and it can change only if the number of protons changes. As mentioned above, if one takes the charge of one proton as the physical unit to be used for charge, then the nuclear charge, denoted by *Z*, can be only an integer value. Indeed, for real molecules, the nuclear charge is "quantized". If this natural physical unit is used for charge, then there is no atomic nucleus in a molecule with a charge of a noninteger value. For example, it is impossible to have an atomic nucleus with a nuclear charge equal to 3.14159, an approximate rounded-off value of pi.

In the first instance, a simple chemical example, easily appreciated, may serve as an illustration, where the approach used in this example for the De-quantization–Re-quantization Paradox has had many useful applications, leading to novel scientific results. In the next section we briefly review this example.

One may use an artificial model, where the nuclear charges are taken as continuous variables, which allows the model to generate formal transformations between very different molecules; for example, "turning" the molecule of carbon monoxide (CO) into the molecule of nitrogen (N_2). Using the above discussed physical unit, the nuclear charges of CO can be listed as (6,8), taken as a formal row-vector, and, in the same way, the nuclear charges of N_2 can be taken as (7,7). By assuming an artificial, *continuous* transformation between these vectors, in this de-quantized model, for any common bond length, one is able turn CO into N_2. What is the possible advantage of doing this? Without going into great detail [see Mezey (1981a, 1981b, 1983, 1985) for explanations of the treatment of CO, N_2, and other complex molecules], one may exploit a formal mathematical feature of quantum mechanics. As one of the founding features of quantum mechanics, various measurable properties of physical systems (such as molecules), where the entire physical system (e.g., the entire molecule) is represented by wavefunctions, these properties can be computed using various mathematical tools, called "operators" (Mezey,). In our case, the relevant mathematical tool, the quantum chemistry electronic energy Hamilton operator, depends linearly on the nuclear charges. By the so-called variational theorem, this implies that the actual electronic energy must be a concave function (concave from below) of any such linear variable of the Hamiltonian, such as the "now de-quantized, hence continuous", real variable Z, the nuclear charge. This simple fact of concavity, and using a formal, continuous, "de-quantized" variable nuclear charge Z, give a rigorous proof that shows the electronic energy of the molecule CO is always less than the electronic energy of the molecule N_2 for any common bond length value. It is noteworthy that the truthfulness of the proof is not affected by using a fictitious "de-quantized" nuclear charge Z. Furthermore, one needs no quantum-chemistry software, nor a computer to arrive at this result on molecular energies—only a pen and less than five lines on paper. Evidently, such an abstract, fictitious exercise of pretending that the number of protons in an atomic nucleus can change continuously can lead to valid, novel, and useful results, which are rigorously proven in this way (Mezey, 1981a, 1981b, 1983, 1985).

Similarly, valid results can be easily obtained for many other molecular comparisons using this paradoxical, artificial, de-quantized, and eventually re-quantized model for nuclear charge. Ambitious applications of the general De-quantization–Re-quantization idea in the molecular context have also been pursued. Given that the model allows the inclusion of zero nuclear charges for a finite number of nuclei for some molecules, many nuclei can be introduced because formal "nuclei" with zero nuclear charge do not affect the physical relations among the "real" charge-bearing nuclei and electrons. Hence, one can imagine an abstract space of very high dimensions, where the coordinates are the individual nuclear charges, represented as "de-quantized", possibly continuously changing real variables, where, depending on the case, there can be any number of zero-charge nuclei. In this extended, high-dimensional "Nuclear Charge Space" Z, relations (e.g., energy relations) can be established among many greatly different molecules; in fact, for as many molecules as one wishes. In this abstract space Z, with the potential of infinitely many "impossible" or "nonexistent" molecules with noninteger nuclear charges, only such comparisons are useful when they lead to results related to real molecules. This reverts to the scenario where we take only integer values for the individual nuclear charges Z. By using and exploiting the introduced "flexibility" of continuity (de-quantization) and then returning to reality by re-introducing the natural restriction of integer nuclear charges (re-quantization), it is possible, in an unexpected way, to take advantage of some of the mathematical features of quantum mechanics. Nevertheless, the energy relations established for real molecules, based on a "de-quantized" continuity in this abstract Nuclear Charge Space Z, but eventually returning to the re-quantized reality, are valid, and can lead to many useful conclusions. That is, by using imaginary molecules with continuously changing nuclear charges, actual energy relations can be established for real molecules.

This Z-space model, the Nuclear Charge Space model, and the $M*Z$ Product Space Model where the Z-space is combined with a more "traditional" space M of the continuously changing nuclear coordinates, has many applications. However, the quantum-chemical origin of the Chemical Space model can lead to an "Extended Chemical Space" approach, where even some additional integer quantities (such as the number of electrons of a given spin in a molecule) can be taken as formal continuous variables. This creates a model that, in principle, includes all actual molecules (infinitely many) in various electronic states. Of course, this extended space also includes infinitely many fictitious molecules with physically impossible characteristics, but provide a useful intermediate role to give new results for real molecules. The "Extended Chemical Space" approach includes a special extreme case, the so-called Universal Molecule Model (Mezey, 2012, 2013), which is a single fictitious entity, where all actual molecules are special cases of the Universal Molecule. In addition, this Universal Molecule also serves as a central model from which, at least in a formal sense, the entire Chemical Space model, the Nuclear Charge Space model, and all possible real molecular models can be formally derived. Another model is the so-called Reference Cluster model (Mezey, 2015a, 2015b), where all nuclei have the same, possibly noninteger nuclear charge that is the formal average of all of the actual nuclear charges for the given molecule. This Reference Cluster model provides interesting symmetry connections, and symmetry-deficiency transformations between seemingly unrelated molecules (Mezey, 2015a, 2015b).

11.4 Localization–Delocalization Paradox

Local and global aspects may appear in many contexts, often in a spatial sense. These may include the surface of the planet Earth or just within a country, or in the context of an infection affecting only a small area of someone's skin or spread all over the body. Further reduction of the spatial size of the problem to a molecular level may involve local molecular ranges relating to a reactive functional group or to the global properties of the complete molecule.

The Localization–Delocalization aspects also appear on a more abstract level, when no spatially defined regions are relevant. Rather, we are referring to abstract domains. Examples of the Localization–Delocalization Paradox may be found in specific areas of science or in others, that are more directly beneficial within an interdisciplinary scientific context. The wisdom of some proverbs shows some of the fundamental aspects of this paradox, emphasizing the local–global mutual influences:

The devil is in the detail ...

A large balloon can be deflated by a very small hole made by a pin ...

11.4.1 Localized–Delocalized Areas of Science: Specific Disciplines and Interdisciplinarity

It is often assumed that to become very efficient in a given field of science, it is necessary to obtain a very deep knowledge in that field, with a strong time-commitment and devotion to the subject. Because of time limitations, this may mean far lesser focus and less time devoted to alternative areas of science and culture. This philosophy assumes that a strong local interest, local to the given field of science, is required. Yet, there are many examples of excellent, ground-breaking scientists who have studied rather broad, even seemingly unrelated topics, and have had a rather interdisciplinary education and background. Many of these scientists boast interdisciplinary achievements that cover a wide range of sciences.

One prominent example of such a scientist is the Nobel Laureate Prof. Kenichi Fukui, a theoretical and mathematical chemist, originally with an engineering background, who, after successes as an engineer, switched fields and had major achievements in quantum chemistry. However, his example also reveals an interesting manifestation of the Localization–Delocalization Paradox. After "localizing" his main scientific activities in the field of quantum chemistry (Fukui, 1970, 1981, 1986; Mezey, 2018; Tachibana & Fukui, 1978), many of his results suggest that his thought processes benefitted from his "delocalized" scientific background, including engineering, as well as cultural aspects of science (Mezey, 2018). Evidently, his most innovative scientific results in his local area of quantum chemistry had been strongly influenced by the more practical, engineering-type approaches he had used in his previous field, and a "delocalized scientific approach", involving a way of phrasing and solving a problem, and reaching new, innovative conclusions had a major role in his successes.

11.4.2 The Localization–Delocalization Paradox of the Location of Discovery and Utilization

New scientific breakthroughs and new technical inventions are typically based on various levels of familiarity with the accumulated knowledge of humanity. In a geographical sense, this knowledge is highly delocalized such that there is a fundamentally *delocalized component* in most scientific discoveries and inventions.

Yet, the actual final step leading to the given discovery or invention typically occurs locally. This may occur in a specific laboratory or in any place where the idea of a solution occurs to a single individual, or, possibly, a team devoted to the topic. This means that there is also a fundamentally *localized component* in the geographical sense where the decisive step of the discovery or invention occurs.

There is yet another twist in this Localization–Delocalization Paradox, namely, the fact that highly successful discoveries and inventions are eventually used globally, unless some politically motivated secrecy or business-motivated patent restriction stands in the way. This *delocalization feature*, which occurs in the geographical sense, is typical for most successful discoveries and inventions.

The actual success achieved when a scientific breakthrough occurs, or when a prototype of a technical invention first functions are typically localized events, yet their effects on human culture is eventually highly delocalized, often global. This is another example of the Localization–Delocalization Paradox.

11.4.3 Three Examples of the Localization–Delocalization Paradox in Chemistry

In thermodynamics, it is customary to discuss actual physical and chemical problems by first identifying what is to be considered as the "system", the main target of the study, and to distinguish it from the rest of the universe, often called the "surroundings". Most investigations of thermodynamics deal with processes within the system, but also consider processes between the system and the surroundings (Arens, 1963; Atkins, 2007; Bazarov, 1964; Glansdorff & Prigogine, 1971; Mezey, 1984, 2015a, 2015b; Stull et al., 1969; von Bertalannfy, 1968; Van Ness, 1969).

One important and general phenomenon that can be observed for most systems that are close to or in equilibrium is their "reaction" to some stress applied from the surroundings. This general principle (Le Chatelier's Principle) states that in such a case, the system readjusts to oppose the effects of the stress applied to it. This usually means only a partial readjustment, and the original state is seldom reconstructed.

This thermodynamics model inherently involves the Localization–Delocalization Paradox, because the system may be regarded as a local entity that is enclosed by the surroundings as the global entity.

For example, consider a chemical equilibrium of a solution of a mixture of compounds that can transform into one another, but some time after mixing them, their relative concentrations do not change; that is, they establish an equilibrium. This is typically a dynamic equilibrium, because individual molecules can change, but as far as overall concentrations are concerned, in such an equilibrium, the opposite changes cancel out. However, if a stress is applied from the surroundings by adding an extra quantity of one of the compounds, then the system is no longer at equilibrium. The system will then readjust to the sudden increase in concentration of one of the compounds by increasing the rate of reactions that convert some of the newly arrived compound to other compounds, thereby re-establishing an equilibrium. In such a case, the new concentration of the added compound may be slightly higher than it was before, but it will always be less than it was right after its addition. Although the original concentrations are not restored, the suddenly increased concentration of the added compound will have been somewhat reduced by this "response" of the system to the "stress".

This example shows that Le Chatelier's principle is related to the Localization–Delocalization Paradox, where the system can be taken as a localized entity, and the interaction with the surroundings as a manifestation of delocalization.

An analogous manifestation of this principle is the so-called Quantum-Chemical Le Chatelier Principle (Mezey, 1984, 2015a, 2015b), applied on a somewhat more fundamental level, where the energy components, such as the nuclear repulsion part E_n and the electronic energy part E_e, making up the total energy E_t of a molecule, show rather characteristic interrelations in some motions of conformational rearrangements or chemical reactions. If one regards the set of nuclei and the set of electrons within a molecule to exist in some "equilibrium", than a sudden change of some nuclear arrangement can be taken as a "stress" on this equilibrium, where this imbalance, or stress, is partially alleviated by a change of the electron distribution. These changes are reflected in changes in the E_n and E_e components of the total energy, and in many cases, the changes in these two energy components have opposite phases, as one change partially compensates for the other (Mezey, 1984, 2015a, 2015b).

Even small local changes that directly affect only some of the nuclear positions can trigger a more distributed, delocalized change in the electron distribution, providing an example of the Localization–Delocalization Paradox.

Another chemical example is also from quantum chemistry: molecular orbitals are used as mathematical representations of the molecular wavefunction, a function that in principle carries all the measurable information about an individual molecule. There is a considerable freedom in choosing such molecular orbital representations, which are quantum-mechanically equivalent, and all fulfill the physical restrictions of validity. However, for interpretative purposes, some representations where the orbitals have some local dominance in one or another region of the molecule may be more suitable. By a variety of criteria, such orbitals are regarded as "localized orbitals"; some of the most used choices are the Boys localization (Foster & Boys, 1960), the so-called Ruedenberg localization (Edmiston & Ruedenberg, 1963), while the scheme most frequently employed in current commercial quantum chemistry software is the Pipek-Mezey localization (Pipek & Mezey, 1989).

In the Pipek-Mezey localization, the Localization–Delocalization Paradox appears in more than one way, because a very closely related approach may also serve to explicitly investigate these orbital representations that are the *most delocalized* within a given molecule. These can provide useful interpretative tools to address questions on how a local influence at one end of a molecule may produce effects at the opposite end of the same molecule (Pipek & Mezey, 1988). Such knowledge in important in the advancement of computer-based molecular design.

11.4.4 A Holographic Principle and the Localization–Delocalization Paradox

Another chemistry example connected to the Localization–Delocalization Paradox, although in a fundamentally different way, is the holographic property of molecular electron density clouds. As has been proven (Mezey, 1999a), if any molecule is taken in its most stable, lowest electronic state, then any small (but positive volume) part of the electron density cloud contains the complete information about all properties of the entire molecule. This concept has been established by the Holographic Electron Density Theorem (Mezey, 1999a). This theorem is only analogous with actual holography, but one common "holographic" feature is important: a part of the holographic recording is suitable to reconstruct the complete three-dimensional image, and for molecules too, any small, positive volume of the electronic cloud is sufficient to deliver the complete molecular information.

Several consequences of this theorem provide tools for computer-based molecular design, for example, in the search for new pharmaceuticals or herbicides (Mezey, 1999b), or in the study of connections between local and global similarities of molecules (Mezey, 2000a, 2000b), which is a direct example of the Localization–Delocalization Paradox. The assurances given by the theorem also have practical uses in the study of molecular recognition (Mezey, 2001a). Examples include the mutual recognition between an enzyme cavity and a complementary pharmaceutical compound, or in the study of potentially exhibited molecular properties, such as electronic excitation generated after a photon of specific energy reaches the molecule (Mezey, 2001b).

11.4.5 Individual Cultures–Multiculturalism

An important example of the Localization–Delocalization Paradox is related to cultural developments and the local and global relations among cultures. This topic deserves a far more detailed study than the comments provided in this short section. This Localization–Delocalization Paradox is one of the oldest problems of humanity and is often a source of great conflict, but it is also the source and motivation for great developments. Two old proverbs illustrate this paradox:

1. The old advice for a visitor to the ancient Rome: In Rome, live like a Roman…
2. The wise man learns from his neighbor.

In today's world of the Internet and instant communication, there are many related issues, and we shall not discuss them here. Mezey (2019) illustrates some of the connections among some of the current cultural trends, and the analogies between the current, generalized environmental issues, involving both the natural and the cultural environments.

11.5 Sharp–Fuzzy Paradox

The Sharp–Fuzzy Paradox is also present in various forms in some of the examples mentioned above, yet an additional, extreme example may provide some interesting insight.

Fuzzy sets provide alternative representations to problems that can also be described using the tools of probability theory, but the fuzzy set approach introduced by Zadeh (1965) has many advantages. The example of the Sharp–Fuzzy Paradox discussed here is that of graph theory (Bollobas, 1998), one where one of the most characteristically discrete mathematics subject, which not only has continuous features, such as labeling the graph vertices or edges with continuously changing variables, but also with fuzzy features. Thereby, a very versatile mathematical model is obtained (Blue et al., 2002; Rosenfeld, 1975), which can rely on the advantages of both discrete and sharp descriptors that is usually beneficial for clarity of concepts and simplifications. However, such a model can also exploit the benefits of continuous, as well as fuzzy variables. The apparent Sharp–Fuzzy Paradox between these two types of features provides a versatility within a single framework that has many advantages.

11.6 Summary

This chapter has illustrated some of the scientific and culture-related paradoxes through various examples, some relating to the history of scientific and technological discoveries and advances, but others relating to more fundamental and often interdisciplinary areas of human activities. Although some of the paradoxes have been sources of negative developments, this chapter has focused on the benefits of paradoxes. The element of surprise and the resulting intellectual curiosity that occur when a paradox is encountered have often had motivating roles in novel discoveries. In fact, paradoxes are often compelling signs that some new way of thinking and new way of problem solving is required. In this role, paradoxes can be seen as forces acting along the evolution of science and many other human activities.

Appendix 1

Benefits of Clearly Defined but Not Well-Known Components as Temporary Tools for Advancement: A Witty Proof of the Theorem of Pythagoras

The theorem of Pythagoras, $c^2 = a^2 + b^2$, for the three sides *a*, *b*, and *c* (the longest side) of a triangle having a right-angle opposite to side *c* has many proofs. However, one proof, with a simple consideration of the fact that the theorem describes a relation among distances, and, if the areas of triangles are considered, then for such areas, distance squared must be the dimension for the results eventually obtained. This proof is particularly witty and demonstrates very clearly that some temporarily used entities in the proof do not appear in the final result. Such entities do not need to be known in any detail as long as they have a clear definition. It is sufficient to know that such an entity exists, even if its function is unknown. After using such an entity as a tool in some intermediate steps of the proof, one may discard it without ever clarifying what it really is.

Consider a triangle T with vertices A, B, and C, with angles at these vertices denoted by α, β, and γ, respectively, where $\gamma = 90°$, and with sides opposite to these vertices denoted by *a*, *b*, and *c*, respectively. Consider a straight line issued from C, perpendicular to *c*, with the intercept on *c* denoted by M. The line segment from C to M is denoted by *m*. Without any loss of generality, we may assume that $\alpha \geq \beta$.

This configuration gives three triangles, each with a right angle. The original triangle T with sides *a*, *b*, and *c*, contains two smaller triangles; one we refer to as T_a, with sides *m*, *a*, and one part of *c*, and the other as T_b with sides *m*, *b*, and the other part of *c*. In T, in T_a, and in T_b, the right angle is opposite to sides *c*, *a*, and *b*, respectively. These three triangles are similar in a geometrical sense in that they are scaled versions of one another.

Evidently, the two small triangles make up the original one, so the areas *E* of these three triangles are related by

$$E(\mathrm{T}) = E(\mathrm{T_a}) + E(\mathrm{T_b}). \tag{11.1}$$

Clearly, the area of each triangle is uniquely determined by its longest side and the angle α. If we know these, all other angles are known and the triangles and their areas are fully defined. Therefore, there must be a function *F* for the area *E* that depends on the longest side *s* and the angle α, and nothing else:

$$E = F(s, \alpha) \tag{11.2}$$

The witty dimension argument then proceeds as follows.

The result for the area *E* must come out as a quantity with dimensions of "distance squared". The angle variable in function $F(s, \alpha)$ cannot produce such a dimension; this must all originate from *s*. That is, the area function must have the structure of

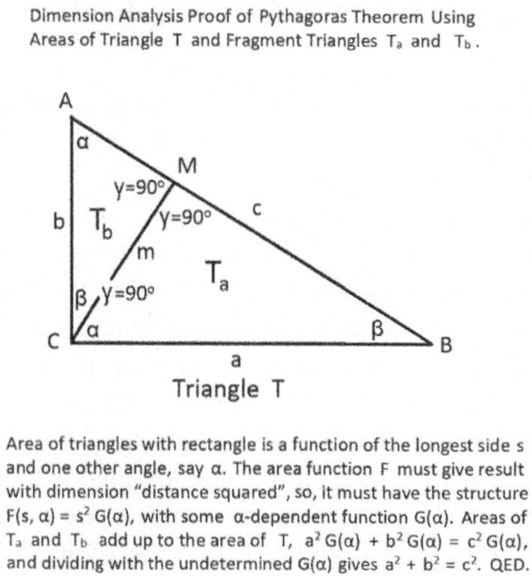

Fig. 11.1 Illustration of the dimension analysis proof of the theorem of Pythagoras

$$E = F(s, \alpha) = s^2 G(\alpha) \tag{11.3}$$

with some function $G(\alpha)$ for the contribution of the angle variable.

Clearly, the side length s cannot produce other dependences like s^3, or s^{17}, or $\sin(s)$, or s^{-2} because the overall result must have the dimension of area, and only s^2 can deliver that.

That is, Eq. (11.1) can be written as

$$c^2 G(\alpha) = a^2 G(\alpha) + b^2 G(\alpha). \tag{11.4}$$

Division with the unknown angle dependence $G(\alpha)$ immediately gives the proof of the theorem of Pythagoras (Fig. 11.1):

$$c^2 = a^2 + b^2. \tag{11.5}$$

It is instructive to realize that in the course of this proof we have used a function, $G(\alpha)$, which had an important role in the proof, yet we could eliminate it and reach a result without knowing the form of the function.

It is likely that similar situations occurring in many different fields beyond mathematics and the natural sciences, where it is worthwhile to use concepts and ideas that are defined but not determined, because their use allows actual conclusions to be reached without detailed specification.

Appendix 2 Useful Complications Leading to Simplifications

It is normal to regard complications as something negative, something to be avoided, but in many situations, a temporary complication does lead to a solution. In some cases, the very complication can be the source of simplification. Paradoxically, by seemingly complicating a problem and the choice of approach, it is often the key to a simple solution. One such complication that can be useful is the extension of dimensions; for example, extending the description of some mathematical problems originally defined on the one-dimensional real axis to the two-dimensional complex plane can allow a solution to be easily found.

More generally, such an apparently paradoxical approach can be summarized as an "Excursion to higher dimensions without looking around, just taking what one needs". In some cases, the only goal is to obtain a proof for a result that cannot be proven in the three-dimensional model; we use higher dimensions only to get some results, without checking out most of the properties of this higher dimensional model. We complete the proof there, and then return to the "normal" three-dimensional space with a result.

A similar dimension extension, which initially may appear as a complication, can provide quick shortcuts to solutions for various problems in physics. Here we briefly describe one such case, relating to the proof of the Holographic Electron Density Theorem, briefly discussed in an earlier section.

This approach has been used in the actual proof of the holographic theorem. Without getting into the details, and only providing a sketch of the main steps, in the ordinary three-dimensional (3D) description, the difficulty is that the electron density, as a function, is defined on a noncompact set (the 3D space; compactness is a desirable property allowing the use of many mathematical tools), and for noncompact sets an important and very useful theorem, the analytic continuation theorem, is not applicable. Yet, the analytical continuation theorem is able to provide strong relations between the part and the whole, the very subject of the Holographic Electron Density Theorem. However, one can move the problem to higher dimensions, define the relevant electron density function there in a consistent way, use a 4D variant of the so-called stereographic projection familiar from various geographical maps, and then re-define electron density on a 3D sphere embedded in 4D. After that, a compactification (the Alexandrov one-point compactification) is applicable to the 3D sphere embedded in 4D without affecting the proper physical definition of the electron density. Hence, for this new higher-dimensional arrangement, the analytic continuation theorem is now applicable, and that leads to the proof of the holographic electron density theorem.

This "dimension-hopping" proof is an example of the local–global paradox, in more than one way: the theorem itself states a relation between local and global aspects of the electron density, but it is also an example of using an intermediate mathematical entity to reach the proof, such as the 3D spherical hypersurface embedded in 4D, without exploring any further how this intermediate representation of the electron density function behaves, except for the very properties we need.

References

Arens, R. (1963). An axiomatic basis for classical thermodynamics. *Journal of Mathematical Analysis and Applications, 6*, 207–229.

Atkins, P. (2007). *Four laws that drive the universe*. Oxford University Press.

Bak, P. (1996). *How nature works: The science of self-organized criticality*. Springer.

Bazarov, I. P. (1964). *Thermodynamics*. Pergamon Press.

Blue, M., Bush, B., & Puckett, J. (2002). *Fuzzy Sets and Systems, 125*(3), 355–368.

Bollobas, B. (1998). *Modern graph theory*. Springer.

Darwin, C. (1859). *The origin of species*. John Murray.

Edmiston, C., & Ruedenberg, K. (1963). Localized atomic and molecular orbitals. *Reviews of Modern Physics, 35*, 457–465.

Fink, T., Bruggesser, H., & Reymond, J.-L. (2005). Virtual exploration of the small-molecule chemical universe below 160 Daltons. *Angewandte Chemie International Edition, 44*, 1504.

Foster, J. M., & Boys, S. F. (1960). Canonical configuration interaction procedure. *Reviews of Modern Physics, 32*, 300–302.

Fukui, K. (1970). A formulation of the reaction coordinate. *Journal of Physical Chemistry, 74*, 4161–4163.

Fukui, K. (1981). The path of chemical reactions—The IRC approach, accounts of chem. *Philosophy and Phenomenological Research, 14*, 363–368.

Fukui, K. (1986). Reminiscences—My forty years study of chemical reactions. In H. G. Herzberg, R. S. Mulliken, K. Fukui, W. Lipscomb, R. Hoffman, H. L. Honolulu, V. H. Smith, Jr., H. F. Schaefer III, & K. Morokuma (Eds.), *Applied quantum chemistry, proceedings of the Hawaii 1985 nobel laureate symposium on applied quantum chemistry* (pp. 1–25), Reidel Publ. Co., Dec 16–21, 1984.

Glansdorff, P., & Prigogine, I. (1971). *Thermodynamic theory of structure, stability and fluctuations*. Wiley.

Kauffman, S. A. (1993). *The origins of order: Self-organization and selection in evolution*. Oxford University Press.

Kirkpatrick, P., & Ellis, C. (2004). Chemical space. *Nature, 432*, 823.

Maynard-Smith, J., & Szathmary, E. (1995). *The major transitions in evolution*. Freeman.

Mezey, P. G. (1981a). Electronic energy inequalities for isoelectronic molecular systems. *Theoretica Chimica Acta, 59*, 321–332.

Mezey, P. G. (1981b). The Isoelectronic and Isoprotonic energy hypersurface and the topology of the nuclear charge space. *International Journal of Quantum CHemistry Symposium, 15*, 279–285.

Mezey, P. G. (1983). A molecular geometry invariant property of energy level set boundaries in Z-space. *International Journal of Quantum CHemistry, 24*, 523–526.

Mezey, P. G. (1984). A General formulation of the "quantum chemical le chatelier principle." *International Journal of Quantum CHemistry, 25*, 853–861.

Mezey, P. G. (1985). A simple relation between nuclear charges and potential surfaces. *Journal of the American Chemical Society, 107*, 3100–3105.

Mezey, P. G. (1999a). The holographic electron density theorem and quantum similarity measures. *Molecular Physics, 96*, 169–178.

Mezey, P. G. (2000a). The T-Hull approach to transformations of discrete point sets to continua and shape transformations between discontinuous objects using alpha hulls. *Journal of Mathematical Chemistry, 27*, 53–60.

Mezey, P. G. (2001a). A uniqueness theorem on molecular recognition. *Journal of Mathematical Chemistry, 30*, 305–313.

Mezey, P. G. (2001b). The holographic principle for latent molecular properties. *Journal of Mathematical Chemistry, 30*, 299–303.

Mezey, P. G. (2015a). Compensation effects in molecular interactions and the quantum chemical le chatelier principle, (Invited paper to the jacopo tomasi festschrift). *Journal of Physical Chemistry A, 119*, 5305–5312.

Mezey, P. G. (1999b). Holographic electron density shape theorem and its role in drug design and toxicological risk assessment. Journal of Chemical Information Computer Science, *39*, 224–230.

Mezey, P. G. (2000b). Local and global similarities of molecules: Electron density theorems, computational aspects, and applications. In *Proceedings of European congress on computational methods in applied sciences and engineering*. ECCOMAS 2000.Barcelona, Sept 11–14, 2000.ISBN-84-89925-70-4, 1–10. Available at: congress.cimne.com/eccomas/eccomas2000/pdf/811.pdf.

Mezey, P. G. (2012). Discrete skeletons of continua in the universal molecule model. In *AIP (American institute of physics) conference proceedings. COMPUTATION IN MODERN SCIENCE AND ENGINEERING: Proceedings of the international conference on computational methods in science and engineering 2009 (ICCMSE 2009)* (Vol. 1504, pp. 725–728).

Mezey, P. G. (2013). On discrete to continuum transformations and the universal molecule model—a mathematical chemistry perspective of molecular families. In *AIP (American Institute of physics) conference proceedings, COMPUTATION IN MODERN SCIENCE AND ENGINEERING: Proceedings of the international conference on computational methods in science and engineering 2007 (ICCMSE 2007)*, (Vol. 963/2, parts A and B, pp. 513–516).

Mezey, P. G. (2015b). Relations between real molecules through abstract molecules: The reference cluster approach (Invited paper to the Peter Surjan Festschrift). *Theoretical Chemistry Accounts, 134*(134), 25–30. https://doi.org/10.1007/s00214-015-1728-1.

Mezey, P. G. (2018). The intrinsic reaction coordinate and a path of ascent to mount Hiei: In memory of professor kenichi fukui, a nobel prize winning chemist, and a pioneer of many new paths in mathematical chemistry. In G. Fang, M. Amini, H. Chen, N. Fukuda, H. Hosoya, M. Kawai, J. E. LeBlanc, P. G. Mezey, I. Naruki, T. Okada, E. Rambo, M. Spivakovsky, S. Takeuchi, K. F. Taylor, H. Wong, S. Yamanaka, M. Yokotani, P. Zizler, & S. Arimoto (Eds.), *Mathematics and chemistry interdisciplinary joint research and the Fukui project XXV* (Vol. 60, pp 37–42). Bulletin of National Institute of Technology, Tsuyama College

Mezey, P. G. (2019). The role of analogies and data structures in cultural, environmental, and scientific developments. In Z. Androvičová, & E. Belaňová (Eds.), *Selected aspects of integrated environmental management: Culture and environment", technical university of Zvolen (Slovakia)*, UNESCO publ. (pp. 9–16) (ISBN 978-80-228-3200-7, ©Technická univerzita vo Zvolene, ©Technical University of Zvolen).

Murase, M. (1992). *The dynamics of cellular motility*. Wiley.

Murase, M. (1996). alzheimer's disease as subcellular 'cancer': The scale-invariant principles underlying the mechanisms of aging. *Progress of Theoretical Physics, 95*(1), 1–36.

Murase, M., & Mursae, T. (2021a). Transdisciplinary study of how to integrate shattered world: The self-nonself circulation principle of "living" wholeness. In S. Hill, T. Yagi, S. Yamash'ta (Eds.), *The Kyoto manifesto II*. Springer

Murase, M., & Mursae, T. (2021b). The self-nonself circulation principle of "living" nature: How to survive shattered world. In S. Hill, T. Yagi, S. Yamash'ta (Eds.), *The Kyoto manifesto II*. Springer

Murase, M. (2000). *Life as history: The construction of self-nonself circulation theory* (in Japanese) (pp. 369–376), Kyoto University Press, Kyoto (in English). https://repository.kulib.kyoto-u.ac.jp/dspace/bitstream/2433/49765/1/Murase2000b2.pdf.

Murase, M. (2008). Environmental pollution and health: an interdisciplinary study of the bioeffects of electromagnetic fields. *SANSAI, An Environmental Journal for the Global Community, 3*, 1–35.

Murase, M. (2018). A self-similar dynamic systems perspective of "living" nature: The self-nonself circulation principle beyond complexity. In Y. Stomu, Y. Tadashi, S. Hill (Eds.), *The Kyoto Manifesto for global economics the platform of community, humanity, and spirituality* (pp. 257–283) Springer.

Nicolis, G., & Prigogine, I. (1977). *Self-organization in nonequilibrium systems: From dissipative structures to order through fluctuations*. Wiley.

Pipek, J., & Mezey, P. G. (1988). Dependence of MO shapes on a continuous measure of delocalization, *International Journal of Quantum Chemistry Symposium, 22*, 1–13.

Pipek, J., & Mezey, P. G. (1989). A fast intrinsic localization procedure applicable for ab initio and semiempirical linear combination of atomic orbital wave functions. *The Journal of Chemical Physics, 90*, 4916.
Planck, M. (1901a). Über das Gesetz der Energieverteilung im Normalspektrum. *Annalen der Physik, 4*; 1, 553–563; 717–727 (PAV).
Planck, M. (1901b). Über die Elementarquanta der Materie und der Elektrizität. *Annalen der Physik, 4;* 1, 564–566; 728–730 (PAV).
Rosenfeld, A. (1975). *Fuzzy sets and their applications* (pp. 77–95). Academic Press.
Smith, J. S., Nebgen, B., Lubbers, N., Isayev, O., & Roitberg, A. E. (2018). Less is more: Sampling chemical space with active learning. *The Journal of Chemical Physics, 148*(24), 241733
Stull, D. R., Westrum, E. F., & Sinke, G. C. (1969). *The chemical thermodynamics of organic compounds Wiley.*
Tachibana, A., & Fukui, K. (1978). Differential geometry of chemically reacting systems. *Theoretica Chimica Acta, 49*, 321–347.
Van Ness, H. C. (1969). *Understanding thermodynamics*. Dover.
von Bertalannfy, L. (1968). *General system theory*. Braziller.
von Lilienfeld, O. A. (2013). First principles view on chemical compound space: Gaining rigorous atomistic control of molecular properties. *International Journal of Quantum Chemistry, 113*, 1676–1689.
Zadeh, L. A. (1965). Fuzzy sets. *Information and Control, 8*, 338–352.

Masatoshi Murase received his Ph.D. degree from The University of Tokyo in 1987. Since 1992, he has been an associate professor at the Yukawa Institute for Theoretical Physics, Kyoto University. In 1987 and 1988 he was a visiting scientist at the Duke University Medical Center, Durham, NC, USA. Since 2010, he has been a member of the Cooperation Promotion Committee of the International Research Unit of Integrated Complex System Science, Kyoto University. Between 2015 and 2020 he was a director of the Research Promotion Strategy Office of the International Research Unit of Advanced Future Studies, Kyoto University.

Paul G. Mezey is a Hungarian-Canadian mathematical chemist. He was the Canada Research Chair in Scientific Modeling and Simulation in the Department of Chemistry at the Memorial University of Newfoundland. Mezey received a Master's degree in chemistry, a Ph.D. in Chemistry, and a Master's degree in mathematics, all from Eötvös Loránd University in Budapest, in the years 1967, 1970, and 1972, respectively. From 1982 to 2003, he was a professor of chemistry and mathematics at the University of Saskatchewan, where he received a D.Sc. in 1985 in mathematical chemistry. He was a faculty member at Memorial University from 2003 to 2018. Mezey is Editor-in-Chief of the *Journal of Mathematical Chemistry*.

Part III
Emergent Dynamics in Complex Social and Physical Sciences: Exploring the Underlying Fluctuations in Collective Modes

Chapter 12
Elucidation of Chaotic Market Hypothesis Based on Ergodic Theory

Ken Umeno

Abstract We develop a method of directly testifying the efficient market hypothesis (EMH) proposed by Fama by using the ergodic theory to connect microscopic price fluctuations with macroscopic behavior. After testing the validity of the method by using exactly solvable chaos, we found a novel periodic structure in the 5-min chart data of the Nikkei averages in 2019 with our developed new correlation function based on the characteristic function. This *directly* denies the EMH. Statistical data of the empirical studies have shown that a stable law well describes the price fluctuations of financial markets as predicted by the most generalized version of the central limit theorem called the universal super generalized central limit theorem (USGCLT) we discovered recently. The concept and proof of the extension of the super generalized central limit theorem (SGCLT) is also given to illustrate the mechanism of *universality* such that a sum of random numbers with *nonidentical* power law distributions converges to a stable law in distribution. With these theoretical facts together with the empirical fact of the data denying the EMH, we propose the chaotic market hypothesis (CMH) based on the ergodic theory to capture the essential characteristics of the financial markets.

12.1 Introduction

Ergodicity is the fundamental concept that connects macroscopic behavior with the microscopic behavior for complex systems such as the financial market. To characterize the financial market, Fama (1970) propose a clear vision that the the financial market is essentially characterized by the random walk by posing basic assumptions of *efficiency* that the equality of opportunities exist for every participants under

K. Umeno (✉)
Graduate School of Informatics, Kyoto University, 36-1 Yoshida-Honmachi, Sakyoku, Kyoto 606-8501, Japan

International Research Unit of Integrated Complex System Science, Kyoto University, Kyoto 606-8501, Japan
e-mail: umeno.ken.8z@kyoto-u.ac.jp

K. Nishimura et al. (eds.), *Creative Complex Systems*, Creative Economy,
https://doi.org/10.1007/978-981-16-4457-3_12

the fair condition that there is no insider trading with privileged information. This assumption has been supported by many empirical studies such that financial time series data such as logarithmic returns of price fluctuations have *no correlation* as firstly observed by Fama (1970). Since then, the efficient market hypothesis (EMH) has prevailed with this view by stating that the financial market is just a *random walk* and no one can predict a future price from information up to the present. Actual financial time series are, however, known to have a short-term memory effect showing *correlation* by practitioners participating in markets. Thus, there is a non-negligible *gap* between EMH and actual financial markets. The purpose of the present study is to clearly characterize the gap by giving the reason on the tail risk behavior characterized by a certain generalized version of the central limit theorem and introducing a new hypothesis called the chaotic market hypothesis (CMH). In the CMH, a small amount of correlation is *acceptable* for financial time series such as a *mixing* behavior representing *chaos*, together with the fundamental concepts: super generalized central limit theorem showing a universality of fat tail behavior. In Sect. 12.2, we review the super generalized central limit theorem (SGCLT) as the basic ingredients of its further generalized central limit theorem showing a universal tail behavior of financial markets, which will be presented in Sect. 12.3. In Sect. 12.4, a novel correlation function based on a characteristic function is introduce to test the EMH by using the empirical data of the 5-min chart data of logarithmic returns of the Nikkei average in 2019. In Sect. 12.5, CMH is proposed to capture the essential characteristics of the financial market behavior as described above. Discussion about the difference between the CMH and the fractal market hypothesis is presented in Sect. 12.6. Section 12.7 summarizes and concludes the chapter.

12.2 Super Generalized Central Limit Theorem and its Generalization

As the central limit theorem is the fundamental law to characterize a limiting distribution for sums of random numbers with finite variance, the generalized central limit theorem (Gnedenko & Kolmogorov, 1954) has a key role in characterizing a limiting power law distributions for sums of random numbers obeying the power law with infinite variance such as Cauchy law. Such a limiting power law can be characterized as the stable law in a *universal* manner, which is the crux of the ubiquitous nature of macroscopic distributions due to the generalized central limit theorem.

Stable Law $S(x;\alpha,\beta,\gamma,\mu)$ is given by the Fourier transform of the characteristic function $\phi(t)$

$$S(x;\alpha,\beta,\gamma,\mu)=\frac{1}{2\pi}\int_{-\infty}^{\infty}\phi(t)e^{-Ixt}dx$$

where characteristic function $\phi(t)$ with the four parameters: α,β,γ and μ has the form:

$$\phi(t) = \exp\{\mathrm{I}\mu t - \gamma^{\alpha}|t|^{\alpha}(1 - \mathrm{I}\beta\, \mathrm{sgn}(t) w(\alpha, t)\} \tag{12.1}$$

with

$$w(\alpha, t) = \begin{cases} \tan(\pi\alpha/2) & \text{if } \alpha \neq 1 \\ -2/\pi \ln|t| & \text{if } \alpha = 1. \end{cases}$$

Here the parameters α, β, γ and μ are the scaling exponent parameter $\alpha \in (0, 2]$ representing the fatness of the tail, the skewness parameter $\beta \in [-1, 1]$, the scaling parameter $\gamma > 0$, and the location parameter $\mu \in \mathbb{R}$, respectively. When $\alpha = 2$, it corresponds to the Gauss law with $\beta = 0$ and when $\alpha = 1, \beta = 0$, it corresponds to the Cauchy law.

Here, we introduce the the **condition for the super generalized central limit theorem** according to Shintani and Umeno (2018):

1. The random variables $C_+ > 0$ and $C_- > 0$ obey respectively the distributions $\mathrm{P}_{c_+}(c)$ and $\mathrm{P}_{c_-}(c)$, and satisfy

$$\mathrm{E}[C_+] < \infty, \text{ and } \mathrm{E}[C_-] < \infty.$$

2. The probability distribution function $f_i(x)$ of the random variables X_i has a following limiting form when $0 < \alpha < 2$:

$$f_i(x) \simeq \begin{cases} c_{+i} x^{-(\alpha+1)} & \text{for } x \to \infty \\ c_{-i} |x|^{-(\alpha+1)} & \text{for } x \to -\infty, \end{cases} \tag{12.2}$$

   where c_{+i} and c_{-i} are samples obtained by C_+ and C_-, respectively.

The following theorem holds (Shintani and Umeno 2018).

Theorem 1 (Super Generalized Central Limit Theorem) *Suppose that Condition 1 and Condition 2 are satisfied. Then the following superposition S_n of independent and nonidentical random variables with power laws converges in density to a unique stable distribution $S(x; \alpha, \beta^*, \gamma^*, 0)$ for $n \to \infty$, where*

$$S_n = \frac{\sum_{i=1}^{n} X_i - A_n}{n^{\frac{1}{\alpha}}} \xrightarrow{d} S(x; \alpha, \beta^*, \gamma^*, 0) \text{ for } n \to \infty,$$

$$A_n = \begin{cases} 0 & \text{if } 0 < \alpha < 1 \\ n \sum_{i=1}^{n} \Im \ln(\varphi_i(1/n)) & \text{if } \alpha = 1 \\ \sum_{i=1}^{n} \mathrm{E}[X_i] & \text{if } 1 < \alpha < 2 \end{cases}$$

with $\varphi_i(t)$ being a characteristic function of X_i as the expected value of $\exp(\mathrm{I}tX_i)$ *and parameters $\beta^*, \gamma^*, \beta_i, \gamma_i$ are expressed as:*

$$\beta^* = \frac{E_{C_+,C_-}[\beta_i \gamma_i^{\alpha}]}{E_{C_+,C_-}[\gamma_i^{\alpha}]}, \gamma^* = \{E_{C_+,C_-}[\gamma_i^{\alpha}]\}^{\frac{1}{\alpha}}, \tag{12.3}$$

$$\beta_i = \frac{c_{+i} - c_{-i}}{c_{+i} + c_{-i}}, \gamma_i = \left\{ \frac{\pi(c_{+i} + c_{-i})}{2\alpha \sin(\pi\alpha/2)\Gamma(\alpha)} \right\}^{\frac{1}{\alpha}},$$

where $E_{C_+,C_-}[X]$ *denotes the expectation value of* X *with respect to the random parameter distributions* $P_{c_+}(c)$ *and* $P_{c_-}(c)$. *Here,* $\Im$ *is an imaginary part of the argument.*

Consider a further generalization of the super generalized central limit theorem. Here, a *generalization* means that we extend the super generalized central limit theorem with a unique power index α to a further generalization of the generalized central limit theorem for superposition of random variables with *different* power indices α_m where

$$0 < \alpha_1 < \alpha_2 < \cdots < \alpha_m < 2$$

for a sum of independent random numbers X_i obeying the probability distribution function f_i such that

$$f_i(x; \alpha_k) \simeq \begin{cases} c_{+i} x^{-(\alpha_k+1)} & \text{for} \quad x \to \infty \\ c_{-i} |x|^{-(\alpha_k+1)} & \text{for} \quad x \to -\infty \end{cases} \tag{12.4}$$

When $m = 1$, it corresponds to the SGCLT with a unique α. Thus, we consider the case $m \geq 2$.
The following conditions are essential:
Condition for the Universal Super Generalized Central Limit Theorem (USGCLT)

1. Each random number X_i satisfies the conditions of the super generalized central limit theorem (SGCLT).
2. The power indices of the probability distribution are in the order that $0 < \alpha_1 < \alpha_2 < \cdots < \alpha_m < 2$ and the equality $\sum_{k=1}^{m} q_{\alpha_k} = 1$ is satisfied where $q_{\alpha_k} (0 < q_{\alpha_k} < 1)$ is a probability that the power index $\alpha_k (1 \leq k \leq m)$ is randomly selected from the m power indices $(\alpha_1, \alpha_2, \cdots, \alpha_m)$. Furthermore, the probabilities q_{α_k} are assumed to be positive constants.

We have the following theorem:

Theorem 2 (Universal Super Generalized Central Limit Theorem *(USGCLT)*) *If independent random variables* $X_1, X_2, \ldots$ *satisfy the* **condition for the universal super generalized central limit theorem (USGCLT)**, *then a superposition*

$$S_n \equiv \frac{\sum_{i=1}^{n} X_i - A_n}{n^{\frac{1}{\alpha_1}}}$$

converges in density to a stable law as

$$S_n \xrightarrow{d} S(x;\alpha_1,\beta^*[\alpha_1],\gamma^*[\alpha_1]\{q_{\alpha_1}\}^{\frac{1}{\alpha_1}},0),$$

where

$$A_n = \sum_{k=1}^{m} A_{nq_{\alpha_k}}[\alpha_k] \tag{12.5}$$

with $A_{nq_{\alpha_k}}[\alpha_k]$'s sum A_n being a sample mean when α_k is selected and the parameters $\beta^*[\alpha_1]$ and $\gamma^*[\alpha_1]$ are a skewness parameter and a scaling parameter, respectively which corresponds to the parameters of a limiting stable distribution with **the minimum characteristic exponent** *α_1 via the Super Generalized Central Limit Theorem when $nq_{\alpha_1} \to \infty$.*

Proof We first note that the following decomposition always holds for $\sum_{i=1}^{n} X_i$:

$$\mathrm{E}\left[\mathrm{Exp}\left(\mathrm{I}\left(\sum_{i=1}^{n} X_i\right)\right)\right] = \prod_{k=1}^{m} \mathrm{E}\left[\mathrm{Exp}\left(\mathrm{I}\left(\sum_{j=1}^{n(k)} X_j[\alpha_k]\right)\right)\right]$$

where $\lim_{n\to\infty} \dfrac{n(k)}{n} = q_{\alpha_k}$ is satisfied according to the condition for the USGLT. Since each stable law is an *infinitely divisible distribution* and its characteristic function $\phi_{\alpha,\beta}(t)$ always satisfies

$$|\phi_{\alpha,\beta}(t)| \le 1,$$

then the decomposition is interchangeable even for $n \to \infty$ as:

$$\begin{aligned} &\mathrm{E}\left[\mathrm{Exp}\left(\mathrm{I}\left(\sum_{i=1}^{n} X_i\right)\right)\right] \\ &= \prod_{k=1}^{m} \mathrm{E}\left[\mathrm{Exp}\left(\mathrm{I}\left(\sum_{j=1}^{n(k)} X_j[\alpha_k]\right)\right)\right] \to \prod_{k=1}^{m} \mathrm{E}\left[\mathrm{Exp}\left(\mathrm{I}\left(\sum_{j=1}^{nq_k} X_j[\alpha_k]\right)\right)\right] \end{aligned} \tag{12.6}$$

as $n \to \infty$. Here we can thus apply the super generalized central limit theorem to a superposition $\sum_{j=1}^{nq_k} X_j[\alpha_k]$ with the power index α_k, then for each $\alpha_k (1 \le k \le m)$, the relation

$$\frac{\sum_{j=1}^{nq_{\alpha_k}} X_j[\alpha_k] - A_{nq_{\alpha_k}}}{\{nq_{\alpha_k}\}^{\frac{1}{\alpha_k}}} \xrightarrow{d} S(x;\alpha_k,\beta^*[\alpha_k],\gamma^*[\alpha_k],0) \tag{12.7}$$

as $n \to \infty$ holds. In particular, for the case when $k = 1$, the relation

$$\frac{\sum_{j=1}^{nq_{\alpha_1}} X_j[\alpha_1] - A_{nq_{\alpha_1}}}{\{nq_{\alpha_1}\}^{\frac{1}{\alpha_1}}} \xrightarrow{d} S(x; \alpha_1, \beta^*[\alpha_1], \gamma^*[\alpha_1], 0) \tag{12.8}$$

as $n \to \infty$ holds. By seeing the obvious relation $\alpha_2(> \alpha_1)$ for example, we have the relations:

$$\frac{\sum_{j=1}^{nq_{\alpha_2}} X_j[\alpha_2] - A_{nq_{\alpha_2}}}{\{nq_{\alpha_1}\}^{\frac{1}{\alpha_1}}} = \frac{\sum_{j=1}^{nq_{\alpha_2}} X_j[\alpha_2] - A_{nq_{\alpha_2}}}{\{nq_{\alpha_2}\}^{\frac{1}{\alpha_2}}} \frac{\{nq_{\alpha_2}\}^{\frac{1}{\alpha_2}}}{\{nq_{\alpha_1}\}^{\frac{1}{\alpha_1}}}$$

$$= \frac{\sum_{j=1}^{nq_{\alpha_2}} X_j[\alpha_2] - A_{nq_{\alpha_2}}}{\{nq_{\alpha_2}\}^{\frac{1}{\alpha_2}}} \cdot \frac{1}{n^{\frac{1}{\alpha_1} - \frac{1}{\alpha_2}}} \cdot \frac{q_{\alpha_2}^{\frac{1}{\alpha_2}}}{q_{\alpha_1}^{\frac{1}{\alpha_1}}} \xrightarrow{d} \delta(x)$$

as $n \to \infty$. Here, we use the SGCLT for α_2 and the obvious fact that $\dfrac{1}{\alpha_1} - \dfrac{1}{\alpha_2} > 0$ and then as $n \to \infty$

$$\frac{1}{n^{\frac{1}{\alpha_1} - \frac{1}{\alpha_2}}} \to 0.$$

This means that the density function of the random variable $\dfrac{\sum_{j=1}^{nq_{\alpha_2}} X_j[\alpha_2] - A_{nq_{\alpha_2}}}{\{nq_{\alpha_1}\}^{\frac{1}{\alpha_1}}}$ approaches to Dirac's delta function $\delta(x)$ as $n \to \infty$ whose characteristic function is just unity.

In the same way, for more general $\alpha_k (k \geq 2)$ by using the obvious relation $\dfrac{1}{\alpha_1} - \dfrac{1}{\alpha_k} > 0$, the relation

$$\frac{\sum_{j=1}^{nq_{\alpha_k}} X_j[\alpha_k] - A_{nq_{\alpha_k}}}{\{nq_{\alpha_1}\}^{\frac{1}{\alpha_1}}} \xrightarrow{d} \delta(x) \quad \text{for} \quad n \to \infty$$

holds. Thus, in the limit $n \to \infty$

$$\frac{\sum_{i=1}^{n} X_i - A_n}{\{nq_{\alpha_1}\}^{\frac{1}{\alpha_1}}} \xrightarrow{d} S(x; \alpha_1, \beta^*[\alpha_1], \gamma^*[\alpha_1], 0).$$

Therefore, we conclude:

$$\frac{\sum_{i=1}^{n} X_i - A_n}{n^{\frac{1}{\alpha_1}}} \xrightarrow{d} S(x; \alpha_1, \beta^*[\alpha_1], \gamma^*[\alpha_1]\{q_{\alpha_1}\}^{\frac{1}{\alpha_1}}, 0).$$

This universal super generalized central limit theorem (USGCLT) proposed and proven here says that the *minimum power index* α_1 corresponding to the *biggest tail-risk* component is dominant factor to characterize a limit theorem for a mixed superposition of power laws. In other words, we can say that the biggest risk represented by the minimum tail power index α_1 is *dominant* in a limiting behavior of macroscopic risk, which is conceptually similar to Gresham's Law that states "bad money drives out good" because "goodness" of money can be measured by tail risk indicator α. In addition to this feature, this USGCLT gives a theoretical explanation about the reason why stable laws appear universally in empirical distributions of price fluctuations.

12.3 Exactly Solvable Chaos and Stable Law to Test Universal Super Generalized Central Limit Theorem

In this section, we introduce the connection between a certain ideal class of chaotic mappings and stable laws by ergodic theory. Umeno (1998) showed that the following chaotic mapping

$$Y = \frac{1}{2}\left(X - \frac{1}{X}\right) \tag{12.9}$$

has the Cauchy distribution with $\alpha = 1$ as an ergodic invariant measure while more generalized map

$$Y = \frac{1}{2}\left||X|^{\alpha} - \frac{1}{|X|^{\alpha}}\right|^{\frac{1}{\alpha}} \cdot \mathrm{SGN}\left[X - \frac{1}{X}\right], \quad (\alpha > 0) \tag{12.10}$$

has the power law with the power index α whose superposition converges to a stable law by the generalized central limit theorem (GCLT) for $0 < \alpha < 2$. All the chaotic mappings can be categorized as *exactly solvable chaos* which has an exact solution and the explicit ergodic invariant measure (Umeno, 1997). Figure 12.1 shows the return maps of the above exactly solvable chaos. Empirical density obtained by 100,000 iterations of the chaotic mapping (12.9) is shown in Fig. 12.2 and it remarkably matches the exact Cauchy probability density. According to Fig. 12.3, empirical density obtained by 100,000 iterations of the chaotic mapping (12.10) at $\alpha = 3/2$ also matches an analytical density

$$\rho(x;\alpha) = \frac{\alpha|x|^{\alpha-1}}{\pi(1 + |x|^{2\alpha})} \tag{12.11}$$

at $\alpha = 3/2$ very well (Umeno, 1998). Thus, we can say that *ergodic theory works well* for characterizing a logical connection between statistics and deterministic chaos, especially for the case of exactly solvable chaos. Note that there are other more

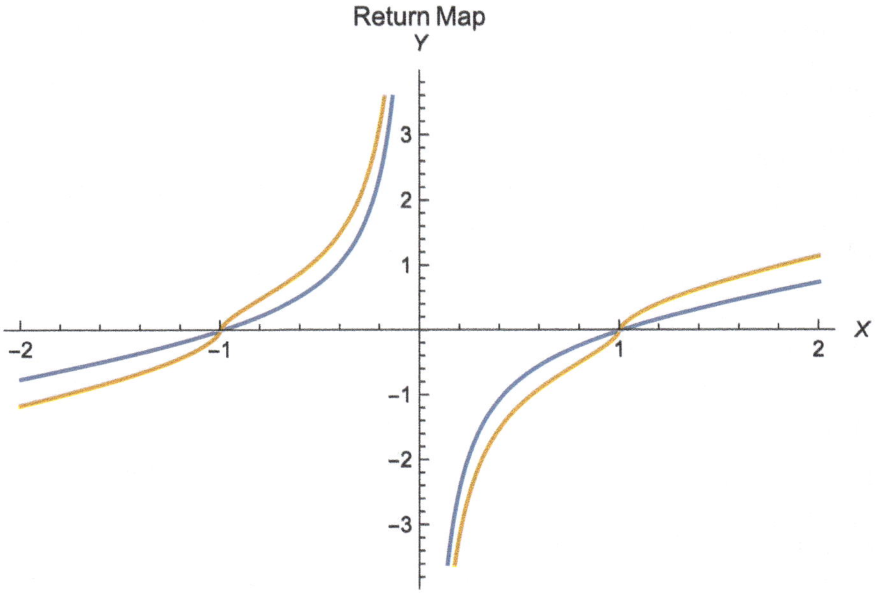

Fig. 12.1 Chaotic maps generating Cauchy distribution $\alpha = 1$ (blue) and Levy's stable law with $\alpha = 3/2$ (orange)

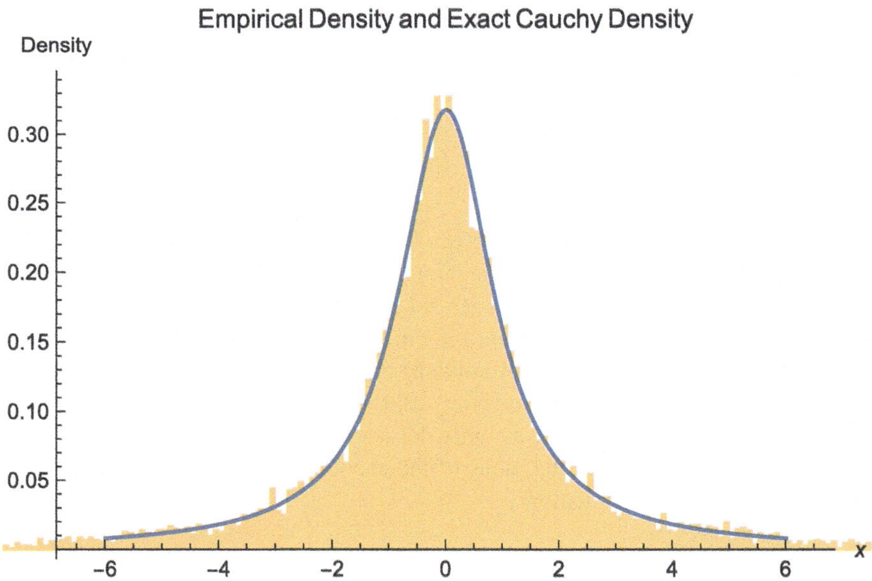

Fig. 12.2 Empirical Density $N = 100{,}000$ and Cauchy Probability Density

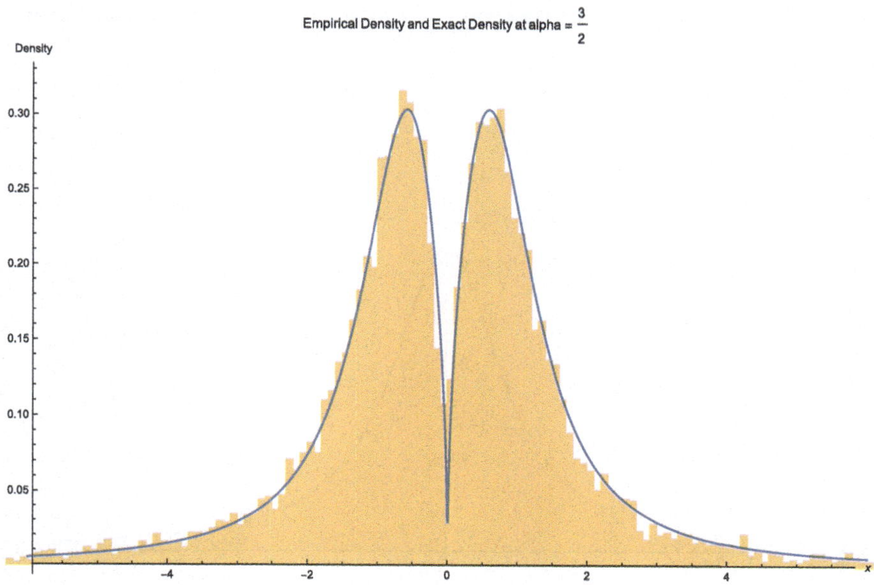

Fig. 12.3 Empirical Density $N = 100{,}000$ and Exact Probability Density at $\alpha = \frac{3}{2}$

chaotic mappings with ergodic Cauchy distribution such as the generalized Boole transformation in Eq. (12.12) with an analytical Lyapunov exponent λ in Eq. (12.13) (Umeno & Okubo, 2016) and the super generalized Boole transformations (Okubo & Umeno, 2018, 2021) that have Cauchy distribution as an ergodic invariant measure.

$$Y = aX - \frac{b}{X}, \quad (0 < a < 1, b > 0) \tag{12.12}$$

$$\lambda = \log\left(1 + 2\sqrt{a(1-a)}\right), \quad (0 \leq a \leq 1). \tag{12.13}$$

Next, we use these exactly solvable chaotic mappings to generate a mixed type of superposition to test the universal super generalized central limit theorem (USG-CLT). Let us consider a simple mixed type of superposition composed of chaotic dynamics with Cauchy distribution ($\alpha = 1$) and chaotic dynamics with the power law ($\alpha = 3/2$). Thus, this example corresponds to the case that $\alpha_1 = 1$ and $\alpha_2 = 3/2$. In Fig. 12.4, a mixed superposition of Cauchy chaos mapping in Eq. (12.9) with $\alpha_1 = 1$ and chaos mapping in Eq. (12.10) with $\alpha_2 = 3/2$ where $q_{\alpha_1} = q_{\alpha_2} = 1/2$ is shown to converge to the Cauchy distribution which is a confirmation of the validity of the USGCLT. Here, Cauchy chaos at $\alpha = 1$ is *dominant* as predicted by the USGCLT.

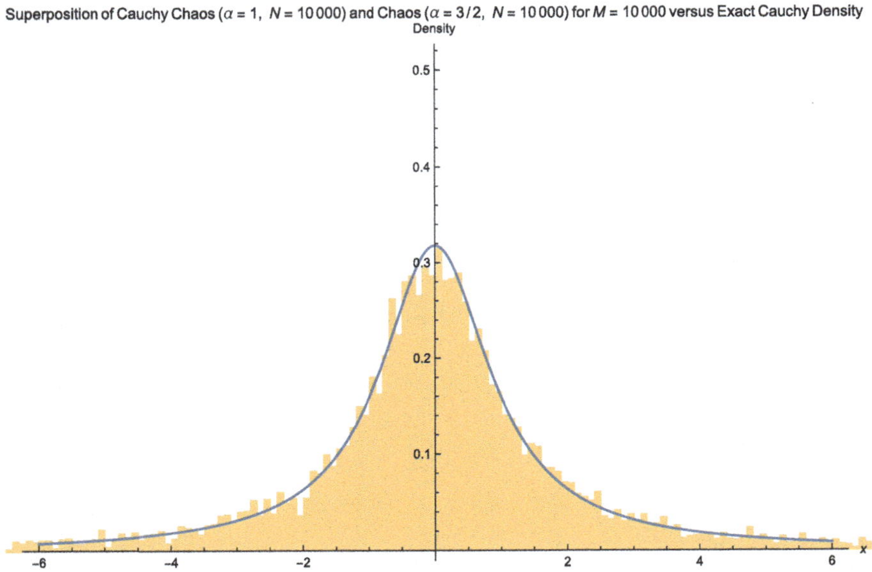

Fig. 12.4 Mixed superposition of Cauchy Chaos ($\alpha = 1, N = 10{,}000$) and Chaos ($\alpha = 3/2, N = 10{,}000$) for $M = 10{,}000$

12.4 Testing Efficient Market Hypothesis and Discovery of Novel Periodic Structure

Now we consider the test of the efficient market hypothesis (EMH) by empirical data. If EMH holds, then there must be *no correlation* in price fluctuations. Thus we investigate a *correlation structure* of empirical fluctuations of financial markets. We use the 5-min (high-frequency) chart of logarithmic returns of the Nikkei Averages in 2019 to test the EMH. Many empirical studies have shown that such high-frequency charts of logarithmic returns show no correlation or random fluctuations as predicted by EMH. Is it true? We solve the very question on EMH. In Figs. 12.5 and 12.6, a normalized standard correlation $C(l) = \text{Const.} \sum_{i=1}^{N} X_i X_{i+l}$ of the 5-min chart of logarithmic returns of the Nikkei averages in 2019 is depicted, which shows that no nontrivial correlation exists and EMH *seems to hold*. The reason why the correlation looks likely to be zero can be explained by the fact that the variance of price fluctuations corresponding to the normalized constant of the correlation is generally quite large due to the fat tail distribution and then the normalized correlation $C(l)$ computed by the division of the variance becomes near zero even if there exist a non-trivial correlation in data. Thus, with this normal correlation, we cannot conclude that EMH holds for the 5-min chart of logarithmic returns of the Nikkei averages in 2019 at this stage because there might be a *non-trivial correlation structure* in data. To investigate a new possibility, we consider an essence of the financial time series data by ergodic theory.

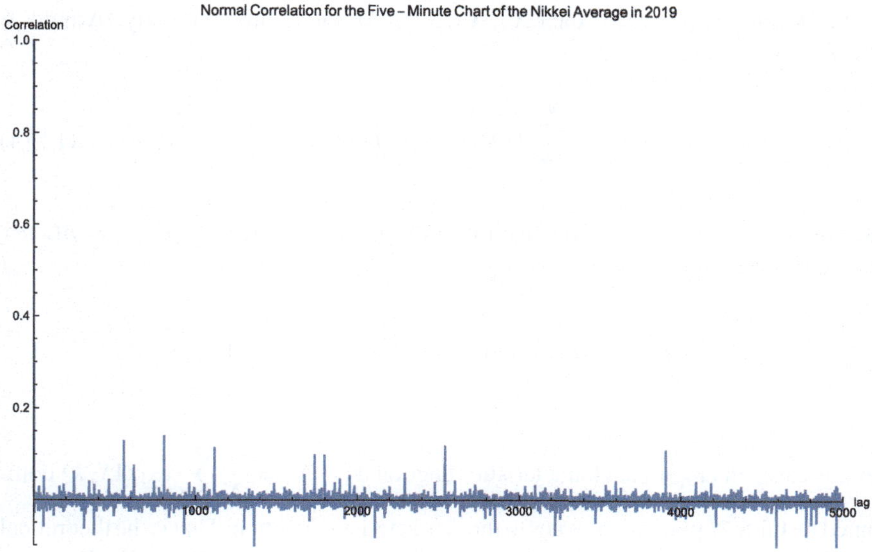

Fig. 12.5 Normal correlation for the 5-min chart of log-returns of the Nikkei averages in 2019

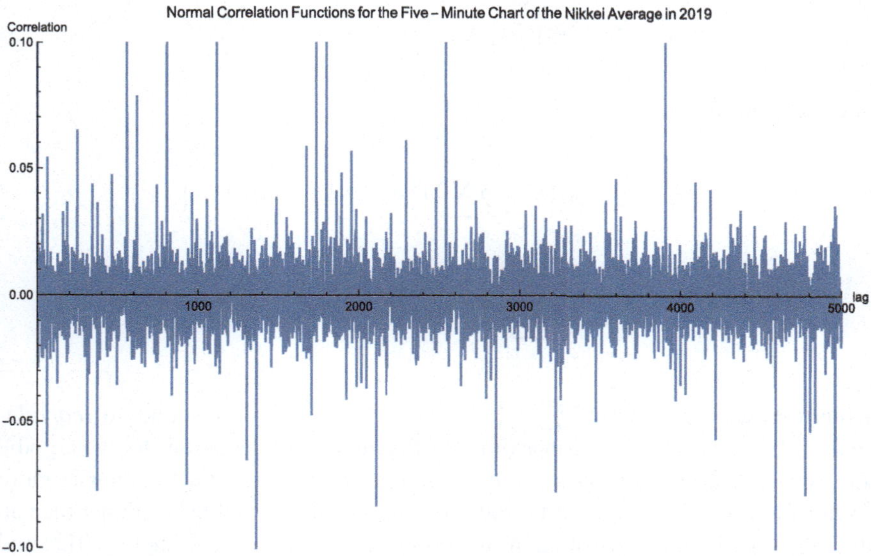

Fig. 12.6 Enlarged figure of normal correlation for the 5-min chart of log-returns of the Nikkei averages in 2019

Consider a sequence of data $X_1, X_2, \ldots \in M$ satisfying ergodicity (Arnold & Avez, 1968) in a sense that

$$\lim_{N\to\infty} \frac{1}{N} \sum_{j=1}^{N} f(X_j) = \int_M f(x)\mu(dx) \quad \text{a.e.} \tag{12.14}$$

By ergodicity, the characteristic function of the probability measure $\mu(dx) = \rho(x)dx$ can thus be computed by

$$\phi(k) = \int_{-\infty}^{\infty} \exp(\mathrm{I}kx)\rho(x)dx = \lim_{N\to\infty} \frac{1}{N} \sum_{n=1}^{N} \exp(\mathrm{I}kX_n) \quad \text{a.e.}$$

In this case, an empirical characteristic function $\phi(k; N) = \frac{1}{N}\sum_{n=1}^{N} \exp(\mathrm{I}X_n k)$ computed by finite N points converges to the characteristic function. That is the theoretical foundation of analysis called Chaos Fourier Transform (Umeno, 2016). Furthermore, we can compute the mean square deviation of the characteristic function by $V_k(N)$ as

$$V_k(N) \equiv E[|\phi(k; N) - \phi(k)^2] = \frac{D_k}{N} + \frac{E_k}{N^2}$$

where D_k and E_k are:

$$D_k = 1 - \phi(k)^2 + 2\sum_{l=1}^{N} \left\{ \langle e^{\mathrm{I}k(X_0 - X_l)} \rangle - |\phi(k)|^2 \right\}$$

$$E_k = -2\sum_{l=1}^{N} l \left\{ \langle e^{\mathrm{I}k(X_0 - X_l)} \rangle - |\phi(k)|^2 \right\}.$$

In this representation, a term $2\sum_{l=1}^{N}\{\langle e^{\mathrm{I}k(X_0-X_l)}\rangle - |\phi(k)|^2\}$ corresponds to a *correlation term* that plays a key role in estimating fluctuations of deviations for the ergodic sum (Umeno, 2000). Because an investigation of characteristic function is more essential to capture a feature of fat-tail distributions such as stable law rather than an investigation of simple distribution function [see Fukunaga and Umeno (2017) and Kakinaka and Umeno (2020a, b) for detailed explanation on this matter], we are now motivated to define a new correlation function $C_k(l)$ as

$$C_k(l) \equiv \langle e^{\mathrm{I}k(X_0 - X_l)} \rangle - |\phi(k)|^2. \tag{12.15}$$

This correlation function can be computed by the following formulae:

$$\hat{C}_k(l; N) \equiv \frac{1}{N}\sum_{n=1}^{N} e^{\mathrm{I}k(X_n - X_{n+l})} - \left|\frac{1}{N}\sum_{n=1}^{N} e^{\mathrm{I}kX_n}\right|^2$$

$$\lim_{N\to\infty} \hat{C}_k(l; N) = C_k(l) \quad \text{a.e.}$$

Thus, we call this novel correlation function $C_k(l)$ characteristic function-based correlation function or CF-based correlation function.

Efficient Market Hypothesis

Now we test the EMH (Fama, 1970) by investigating whether there exist a nontrivial correlation structure. If the efficient market hypothesis holds, then

$$C_k(l) = 0 \quad \text{for} \quad \text{any} \quad l(\neq 0).$$

Thus, to deny the EMH, it is sufficient to detect some correlation such that

$$C_k(l) \neq 0 \quad \text{for} \quad \text{some} \quad l(\neq 0) \text{ and } k(\neq 0).$$

Note that $C_0(l) = 0$ for any l. In Fig. 12.7 the absolute value of characteristic function-based correlation at $k = 10$ defined by Eq. (12.15) is depicted for the 5-min chart of logarithmic returns of Nikkei averages in 2019, which shows no clear correlation supporting EMH. However, in Fig. 12.8, a *nontrivial periodic structure* (small correlation) is shown to exist for the absolute value of characteristic function-based

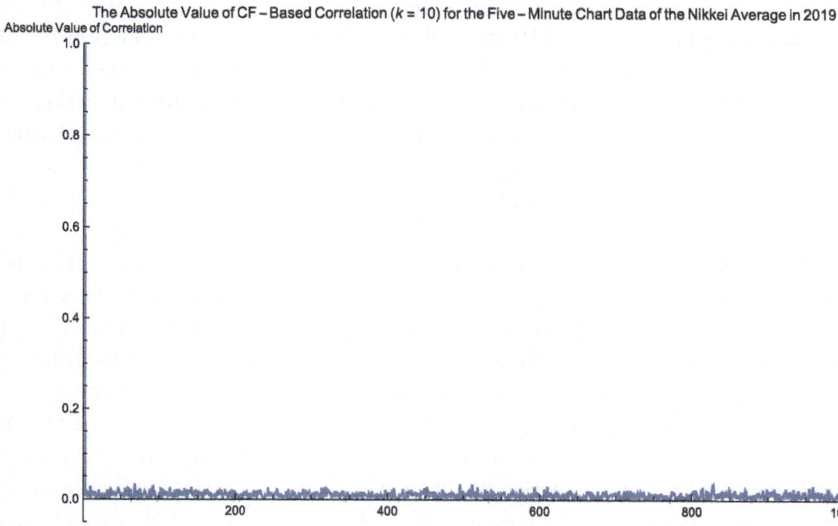

Fig. 12.7 The absolute value of characteristic function-based correlation ($k = 10$) for the 5-min chart of log-returns of the Nikkei averages in 2019

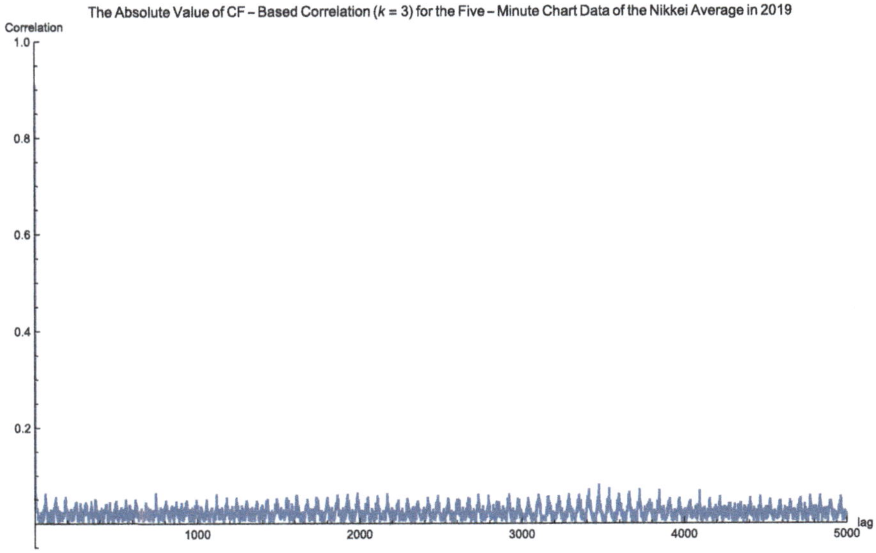

Fig. 12.8 The absolute value of characteristic function-based correlation ($k = 3$) for the 5-min chart of log-returns of the Nikkei averages in 2019

correlation at $k = 3$ with the 5-min chart of logarithmic returns of Nikkei averages in 2019, which clearly *deny* the EMH in a direct manner. A clear periodic structure with 5-h periodicity is shown in Fig. 12.9, which is an enlarged version of Fig. 12.8. This 5-hour periodicity is exactly the same as the trading duration per day in the Nikkei stock market in 2019 (the daily trading time is: 9:00–11:30 and 12:30–15:00 in Japan standard time). In Fig. 12.10, the real part of the characteristic function-based correlation at $k = 3$ with the 5-min chart of logarithmic returns of Nikkei averages in 2019 is depicted, which shows clear periodic structure in correlation with the *5-h periodicity* which corresponds to a 60-lag periodicity satisfying the simple relation

$$\text{Period} = 60 \text{ (lags)} \times 5 \text{ min} = 5 \text{ h (trading time duration)}. \qquad (12.16)$$

On the contrary, no correlation structure is detected in the imaginary part of the characteristic function-based correlation at $k = 3$ as depicted in Fig. 12.11. This novel periodicity according to the trading time duration also appears in the absolute value of characteristic function-based correlation at $k = 1$ for the 5-min chart of logarithmic returns of Nikkei averages in 2019 as depicted in Fig. 12.12. Thus, this periodicity detected with CF-based correlation is NOT a peculiar feature at special $k(\neq 0)$ but rather a *universal* phenomenon capturing the market periodicity corresponding to the daily periodicity with the market. Such an exploration of novel periodic structure of financial markets such as the commodity market is now extensively investigated in our group (Shiihashi, 2020).

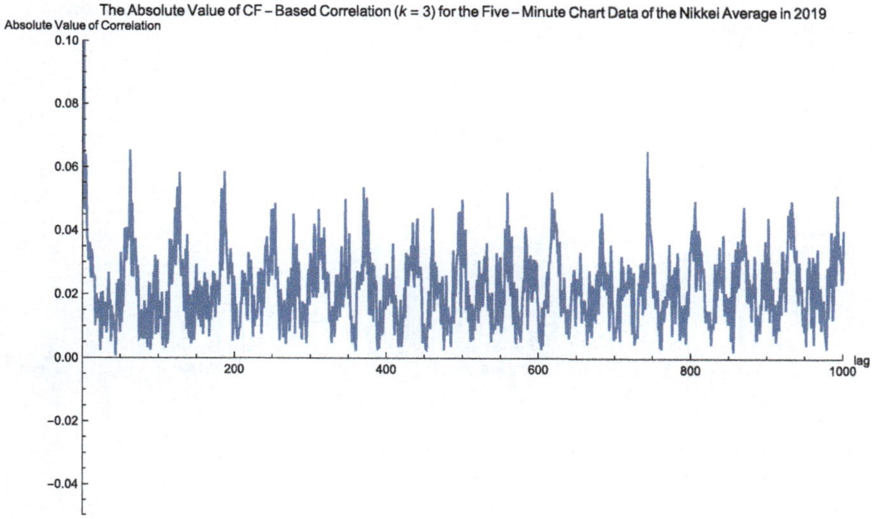

Fig. 12.9 Enlarged figure of the absolute value of characteristic function-based correlation ($k = 3$) for the 5-min chart of log-returns of the Nikkei averages in 2019

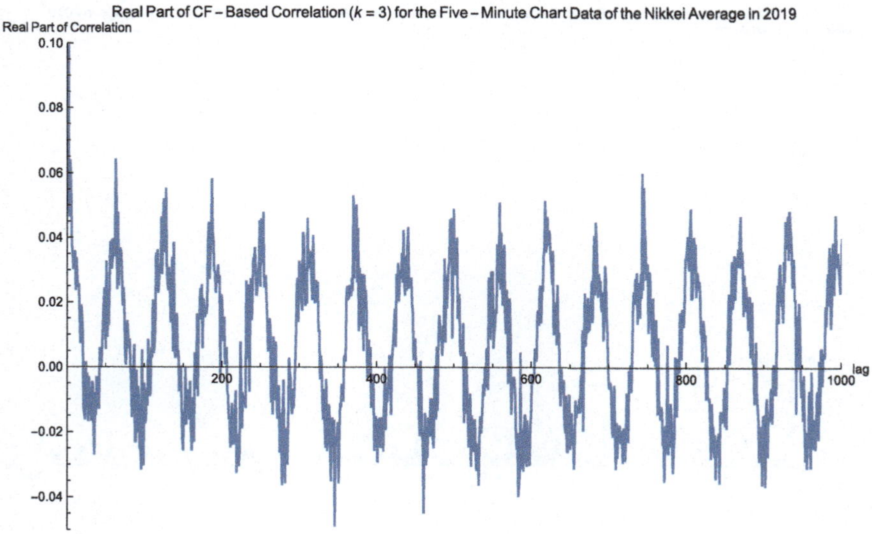

Fig. 12.10 Real part of characteristic function-based correlation ($k = 3$) for the 5-min chart of log-returns of the Nikkei averages in 2019

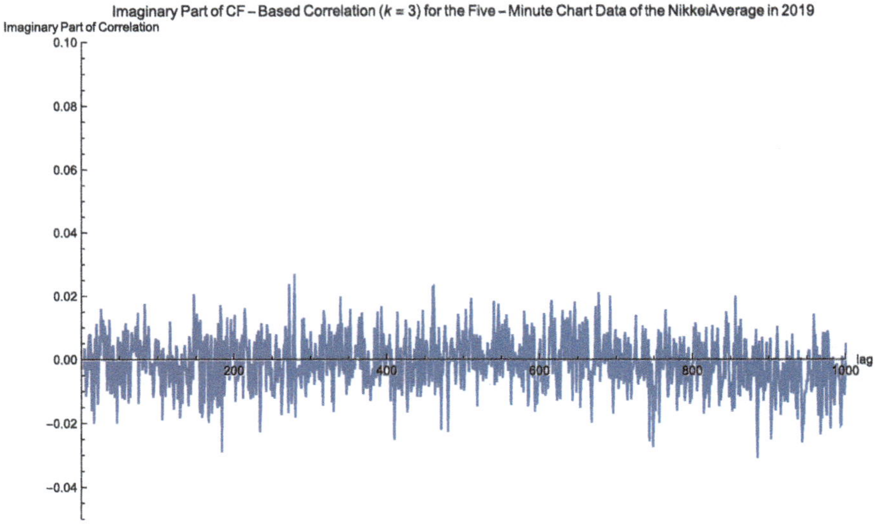

Fig. 12.11 Imaginary part of characteristic function-based correlation ($k = 3$) for the 5-min chart of log-returns of the Nikkei averages in 2019

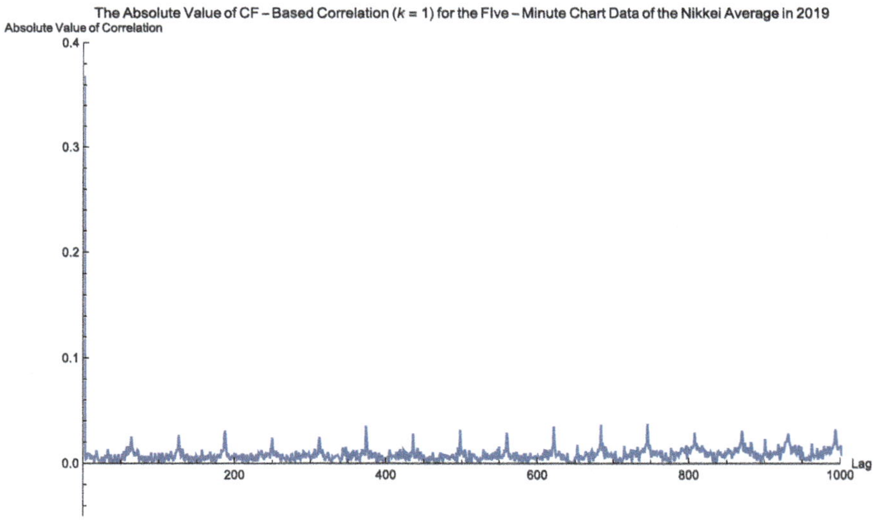

Fig. 12.12 The absolute value of characteristic function-based correlation ($k = 1$) for the 5-min chart of log-returns of the Nikkei averages in 2019

12.5 Elucidation of Chaotic Market Hypothesis

We are now approaching the construction of new model of financial markets based on our findings about the universal super generalized central limit theorem (USGCLT) and nontrivial periodic structure in CF-based correlation functions as discussed in the previous sections. While the efficient market hypothesis (EMH) well captures the random structure of financial markets, we are now focused on the detailed distribution structure and its mechanism to show the universality of stable law and nontrivial correlation structure, both of which cannot be captured by EMH. In Fig. 12.13, the whole time series of the rescaled logarithmic returns for the Nikkei Averages in 2019 is depicted, which clearly shows randomness. Thus, a new model should also possess a capability of explaining randomness of the financial markets in addition to the capabilities of explaining the universality of stable law and nontrivial correlation structure. In Fig. 12.14, an empirical probability density of rescaled logarithmic returns of the Nikkei Averages in 2019 is compared to the stable law with $\alpha = 1.4$, which shows that the USGCLT seems to hold such that stable law at $\alpha = 1.4$ matches the empirical probability density for the Nikkei Averages in 2019. After seeing the evidence to support USGCLT and nonrandom time correlation structure in randomness look-like time series of price fluctuaions, we are now motivated to propose the following chaotic market hypothesis capturing these properties as follows.

Chaotic Market Hypothesis:

For financial market we say that *chaotic market hypothesis* (CMH) holds (1) if stable law can *universally* capture the averaged data (index data) of price fluctuations by

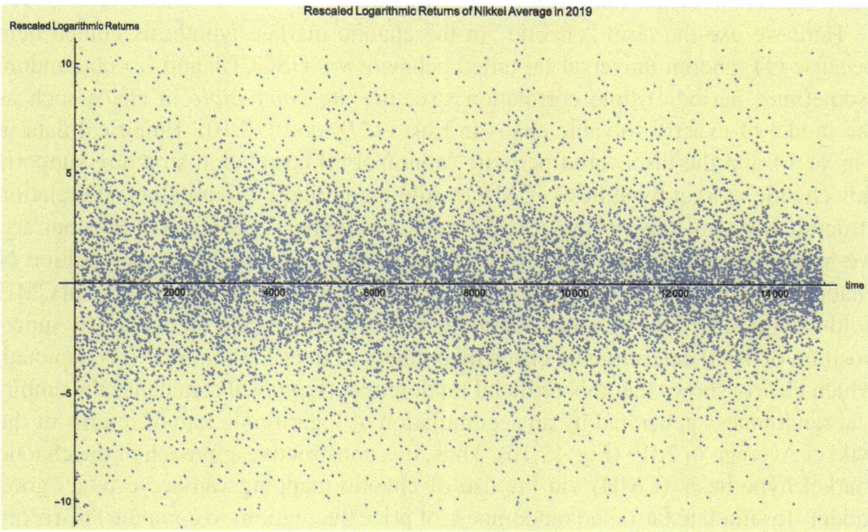

Fig. 12.13 The whole time series of the rescaled logarithmic returns of the Nikkei averages in 2019

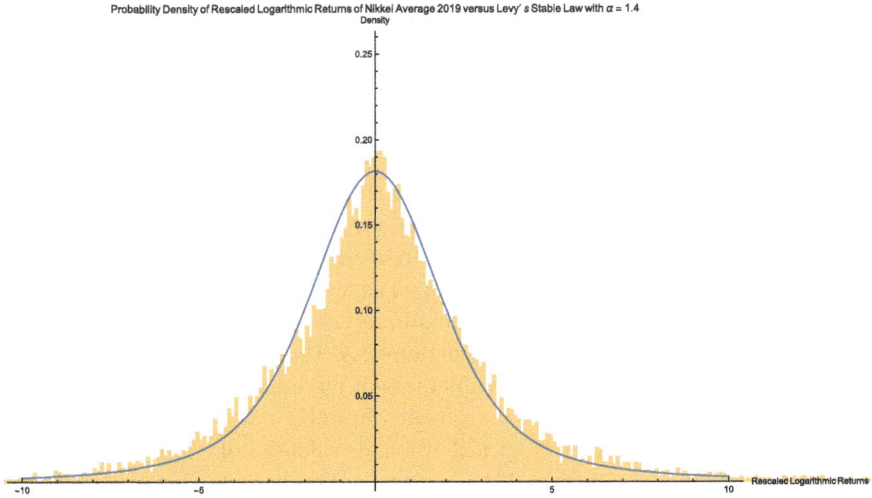

Fig. 12.14 Empirical probability fensity of rescaled logarithmic returns of the Nikkei averages in 2019 versus Levy's stable law with $\alpha = 1.4$

the universal super generalized central limit theorem (USGCLT) -*random fat-tailed* behavior- and (2) if the characteristic function based correlation functions show non-zero (sometimes periodic) correlation structure as

$$C_k(l) = \langle e^{Ik(X_0 - X_l)} \rangle - |\phi(k)|^2 \neq 0 \text{ for some } l(\neq 0) \text{ and } k(\neq 0)$$

-*nonrandom (sometimes periodic)* time structure.

Here we use the term "chaotic" in the chaotic market hypothesis coined here because (1) random universal fat-tailed behavior via USGCLT and (2) nonrandom (sometimes periodic) time correlation structure are *compatible* in chaos such as the model of exactly solvable chaos in Eqs. (12.9) and (12.10). Empirical data in Fig. 12.14 and the discovered periodic (nonrandom) correlation structure supports this chaotic market hypothesis (CMH) while the periodic (nonrandom) correlation structure does not support the efficient market hypothesis (EMH). On the contrary, we now see how artificial market price fluctuations generated by superposition of chaotic dynamics mimic the empirical market fluctuations and can support that CMH holds. In Fig. 12.15, Levy walk (deterministic diffusion process) generated by superposition of 1000 deterministic chaos mappings in Eq. (12.10) at $\alpha = \frac{3}{2}$ is depicted, which clearly shows that this artificial randomness (Levy walk) matches the empirical randomness generated by an accumulation of logarithmic rate of returns of the Nikkei Average in 2019 (Fig. 12.16). Thus, this constructive approach of the chaotic market hypothesis (CMH) via the use of chaotic mapping can serve us a "good model" to simulate fat-tailed randomness of price fluctuations to compute tail-risk in the financial market. Because chaotic dynamics generally has a mixing property such that correlation has an *exponential decay* structure whose decay rate is characterized

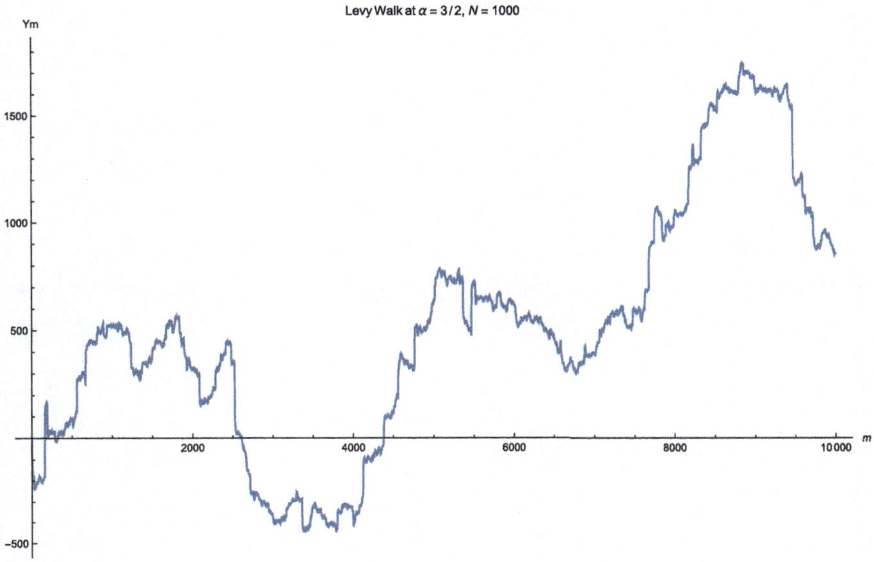

Fig. 12.15 Levy walk generated by superposition of chaos $\alpha = 3/2$, $N = 1000$

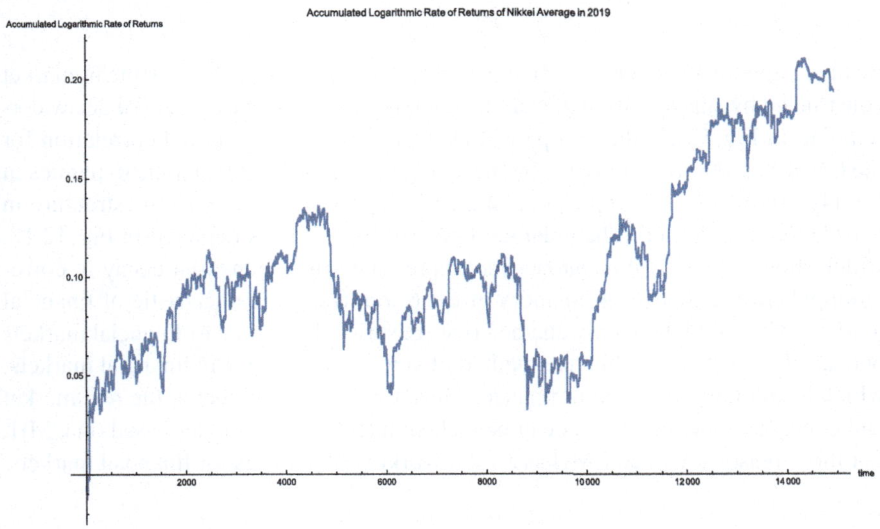

Fig. 12.16 Accumulated logarithmic rate of returns of the Nikkei averages in 2019

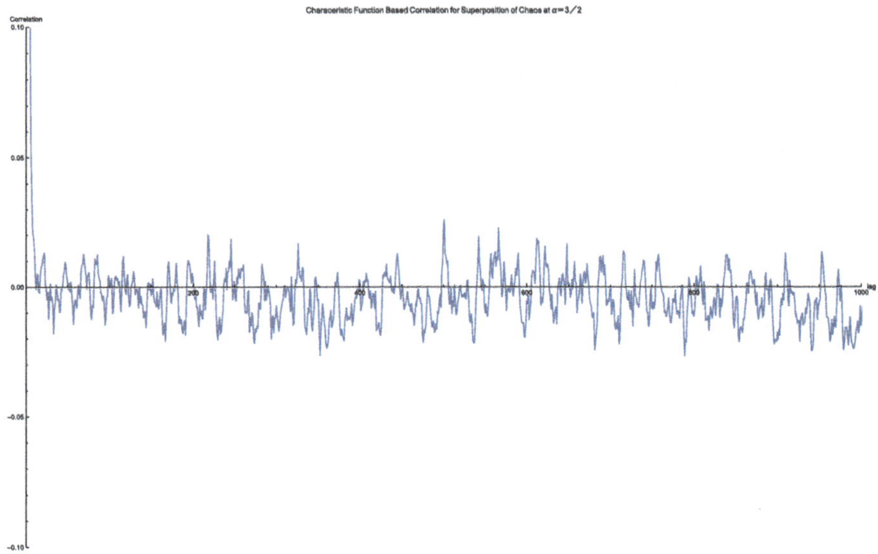

Fig. 12.17 Real part of characteristic function-based correlation for the superposition (N = 10,000) of chaos at $\alpha = 3/2$. Exponential decay of the correlation function can be observed

by the Lyapunov exponent $\lambda > 0$, it is highly expected that such an artificial market constructed by superposition of chaotic dynamics also has an exponential decay correlation. In Fig. 12.17, the real part of characteristic function-based correlation for the CMH-based artificial market by the superposition of 10,000 chaotic dynamics in Eq. (12.10) at $\alpha = \frac{3}{2}$ is depicted, and clearly shows *exponential decay* structure in correlation as expected. The enlarged figure of Fig. 12.17 is depicted in Fig. 12.18, which shows *short-term memory effect* represented by exponential decay in correlation where the short-term memory effect is a universal characteristic of financial markets. Thus, CMH not only characterizes the price fluctuation of financial markets but can also work as an effective method of simulating tail-risk in financial markets, which is an important topic in finance. Thus, CMH can well match the real market and conversely the real market can be well simulated by the model based on CMH. For those reasons, we see CMH as a valid working hypothesis for financial markets.

12.6 Discussions

The proposed chaotic market hypothesis (CMH) seems similar to the fractal market hypothesis (FMH) proposed by Peter (1991) which is also based on chaos theory. It is known that FMH can also capture the market behavior, in particular persistent behavior with long memory ("trends") *better* than the efficient market hypothesis (EMH). What is the difference between the CMH and FMH? While the FMH *assumes* self-

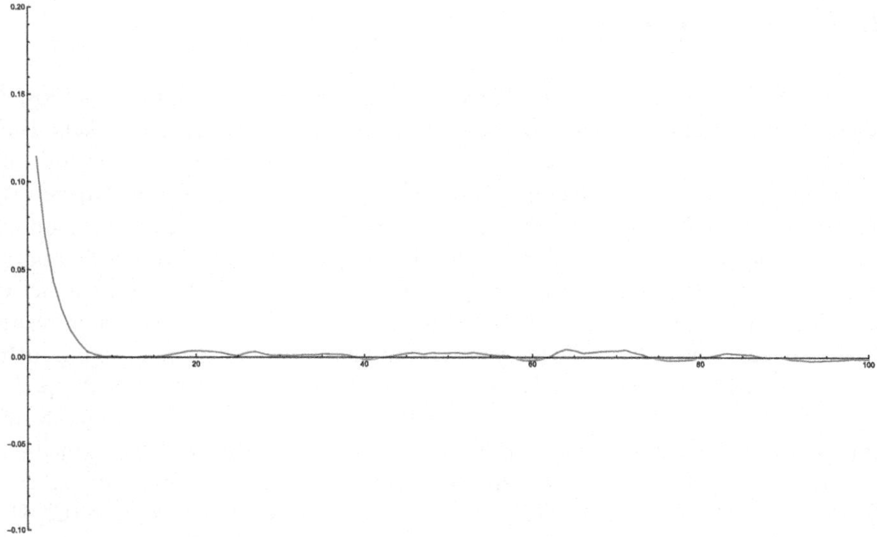

Fig. 12.18 Enlarged figure of the real part of characteristic function-based correlation for the superposition ($N = 10{,}000$) of chaos at $\alpha = 3/2$. *Exponential decay* of the correlation function can be observed, which shows the essential characteristic of chaotic nature. The horizontal axis corresponds to a lag while the vertical axis corresponds to the real part of characteristic function-based correlation

similarity (fractal property) of a price time-series quantified by the Hurst exponent or the fractal dimension, CMH does NOT assume self-similar structure in the time-series but assume a power-law with an index α in a distribution of price fluctuations, which is theoretically founded by the universal super generalized central limit theorem (USGCLT). Although the power index α in the CMH is related to the Hurst exponent in the FMH, the α is but one of four parameters α, β, γ, and μ to characterize a stable law in CMH via the USGCLT. Furthermore, a skewness parameter β can also be important in characterizing the financial market (Fukunaga & Umeno, 2017) as in CMH while there is no similar counterpart parameter such as a skewness parameter in the FMH. Furthermore, a discovered *periodic* structure in CF-based correlation in Sect. 12.4 cannot be explained by the FMH where a corresponding correlation structure must also have self-similar structure in FMH and CF-based correlation function is not provided in the theoretical framework of FMH. Thus, the self-similarity assumption in FMH does not match the observed periodicity in CF-based correlation in the empirical study. Because the self-similarity assumption is the crux of FMH, we can argue that the crux of FMH can miss important trading characteristics of financial markets such as daily trading time duration.

12.7 Conclusions

A generalization of the super generalized central limit theorem (SGCLT) is proposed to show a universality of stable laws in the price fluctuations in financial markets. The validity of the USGCLT is confirmed by numerical simulations using superposition of different kinds of exactly solvable chaotic mappings with power laws with different power indices. Then by using ergodic theory, we propose CF-based correlation to test the efficient market hypothesis (EMH) and directly deny EMH by finding a novel periodic (non-random) structure in CF-based correlation whose periodicity estimated 5 h is exactly the same as the trading time duration in the 5-min chart Nikkei averages in 2019, while the normal correlation function cannot capture such kind of periodic structure because of the general fat tail distribution of the price fluctuations making the variance (a normalized constant) divergent to output no correlation. Thus, we conclude that the very CF-based correlation function method based on ergodic theory correctly captures a fine time structure in financial markets. Based on these findings and USGCLT, we propose chaotic market hypothesis (CMH) to capture the essential characteristics of financial market that EMH and other models like the fractal market hypothesis fail to capture. Furthermore, this CMH can be a good model or good simulation method to measure a tailed risk in finance because a constructive approach to modeling of financial markets is easy by direct applications of USGCLT. To conclude, the CMH is based on ergodic theory and is a valid working hypothesis for modeling, featuring and simulating financial markets with theoretical foundation of the universal super generalized central limit theorem (USGCLT).

References

Arnold, V. I., & Avez, A. (1968). *Ergodic problems of classical mechanics*. Benjamin.

Fama, F. (1970). Efficient capital markets: A review of theory of empirical work. *The Journal of Finance*, *25*, 383–417.

Fukunaga, T., & Umeno, K. (2017). Universal Levy's stable law of stock market and its characterization. arXiv:1709.06279

Gnedenko, B. V., & Kolmogorov, A. N. (1954). *Limit distributions for sums of independent random variables*. Addison-Wesley.

Kakinaka, S., & Umeno, K. (2020a). Characterizing cryptocurrency market with Levy's stable distributions. *Journal of the Physical Society of Japan*, *024802*. https://doi.org/10.7566/JPSJ.89.024802

Kakinaka, S., & Umeno, K. (2020b). Flexible two-point selection approach for characteristic function-based parameter estimation of stable laws. *Chaos*, *073128*. https://doi.org/10.1063/5.0013148

Okubo, K., & Umeno, K. (2018). Universality of the route to chaos: Exact analysis. *Progress of Theoretical and Experimental Physics*, *103A01*. https://doi.org/10.1093/ptep/pty094

Okubo, K., & Umeno, K. (2021). Infinite ergodicity that preserves the Lebesgue measure. *Chaos*, *033135*. https://doi.org/10.1063/5.0029751

Peter, E. (1991). *Fractal market analysis: Applying chaos theory to investment and economics* (p. 336). Wiley Finance: John Wiley Science.

Shiihashi, K. (2020). Graduation thesis, Undergraduate School of Informatics and Mathematical Science. Kyoto University (in Japanese).
Shintani, M., & Umeno, K. (2018). Super generalized central limit theorem -limit distributions for sums of non-identical random variables with power laws. *Journal of the Physical Society of Japan, 87*. https://doi.org/10.7566/JPSJ.87.043003
Umeno, K., & Okubo, K. (2016). Exact Lyapunov exponents of the generalized Boole transformations. *Progress of Theoretical and Experimental Physics, 21A01*. https://doi.org/10.1093/ptep/ptv195
Umeno, K. (1997). Method of constructing exactly solvable chaos. *Physical Review E, 55*, 5280–5284.
Umeno, K. (1998). Superposition of chaotic processes with convergence to Levy's stable law. *Physical Review E, 58*, 2644–2647.
Umeno, K. (2000). Chaotic Monte Carlo computation: A dynamical effect of random number generations. *Japanese Journal of Applied Physics, 39*, 1442–1456.
Umeno, K. (2016). Ergodic transformations on R preserving Cauchy laws. *NOLTA, IEICE, 7*, 14–20.

Ken Umeno graduated from the Department of Physics, University of Tokyo and obtained a Ph.D. in 1995. He has been a professor at the department of Applied Mathematics and Physics, Graduate School of Informatics, Kyoto University. Umeno has studied chaos theory and complex systems and developed their applications for more than 30 years. His major discoveries are: theory of exactly solvable chaos (1997), chaotic Monte Carlo computation and super efficiency (2000), chaos-CDMA (1999), brain-signal processing method to separate chaotic mixture (2006), chaos-Fourier transform (2016), super generalized central limit theorem (SGCLT) (2018) and chaotic market hypothesis (CMH) (present chaper).

Chapter 13
Itinerant-Electron Magnetism and Spin Fluctuations—Aspects of Theories and Experiments

Kazuyoshi Yoshimura

Abstract This chapter reviews the history of magnetism, which is one of the main streams of solid-state physics and quantum physics, and developments of itinerant-electron magnetism and exotic superconductivity in terms of spin fluctuations are discussed from theoretical and experimental points of view. Spin-fluctuation theories beyond mean-field approximations are important to explain static magnetic properties as well as the dynamical properties of itinerant magnets. It was found that electron-pair formation due to magnetic spin fluctuations can be realized in exotic superconductors. Problems in itinerant-electron systems continue to attract our interest.

Keywords Spin fluctuations · Itinerant electrons · Collective modes of electrons · Itinerant magnetism · Exotic superconductivity · Itinerancy (Delocalization) · Quantum physics · Solid-state physics

13.1 Brief History of Magnetism

The research field of magnetism has been developed since Pierre Curie discovered the Curie law experimentally in 1895 (Curie, 1895). The Curie law for paramagnets can be represented as the temperature dependence of the magnetic susceptibility as $\chi^{-1} = T/C$, where C is the Curie constant, and was explained theoretically by Langevin as (1905)

K. Yoshimura (✉)
Department of Chemistry, Graduate School of Science, Kyoto University, Kitashirakawa, Oiwake-Cho, Sakyo-ku, Kyoto 606-8502, Japan
e-mail: yoshimura.kazuyoshi.8e@kyoto-u.ac.jp

Research Center for Low Temperature and Material Sciences, Kyoto University, Kyoto 606-8501, Japan

International Research Unit of Integrated Complex System Science, Kyoto University, Kyoto 606-8501, Japan

K. Nishimura et al. (eds.), *Creative Complex Systems*, Creative Economy,
https://doi.org/10.1007/978-981-16-4457-3_13

$$\chi = \frac{C}{T} = \frac{N_0\,\mu_{eff}^2}{3k_\mathrm{B}T}, \tag{13.1}$$

where N_0 is the number of magnetic atoms or ions, and k_B is the Boltzmann constant. Here, the effective paramagnetic moment, μ_{eff}, can be represented by quantum theory (Brillouin, 1922, 1930) with the Bohr magneton, μ_B, as

$$\mu_{eff} = 2\mu_\mathrm{B}\sqrt{S(S+1)} \tag{13.2}$$

in the quantum spin (S) system. Here, the μ_{eff} is independent of temperature in the classical quantum theory of magnetism with spin quantum number S.

The theory of ferromagnetism has been developed by the mean-field theory (molecular-field theory) proposed by Weiss (1907), and modified with the development of the quantum mechanics in the twentieth century. Thus, Weiss's molecular-field theory of ferromagnetism was able to explain the ferromagnetic phase transition very well, and the paramagnetic susceptibility was derived together with the same μ_{eff} by Eq. (13.2) as follows:

$$\chi = \frac{N_0\,\mu_{eff}^2}{3k_B(T-\theta_\mathrm{W})}, \tag{13.3}$$

which is the so-called Curie–Weiss law of magnetic susceptibility of the ferromagnet above the ferromagnetic transition temperature (i.e., the Curie temperature T_C). Here, the magnetic interaction term is normalized by the temperature scale shift of the Weiss temperature θ_W. Since Eq. (13.3) shows divergent behavior at θ_W, T_C is equal to θ_W, and is represented in the molecular-field theory as

$$T_\mathrm{C} = \theta_\mathrm{W} = C\gamma_m = \frac{N_0\mu_{eff}^2}{3k_\mathrm{B}}\gamma_m = \frac{2J}{3k_\mathrm{B}}zS(S+1), \tag{13.4}$$

where γ_m is the molecular-field coefficient. Furthermore, in Eq. (13.4), the Weiss molecular-field theory is modified by Heisenberg's quantum model (Heisenberg, 1928) of the ferromagnetic exchange interaction, and J here is Heisenberg's exchange coupling constant between neighboring spins in the Heisenberg Hamiltonian as

$$H = -2J\sum_{i>j} S_i \cdot S_j, \tag{13.5}$$

where indices i and j represent the neighboring spin sites with the coordination number z. Experimentally, θ_W is always higher that T_C in general because of the existence of short-range ordering just above T_C.

The Weiss molecular-field theory for ferromagnets was expanded and applied to the cases of antiferromagnets and ferrimagnets by Néel (1948), leading to complete understanding of the static behaviors of ferro-, antiferro- and ferrimagnetism of localized moment (spin) systems.

In the next step, we should discuss the magnetic excitation, or spin excitation (so-called magnon), which is important to understand ferromagnetic behaviors at finite temperatures and even at low temperatures. The magnon excitations can be well described by the Holstein-Primakoff model (Holstein & Primakoff, 1940), in which only the transverse spin excitations (i.e., spin fluctuations) are considered in localized moment (spin) systems. These magnons are virtual bosons, like energy quanta in quantum mechanics. In magnon theory, the total spin operators, S_x, S_y and S_z, which give the amplitudes of spin components when they operate on the wave function φ are described by the bosonic creation and annihilation operators b^+ and b for magnons as

$$\begin{cases} S_+ = S_x + iS_y = \hbar\sqrt{2s}\sqrt{1 - \frac{b^+b}{2s}}\,b \\ S_- = S_x - iS_y = \hbar\sqrt{2s}b^+\sqrt{1 - \frac{b^+b}{2s}} \end{cases} \tag{13.6}$$

and

$$S_z = \hbar\left(s - b^+b\right), \tag{13.7}$$

where S_z is the projection of spin to the quantized axis z and S_+ and S_- are spin-raising and spin-lowering operators. Here, the unit of angular momentum is expressed as $\hbar = h/2\pi$ with the Planck constant h. Thus, the magnetically excited state can be expressed by introducing virtual bosons like energy quanta in the magnetic state. The spin wave state corresponds to that with magnons created in the system. When a magnon is created, the magnetization is lowered by $\hbar$ in S_z. decreasing the magnetization in the ferromagnetically ordered state. Finally, n magnons are excited, and the spin projection to z, $\langle S_z \rangle$, becomes

$$\langle S_z \rangle = S_z\varphi = \hbar\left\{s - \left(b^+b\right)^n\right\}|0\rangle_{\mathrm{m}}, \tag{13.8}$$

where $|0\rangle_m$ is the ferromagnetic state with no magnons. The spin-wave (magnon) thoery gives $T^{3/2}$ dependence of the magnetization in ferromagnets at low temperatures, which agree well with many experimental results (Ashcroft & Mermin, 1976; Holstein & Primakoff, 1940; Holtzberg et al., 1964; Kittel, 1953). Spin-waves (i.e., magnons) are elementary spin excitations and are also treated as virtual (pseudo-) bosons. Here, it is very important that the square amplitude of spin momentum, $\langle S^2 \rangle$, is completely conserved even in the quantum state:

$$\left\langle S^2 \right\rangle = \left\langle S_x^2 + S_y^2 + S_z^2 \right\rangle = \hbar^2 S(S+1). \tag{13.9}$$

Thus, even in the case with spin fluctuations (here, magnon excitations), the magnetism of the localized moment (spin) systems has been completely understood except for some complicated problems as quantum-spin effects due to spin frustration, and so on (e.g., Kosterlitz & Thouless, 1973).

On the other hand, understanding of the magnetism of metals and metallic compounds is much more complicated because the collective modes of band electrons (so called the band magnetism and/or the itinerant-electron magnetism) should be considered when discussing magnetic states. Hereafter, itinerant-electron magnetism is reviewed and discussed.

13.2 Itinerant-Electron Magnetism

The gas or liquid state of band electrons can be written as (Landau & Lifshitz, 1980)

$$\phi_{\mathrm{F}} = \prod_{k \le k_F} a^{+}_{-k\downarrow} a^{+}_{k\uparrow} \varphi_0 = \prod_{k \le k_F} B^{+}_{k} \varphi_0 = \prod_{k \le k_F} B^{+}_{k} |0\rangle, \tag{13.10}$$

where ↑ and ↓ represent the up and down spins of electrons, respectively, k is the wave number (the momentum $\boldsymbol{p}$ of the electron is written as $p = \hbar k$), k_{F} is the electron wave number at the Fermi level. Here, $B^{+}_{k}$and B_k are the bosonic creation and annihilation operators of an electron pair with $-k \downarrow$ and $k \uparrow$ represented as

$$B^{+}_{k} = a^{+}_{-k\downarrow} a^{+}_{k\uparrow}\,, \quad B_k = a_{k\uparrow} a_{-k\downarrow}. \tag{13.11}$$

This state of the electrons in Eq. (13.10) corresponds to the Pauli paramagnetic state of metals, in which electron pairs with opposite wave number vectors and opposite spins as $-k \downarrow$ and $k \uparrow$ are contained in the vacuum state with no electrons, $\varphi_0 = |0\rangle$, and the electron band is filled up to the Fermi energy, $E_{\mathrm{F}} = \hbar^2 k_{\mathrm{F}}^2/2m$ (Ashcroft & Mermin, 1976; Kittel, 1953; Pauli, 1925, 1927). This electronic situation arises from Pauli's exclusion principle for fermions (Pauli, 1925, 1927). The Pauli paramagnetic state gives its magnetic susceptibility χ_{P} as

$$\chi_{\mathrm{P}} = 2\mu_{\mathrm{B}}^2 D(E_F)\left[1 + \frac{\pi^2}{6} k_{\mathrm{B}}^2 \left\{\frac{D''}{D} - \left(\frac{D'}{D}\right)^2\right\} T^2 + \cdots\right], \tag{13.12}$$

where $D(E_{\mathrm{F}})$ is the density of states of electrons at the Fermi level, and D' and D'' are its first and second derivatives at the Fermi level. In general, the coefficient of T^2 in Eq. (13.12) is not so large that χ_{P} is almost independent of T, which is completely different from the Curie and Curie–Weiss laws of the paramagnetic susceptibilities in localized moment (spin) magnets. The T-dependence of χ_{P} is due to the T-dependence of the Fermi energy, that is, strictly speaking, that of the chemical potential of electrons.

Here, it is important to explain the correlation effects between electrons in the metallic and Pauli paramagnetic states and to understand how the insulating state of electrons becomes metallic and Pauli paramagnetic states. The following Hubbard Hamiltonian gives answers for these questions (Hubbard, 1963, 1964a, 1964b):

$$H = -t \sum_{\langle i,\, j\rangle,\, \sigma} \left(a_{i,\,\sigma}^{+} a_{j,\,\sigma} + a_{j,\,\sigma}^{+} a_{i,\,\sigma} \right) + U \sum_{i=1}^{N} a_{i\uparrow}^{+} a_{i\uparrow} \cdot a_{i\downarrow}^{+} a_{i\downarrow}, \qquad (13.13)$$

where t is the value of the transfer integral, U is the strength of the onsite Coulomb interaction, σ is spin ↑ or ↓, and i and j stand for the atomic sites to which the electrons belong. The electron momentum $p = \hbar k$ is $\sim \hbar k_{\mathrm{F}}$ because the electrons with $k \ll k_{\mathrm{F}}$ are confined within the Brillouin zone and cannot move. The first term describes the energy gain when the electrons move from sites i to j. The second term describes the energy loss because of the Coulomb repulsion when electrons with ↑ and ↓ spins occupy the same site i. Therefore, when $t > U$, an electron can move from site i to j, leading to the metallic electronic state in the Hubbard model. When $t < U$, an electron cannot move, leading to the insulating electronic state in the Hubbard model, which is important for discussing and explaining strong-correlation behaviors of electrons like the metal–insulator Mott transition (Mott, 1949, 1968). The Heisenberg Hamiltonian [Eq. (13.5)] in the itinerant system can be derived from the Hubbard model, obtaining J of the exchange interaction simply as

$$J = \frac{4t^2}{U}. \qquad (13.14)$$

Thus, the Hubbard model (Hubbard, 1963, 1964a, 1964b) is very useful and important in discussing itinerancy and/or localization of electrons in magnetic and strongly correlated electron systems.

In trying to explain band ferromagnetism (itinerant ferromagnetism), molecular-field approximation was adapted to the Pauli paramagnetic state of the free-electron model by Stoner (Stoner, 1938, 1939) and later by Wohlfarth (Rhodes & Wohlfarth, 1963; Wohlfarth, 1978) in the 1940s to the 1970s in work that is known as the the Stoner-Wohlfarth (SW) theory of band magnetism (Rhodes & Wohlfarth, 1963; Stoner, 1938, 1939; Wohlfarth, 1978). When the molecular field, $H_m = \gamma_m M$, and the external field H_{ext} is added to the Pauli paramagnetic state, the magnetization (magnetic moment) M is induced as

$$M = \chi_P (H_{ext} + H_m) = \chi_P (H_{ext} + \gamma_m M). \qquad (13.15)$$

Therefore, the magnetic susceptibility ($\chi = M/H_{ext}$) is enhanced as

$$\chi = M/H_{ext} = \frac{\chi_P}{1 - \gamma_m \chi_P} = \frac{\chi_P}{1 - 2\gamma_m \mu_B^2 D(E_F)}, \tag{13.16}$$

leading to the exchange-enhanced Pauli paramagnetism, brcause γ_m is caused by the exchange interaction like Eq. (13.4). Here, the factor $1/(1-2\gamma_m \mu_B^2 D(E_F))$ is called as the "Stoner enhancement factor." If $\gamma_m \chi_P = 1$, χ shows divergence in Eq. (13.16). Therefore, the condition,

$$\gamma_m \chi_P = 2\gamma_m \mu_B^2 D(E_F) = 1, \tag{13.17}$$

is called the "Stoner criterion" for ferromagnetism. When the exchange interaction becomes larger and exceeds the Stoner criterion [Eq. (13.17)], $\gamma_m \chi_P > 1$ is realized, implying that the ferromagnetic state can occur in this itinerant-electron system. In this situation, spontaneous band-splitting takes place, leading to the appearance of the ferromagnetic spontaneous magnetic moment, μ_s. Although this static mean-field SW theory for band electrons can explain the ground state properties (e.g., μ_s at 0 K which corresponds to the strength of the band splitting caused by the exchange interaction), it was found that SW theory could not explain the typical behaviors of itinerant-electron ferromagnets at finite temperatures. One of the most serious matters is that SW theory could not explain the Curie–Weiss law of the paramagnetic susceptibility like Eq. (13.2) which can be seen in all the ferromagnets including itinerant ones. Instead, SW theory always gives the T^2-dependence of the inverse paramagnetic susceptibility χ^{-1}, essentially because of the same reason that the Pauli paramagnetic susceptibility indicates slight T^2-dependence shown in Eq. (13.12). This T^2-dependence is scarcely observed in the paramagnetic susceptibility above T_C in experiments.

Furthermore, typical weakly itinerant ferromagnets such as MnSi, Sc_3In, $ZrZn_2$ and Ni_3Al were discovered by Mathias et al. and by other groups as by-products of searches for superconducting compounds (Bloch et al., 1975; Boer, 1969; Boer et al., 1969; Mathias et al., 1963; Ogawa, 1976; Shinoda & Asanabe, 1966; Takeuchi & Masuda, 1979; Williams et al., 1966). Subsequently, the Laves-phase system, $Y(Co_{1-x}Al_x)_2$, was also discovered to show weakly itinerant ferromagnetism in 1985 (Yoshimura & Nakamura, 1985; Yoshimura et al., 1987, 1988a). They all showed common characteristic features: (1) very low T_C (usually less than 50 K), (2) very small value of spontaneous ordered moment μ_s in the ground state (far less than $1\mu_B$), and (3) the CW law of the paramagnetic susceptibility (see Fig. 13.1) showed a larger μ_{eff} than the small ordered moment, μ_s (Bloch et al., 1975; Boer, 1969; Boer et al., 1969; Mathias et al., 1963; Ogawa, 1976; Shinoda & Asanabe, 1966; Takeuchi & Masuda, 1979; Williams et al., 1966; Yoshimura & Nakamura, 1985; Yoshimura et al., 1987, 1988a). The small μ_s can be explained by SW theory with a moderately small γ_m. However, it could not explain the other two properties for the CW law mentioned above. Furthermore, SW theory could not explain even very low T_C, because SW theory is a static mean-field theory so that T_C calculated by SW theory becomes the same order of Fermi energy ~10^4 K. On the other hand, the observed

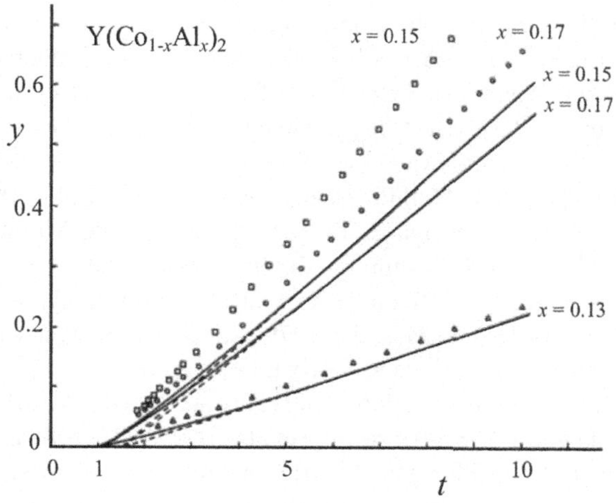

Fig. 13.1 Inverse magnetic susceptibility y plotted against reduced temperature, $t = T/T_C$ for $Y(Co_{1-x}Al_x)_2$ with $x = 0.13$ (triangles), 0.15 (squares) and 0.17 (circles) (Yoshimura et al., 1987). Solid curves represent theoretical calculations: broken lines represent extrapolations of calculated y from high T

values of T_C of these weakly itinerant ferromagnetic compounds are usually less than 50 K (Bloch et al., 1975; Boer, 1969; Boer et al., 1969; Mathias et al., 1963; Ogawa, 1976; Shinoda & Asanabe, 1966; Takeuchi & Masuda, 1979; Williams et al., 1966; Yoshimura & Nakamura, 1985; Yoshimura et al., 1987, 1988a): for example, 26 K (the maximum) ($\theta_W = 40$ K) for $Y(Co_{1-x}Al_x)_2$ with $x = 0.15$ (Yoshimura & Nakamura, 1985; Yoshimura et al., 1987, 1988a).

Rhodes and Wohlfarth (1963; Wohlfarth, 1978) gave us an important suggestion for itinerant ferromagnets as follows. Here, we can estimate the value of $2S$ from μ_{eff} by using Eq. (13.2) and define the paramagnetic moment, μ_C, as $\mu_C = 2S$. In the localized moment system, μ_C is naturally equal to μ_s at 0 K, which is guaranteed by quantum mechanics. However, in cases of itinerant ferromagnets, they suggested that this simple formula, $\mu_C = \mu_s$, is not valid anymore, and that μ_C is always larger than μ_s in itinerant systems. If we take the ratio μ_C/μ_s, which is called as the Rhodes-Wohlfarth ratio (RWR), the value of RWR always comes to be more than unity. Rhodes and Wohlfarth plotted RWR against T_C (Rhodes-Wohlfarth plot: Rhodes & Wohlfarth, 1963; Wohlfarth, 1978) and found that the values of RWR for various itinerant ferromagnets distribute widely above the unity line and may be less than some function that decreases with T_C. However, they could not obtain this function. This plot is very informative for studying the itinerant magnetism, implying its importance in the concept of the itinerant magnetism. This relation between RWR and T_C was lately solved by Takahashi's theory of spin fluctuations (Takahashi, 1986).

In the several decades since the emergence of SW theory, many important theoretical and experimental approaches have been conducted to understand itinerant-electron magnetism (Anderson, 1959, 1961; Boer, 1969; Boer et al., 1969; Bloch et al., 1975; Chen et al., 2010; Gat-Malureanu et al., 2011; Goodenough, 1971; Imai et al., 2014, 2015; Ishikawa et al., 1985; Kiyama et al., 1996, 1998; Kontani et al., 1975; Masuda, 1983; Mathias et al., 1963; Michor et al., 2004; Moriya & Kawabata, 1973a, 1973b; Moriya & Takahashi, 1978, 1995; Murata & Doniach, 1972; Ogawa, 1976; Ohta & Yoshimura, 2009a, 2009b; Ohta et al. 2010a, 2010b, 2011; Sakakibara et al., 1986; Shinoda & Asanabe, 1966; Sugiyama et al., 2015; Takahashi, 1986; Takeuchi & Masuda, 1979; Takahashi & Moriya, 1983, 1985; Takaigawa & Yasuoka, 1982; Ueda & Moriya, 1974; Uemura et al., 2006; Umemura & Masuda, 1983; Williams et al., 1966; Yang et al., 2011, 2013; Yasuoka et al., 1978; Yoshimura & Nakamura, 1985; Yoshimura et al., 1984, 1987, 1988a, 1988b, 1988c, 1999; Ziebeck et al., 1982; Zhang et al., 2018; and references therein). Many review articles have been published on itinerant-electron magnetism (e.g., Moriya, 1979, 1981, 1985, 1987; Takahashi, 2013, 2017a, 2017b; Takahashi & Yoshimura, 2012; Yoshimura, 2017, 2020). First, dynamical mean-field theory, also called random phase approximation (RPA) theory, was constructed by considering a spin fluctuation only with the same phase (Murata & Doniach, 1972), which can explain even dynamical properties of widely distributed magnets from weak itinerant regime through an intermediate to localized regime. However, RPA theory was found to be valid only at low temperatures just above 0 K because no mode–mode coupling of spin fluctuations was considered in RPA theory. Soon after, an epoch-making development was achieved by Toru Moriya and his coworkers, which was the great success of spin-fluctuation theory for weakly itinerant ferromagnets and antiferromagnets based upon the self-consistent renormalization (SCR) of spin fluctuations to magnetic free energy (SCR theory) (Moriya & Kawabata, 1973a, 1973b; Moriya, 1979, 1981, 1985, 1987), proceeding beyond static mean-field SW theory (Rhodes & Wohlfarth, 1963; Stoner, 1938, 1939; Wohlfarth, 1978) and dynamical mean-field RPA theory (Murata & Doniach, 1972). In SCR theory of spin fluctuations, the CW law can be explained in completely different terms as a new concept from former theories of magnetism as:

$$\frac{1}{\chi} = \frac{(1-\alpha)}{\chi_0} + \frac{5}{3} g N_0^2 \langle S(T)^2 \rangle = \left[4 N_0 I^2 \langle S(T_C)^2 \rangle / 3 T_C T_0 \right] (T - T_C), \quad (13.18)$$

where the first term corresponds to the inverse of the exchange-enhanced Pauli paramagnetic susceptibility (α and I are related to the exchange interaction of itinerant electrons) and the second term is due to spin fluctuations with the spin-fluctuation coupling, g. Here, the average square amplitude of total local spin, $\langle S(T)^2 \rangle$, which is due to the dynamical band splitting and thus contains the longitudinal spin fluctuation together with the transverse one, is dependent on temperature and is linear with T above T_C. T_0 is a very important spin-fluctuation parameter that corresponds to the energy width of the spin-fluctuation spectrum, that is, the dynamical susceptibility,

$\chi(q, \omega)$, with its wave vector, q, and frequency, ω, of spin fluctuation, which is explained in the next section.

Afterwards, spin-fluctuation theory has been developed toward the unified theory between the weakly itinerant regime (Moriya & Kawabata, 1973a, 1973b) and the localized moment regime (Anderson, 1959, 1961) in metallic magnets using a phenomenological method by Moriya and Takahashi (1978). Furthermore, SCR theory was successfully extended to explain the characteristic magnetic behaviors of heavy-fermion systems (Moriya & Takimoto, 1995). Then, SCR theory was developed and rearranged in a quantitative way by Takahashi and Moriya in 1985 (Quantitative Aspect of Spin Fluctuations) (Takahashi, 1986; Takahashi & Moriya, 1985), by which we can compare the experiments and the SCR theory quantitatively by means of a set of (several numbers of) spin-fluctuation parameters (Takahashi, 1986, 2013; Takahashi & Moriya, 1985; Takahashi & Yoshimura, 2012; Yoshimura, 2017, 2020; Yoshimura et al., 1987).

13.3 Quantitative Aspects of SCR Theory of Spin Fluctuations

In the SCR theory of spin fluctuations for itinerant-electron magnetic systems (Moriya, 1979, 1981, 1985, 1987; Moriya & Kawabata, 1973a, 1973b; Moriya & Takahashi, 1978; Takahashi & Moriya, 1983, 1985; Takahashi & Yoshimura, 2012), the definitions of spin fluctuations, $\langle S^2\rangle = \langle S^2\rangle_T + \langle S^2\rangle_{Z.P.}$, which are magnetic excitations and band fluctuations in itinerant-electron magnetic systems, are

$$\begin{cases} \langle S^2\rangle_T = \dfrac{6}{N_0^2}\displaystyle\sum_q \int_0^\infty \dfrac{d\omega}{\pi} n(\omega) Im\chi(q, \omega) \\ \langle S^2\rangle_{Z.P.} = \dfrac{3}{N_0^2}\displaystyle\sum_q \int_0^\infty \dfrac{d\omega}{\pi} Im\chi(q, \omega) \end{cases} \quad (13.19)$$

where $\langle S^2\rangle_T$ is the thermal spin fluctuations, $\langle S^2\rangle_{Z.P.}$ is the zero-point spin fluctuations, q is the wave vector of the spin fluctuation ($\hbar q$is the momentum of the spin fluctuation), ω is the frequency of the spin fluctuation ($\hbar\omega$is the energy of the spin fluctuation), N_0 is the number of magnetic atoms in the system, and Im χ (q, ω) is the imaginary part of the dynamical magnetic susceptibility as a function of q andω. Furthermore, $n(\omega)$ is the bosonic factor, namely the boson number operator written as

$$n(\omega) = b^+ b\phi(\omega). \quad (13.20)$$

The difference between $\langle S^2\rangle_T$ and $\langle S^2\rangle_{Z.P.}$ is only the presence of $2n(\omega)$ Takahashi, 1986, 2013, 2017a, 2017b; Takahashi & Yoshimura, 2012).

SCR theory of spin fluctuations, which is the mode–mode coupling theory between different wave q-vectors of spin fluctuations, has successfully explained many experimental magnetic properties of the itinerant ferromagnets and antiferromagnets, including the low ferromagnetic transition temperature, the Curie temperature T_{C}, and the low antiferromagnetic transition temperature (Néel temperature T_{N}) in itinerant-electron systems and the dynamical measurements of spin dynamics in itinerant magnets (Moriya, 1979, 1981, 1985, 1987; Moriya & Kawabata, 1973a, 1973b; Takahashi, 2013, 2017a, 2017b; Takahashi & Moriya, 1985; Takahashi & Yoshimura, 2012; Yoshimura, 2017, 2020). The Curie–Weiss law in χ above T_{C} was successfully explained by the T-linear increase of $\langle S^2 \rangle_T$ in SCR theory (Moriya, 1979, 1981, 1985, 1987; Moriya & Kawabata, 1973a, 1973b; Takahashi & Moriya, 1985) as explained in the previous section. Hereafter, the quantitatively developed SCR theory is reviewed in this section.

In SCR theory, Im χ (q,ω) is usually written by the following double Lorentzian formula as (oriya, 1979, 1981, 1985, 1987; Takahashi, 1986, 2013, 2017b; Takahashi & Moriya, 1985; Takahashi & Yoshimura, 2012; Yoshimura, 2017, 2020)

$$\begin{cases} Im\chi(q,\omega) = \frac{\chi(0,0)}{1+q^2/\kappa^2} \cdot \frac{\omega\Gamma_q}{\omega^2+\Gamma_q^2} \\ \Gamma_q = \Gamma_0 q\left(\kappa^2+q^2\right), \quad \kappa^2 = \frac{1}{2\overline{A}} \cdot \frac{N_0}{\chi} \end{cases}, \tag{13.21}$$

where κ and Γ_q represent the q- and ω-widths of the spin-fluctuation spectrum (dynamical magnetic susceptibility), and κ^2 corresponds to the inverse susceptibility, $1/\chi$. In SCR theory, κ and Γ_q are important spin-fluctuation parameters for expressing magnetic quantities. Furthermore, Γ_0 and $\overline{A}$ are the parameters representing ω-width and q-damping width of the double Lorentzian form. Here, by utilizing a set of spin-fluctuation parameters:

$$p_s, \quad \overline{F_1}, \quad T_0, \quad T_A. \tag{13.22}$$

Here, p_s is the spontaneous Bohr magneton number ($=\mu_{\mathrm{s}}/\mu_{\mathrm{B}}$), $\overline{F_1}$ corresponds to g in Eq. (13.18), and T_0 and T_A are the characteristic temperatures corresponding to the q- and ω-widths of the spin-fluctuation spectrum deduced from Γ_0 and $\overline{A}$ as

$$\begin{cases} T_0 = \Gamma_0 q_{\mathrm{B}}^3/2\pi \\ T_A = \overline{A} q_{\mathrm{B}}^2 \end{cases}, \tag{13.23}$$

where q_{B} is the wave vector q at the Brillouin zone boundary. The values of T_0 and T_A can be obtained experimentally and directly from neutron-scattering experiments (Ishikawa et al., 1985; Ziebeck et al., 1982).

Furthermore, $\overline{F_1}$ corresponds to the coefficient of the M^4 term in the Landau expansion of magnetic free energy (Landau,) as:

$$F(M, H) = F(0, 0) + \frac{1}{2}aM^2 + \frac{1}{4}bM^4 + \frac{1}{6}cM^6 + \ \ldots \ - M \cdot H, \quad (13.24)$$

where *a*, *b*, *c*, … are the Landau expansion coefficients, and are represented by spin-fluctuation parameters as (Takahashi, 1986, 2013, 2017b; Takahashi & Moriya, 1985; Takahashi & Yoshimura, 2012; Yoshimura, 2020)

$$\begin{cases} b = \frac{\overline{F}_1}{2\mu_B} \\ c = \left[2\left\{3\pi\left(2+\sqrt{5}\right)\right\}^2 N_0 \mu_B^3 \frac{T_C^2}{T_A^3 p_s^3}\right]^{-1} \end{cases}. \quad (13.25)$$

The coefficient *a* below T_C corresponds to *T*-dependence of ordered spontaneous moment, $\mu_{s,}$, and is written as

$$a = -\frac{\mu_s(T)^2\overline{F}_1}{2\mu_B} = -\frac{p_s(T)^2\overline{F}_1\,\mu_B}{2}. \quad (13.26)$$

The value of $\overline{F_1}$ can be obtained experimentally from the Allott-plot [M^2 vs. *H/M* plot (*M* being magnetization and *H* being magnetic field)] of magnetizations *M* as:

$$2\mu_B(H/M) = \overline{F}_1\left[M(T, H)^2 - M(T, 0)^2\right], \quad (13.27)$$

which can be deduced as the equilibrium condition ($\partial F/\partial M = 0$) of the Landau expansion [Eq. (13.24)] considered up to the M^4 term.

In quantitatively moderated SCR theory, magnetic quantities, such as T_C, and the temperature dependence of the magnetic susceptibility χ above T_C, can be deduced in quantitative formulae by using Eq. (13.21) with a set of spin-fluctuation parameters, p_s, $\overline{F_1}$, T_0 and T_A [Eq. (13.22)] (Takahashi, 1986, 2013, 2017b; Takahashi & Moriya, 1985; Takahashi & Yoshimura, 2012; Yoshimura, 2017, 2020), For instance, T_C is given through the following SCR relation among spin-fluctuation parameters as

$$p_s^2/4 \cong \frac{15cT_0}{T_A}\eta^4 = \frac{15cT_0}{T_A}\left(\frac{T_C}{T_0}\right)^{4/3}, \quad (13.28)$$

where *c* is the constant being 0.3353… (Takahashi & Moriya, 1985).

The inverse magnetic susceptibility, *y*, in SCR theory can be deduced by an integral form of digamma function with spin-fluctuation parameters as (Takahashi & Moriya, 1985)

$$y = \frac{N_0}{2T_A\eta^2}\chi^{-1} \cong \frac{\overline{F}_1 p_s^2}{8T_A\eta^2}\left\{-1 + \frac{1}{c}\int_0^{1/\eta} dzz^3\left[lnu - \frac{1}{2u} - \psi(u)\right]\right\}, \quad (13.29)$$

with

$$u = z(y + z^2)/t, \quad t = T/T_C, \quad \eta = (T_C/T_0)^{1/3}, \quad \psi : \text{ digamma function.} \tag{13.30}$$

Furthermore, the nuclear spin–lattice relaxation rate, $1/T_1$, can be obtained from SCR theory. In general, $1/T_1$ can be expressed by the imaginary part of the dynamical susceptibility as

$$\frac{1}{T_1} = 2\gamma_N^2 k_B T \sum_q A_q^2 \frac{Im\chi(q, \omega_0)}{\omega_0}, \tag{13.31}$$

where γ_N is the nuclear gyromagnetic ratio and A_q is the q-dependent hyperfine coupling constant in nuclear magnetic resonance (NMR) experiments. From Eq. (13.31), $/T_1$ in the paramagnetic state above T_C can be obtained by using the SCR expression of Eq. (13.21) as (Kontani et al., 1975; Moriya, 1979, 1981, 1985, 1987; Takahashi, 1986, 2013, 2017b; Takahashi & Moriya, 1985; Takahashi & Yoshimura, 2012; Ueda & Moriya, 1974; Yasuoka et al., 1978; Yoshimura, 2017, 2020)

$$\frac{1}{T_1} = \gamma_N^2 A_{hf}^2 T \frac{\chi(q=0)}{4\pi^2 \mu_B \Gamma_0} = \left(\gamma_N A_{hf}\right)^2 \frac{p_{eff}^2}{8\pi T_0} \frac{T}{T - T_C}. \tag{13.32}$$

The important spin-fluctuation parameter T_0 can also be obtained experimentally from $1/T_1$ measurements by NMR (Takahashi, 1986; Takahashi & Moriya, 1985) instead of neutron- scattering experiments. Therefore, NMR experiments have been used to characterize the itinerant magnets (Chen et al., 2010; Imai et al., 2015; Kontani et al., 1975; Masuda, 1983; Ohta et al. 2010b; Takaigawa & Yasuoka, 1982; Umemura & Masuda, 1983; Yoshimura et al. 1984, 1987, 1988a, 1988b, 1988c, 1999; Yasuoka et al., 1978).

In Fig. 13.1, the experimental and calculated inverse magnetic susceptibilities y for $Y(Co_{1-x}Al_x)_2$ are indicated against reduced temperature, $t = T/T_C$ (Yoshimura et al., 1987). Here, the triangular, square and circle symbols represent experimental data for $x = 0.13$, 0.15 and 0.17. The solids lines indicate values calculated by SCR theory ([Eq. (13.29)] with a set of spin-fluctuation parameters [Eq. (13.22)] obtained from experimental data [p_s and $\overline{F_1}$ from magnetic measurements; T_0 from $1/T_1$ measurements (NMR), T_A from using the SCR relation of Eq. (13.28)]. The real values of spin fluctuations were taken from the original papers (Yoshimura et al. 1987, 1988a). For example, $T_0 = 2119$ K and $T_A = 6340$ K for $x = 0.15$ (Yoshimura et al., 1987), which are ~10^2–10^3 times larger than $T_C = 26$ K for $x = 0.15$. Here, we observe satisfactorily good agreements between experimental data and theoretical calculations, because the SCR calculations are simply performed by using Eq. (13.29) with experimentally obtained parameters and there are no intentional fittings between experiments and calculations. Similar good agreements can also be observed in many typical cases of weakly itinerant ferromagnets such as MnSi, Sc_3In, $ZrZn_2$ and Ni_3Al

(Takahashi & Moriya, 1985), (Sr-Ca)RuO_3 (Kiyama et al., 1996; Yoshimura et al., 1999), LaCoAsO (Ohta & Yoshimura, 2009a; Ohta et al. 2010b), ACo_2Se_2 (A: alkaline metal) (Yang et al., 2013), BCo_2P_2 (B: alkaline-earth metal) (Imai et al., 2014, 2015), Fe($Ga_{1-x}Ge_x$)$_3$ (Zhang et al., 2018), etc. Consequently, it can be concluded that SCR theory successfully explains weakly itinerant ferromagnetism evaluated from such kinds of comparisons with various experiments (Moriya, 1979, 1981, 1985, 1987; Takahashi, 1986, 2013, 2017b; Takahashi & Moriya, 1985; Takahashi & Yoshimura, 2012; Yoshimura, 2017, 2020).

13.4 Takahashi's Spin-Fluctuation Theory

After providing quantitative evidence for SCR theory in 1985 (Takahashi & Moriya, 1985), Takahashi has developed spin-fluctuation theory (Takahashi, 1986) in different approaches with two intrinsic assumptions: total spin-fluctuation amplitude conservation (TAC) and global consistency (GC) for a few decades (~1986 to present) (; Takahashi, 1986, 2013; Takahashi & Yoshimura, 2012; Yoshimura, 2017, 2020)), which led us to a new unified picture of metallic magnetism with a wide variety of itinerant-electron magnets based upon spin-fluctuation approaches (Takahashi, 2013, 2017a, 2017b; Takahashi & Yoshimura, 2012; Yoshimura, 2017, 2020).

Here, we stress on TAC hypothesis, because it is directly concerned with the unified picture of itinerant magnetism. In Takahashi's theory, the following TAC Eq. (13.33) is assumed to be valid even in explaining magnetic properties of itinerant electrons at finite temperatures (Takahashi, 1986). Therefore, the total spin fluctuation, namely, the average total square amplitude of the local spin fluctuation is constant and conserved even in the itinerant system as (Takahashi, 1986, 2013, 2017a, 2017b; Takahashi & Yoshimura, 2012; Yoshimura, 2017, 2020)

$$\langle S_{loc}^2 \rangle = \langle S_{loc}^2 \rangle_T + \langle S_{loc}^2 \rangle_{Z.P.} + \sigma_s^2 = const. \tag{13.33}$$

where σ_s^2 is the spin-fluctuation term in the ordered state. Equation (13.33) is naturally satisfied in the localized moment system as mentioned in Sect. 13.1. Takahashi assumed that Eq. (13.33) is valid even in an itinerant system, although that is not intuitive in the itinerant system. From this TAC assumption [Eq. (13.33)], Takahashi obtained the next important relation between spin-fluctuation parameters as

$$\overline{F_1} = \frac{4}{15} \frac{k_B T_A^2}{T_0}. \tag{13.34}$$

By using the relation of Eq. (13.34) combined with SCR relations as Eqs. (13.28), (13.29) and (13.32), the magnetic properties at finite temperatures can be reproduced and explained (Takahashi, 1986, 2013, 1988a,2017a,2017b; Takahashi & Yoshimura,

2012; Yoshimura, 2017, 2020; Yoshimura et al., 1987). This allows the new unification of the itinerant-electron magnetism from the exchange-enhanced Pauli paramagnetic weak limit (Moriya & Kawabata,) to the localized-moment limit in the metallic state (Anderson, 1959, 1961), leading to the unified relation between p_{eff} $/p_s$ (similar to RWR) and T_C $/T_0$ independent of magnetic materials by combination of Eqs. (13.28), (13.29) and (13.34) as

$$p_{eff}/p_s \approx 1.4 \times (T_C/T_0)^{-2/3}, \tag{13.35}$$

which gives the new universal picture (the universal p_{eff} $/p_s$ vs T_C/T_0 plot) among various ferromagnetic materials (Imai et al., 2014, 2015; Takahashi, 1986, 2013, 2017b; Takahashi & Yoshimura, 2012; Yang et al., 2013; Yoshimura, 2017, 2020; Zhang et al., 2018), giving physical meanings to phenomenological Rhodes-Wohlfarth plot (p_C/p_s vs. T_C) (Rhodes & Wohlfarth, 1963; Wohlfarth, 1978). Here, the universal p_{eff} $/p_s$ vs T_C/T_0 plots (Takahashi plot) for various itinerant ferromagnets including recent planar 2D ferromagnets as LaCoAsO (Ohta & Yoshimura, 2009a; Ohta et al. 2010b), ACo_2Se_2 (A: alkaline metal) (Yang et al., 2013) and BCo_2P_2 (B: alkaline-earth metal) (Imai et al., 2014, 2015) are shown in Fig. 13.2. The solid straight line represents Takahashi's relation, Eq. (13.35). In the spin-fluctuation theories for 2D itinerant ferromagnets (Hatatani & Moriya, 1995; Takahashi, 1997) have indicated that the relation of Eq. (13.35) shifts downward in Takahashi plots of Fig. 13.2. Even if we consider this situation in Fig. 13.2 (coexisting of 2D and 3D itinerant ferromagnets in Fig. 13.2), it can be seen that almost all the itinerant ferromagnets follow Takahashi's relation, Eq. (13.35) in Fig. 13.2, resulting

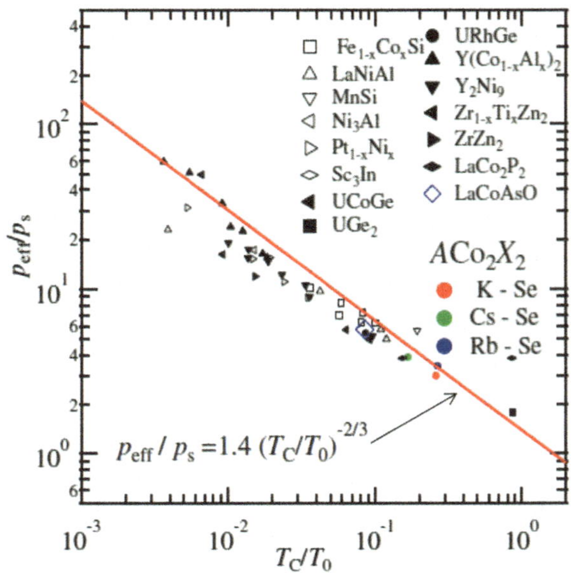

Fig. 13.2 Generalized Rhodes-Wohlfarth plots (Takahashi plots: p_C/p_s vs. T_C) in logarithmic scales (for references, see text)

in the fact that the Takahashi (p_{eff}/p_s vs. T_C/T_0) plots realize the universal Rodes-Wohlfarth plots for itinerant ferromagnetic systems. Therefore, we prefer to conclude that Takahashi's theory of spin fluctuations possesses the unified dynamical picture of itinerant magnetism due to the TAC hypothesis (Takahashi, 1986, 2013, 2017a, 2017b; Takahashi & Yoshimura, 2012; Yoshimura, 2017, 2020). However, the TAC assumption completely contradicts the concept of the SCR theory of spin fluctuations [Eq. (13.18)], which is now discussed. How to treat the zero-point spin fluctuation in Eq. (13.19) has not been completely solved, and is still an open question.

13.5 Exotic Superconductivity and Spin Fluctuations

This section discusses superconductivity which is another typical example of a collective mode of electrons in solid-state physics. Superconductivity was discovered by Kamerlingh Onnes in 1911 in metallic Hg (Onnes, 1911), and theoretically explained in 1957 by the Bardeen-Cooper-Schrieffer (BCS) theory of superconductivity by utilizing second quantization methods, such as the quantum field theory of quantum mechanics (Bardeen et al., 1957). Recently, there have been great developments in this field including the remarkable discovery of high-T_c cuprate (T_c is the superconducting critical temperature) by Bednorz and Müller (1986). Next, BCS theory (Bardeen et al., 1957) is introduced and discussed to demonstrate another collective mode of itinerant electrons.

In BCS theory, the BCS electronic state is considered based upon Eq. (13.10) as

$$\varphi_\theta = \Pi_k\left(u_k + exp(i\theta)v_k B_k^+\right)\varphi_0 = \Pi_k\left(u_k + exp(i\theta)v_k B_k^+\right)|0\rangle, \tag{13.36}$$

where $\exp(i\theta)$ is the phase factor in the phase description of the electron-pair wave function. Here, the electron pair is called a Cooper pair in superconductivity. Furthermore, u_k and v_k are parameters, that are related by

$$u_k^2 + v_k^2 = 1, \\ u_k = \begin{cases} 0\ (k << k_F) \\ 1\ (k >> k_F) \end{cases}, \ v_k = \begin{cases} 1\ (k << k_F) \\ 0\ (k >> k_F) \end{cases} \tag{13.37}$$

The superconducting state is assumed to be the Bose–Einstein condensation (BEC) state of Cooper pairs of electrons, where the Cooper pairs have the phase θ, so that they move in phase in the superconducting BEC state with the same value of $exp(i\theta)$.

The BCS Hamiltonian is (Bardeen et al., 1957).

$$H = \sum_{k,\sigma}\left(\frac{\hbar^2k^2}{2m} - \frac{\hbar^2k_F^2}{2m}\right)a_{k\sigma}^+\ a_{k\sigma} - \frac{g}{V}\sum_k{}'\sum_l{}' B_k^+ B_l, \tag{13.38}$$

where the first term is the relative kinetic energy compared with the Fermi Energy, E_F, the second term is the interaction between the Cooper pairs, g/V is the strength of their interaction per volume (V), and $\sum'$ is the summation of the wave vector, $\boldsymbol{k}$ or $\boldsymbol{l}$, near $\boldsymbol{k}_F$. By the BCS state [Eq. (13.36)] operating on the BCS Hamiltonian [Eq. (13.38)], we obtain the relative energy compared with the non-superconducting (normal) state. When g is positive, the interaction between the Cooper pairs is attractive, and the superconducting state is stabilized. In BCS theory, this positive interaction is attributed to the electron–phonon (energy quantum of lattice vibration) interaction. In other words, the phonon scattering of the electrons is the origin of the attraction of the Cooper pairs. Therefore, in the second term of the Eq. (13.38), when the electron pair with wave vector l is scattered by the phonon to become the electron pair with wave vector k, the system gains the energy of g per V (Bardeen et al., 1957). However, the Hamiltonian of Eq. (13.38) cannot be rigorously solved because it includes the many-body problem. Therefore, BCS theory can be modified to give the mean-field theory via the one-body approximation as

$$H_m = \sum_{k,\sigma} \left(\frac{\hbar^2 k^2}{2m} - \frac{\hbar^2 k_F^2}{2m} \right) a_{k\sigma}^+ a_{k\sigma} - \sum_k{}' \left(\Delta^* B_k + \Delta B_k^+ \right) + \frac{V}{g} |\Delta|^2, \quad (13.39)$$

where $\boldsymbol{H}_m$ is the Hartree–Fock-Gor'kov Hamiltonian (Gor'kov, 1959). The superconducting gap energy Δ and its conjugate Δ^* are introduced, where Δ can be written as:

$$\Delta = \frac{g}{V} \sum_k{}' \langle \varphi_\theta | a_{k\uparrow} a_{-k\downarrow} | \varphi_\theta \rangle = \frac{g}{V} exp(i\theta) \sum_k{}' u_k v_k. \quad (13.40)$$

The second term in Eq. (13.39) can be written with one bosonic creation or annihilation operator, whereas this term is given by the product of two operators in Eq. (13.38). The third term of Eq. (13.39) is a correction of the mean-field approximation. Although the mean-field Hamiltonian of Eq. (13.39) by Gor'kov was not diagonalized, the perfect formalism of the diagonalization of the BCS mean field Hamiltonian was done by Bogoliubov by means of introducing the Bogoliubov transformation of creation and annihilation operators for fermions (Bogoliubov, 1958a, 1958b, 1958c).

The formalism of BCS mean-field theory should be valid essentially even in high-T_c cuprates (Bednorz & Müller, 1986; Wu et al., 1987) and iron pnictides (Kamihara et al., 2006, 2008) superconductors, as well as other strongly correlated electron superconductors, such as heavy-fermion superconductors (Steglich et al., 1979) and organic superconductors (Lebed, 2008), although the mediation mechanism of Cooper pairs should be different from that of BCS theory (Anderson, 1973, 2013; Lee & Nagaosa, 1992; Moriya, 2006; Moriya & Ueda, 2003; Moriya et al., 1990, 1992; Nagaosa & Lee, 1990; Nomura & Yamada, 2003; Pines, 2013). In high-T_c cuprates, microscopic experiments have shown that the magnetic excitations were crucial (Asayama et al., 1996a, 1996b; Imai, 1990; Imai et al. 1993a, 1993b; Kitaoka et al., 1990, 1992), leading to the possible mechanism involving magnetic interaction-mediated Cooper pairs (Anderson, 1973, 2013; Lee & Nagaosa, 1992; Moriya, 2006; Moriya & Ueda, 2003; Moriya et al., 1990, 1992; Nagaosa & Lee, 1990; Nomura & Yamada, 2003; Pines, 2013). BEC of fermion pairs was also discovered in liquid ^{3}He as part of discovery of its superfluidity (Osheroff et al., 1972), and has become more and more important in quantum physics.

Given that novel superconductors have been discovered in strongly correlated electron systems, such as heavy-fermion compounds and intermetallics (Steglich et al., 1979), the organic systems (Lebed, 2008), high-T_c cuprates (Bednorz & Müller, 1986; Wu et al., 1987), pyrochlore compounds (Hanawa et al., 2001; Sakai et al., 2001), triangular-lattice layered Co-oxides (Takada et al., 2003) and Fe pnictides (Kamihara et al., 2006, 2008), 3–4–13 cluster compounds (Klintberg et al., 2012; Yang et al., 2010; Yu et al., 2015) and others, the electron–electron correlations and interplays between itinerant magnetism and novel superconductivity, called exotic superconductivity, have been important. Itinerant-electron characteristics have recently become one of the most difficult and important problems in solid-state sciences (Anderson, 1973, 2013; Lee & Nagaosa, 1992; Moriya, 2006; Moriya & Ueda, 2003; Moriya et al., 1990, 1992; Nagaosa & Lee, 1990; Nomura & Yamada, 2003; Pines, 2013). The formalism of BCS mean field theory (Bardeen et al., 1957; Bogoliubov, 1958a, 1958b, 1958c; Gor'kov, 1959) should be valid even in high-T_c cuprate and iron pnictide superconductors with high T_c, as well as other strongly correlated electron superconductors, such as heavy-fermion superconductors and organic superconductors etc., although the mediation mechanism of Cooper pairs may be different from that of BCS theory (Bardeen et al., 1957; Bogoliubov, 1958a, 1958b, 1958c; Gor'kov, 1959). In high-T_c cuprates, microscopic dynamical experiments (Asayama et al., 1996a, 1996b; Imai, 1990; Imai et al. 1993a, 1993b; Kitaoka et al., 1990, 1992) have shown that the magnetic excitations were crucial, leading to the possible mechanism involving magnetic interaction-mediated Cooper pairs (Anderson, 1973, 2013; Lee & Nagaosa, 1992; Moriya, 2006; Moriya & Ueda, 2003; Moriya et al., 1990, 1992; Nagaosa & Lee, 1990; Nomura & Yamada, 2003; Pines, 2013). In spin-fluctuation theory, the superconducting transition temperatures T_c are found to be universally scaled by the characteristic temperature corresponding to spin-fluctuation cut-off energy which is related to the characteristic energy width of the spin-fluctuation spectrum, T_0, in exotic superconductors (Moriya, 2006; Moriya & Ueda, 2003; Moriya et al., 1990, 1992). In Fig. 13.3 the values of T_c of exotic superconductors such as high-T_c cuprates and heavy fermion superconductors are plotted against T_0. Here, plots are also shown for the triangular superconductor, $Na_xCoO_2 \cdot yH_2O$, and the cluster superconductor, $Ca_3Ir_4Sn_{13}$, with T_0 estimated from NMR data for $1/T_1$ (Chen et al., 2015; Ihara et al., 2008; Kato et al., 2006; Yoshimura et al., 2007). Both species have ferromagnetic spin fluctuations in their normal states. It can be seen from the plots of T_c against T_0 in Fig. 13.3 that T_c is correlated to T_0 in every exotic superconductors as a universal linear line, implying that the superconducting T_c is dominated by the antiferromagnetic or ferromagnetic spin fluctuation,

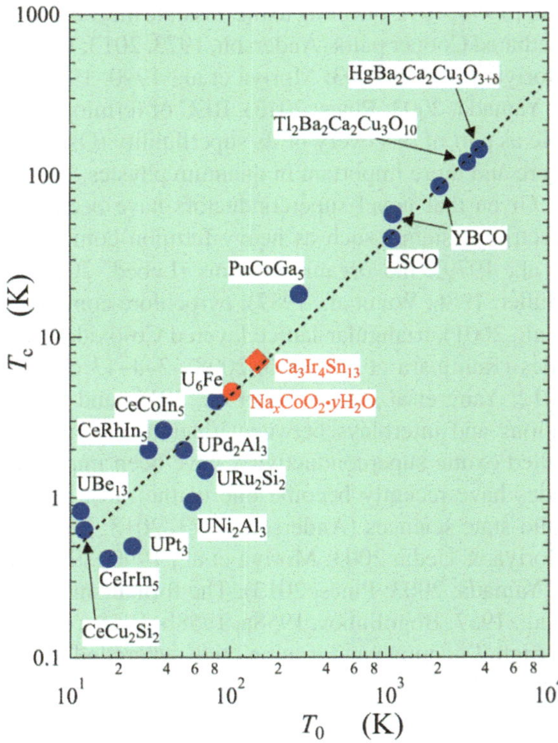

Fig. 13.3 Superconducting transition temperature T_c plotted against the characteristic temperature, T_0, corresponding to the ω-width of the spin-fluctuation spectrum for several exotic superconductors using logarithmic scales (for references, see text)

and that the mediating mechanism of electron pairs is related to magnetic interaction, that is, spin fluctuations. A similar theory of antiferromagnetic spin correlation was published by D. Pines (2013). Of course, there exist the different theories with different ideas of antiferromagnetic interactions (Anderson, 1973, 2013; Lee & Nagaosa, 1992; Nagaosa & Lee, 1990) as well as Fermi-liquid theory with antiferromagnetic interactions (Nomura & Yamada, 2003). Discussion will be ongoing in pursuit of superconducting mechanisms of exotic superconductors. The dynamical theory beyond the mean-field approximation should be used to explain many problems in exotic superconductors such as spin-gap (pseudo-gap) behaviors from far above T_c and extremely high values of T_c itself.

Final Remarks and Acknowledgement

In solid-state quantum physics, P.W. Anderson's famous words were summarized in the title of his famous paper (Anderson, 1972)): "more is different." Many fascinating phenomena exist in this field, in which the many-particle problem itself is essential. In other words, it is impossible for a single particle (electron) or small number of particles (electrons) to show any such phenomena. Among them, magnetism and superconductivity, which are main themes of this chapter, are the typical and outstanding examples of "more is different." These phenomena are caused by the collective modes of itinerant electrons. This chapter has reviewed the recent developments in itinerant-electron magnetism and exotic superconductivity. Spin-fluctuation theories beyond the mean-field theories are so important to explain even static magnetic properties as well as dynamical properties of itinerant magnets. For example, the RW plot can be explained universally for the first time by spin-fluctuation theory (Takahashi theory). Furthermore, it was found that the electron-pair formation due to magnetic spin fluctuations would be realized in exotic superconductors, in which the superconducting transition temperature T_c is scaled with the characteristic temperature T_0 corresponding to the energy width of spin fluctuations. The concepts and theories beyond the mean-field approximations are very important in itinerant systems, and itinerant-electron problems are still currently attracting our interest.

The author (K.Y.) thanks Professor Emeritus Masashi Takigawa, the Institute of Solid State Physics (ISSP) , University of Tokyo, and Professor Emeritus Hiroshi Yasuoka, ISSP, University of Tokyo, and the late Professor Emeritus Charles P. Slichter, University of Illinois at Urbana-Champaign, for guidance into the research field of solid-state physics with utilization of NMR techniques and for their great interest and knowledge in microscopic magnetic measurements. Professor Masashi Takigawa was a student and coworker of Professor Yasuoka, and is also a good friend of K.Y. Professor Yasuoka was a supervisor of K.Y. when he was a Ph.D. student, and is still very active as a special invited researcher at Los Alamos National Laboratory, USA, and also at the Max Planck Institute Dresden, Germany. Professor Charles P. Slichter was a famous professor owing to his famous NMR textbook, *Principles of Magnetic Resonance* (Springer, 1978, 1998). He was also a classmate of the late Professor Philip W. Anderson in the laboratory of Professor John H. van Vleck, Department of Physics, Harvard University, when he was a student there, and was a colleague of the late Professor David Pines at Department of Physics, University of Illinois at Urbana-Champaign. When K.Y. was staying there, both Charlie and David were still very active. The author (K.Y.) feels greatly honored to be their student and coworker in solid-state physics research.

The author is also grateful to Professor Emeritus Yoshinori Takahashi, University of Hyogo Prefecture, and Professor Emeritus Toru Moriya, University of Tokyo, for kindly teaching the theory of spin fluctuations and for guidance in theoretical approaches of itinerant-electron magnetism, especially in terms of spin fluctuations. Professor Takahashi was the last coworker of Professor Moriya at ISSP, University of Tokyo. Finally, they were struggling with each other to develop the theories of spin fluctuations. Such behavior is very important in achieving breakthroughs in scientific research. The author (K.Y.) is honored to have been there.

References

Anderson, P. W. (1959). New approach to the theory of superexchange interactions. *Physical Review, 115*, 2–13.

Anderson, P. W. (1961). Localized magnetic states in metals. *Physical Review, 124*, 41–53.

Anderson, P. W. (1972). More is different. P. W. Anderson. *Science, 117*, 393–396.
Anderson, P. W. (1973). Resonating valence bonds: A new kind of insulator? *Material Research Bulletin, 8*, 153–160.
Anderson, P. W. (2013). Twenty-five years of high-temperature superconductivity -A personal review. *Journal of Physics Conference Series, 449*, 012001/1–10.
Asayama, K., Kitaoka, Y., Zheng, G.-q, & Ishida, K. (1996a). NMR studies of high T_c superconductors. *Progress in Nuclear Magnetic Resonance Spectroscopy, 28*, 221–253.
Asayama, K., Kitaoka, Y., Zheng, G.-q, Ishida, K., & Magishi, K. (1996b). NMR study of high-T_c superconductors. *Physica B: Condensed Matter, 223–224*, 478–483.
Ashcroft, N. W., & Mermin, N. D. (1976). *Solid state physics*. Harcourt.
Bardeen, J., Cooper, L., & Schrieffer, J. R. (1957). Theory of superconductivity. *Physical Review, 108*, 1175–1204.
Bednorz, J. G., & Müller, K. A. (1986). Possible highT_c superconductivity in the Ba–La–Cu–O system. *Zeitschrift Für Physik B Condensed Matter, 64*, 189–193.
Brillouin, L. (1922). La Théorie des Quanta et l'Atome de Bohr. *Presse Universitaires De France, 1922*, 1931.
Brillouin, L. (1930). *Les statistiques quantiques Et leurs applications 2 Vols*. Presse Universitaires de France.
Bloch, D., Voiron, J., Jaccarino, V., & Wermick, J. H. (1975). The high field-high pressure magnetic properties of MnSi. *Physics Letters, 51A*, 259–261.
Boer, F. R. de. (1969). *Itinerant electron magnetism in the Ni_3Al and Ni_3Ga phases*. Ph.D Thesis (University of Amsterdam).
Bogoliubov, N. N. (1958a). New Mwthod in the Theory of Superconductivity. *I Soviet Physics JETP, 34*, 41–46.
Bogoliubov, N. N. (1958b). New Mwthod in the Theory of Superconductivity. *II Soviet Physics JETP, 34*, 46–50.
Bogoliubov, N. N. (1958c). New method in the theory of superconductivity. *III Soviet Physics JETP, 34*, 51–55.
Chen, B., Ohta, H., Michioka, C., Itoh, Y., & Yoshimura, K. (2010). ^{27}Al NMR studies of itinerant electron ferromagnetic Ni_3AlC_x. *Physical Review, B 81*, 134416/1–8.
Chen, B., Yang, J., Guo, Y., & Yoshimura, K. (2015). ^{119}Sn NMR study in the normal state of the superconductor $Ca_3Ir_4Sn_{13}$. *EPL, 111*, 17005/1–5.
Curie, P. (1895). Proprietes magnetiques des corps a diverses temperatures. *Journal de Physique, 4*, 197–212 and 263–272; *Annales de Chimie et de Physique* [7], *5*, 232–334.
de Boer, F. R., Schinkel, C. J., Biesterbos, J., & Proost, S. (1969). Exchange-enhanced paramagnetism and weak ferromagnetism in the Ni_3Al and Ni_3Ga phases; giant moment inducement in Fe-Doped Ni_3Ga. *Journal of Applied Physics, 40*, 1049.
Gat-Malureanu, I. M., Carlo, J. P., Goko, T., Fukaya, A., Ito, T., Kyriakou, P. P., Larkin, M. I., Luke, G. M., Russo, P. L., Savici, A. T., Wiebe, C. R., Yoshimura, K., & Uemura, Y. J., (2011). Muon spin relaxation and susceptibility measurements of an itinerant-electron system $Sr_{1-x}Ca_xRuO_3$: Quantum evolution from ferromagnet to paramagnet. *Physical Review B, 84*, 224415/1–12.
Goodenough, J. B. (1971). Metallic oxides. *Progress in Solid State Chemistry, 5*(1971), 145–399.
Gor'kov, L. P. (1959). Microscopic derivation of the Ginzburg-Landau equations in the theory of superconductivity. *Soviet Physics JETP, 9*, 1364–1367.
Hanawa, M., Muraoka, Y., Tayama, T., Sakakibara, T., Yamaura, J., & Hiroi, Z. (2001). Superconductivity at 1 K in $Cd_2Re_2O_7$. *Physical Review Letters, 87*, 187001/1–4.
Hatatani, M., & Moriya, T. (1995). Ferromagnetic spin fluctuations in two-dimensional metals. *Journal of the Physical Society of Japan, 64*, 3434–3441.
Heisenberg, W. (1928). Zur theorie des ferromagnetismus. *Zeitschrift Für Physik, 49*, 619–636.
Holstein, H., & Primakoff, H. (1940). Field dependence of the intrinsic domain magnetization of a ferromagnet. *Physical Review, 58*, 1098–1113.
Holtzberg, F., McGuire, T. R., Methfessel, S., & Suits, J. C. (1964). Ferromagnetism in rare-earth group VA and VIA compounds with Th_3P_4 structure. *Journal of Applied Physics, 35*, 1033–1038.

Hubbard, J. H. (1963). Electron correlations in narrow energy bands. *Proceedings of the Royal Society, A276*, 238–257.
Hubbard, J. H. (1964a). Electron correlations in narrow energy bands. II. The degenerate band case. *Proceedings of the Royal Society, A277*, 237–259.
Hubbard, J. H. (1964b). Electron correlations in narrow energy bands. III an improved solution. *Proceedings of the Royal Society, A281*, 401–419.
Ihara, Y., Ishida, K., Ohta, H., & Yoshimura, K. (2008). Co-NMR measurements on crystalline sample of the bilayer hydrate $Na_x(H_3O)_zCoO_2{\cdot}yH_2O$. *Journal of the Physical Society of Japan, 77*, 07302/1–4.
Imai, M., Michioka, C., Ohta, H., Matsuo, A., Kindo, K., Ueda, H., & Yoshimura, K. (2014). Anomalous itinerant-electron metamagnetic transition in the layered $Sr_{1-x}Ca_xCo_2P_2$ system. *Physical Review B, 90*, 014407/1–6.
Imai, M., Michioka, C., Ueda, H., & Yoshimura, K. (2015). Static and dynamical magnetic properties of an itinerant ferromagnet $LaCo_2P_2$. *Physical Review B, 91*, 181144/1–7.s
Imai, T. (1990). Analysis of nuclear relaxation experiments in high T_c oxides based on mila-rice hamiltonian. *Journal of the Physical Society of Japan, 59*, 2508–2521.
Imai, T., Slichter, C. P., Yoshimura, K., & Kosuge, K. (1993a). Low frequency spin dynamics in undoped and Sr-doped La_2CuO_4. *Physical Review Letters, 70*, 1002–1005.
Imai, T., Slichter, C. P., Yoshimura, K., Kato, M., & Kosuge, K. (1993b). Spin-spin correlation in the quantum critical regime of La_2CuO_4. *Physical Review Letters, 71*, 1254–1257.
Ishikawa, Y., Uemura, Y., Majzak, C. P., Shirane, G., & Noda, Y. (1985). Paramagnetic spin fluctuations in the weak itinerant-electron ferromagnet MnSi. *Physical Review B, 31*, 5884–5893.
Kato, M., Michioka, C., Waki, T., Itoh, Y., Yoshimura, K., Ishida, K., Sakurai, H., Takayama-Muromachi, E., Takada, K., & Sasaki, T. (2006). Possible spin triplet superconductivity in $Na_xCoO_2{\cdot}{}_yH_2O$-^{59}CoNMR Studies. *Journal of Physics: Condensed Matter 669*, 669–682.
Kamerlingh Onnes, H. (1911). The superconductivity of mercury. *Communications Physics Laboratory University Leiden, 122*, 124.
Kamihara, Y., Hiramatsu, H., Hirano, M., Kawamura, R., Yanagi, H., Kamiya, T., & Hosono, H. (2006). Iron-based layered superconductor: LaOFeP. *Journal of the American Chemical Society, 128*, 10012–10013.
Kamihara, Y., Watanabe, T., Hirano, M., & Hosono, H. (2008). Iron-based layered superconductor $La[O_{1-x}F_x]FeAs$ ($x = 0.05-0.12$) with $T_c = 26$ K. *Journal of the American Chemical Society, 130*, 3296–3297.
Kitaoka, Y., Ohsugi, S., Ishida, K., & Asayama, K. (1990). Antiferromagnetic spin correlation and superconductivity in $La_{2-x}Sr_xCuO_4$: Cu NQR study. *Physica C, 170*, 189–194.
Kitaoka, Y., Ishida, K., Ohsugi, S., Fujiwara, K., Zheng, G. -Q., & Asayama, K. (1992). NMR and NQR studies of spin dynamics and superconductivity in High-T_c oxides. *Journal of Applied Magnetic Resonance, 3*, 549–596.
Kittel, C. (1953). *Introduction to solid state physics.* Wiley
Kiyama, T., Yoshimura, K., Kosuge, K., Ikeda, Y., & Bando, Y. (1996). Invar effect of $SrRuO_3$: Itinerant electron magnetism of ru 4*d* electrons. *Physical Review B, 54*, 756–759.
Kiyama, T., Yoshimura, K., Kosuge, K., Michor, H., & Hilscher, G. (1998). Specific heat of (Sr-Ca)RuO_3. *Journal of the Physical Society of Japan, 67*, 307–311.
Klintberg, L. E., Goh, S. K., Alireza, P. L., Saines, P. J., Tompsett, D. A., Logg, P. W., Yang, J., Chen, B., Yoshimura, K., & Grosche, F. M. (2012). Pressure- and composition-induced structural quantum phase transition in the cubic superconductor (Sr, Ca)$_3Ir_4Sn_{13}$. *Physical Review Letters, 109*, 237008/1–5.
Kontani, M., Hioki, T., & Masuda, Y. (1975). NMR studies on magnetic properties of $ZrZn_2$. *Journal of the Physical Society of Japan, 39*, 665–671.
Kosterlitz, J. M., & Thouless, D. J. (1973). Ordering, metastability and phase transitions in two-dimensional systems. *Journal of Physics C, 6*, 1181–1203.
Landau, L. D. (1937a). On the theory of phase transitions. *Zhurnal Éksperimental'noĭ i Teoreticheskoĭ Fiziki, 7*, 19–32 (*Physics Z Sowjetunion, 11*, 26–47 (1937)).

Landau, L. D. (1937b). On the theory of phase transitions. II. *Zhurnal Éksperimental'noĭ i Teoreticheskoĭ Fiziki, 7*, 627 (*Physics Z Sowjetunion, 11*, 545–555 (1937)).
Landau, L. D., & Lifshitz, E. M. (1980). *Statistical physics, part 2*. Pergamon.
Langevin, P. (1905). Magnétisme et théorie des électrons. *Annales De Chimie De Physique, 5*, 70–1207.
Lebed, A. G. (Ed.) (2008). *The physics of organic superconductors and conductors* (Vol. 110). Springer, Springer Series in Materials Science.
Lee, P. A., & Nagaosa, N. (1992). Gauge theory of the normal state of high-Tcsuperconductors. *Physical Review B, 46*, 5621–5639.
Masuda, Y. (1983). Spin fluctuations in weakly ferro- and antiferromagnetic metals and alloys. *Journal of Magnetism and Magnetic Materials, 31–34*, 259–264.
Mathias, B. T., Geballe, T. H., & Compton, V. B. (1963). Superconductivity. *Reviews of Modern Physics, 35*, 1–22.
Michor, H., El-Hagary, M., Della Mea, M., Pieper, M. W., Reissner, G. Hilscher, M., Khmelevskyi, S., Mohn, P., Schneider, G., Giester, G. & Rogl, P. (2004). Itinerant electron metamagnetism in $LaCo_9Si_4$. *Physical Review B, 69*, 081404(R)/1–4.
Moriya, T., & Kawabata, A. (1973a). Effect of spin fluctuations on itinerant electron ferromagnetism. *Journal of the Physical Society of Japan, 34*, 639–651.
Moriya, T., & Kawabata, A. (1973b). Effect of spin fluctuations on itinerant electron ferromagnetism. *II Journal of the Physical Society of Japan, 35*, 669–676.
Moriya, T., & Takahashi, Y. (1978). Spin fluctuation theory of itinerant electron ferromagnetism –a unified picture. *Journal of the Physical Society of Japan, 45*, 397–408.
Moriya, T. (1979). Recent progress in the theory of itinerant electron magnetism. *Journal of Magnetism and Magnetic Materials, 14*, 1–46.
Moriya, T. (1981). *Electron correlation and magnetism in narrow-band systems*. Springer.
Moriya, T. (1985). *Spin fluctuations in itinerant electron magnetism*. Springer, Springer Series in Solid-State Sciences 56.
Moriya, T. (1987). A unified picture of magnetism. In H. Capellmann (ed.) *Metallic magnetism*. Springer.
Moriya, T., Takahashi, Y., & Ueda, K. (1990). Antiferromagnetic spin fluctuations and superconductivity in two-dimensional metals—A possible model for high T_c oxides. *Journal of the Physical Society of Japan, 59*, 2905–2915.
Moriya, T., Takahashi, Y., & Ueda, K. (1992). Antiferromagnetic spin fluctuations and superconductivity in high T_c oxides. *Journal of Magnetism and Magnetic Materials, 104–107*, 456–460.
Moriya, T., & Takimoto, T. (1995). Anomalous properties around magnetic instability in heavy electron systems. *Journal of the Physical Society of Japan, 64*, 960–969.
Moriya, T., & Ueda, K. (2003). Antiferromagnetic spin fluctuation and superconductivity. *Reports on Progress in Physics, 66*, 1299–1341.
Moriya, T. (2006). Developments of the theory of spin fluctuations and spin fluctuation-induced superconductivity. *Proceedings of the Japan Academy, Ser. B, 82*, 1–16.
Mott, N. F. (1949). The basis of the electron theory of metals, with special reference to the transition metals. *Proceedings of the Physical Society (london), A62*, 416–422.
Mott, N. F. (1968). Metal-insulator transition. *Reviews of Modern Physics, 40*, 677–683.
Murata, K. K., & Doniach, S. (1972). Theory of magnetic fluctuations in itinerant ferromagnets. *Physical Review Letters, 29*, 285–288.
Nagaosa, N., & Lee, P. A. (1990). Normal-state properties of the uniform resonating-valence-bond state. *Physical Review Letters, 64*, 2450–2453.
Néel, L. (1948). Propriétés magnétiques des ferrites, in: ferrimagnétisme et antiferromagnétisme. *Annals of Physics, (Paris)3*, 137–198.
Nomura, T., & Yamada, K. (2003). Higher order perturbation expansion for pairing interaction in repulsive hubbard model. *Journal of the Physical Society of Japan, 72*, 2053–2063.

Ogawa, S. (1976). Electrical resistivity of weak itinerant ferromagnet $ZrZn_2$. *Journal of the Physical Society of Japan, 40*, 1007–1009.

Ohta, H., & Yoshimura, K. (2009a). Anomalous magnetization in the layered itinerant ferromagnet LaCoAsO. *Physical Review B, 79*, 184407/1–5.

Ohta, H., & Yoshimura, K. (2009b). Magnetic properties of *L*CoAsO (*L*=La–Gd). *Physical Review B, 80*, 184409/1–6.

Ohta, H., Michioka, C., Matsuo, A., Kindo, K., & Yoshimura, K. (2010a). Magnetic study of SmCoAsO showing a ferromagnetic-antiferromagnetic transition. *Physical Review B, 82*, 054421/1–8.

Ohta, H., Michioka, C., & Yoshimura, K. (2010b). Spectroscopic study of ^{75}As and ^{139}La NMR on layered structure ferromagnet LaCoAsO. *Journal of the Physical Society of Japan, 79*, 054703/1–5.

Ohta, H., Michioka, C., & Yoshimura, K. (2011). Large magnetoresistance effects in *Ln*CoAsO (*Ln* = Nd, Sm) with a ferromagnetic- antiferromagnetic transition. *Physical Review B, 84*, 134411/1–5.

Osheroff, D. D., Richardson, R. C., & Lee, D. M. (1972). Evidence for a new phase of solid He3. *Physical Review Letters, 28*, 885–888.

Pauli, W. (1925). Über den zusammenhang des abschlusses der elektronengruppen im atom mit der komplexstruktur der spektren. *Zeitschrift Für Physik, 31*, 765–783.

Pauli, W. (1927). Über gasentartung und paramagnetismus. *Zeitschrift Für Physik, 41*, 81–102.

Pines, D. (2013). Finding new superconductors: the spin-fluctuation gateway to high T_c and possible room temperature superconductivity. *The Journal of Physical Chemistry B, 117*, 13145–13153.

Rhodes, P., & Wohlfarth, E. P. (1963). The effective Curie-Weiss constant of ferromagnetic metals and alloys. *Proceedings of the Royal Society A, 273*, 247–258.

Sakai, H., Yoshimura, K., Ohno, H., Kato, H., Kambe, S., Walstedt, R. E., Matsuda, T. D., Haga, Y., & Onuki, Y. (2001). Superconductivity in a pyrochlore oxide, $Cd_2Re_2O_7$. *Journal of Physics: Condensed Matter, 13*, L785–L790.

Sakakibara, T., Goto, T., Yoshimura, K., Shiga, M., & Nakamura, Y. (1986). Itinerant electron metamagnetism in $Y(Co_{1-x}Al_x)_2$. *Physics Letters A, 117*, 243–246.

Shinoda, D., & Asanabe, S. (1966). Magnetic Properties of Silicides of Iron Group Transition Elements. *Journal of the Physical Society of Japan, 21*, 555.

Steglich, F., Aarts, J., Bredl, C. D., Lieke, W., Meschede, D., Franz, W., & Schäfer, H. (1979). Superconductivity in the presence of strong pauli paramagnetism: $CeCu_2Si_2$. *Physical Review Letters, 43*, 1892–1896.

Stoner, E. C. (1938). Collective electron ferromagnetism. *Proceedings of the Royal Society a: Mathematical, Physical and Engineering Sciences, 165*, 372–414.

Stoner, E. C. (1939). Collective electron ferromagnetism II. energy and specific heat. *Proceedings of the Royal Society a: Mathematical, Physical and Engineering Sciences, 169*, 339–371.

Sugiyama, J., Nozaki, H., Umegaki, I., Harada, M., Higuchi, Y., Miwa, K., Ansaldo, E. J., Brewer, J. H., Imai, M., Michioka, C., Yoshimura, K., & M°ansson, M. (2015). Magnetic phase diagram of $Sr_{1-x}Ca_xCo_2P_2$ determined with μ^+SR. *Physical Review B, 91*, 144423/1–5.

Takada, K., Sakurai, H., Takayama-Muromachi, E., Izumi, F., Dilanian, R. A., & Sasaki, T. (2003). Superconductivity in two-dimensional CoO_2 layers. *Nature, 422*, 53–55.

Takahashi, Y., & Moriya, T. (1983). On the spin fluctuations in weak itinerant ferromagnets. *Journal of the Physical Society of Japan, 52*, 4342–4348.

Takahashi, Y., & Moriya, T. (1985). Quantitative aspects of the theory of weak itinerant ferromagnetism. *Journal of the Physical Society of Japan, 54*, 1592–1598.

Takahashi, Y. (1986). On the origin of the curie-weiss law of the magnetic susceptibility in itinerant electron ferromagnetism. *Journal of the Physical Society of Japan, 55*, 3553–3573.

Takahashi, Y. (1997). Spin-fluctuation theory of quasi-two-dimensional itinerant-electron ferromagnets. *Journal of Physics: Condensed Matter, 9*, 10359–10372.

Takahashi, Y., & Yoshimura, K. (2012). *Itinerant Electron Magnets and Spin Fluctuations*, 1st ed., Uchida Rokakuho.

Takahashi, Y. (2013). *Spin fluctuation theory of itinerant electron magnetism*. Springer, Springer Tracts in Modern Physics 253.

Takahashi, Y. (2017a). Current status of spin fluctuation theory of magneto-volume effects in itinerant electron magnets. *Journal of the Japan Society of Powder and Powder Metallurgy, 64*, 159–162.

Takahashi, Y. (2017b). Theoretical development in itinerant electron ferromagnetism. *Journal of Physics: Conference Series, 868*, 012002/1–14.

Takeuchi, J., & Masuda, Y. (1979). Low temperature specific heat of itinerant electron ferromagnet Sc_3In. *Journal of the Physical Society of Japan, 46*, 468–474.

Takigawa, M., & Yasuoka, H. (1982). Nuclear magnetic relaxation in palladium. *Journal of the Physical Society of Japan, 51*, 787–793.

Ueda, K., & Moriya, T. (1974). Nuclear magnetic relaxation in weakly ferro- and antiferromagnetic metals. *Solid State Communications, 15*, 169–172.

Uemura, Y. J., Goko, T., Gat-Malureanu, I. M., Carlo, J. P., Russo, P. L., Savici, A. T., Aczel, A., MacDougall, G. J., Rodriguez, J. A., Luke, G. M., Dunsiger, S. R., McCollam, A., Arai, J., Pfleiderer, C., Boni, P., Yoshimura, K., Baggio-Saitovitch, E., Fontes, M. B., Larrea, J., … Sereni, J. (2006). Phase separation and suppression of critical dynamics at quantum phase transitions of MnSi and $(Sr_{1-x}Ca_x)RuO_3$. *Nature Physics, 3*, 29–35.

Umemura, T., & Masuda, Y. (1983). Nuclear magnetic resonance and relaxation in itinerant electron ferromagnet Ni_3Al. *Journal of the Physical Society of Japan, 52*, 1439–1445.

Weiss, P. (1907). L'hypothèse du champ moléculaire et la propriété ferromagnétique. *Journal of Theoretical and Applied Physics, 6*(1), 661–690.

Williams, H. J., Wernick, J. H., Sherwood, R. C., & Wertheim, G. K. (1966). Magnetic properties of the monosilicides of some 3*d* transition elements. *Journal of Applied Physics, 37*, 1256.

Wohlfarth, E. P. (1978). Magnetic properties of crystalline and amorphous alloys: A systematic discussion based on the Rhodes-Wohlfarth plot. *Journal of Magnetism and Magnetic Materials, 7*, 113–120.

Wu, M. K., Ashburn, J. R., Torng, C. J., Hor, P. H., Meng, R. L., Gao, L., Huang, Z. J., Wang, Y. Q., & Chu, C. W. (1987). Superconductivity at 93 K in a new mixed-phase Y-Ba-Cu-O compound system at ambient pressure. *Physical Review Letters, 58*, 908–910.

Yang, J., Chen, B., Michioka, C., & Yoshimura, K. (2010). Coexistence of superconductivity and ferromagnetic spin-fluctuation in $Ca_3Ir_4Sn_{13}$ single crystal. *Journal of the Physical Society of Japan 79*, 113705/1–4 (letter).

Yang, J., Chen, B., Ohta, H., Michioka, C., Yoshimura, K., Wang H., & Fang, M. (2011). Spin fluctuations on the verge of a ferromagnetic quantum phase transition in $Ni_3Al_{1-x}Ga_x$. *Physical Review B, 83*, 134433/1–9.

Yang, J., Chen, B., Wang, H., Mao, Q., Imai, M., Yoshimura, K., & Fang, M. (2013). Magnetic properties in layered ACo_2Se_2 (A= K, Rb, Cs) with the $ThCr_2Si_2$-type structure. *Physical Review B, 88*, 064406/1–9.

Yasuoka, H., Jaccarino, V., Sherwood, R. C., & Wernick, J. H. (1978). NMR and susceptibility studies of MnSi above T_c. *Journal of the Physical Society of Japan, 44*, 842–849.

Yoshimura, K., Shimizu, T., Takigawa, M., Yasuoka, H., & Nakamura, Y. (1984). Nuclear magnetic relaxation in nearly ferromagnetic YCo_2. *Journal of the Physical Society of Japan, 53*, 503–506 (Letter).

Yoshimura, K., & Nakamura, Y. (1985). New weakly itinerant ferromagnetic system $Y(Co_{1-x}Al_x)_2$. *Solid State Communications, 56*, 767–771.

Yoshimura, K., Takigawa, M., Takahashi, Y., Yasuoka, H., & Nakamura, Y. (1987). NMR study of weakly itinerant ferromagnetic $Y(Co_{1-x}Al_x)_2$. *Journal of the Physical Society of Japan, 56*, 1138–1155.

Yoshimura, K., Mekata, M., Takigawa, M., Takahashi, Y., & Yasuoka, H. (1988a). Spin fluctuations in $Y(Co_{1-x}Al_x)_2$: A transition system from nearly to weakly itinerant ferromagnetism. *Physical Review B, 37*, 3593–3602.

Yoshimura, K., Yamada, M., Mekata, M., Shimizu, T. and Yasuoka, H. (1988b). Nuclear magnetic relaxation in the metallic localized moment system Ir_2MnGa. *Journal of the Physical Society of Japan, 57*, 409–412 (Letter).

Yoshimura, K., Yoshimoto, Y., Mekata, M., Fukamichi, K., & Yasuoka, H. (1988c). Nuclear spin-lattice relaxation in the invar-type itinerant ferromagnet $Lu(Co_{1-x}Al_x)_2$ (x=0.125, 0.15). *Journal of the Physical Society of Japan, 57*, 2651–2654.

Yoshimura, K., Imai, T., Kiyama, T., Thurber, K. R., Hunt, A. W., & Kosuge, K. (1999). ^{17}O NMR observation of universal behavior of ferromagnetic spin fluctuation in the itinerant magnetic system $Sr_{1-x}Ca_xRuO_3$. *Physical Review Letters, 83*, 4397–4400.

Yoshimura, K., Ohta, H., Michioka, C., & Itoh, Y. (2007). In-plane ferromagnetic spin fluctuations in novel superconducting system $Na_xCoO_2 \cdot yH_2O$. *Journal of Magnetism and Magnetic Materials, 310*, 693–695.

Yoshimura, K. (2017). Theme of the workshop on itinerant-electron magnetism. *Journal of Physics: Conference Series, 868*, 012001/1–7.

Yoshimura, K. (2020). Study of novel magnets and superconductors and their physical properties. *Journal of the Japan Society of Powder and Powder Metallurgy, 67*, 59–71.

Yu, W. C., Cheung, Y. W., Saines, P. J., Imai, M., Matsumoto, T., Michioka, C., Yoshimura, K., & Goh, S. K. (2015). Strong coupling superconductivity in the vicinity of the structural quantum critical point in $(Ca_xSr_{1-x})_3Rh_4Sn_{13}$. *Physical Review Letters, 115*, 207003/1–5.

Zhang, Y., Chen, J.-S., Ma, J., Ni, J., Imai, M., Michioka, C., Hadano, Y., Avila, M. A., Takabatake, T., & Yoshimura, K. (2018). Transitions from a Kondo-like diamagnetic insulator into a modulated ferromagnetic metal in $FeGa_{3-y}Ge_y$. *PNAS, 115*, 3273–3278.

Ziebeck, K. R. A., Capellmann, H., Brown, P. J., & Booth, J. G. Z. (1982). Fluctuations in both the ordered and paramagnetic phases of MnSi ! MnSi a heavy fermi liquid? *Zeitschrift Für Physik B, 48*, 241–250.

Kazuyoshi Yoshimura has been Professor of the Graduate School of Science of Kyoto University since 2002. He has been Director of the Research Center for Low Temperature and Material Sciences, Kyoto University since 2013, Director of the International Research Unit of Integrated Complex System Science, Kyoto University since 2015, and Vice Director of the Agency for Health, Safety and Environment, Kyoto University since 2016. Professor Yoshimura received a Ph.D. in engineering from Kyoto University in 1987. Since then, he has held teaching positions in the Faculty of Engineering, Fukui University from 1986 to1988 and in the Department of Chemistry, Faculty of Science, Kyoto University from 1988 to the present. He has consistently studied magnetism and superconductivity in transition-metal compounds and alloys at Kyoto University since 1980 as well as at the Institute for Solid State Physics, University of Tokyo (1983–2000), Technical University Wien (1993), Massachusetts Institute of Technology (1996 and 1998), and University of Illinois at Urbana-Champaign (1998).

Chapter 14
Quantum Size Effect Probed by NMR Measurements

Tomonori Okuno, Shunsaku Kitagawa, Kenji Ishida, Kohei Kusada, and Hiroshi Kitagawa

Abstract Nanoparticle systems are important in both applied and fundamental studies. In 1962, the theory of nanoparticles by R. Kubo predicted that the discrete energy levels in nanoparticles were realized and that their physical quantities behave differently from that of bulk at low temperatures where the thermal fluctuations fall down the energy spacing of the discrete energy levels. In addition, the behavior of physical quantities depends on the parity of the number of the electrons in a nanoparticle. This phenomenon is termed the quantum size effect (QSE). Although various studies on nanoparticles have been conducted to date, QSEs are not well separated from surface effects, which originate from the difference between surface and interior regions of nanoparticles. We succeeded in the separation of QSEs and surface effects, and found novel magnetic fluctuations related to QSEs in Pt nanoparticles. The magnetic fluctuations at low temperatures are not caused by surface effects or magnetic order, and their energy scale seems to be independent of electronic correlation. Further studies on such magnetic fluctuations are in progress.

Keywords Nano particle · Quantum size effect · Kubo effect · NMR · Knight shift

14.1 Introduction

14.1.1 Nanoparticles

Materials with a particle size of several hundred nanometers or less are called nanoparticles. In terms of application, nanoparticles are frequently used as catalysts (Hayashi et al., 2019) and as components for making nanostructures (Sakai et al.,

T. Okuno · S. Kitagawa · K. Ishida (✉)
Department of Physics, Graduate School of Science, Kyoto University, Kitashirakawa, Oiwake-Cho, Sakyo-ku, Kyoto 606-8502, Japan
e-mail: kishida@scphys.kyoto-u.ac.jp

K. Kusada · H. Kitagawa
Department of Chemistry, Graduate School of Science, Kyoto University, Kitashirakawa, Oiwake-Cho, Sakyo-ku, Kyoto 606-8502, Japan

K. Nishimura et al. (eds.), *Creative Complex Systems*, Creative Economy,
https://doi.org/10.1007/978-981-16-4457-3_14

2016). From a chemical viewpoint, it is certainly interesting when it is possible to synthesize metal alloys that do not mix the bulk with nanoparticles (Kusada et al., 2014) and those with crystal structures are not stable in the bulk (Kusada et al., 2013). From a physics viewpoint, as is described later, it is difficult to handle both theoretically and experimentally, and it can be considered to be an underdeveloped and interesting field. Nanoparticles can be located between atoms, molecules, and bulk materials. Theoretically, the atomic and molecular approaches have the multi-body difficulty symbolized by the proposition presented by Poincaré that "more than three-body problems are generally not exactly solvable." On the other hand, in bulk matter, which is an assembly of a large number of atoms on the scale of Avogadro's constant, the physical behavior of the system can be described "almost certainly" in the strict sense of probability theory, which is known as statistical mechanics developed by Boltzmann, Maxwell, Gibbs, and others. However, nanoparticles are a "halfway" multisystem and are difficult to handle statistically. As described in the next section, Kubo (1962) proposed a theory for a population of nanoparticles using probability theory. In recent years, the development of computers has made it possible to perform calculations on halfway multisystems (Nanba et al., 2017). Experimentally, it is considered that the theory has not been sufficiently verified because of problems such as the preparation of a good quality sample, an inability to measure contact, surface effects, and competition with superconductivity. In recent years, improvements in chemical synthesis technology and noncontact measurement have made it possible to measure high-quality samples, which now provides an opportunity to elucidate the physics of nanoparticles.

14.1.2 Quantum Size Effect

We consider the electrons in nanoparticles as particles in a simple spherically symmetric well-shaped potential using quantum mechanics. Each eigenenergy is discrete, and its eigenfunction has a large amplitude near the center and decays toward the edge of the well. The electrons in general nanoparticles are considered to be in a similar state. The effect caused by the discreteness of the intrinsic energy is called the "quantum size effect," and the effect caused by the difference in the wave function between the nanoparticle center and the surface layer is called the "surface effect." In ordinary bulk material, the quantum size effect results in a quasi-continuous energy band. The surface effect is not usually discussed as being negligible compared to bulk physical properties. As can be understood from this comparison, the phenomena peculiar to nanoparticles are the quantum size effect and the surface effect. In this chapter, we focus on the quantum size effect.

As mentioned in the previous section, it is difficult to study nanoparticles from a physical point of view, but such work had already begun in the 1960s. It is said that nanoparticle research began with the need for samples smaller than the penetration depth for nuclear magnetic resonance (NMR) experiments (Kobayashi, 1983) after the development of BCS theory in 1957 (Bardeen et al., 1957). After the theoretical points made by Kubo (1962), research began in earnest. However, there have been very few reports since the discovery of high-temperature superconductivity in 1986. Curiously, it can be said that the first study of nanoparticle physics began with superconductivity and ended with superconductivity.

The quantum size effect was first theoretically pointed out by Kubo (1962). For this reason, the quantum size effect is also called the Kubo effect. Statistical mechanics can be applied to multisystems of nanoparticles. In addition, nanoparticles are not completely uniform and have different boundary conditions, so the interval between one-electron energy levels Δ (named "spacing" in Kubo's paper) was considered to be a random variable. The average δ was estimated to be the reciprocal of the density of states $D(\varepsilon_F)$ at the Fermi energy ε_F per atom in the bulk material multiplied by the average atomic number N of the nanoparticles $\left[\delta = (ND(\varepsilon_F))^{-1}\right]$. It was shown that the effect of the discrete energy level does not appear at high temperature, and the quantum size effect appears at low temperature where the thermal fluctuation is smaller than the energy interval. It is interesting that the behavior of physical quantities at low temperature differs significantly depending on the evenness and oddness of the number of electrons in one nanoparticle. An example of magnetic susceptibility is easy to understand. When the number of electrons is odd, one electron spin configuration survives in the ground state. Therefore, it is considered that the Curie magnetic susceptibility is inversely proportional to the temperature at low temperature. On the other hand, in the case of even-numbered electrons, all electron spins are considered to be in the spin singlet state in the ground state, and their excitation requires energy to destroy the singlet. In other words, it becomes a spin gap. Therefore, the magnetic susceptibility decays exponentially at low temperatures.

Kubo considered the probability distribution of energy intervals to be a completely random Poisson process (energy corresponds to time) and assumed that it follows a Poisson distribution, but Gor'kov and Eliashberg (1965) pointed out that it does not follow the distribution due to the symmetry of the Hamiltonian. Given that a one-electron Hamiltonian includes the kinetic energy, Zeeman energy, and spin–orbit interaction, the temperature dependence of the power of physical quantities differs depending on which of these is dominant. Kawabata et al. (1966) discussed the plasmon mode, and Sone (1977) suggested that the quantum size effect on magnetic susceptibility is weakened by spin–orbit interaction. Shiba (1976) discussed the competition between the nanostructure and superconductivity.

Although various attempts have been made experimentally, here we introduce previous research on copper (Cu) nanoparticles. Magnetic susceptibility and NMR have been measured by Kobayashi et al. (1971). Unfortunately, it is considered that magnetic susceptibility cannot be measured properly because of surface oxidation. In the NMR spectrum, it was observed that the line width increased toward the high Knight shift side, which is considered to increase the Curie magnetic susceptibility of nanoparticles with an odd number of electrons. In addition, at the nuclear spin–lattice relaxation rate $1/T_1$, metallic contributions and temperature-independent constant terms were observed. The latter is thought to be the effect of spin diffusion due to surface oxides. Later, they reported that $1/T_1$ in other nanoparticles showed a much slower relaxation behavior than the bulk (Goto et al., 1989). In contrast, Yee and Knight (1975) reported that at low temperatures, the Knight shift was reduced compared to the bulk, and the reduction was greater for nanoparticles with smaller particle sizes. This is thought to be a decrease in the magnetic susceptibility of nanoparticles with an even number of electrons. It seems that the consistency of experiments by the two groups is not always good.

14.1.3 Previous Pt-NMR Studies

Platinum nanoparticles are promising as catalysts, and the surface effects have been energetically studied since the 1980s. On the other hand, there are few experiments focusing on the quantum size effect. Here we pick up several previous studies on the effects of surface effects and adsorption. Rhodes and coworkers investigated the surface effect using NMR spectroscopy (Rhodes et al., 1982a, b; Stokes et al., 1982). Specifically, the proportion of atoms on the surface (referred to as "dispersion" in the paper) was evaluated by hydrogen adsorption, and the dispersion dependence was investigated by NMR measurement. Regarding the NMR spectrum, it has been reported that the signal near Knight shift $K \sim 0$ increases as the dispersion increases (Rhodes et al., 1982a). It has also been reported that the nuclear spin–lattice relaxation time T_1 becomes longer where the Knight shift is small, but this cannot be explained by a simple Korringa law (Rhodes et al., 1982b). Nuclear spin–spin relaxation oscillates for two pulse intervals in the spin-echo measurements, attributed to different atomic coordination numbers on the surface (Stokes et al., 1982). It has been directly confirmed with the spin-echo double resonance experiment by Makowka and Slichter (1985) that the signal at $K \sim 0$ is a surface signal. Furthermore, changes in the NMR spectrum caused by adsorption of hydrogen and oxygen have also been discussed (Makowka and Slichter 1985; Bucher et al., 1989). It was shown that the local density of states is smaller toward the surface, and it is further reduced by adsorption.

14.2 Experimental Results and Discussion

Our present study focused on the quantum size effect and the discrete energy levels in the electron density of states, which are experimentally separated from the surface effect. Parts of our studies have already been published in the references (Okuno et al., 2020a, b).

14.2.1 Pt-Nanoparticle Sample

We prepared various particle-sized Pt nanoparticles smaller than 10 nm with surfaces covered with polyvinylpyrrolidone (PVP). The Pt nanoparticles with PVP were prepared by reduction of metal ions with alcohol under aqueous conditions without the requirement of any harmful organic reagent. About 3.0 mmol of K_2PtCl_4 was dissolved into deionized water. Then, PVP and reducing agent [ethylene glycol (EG) or triethylene glycol (TEG)] were added to the K_2PtCl_4 solution. The solution was heated to a known temperature for 1.5 h. The precursor solution then slowly changed to black, indicating the formation of Pt nanoparticles. After the reaction finished, the prepared nanoparticles were separated by centrifugation. Size control was achieved by adjusting the concentrations of the reagents and temperature used for the synthesis, as denoted in Table 14.1. Transmission electron microscopy (TEM) images of the PVP-coated nanoparticles are shown in Fig. 14.1a. From the TEM images, we estimated the average and variance of the diameter of the nanoparticle samples, which are shown in histograms in Fig. 14.1b. The distribution of the diameter was about

Table 14.1 Reaction conditions for the synthesis of Pt nanoparticles

Particle diameter (nm)	H_2O (ml)	Reducing agent (ml)	PVP (mmol)	Temperature (°C)
2.5	40	EG/100	9	120
4.0	40	TEG/125	9	120
7.4	3	TEG/75	9	200
9.8	1.5	TEG/75	15	200

PVP polyvinylpyrrolidone; *EG* ethylene glycol; *TEG* triethylene glycol

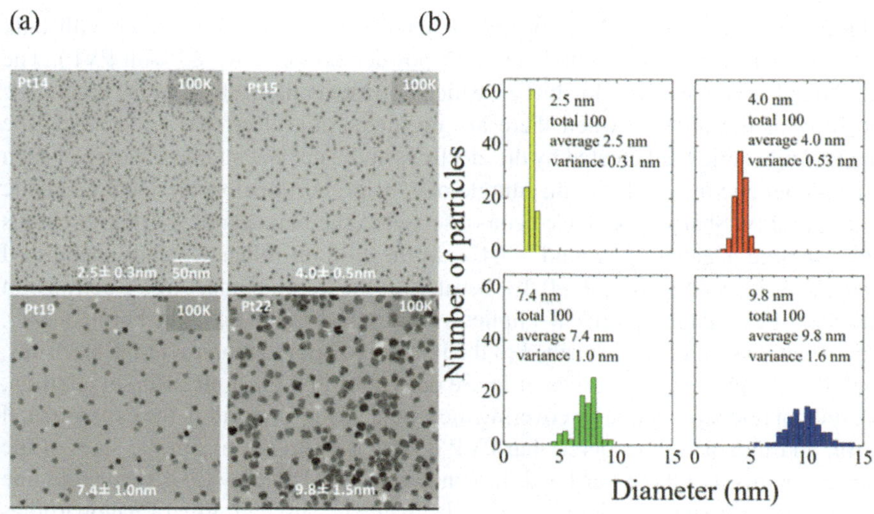

Fig. 14.1 **a** TEM image of the PVP-coated Pt nanoparticle. **b** Histogram of each nanoparticle

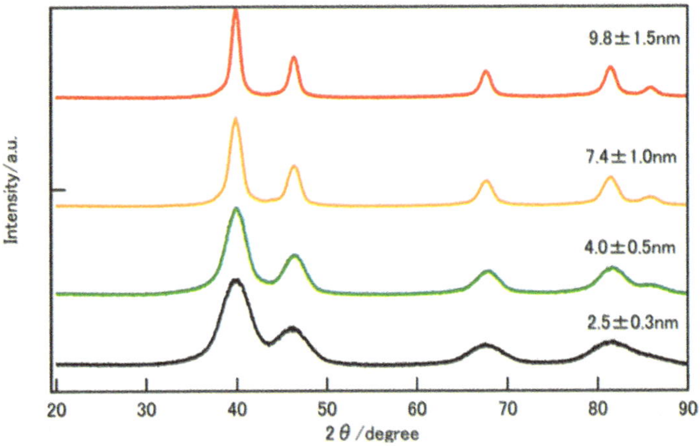

Fig. 14.2 XRD patterns of the nanoparticles

15% in these samples. To investigate the crystal structure, X-ray diffraction (XRD) measurements were performed at room temperature, and the results are shown in Fig. 14.2.

14.2.2 Pt-NMR Measurements

14.2.2.1 NMR Spectroscopy: Separation Between Surface and Interior of Nanoparticles

Figure 14.3 shows the NMR spectra of nanoparticle samples together with bulk Pt samples (a Pt powder sample and a Pt powder sample covered with PVP). The horizontal axis is the Knight shift K, which represents the shift from the resonance position of the nucleus when there are no interacting electrons, and reflects the internal magnetic field created by the electrons at the nuclear site. In the case of such a paramagnetic material, the magnitude of K is proportional to the local magnetic susceptibility. Sharp peaks have been observed in the bulk samples. Bulk Pt metals show a large negative K around -3.4%. In the case of ordinary metals such as Al and Cu, K is usually around $+0.2\%$ because of the Fermi contact of s electrons. In the case of Pt, a large negative K implies a large density of states at the Fermi energy. The negative sign of K is ascribed to the inner core polarization by Pt-5d electrons, and thus the physical properties of Pt are determined by the d electron. In addition, we did not find any change by covering the surface with PVP within the experimental error, and thus it is considered that PVP does not affect the electronic state of the surface atoms. On the other hand, in nanoparticles, a signal with a very wide line width was observed. This indicates that the magnetic susceptibility of nanoparticles is not uniform and that the local magnetic susceptibility is widely distributed. In

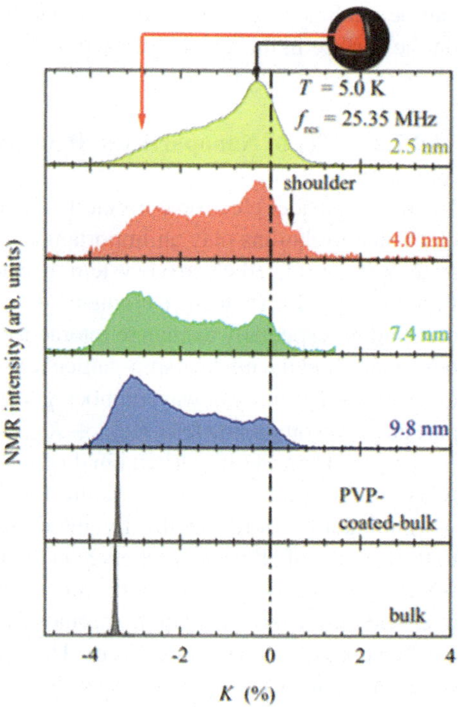

Fig. 14.3 NMR spectra measured at 5.0 K displayed against Knight shift on the horizontal axis. The red and black arrows represent the surface and internal signals. The arrow indicating the shoulder of the 4.0-nm sample is attributed to impurities [$H_2Pt(OH)_6$] (Okuno et al., 2020a)

addition, a signal of $K > 0$, which is shown as a shoulder in Fig. 14.3, was observed in the 4.0 nm sample, and is considered to be an impurity signal. In fact, when a 2.5 nm sample was accidentally left at room temperature, a corresponding signal appeared. It can be seen that the surface of the nanoparticles reacts very sensitively with the outside air, even though it is a precious metal Pt covered with PVP. The signal of this impurity has also been observed in previous studies and is thought to be $H2Pt(OH)_6$.

We demonstrate the separation of the NMR spectra from the surface and from the interior of the nanoparticles with the particle-size dependence of the NMR spectra. In general, as the particle size of nanoparticles becomes smaller, the number of atoms in the surface increases with respect to the number of nuclei inside. As seen in Fig. 14.3, as the particle size of the nanoparticles decreases, the intensity of the signal at $K \sim 0\%$ increases with respect to the signal at $K \sim -3\%$. From this dependence, the signal of $K \sim 0\%$ corresponds to the signal of the surface of the nanoparticles, and the signal at $K \sim -3\%$ corresponds to the signal of the interior of the nanoparticles. The size-dependent results and spectral identification are in good agreement with previous studies. From these results, it seems that the influence of PVP is negligibly small. The K reflects the local electron spin susceptibility or the local density of states, and thus can be simply thought of as reflecting the local electron number density. Therefore, the number of electrons is small in the surface of nanoparticles, because K is small, and the number of electrons is large in the interior of the nanoparticles. This can be qualitatively understood to reflect the small amplitude of the wavefunction on the surface in the spherically symmetric well model.

14.2.2.2 1/T$_1$ of Nanoparticles: Detecting the Quantum Size Effect

Next, we consider electron correlations in the Pt nanoparticles. We consider that the electron correlations play an important role in the Pt nanoparticles as in the case of a strongly correlated electron system, because the Pt-5*d* electron state resides at the Fermi energy. In NMR measurements of metals, parameter $\kappa(\alpha)$ expresses electron correlation (especially exchange interaction). This is obtained from the relationship between *K* and the nuclear spin–lattice relaxation rate $1/T_1$, because *K* reflects static ($\omega = 0$) susceptibility at wavenumber $q = 0$ component. This is separated into a spin part and an orbital part: $K = K_{\text{spin}} + K_{\text{orb}}$. In the present study, we determined K_{orb} $= +0.46\%$ from the impurity signal of insulating H2Pt(OH)$_6$. Nuclear spin–lattice relaxation rate divided by temperature $1/T_1T$ reflects the spin susceptibility with any wavenumber *q*. More specifically, the dynamical spin susceptibility is proportional to the average of all wavenumbers *q* on the Fermi surface of the imaginary part at the NMR resonance frequency (MHz order). Given that the NMR resonance frequency is overwhelmingly slower than the Larmor frequency (GHz order) of electronic spin, it can be approximated to almost zero. The exception is when the Larmor frequency of electronic spin is small, such as near magnetic order. The relationship between K_{spin} and $1/T_1T$ is expressed within the random-phase approximation as follows (Moriya, 1963; Narath & Weaver, 1968):

$$\frac{1}{T_1 T K_{\text{spin}}^2} = \frac{\hbar}{4\pi k_B}\left(\frac{\gamma_n}{\gamma_e}\right)^2 \kappa(\alpha), \kappa(\alpha) = \frac{(1-\alpha_0)^2}{\left\langle \left(1-\alpha_q\right)^2 \right\rangle_F},$$

where k_B is the Boltzmann constant, $\hbar$ is the reduced Planck constant, and γ_n and γ_e are the gyromagnetic ratios of the nucleus and electron, respectively. $\alpha\boldsymbol{q}$ is an electron-correlation constant of the wave number $\boldsymbol{q}$ normalized by the density of states. Also, $\langle \cdots \rangle_F$ means the average for all wave numbers connecting on the Fermi surface. Simply speaking, $\kappa(\alpha)$ is a parameter that indicates the strength of the ferromagnetic or antiferromagnetic electron correlation. If it is smaller (larger) than 1, it means that the electron correlation is ferromagnetic (antiferromagnetic). Let us explain the physical meaning of this relational expression. The static spin susceptibility is enhanced by ferromagnetic electron correlation. In addition, the imaginary part of the dynamical spin susceptibility is enhanced twice by antiferromagnetic electron correlation. Because the imaginary part of the dynamical susceptibility reflects energy dissipation, it is enhanced by both "friction" and "amplitude" components of electron correlations. By taking the ratio of these static and dynamical spin susceptibilities, information about the electron correlation $\kappa(\alpha)$ can be extracted. This equation is for metals, but it is considered that this equation holds even for nanoparticles at sufficiently high temperatures. The meaning of sufficiently high temperature will be described later.

Figure 14.4a shows $(1/T_1T)^{1/2}$ measured at each point of the spectrum at a sufficiently high temperature. While relaxation is fast inside the nanoparticles, it is slow toward the surface. This is consistent with the picture that the number of electrons is smaller at the surface, and relaxation is less likely to occur, but as will be described later, this alone cannot be explained quantitatively, and a modification of electron correlations is required. $(1/T_1T)^{1/2}$ at a sufficiently high temperature does not show a large size dependence. The connection to the bulk sample is also continuous. It is also consistent with previous studies. Figure. 14.4b shows the K dependence of $\kappa(\alpha)$ estimated with K_{spin} and $(1/T_1T)^{1/2}$. The κ (α) of bulk Pt is less than 1, indicating that bulk single Pt is close to ferromagnetism. It is considered that the ferromagnetic electronic correlation becomes weaker from the interior to the surface layer. Because $\kappa(\alpha)$ is proportional to ${K_{\text{spin}}}^{-2}$ as discussed above, there is a large error at K ~ 0. Therefore, $\kappa(\alpha) > 1$ may not be essential.

Figure 14.5 shows the temperature dependence of $1/T_1$ measured on the surface and interior of a 4.0-nm nanoparticle sample. The most important point to note is that the behavior on the surface and interior is almost the same (of course, as mentioned earlier, the values themselves are different). This indicates that the behavior of $1/T_1$ is not a surface effect. If this were a surface effect, anomalies would appear only in the behavior of $1/T_1$ of the signal from the surface, and $1/T_1$ of the signal from the

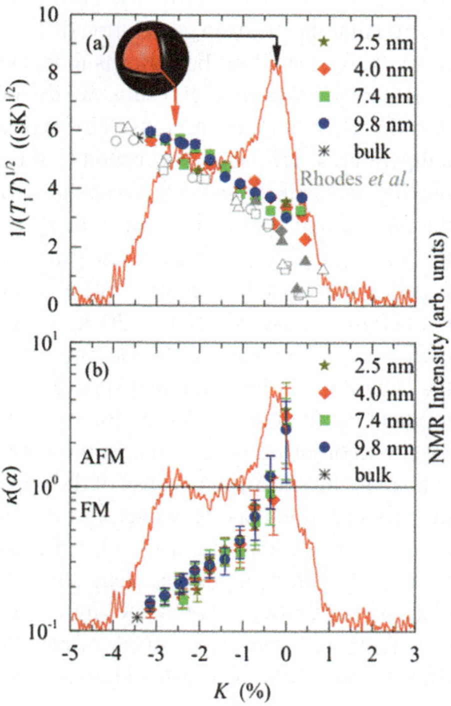

Fig. 14.4 **a** $(1/T_1T)^{1/2}$ measured in the metallic region. For reference, the spectrum of 4.0 nm is shown in the background. The gray symbol is a previous study. **b** Converted to $\kappa(\alpha)$ with K and $1/T_1T$ (see text; Okuno et al., 2020a)

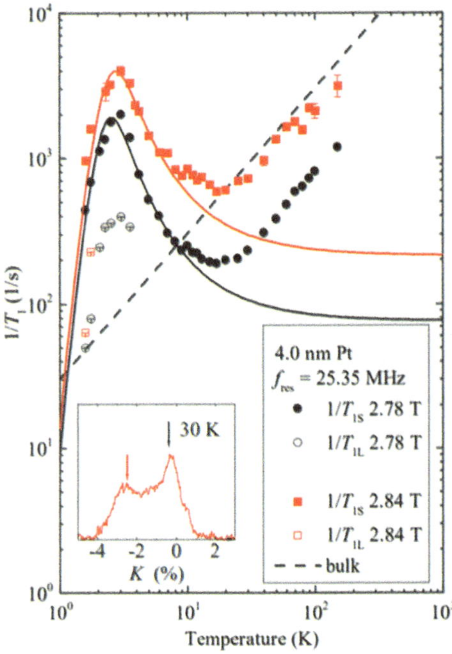

Fig. 14.5 Temperature dependence of $1/T_1$ measured on the surface and interior. From around the temperature at which $1/T_1$ shows the maximum value, $1/T_1$ has multiple components. The fastest and slowest components are displayed. The solid line is the fitting result by the BPP model (see the text). The dotted line is bulk. Inset: Arrows indicate the location of the $1/T_1$ measured spectrum (Okuno et al., 2020a).

interior would behave normally. Furthermore, when we combine with the analysis of $\kappa(\alpha)$ that the electron correlation differs between the nanoparticle surface and the inside, this anomalous behavior is independent of the electron correlation.

Based on experimental results, we divided the characteristics of the $1/T_1$ behavior into three temperature regions. This will be discussed in order below. (1) At high temperatures, $1/T_1$ is in proportion to the temperature. This is the behavior seen in metals, and bulk Pt sample also behaves in the same way. This is consistent with the quantum size effect, because the discrete energy levels are blurred by thermal fluctuation, and the behavior is the same as that of the bulk. This temperature region is called the "metallic region." We evaluated $\kappa(\alpha)$ with experimental values in this temperature range. (2) Below 30 K, $1/T_1$ begins to increase significantly. Such an increase can be considered as an increase in magnetic fluctuation. This is because the ^{195}Pt nucleus has a nuclear spin of 1/2 and does not have an electric quadrupole moment, so there is no electric interaction. Similar $1/T_1$ behavior is observed when magnetic ordering occurs, but in the case of magnetic ordering, there should be a change in the spectrum because of the appearance of an internal magnetic field from the ordered moments. However, in the case of nanoparticles, there is no change in spectrum at low temperatures within the range of experimental accuracy. In addition, the energy scale (temperature dependence) of the $1/T_1$ increase is independent of the electron correlation. (3) On cooling further, $1/T_1$ shows a peak and then decreases rapidly. In addition, the recovery of nuclear magnetization after saturation becomes a multicomponent from around the temperature at which $1/T_1$ reaches the maximum value at low temperatures. In Fig. 14.5, we show the fastest and slowest relaxation components. Given that the surface and the interior in the nanoparticle are separated by Knight shift, it is obvious that his multicomponent relaxation is not caused by the surface effect, but occurs at both the surface and in the interior of the nanoparticle.

The anomaly of $1/T_1$ is an interesting magnetic fluctuation, which is not a surface effect or magnetic order, and is peculiar to nanoparticles not found in the bulk. Predicting that this magnetic fluctuation is a quantum size effect, the size dependence of the nanoparticles of the magnetic fluctuation is very intriguing. Figure 14.6 shows the temperature dependence of $1/T_1$ measured by the interior signal for the samples of various particle sizes. The smaller the particle size, the higher the temperature below which magnetic fluctuation is enhanced. The degree of the enhancement at low temperature does not increase as the particle size decreases. We considered the temperature T^* that characterizes this anomaly to be the temperature at which $1/T_1$ deviates from the metallic behavior indicated by the arrow (Fig. 14.6). The inset of Fig. 14.6 is a plot of T^* against the reciprocal of particle size. The inset also shows $\delta/k_B = [k_B\, N\, D(\varepsilon_F)]^{-1}$, where δ is the mean value of the magnitude of the energy

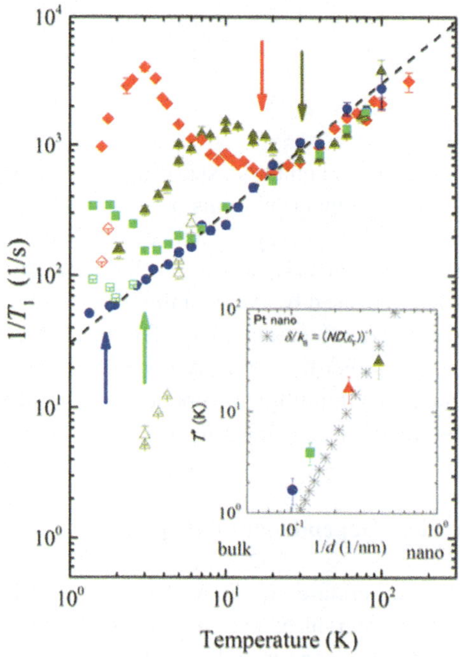

Fig. 14.6 The temperature dependence of $1/T_1$ was measured for samples of each particle size. $1/T_1$ was measured at the interior signal of the nanoparticles. Blue: 9.8 nm, green: 7.4 nm, red: 4.0 nm, yellow: 2.5 nm. The white data shows the slowest component of T_1 when $1/T_1$ is distributed at low temperature. The dotted line is $1/T_1$ of bulk Pt. Inset: Plot of the characteristic temperature T^* where $1/T_1$ deviates from the metallic behavior against the reciprocal of the particle size. Asterisk indicates the value of δ/k_B with respect to the radius of the nanoparticles. Because the radii of nanoparticles take discrete values, δ/k_B also take discrete values (Okuno et al., 2020a)

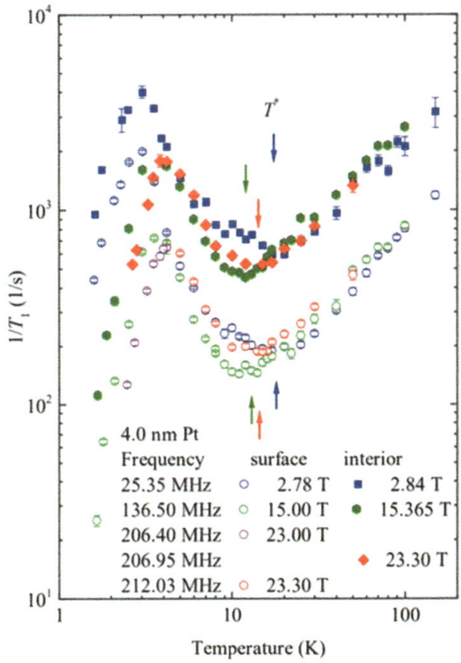

Fig. 14.7 Temperature dependence of $1/T_1$ measured up to a high magnetic field of 23.3 T. Arrow indicates the temperature at which $1/T_1$ deviates from metallic behavior (Okuno et al., 2020b)

interval predicted by the theory of quantum size effect converted to temperature. N is an approximate value of the number of atoms per nanoparticle, $D(\varepsilon_F)$ is the density of states of bulk Pt at the Fermi energy ε_F, and k_B is the Boltzmann constant. The characteristic temperature T^* and δ/k_B are very well scaled, strongly supporting that this magnetic fluctuation is caused by the quantum size effect. The enhancement in magnetic fluctuation as a size effect is a completely new phenomenon that has not been reported previously, neither theoretically nor experimentally. In addition, it is considered that the relaxation of multiple components was visible at low temperature because of the distribution of energy intervals of the quantum size effect.

14.2.2.3 Magnetic-Field Dependence of $1/T_1$

It is suggested that the temperature at which $1/T_1$ deviates from metallic behavior corresponds to the energy interval of the quantum size effect. To show this more directly, we investigated the magnetic field dependence of $1/T_1$. This is because we considered that magnetic field can tune the energy interval related to the size effect through the Zeeman effect. In other words, in even-numbered nanoparticles, the energy gap with the first excited state closes as the magnetic field increases. On the other hand, in odd-numbered nanoparticles, the energy gap with the first excited state opens. If the Zeeman energy is larger than the original energy interval, the opening and closing of the energy gap is considered to be reversed. Figure 14.7

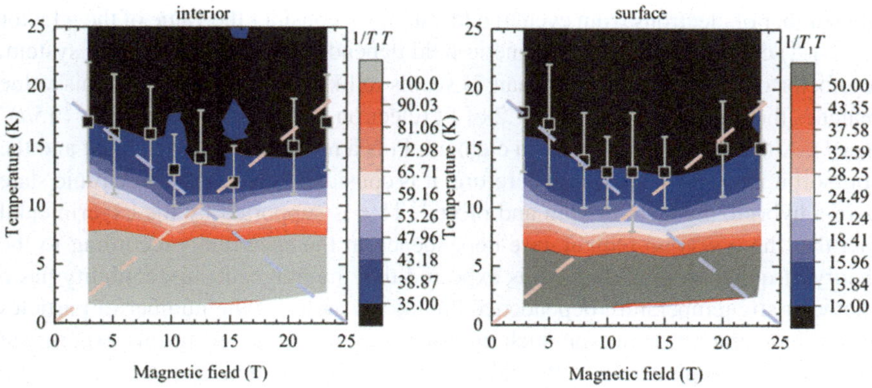

Fig. 14.8 Magnetic field dependence of the temperature at which the nuclear spin-lattice relaxation rate $1/T_1$ in the 4.0-nm nanoparticle deviates from metallic behavior. The color plot is the temperature and magnetic field dependence of $1/T_1T$. The dotted line is the magnetic field dependence of the interval d between two discrete energy levels through the Zeeman effect (light blue: $k_BT = \Delta$ - $g\mu_B\mu_0H$, light red: $k_BT = g\mu_B\mu_0H$), where $\Delta = 20$ K and $g = 1.2$ (Okuno et al., 2020b)

shows the temperature dependence of $1/T_1$ of the surface and interior of the nanoparticles measured in each magnetic field. Both show the fastest component of $1/T_1$. The measurement was performed up to 23.3 T using a cryo-free superconducting magnet at the Institute for Materials Research, Tohoku University. As the magnetic field increased, the temperature at which the behavior that deviated from metallic behavior decreased and then increased. On the other hand, the maximum temperature continued to increase. In NMR studies, magnetic fluctuations are observed through the NMR resonance frequency ($_\mathrm{N}$). Therefore, when the magnetic fluctuation slows down continuously, $1/T_1$ shows a maximum when the characteristic frequency of the magnetic fluctuation becomes equal to ~$_\mathrm{N}$. Therefore, the latter can be explained by the shift of the temperature at which $1/T_1$ takes the maximum to the high temperature side because $_\mathrm{N}$ increases in proportion to the magnetic field. On the other hand, the former is considered to be a quantum size effect. The magnetic field μ_0H dependence of the characteristic temperature T^*, below which $1/T1$ deviates from the metallic behavior, is plotted in Fig. 14.8. The blue dotted line shows the energy gap between the even-numbered electron nanoparticles and the first excited state ($k_BT = \Delta - g\mu_B\mu_0H$) as the magnetic field increases, and the red dotted line shows the increase in the magnetic field in the case of odd-numbered electrons. It shows how the energy gap with the first excited state becomes large ($k_BT = g\mu_B\mu_0H$). Here, μ_B is a Bohr magneton, and μ_0 is the magnetic permeability of vacuum. The level interval Δ was set to $\Delta/k_B = 20$ K, and the g factor was set to $g = 1.2$. For reference, the color plot of $1/T_1T$ is also displayed. The coincidence with these dotted lines supports the realization of discrete energy levels due to the quantum size effect. The smaller the nanoparticle size, the larger the gap should be, so a larger gap gives the faster component of $1/T_1$ from the size dependence of the $1/T_1$ nanoparticles. Therefore, it is considered that the nanoparticles that give a fast component of $1/T_1$ have changed the number of electrons from even to odd. Next, we consider the value of the g factor ($g = 1.2$) estimated from the magnetic field dependence. In the d-electron system, a g factor of 1.2 (much smaller than 2) seems unlikely, but interestingly this value matches the Landé g factor $g_J = 1.2$ of Pt (electron configuration [Xe] $(4f)^{14}$ $(5d)^9$ $(6\,s)^1$). Although Pt is a d-electron element, Pt is a relatively heavy element and the spin–orbit interaction is large. Therefore, it is considered that the Pt electronic state is in a hybridizing state of spin and orbital. This is supported by the experimental fact that there was no temperature dependence in the spectrum. According to the theory of quantum size effect, it is expected that the magnetic susceptibility has a characteristic temperature dependence due to the parity of the number of particles at low temperature. However, such an effect was not observed in this experiment because of the suppression of the effect by spin–orbit interaction.

14.2.2.4 Phenomenological Analysis of Low-Temperature Behavior of 1/T$_1$

We consider the low-temperature behavior of $1/T_1$ at low temperatures. We tried to phenomenologically explain this enhancement in $1/T_1$ by the Bloembergen-Purcell-Pound (BPP) model (Bloenbergen et al., 1948). This model was originally developed to explain the relaxation of nuclear magnetization during NMR analysis of molecules. In this model, nuclear-magnetization relaxation occurs due to fluctuations of the local magnetic field, and correlation of the fluctuating $H(t)$ was assumed to be expressed as $\langle H(t)H(0)\rangle = \langle V^2\rangle e^{-|t|/\tau}$ with a single exponential function of correlation time τ. In this case, $1/T_1$ is expressed as

$$\frac{1}{T_1} = \frac{2\langle V^2\rangle}{1 + (\omega_n \tau)^2}.$$

This analysis is effective as a first approximation for fluctuations in the local magnetic field derived from electrons. The constant of proportionality $\langle V^2\rangle$ gives an index of the magnitude of the hyperfine magnetic field. This model is valid when the local magnetic field fluctuates between the two levels, the activation energy of which is Δ, and in this case the inverse of the correlation time is expressed as $\tau^{-1} = \tau_0^{-1} e^{-\Delta/k_B T}$, where T is the temperature of the system. Analysis of $1/T_1$ by this BPP model was performed on a 4.0-nm sample. The solid lines in Fig. 14.5 are the fitting result and reproduce the behavior at low temperature very well. It is noted that the behavior at high temperatures is outside the scope of this model. This is because the BPP mode cannot be applied at a high temperature where a large number of levels are involved due to thermal fluctuation, but is valid at a sufficiently low temperature where only two energy levels are involved in the thermal fluctuation. Interestingly, the activation energy Δ/k_B in each diameter nanoparticle is close to the energy interval of the quantum size effect. However, there are problems that cannot be explained within this model. For example, the magnitude of the hyperfine magnetic field estimated from $\langle V^2\rangle$ is too small compared to that due to inner core polarization. In addition, the result of the 2.5-nm nanoparticle cannot be explained because the peak structure of $1/T_1$ is too wide. Understanding the relaxation process at low temperatures where $1/T_1$ deviates from the metallic behavior will be an important future task.

14.3 Summary

Nanoparticle systems are important research subjects in both basic and applied research. From the viewpoint of physics, it is considered to be an underdeveloped area. Although theoretical research has progressed considerably in the latter half of the twentieth century, experiments are still in progress. We performed NMR measurements on Pt nanoparticles to observe the intrinsic size effect separated from the surface effect. We succeeded in separating the surface effect and the size effect, and discovered a novel magnetic fluctuation derived from the diameter-size dependence of $1/T_1$. The magnetic fluctuations related to the size effect have a remarkable feature that it hardly depends on electronic density of states or electron correlation. Further theoretical and experimental studies are desired to elucidate the origin of the magnetic fluctuations and the relaxation mechanism of $1/T_1$ of nanoparticles.

Acknowledgements NMR measurements conducted at higher field than 15 T were part of a joint research program with Assistant Professor Tomohiro Hirata and Professor Takahiko Sasaki of the Institute for Materials Research, Tohoku University. Mr. Satoshi Matsuzaki and Mr. Yuta Kinoshita provided assistance in the experiments. The authors thank Shingo Yonezawa and Yoshiteru Maeno for valuable discussions. This research was supported by the Toyota Physical and Chemical Research Institute. The authors thank the staff at the LTM Center for supplying the cryogen, which is indispensable for the experiments.

References

Bardeen, J., Cooper, L., & Schrieffer, J. R. (1957). *Physical Review, 108*, 1175.
Bloenbergen, N., Purcell, E. M., & Pound, R. V. (1948). *Physical Review, 73*, 679.
Bucher, J. P., Buttet, J., & Van Der Klink, J. J. (1989). *Surface Science, 214*, 347.
Gor'kov, L. P., & Eliashberg, G. M. (1965). *Soviet Phyics. JTEP, 21*, 1407.
Goto, T., Komoi, F., & Kobayashi, S.-I. (1989). *Journal of the Physical Society of Japan, 58*, 3788.
Hayashi, E., Yamaguchi, Y., Kamata, K., Tsunoda, N., Kumagai, Y., Oba, F., & Hara, M. (2019). *Journal of the American Chemical Society, 141*, 890.
Kawabata, A., & Kubo, R. (1966). *Journal of the Physical Society of Japan, 21*, 1765.
Kobayashi, S., Takahashi, T., & Sasaki, W. (1971). *Journal of the Physical Society of Japan, 31*, 1442.
Kobayashi, S.-I. (1983). Bussei Kenkyu 39, 197.
Kubo, R. (1962). *Journal of the Physical Society of Japan, 17*, 1975.
Kusada, K., Kobayashi, H., Yamamoto, T., Matsumura, S., Sumi, N., Sato, K., Nagaoka, K., Kubota, Y., & Kitagawa, H. (2013). *Journal of the American Chemical Society, 135*, 5493.

Kusada, K., Kobayashi, H., Ikeda, R., Kubota, Y., Takata, M., Toh, S., Yamamoto, T., Matsumura, S., Sumi, N., Sato, K., Nagaoka, K., & Kitagawa, H. (2014). *Journal of the American Chemical Society, 136*, 1864.
Makowka, C. D., & Slichter, C. P. (1985). *Physical Review B, 31*, 5663.
Moriya, T. (1963). *Journal of the Physical Society of Japan, 18*, 516.
Nanba, Y., Ishimoto, T., & Koyama, M. (2017). *Journal of Physical Chemistry C, 121*, 27445.
Narath, A., & Weaver, H. T. (1968). *Physical Review, 175*, 373.
Okuno, T., Manago, M., Kitagawa, S., Ishida, K., Kusada, K., & Kitagawa, H. (2020a). *Physical Review B, 101*, 121406 (R).
Okuno, T., Kinoshita, Y., Matsuzaki, S., Kitagawa, S., Ishida, K., Hirata, M., Sasaki, T., Kusada, K., & Kitagawa, H. (2020b). *Journal of Physical Society Japan, 89*, 095002.
Rhodes, H. E., Wang, P.-K., Stokes, H. T., Slichter, C. P., & Sinfelt, J. H. (1982a). *Physical Review B, 26*, 3559.
Rhodes, H. E., Wang, P.-K., Makowka, C. D., Rudaz, S. L., Stokes, H. T., Slichter, C. P., & Sinfelt, J. H. (1982b). *Physical Review B, 26*, 3569.
Sakai, D., Harada, K., Hara, Y., Ikeda, H., Funatsu, S., Uraji, K., Suzuki, T., Yamamoto, Y., Yamamoto, K., Ikutame, N., Kawaguchi, K., Kaiju, H., & Nishii, J. (2016). *Science and Reports, 6*, 27767.
Shiba, H. (1976). *Journal of Low Temperature Physics, 22*, 105.
Sone, J. (1977). *Journal of the Physical Society of Japan, 42*, 1457.
Stokes, H. T., Rhodes, H. E., Wang, P.-K., Slichter, C. P., & Sinfelt, J. H. (1982). *Physical Review B, 26*, 3575.
Yee, P., & Knight, W. (1975). *Physical Review B, 11*, 3261.

Tomonori Okuno was a former graduate student. He obtained his Master degree from Kyoto University in 2019.

Shunsaku Kitagawa has been Assistant Professor of Graduate School of Science of Kyoto University since 2016. He obtained his Ph.D. degree from Kyoto University in 2012.

Kenji Ishida has been Professor of Graduate School of Science of Kyoto University since 2007. He obtained his Ph.D. degree from Osaka University in 1991.

Kohei Kusada has been Associate Professor of Graduate School of Science of Kyoto University since 2021. He obtained his Ph.D. degree from Kyoto University in 2014.

Hiroshi Kitagawa has been Professor of Graduate School of Science of Kyoto University since 2009. He obtained his Ph.D. degree from Kyoto University in 1990.

Chapter 15
Recent Topics on Organic Spin Liquid Candidates

Mitsuhiko Maesato

Abstract This chapter discusses the use of organic conductors as a platform for investigating a variety of phenomena such as (super)conductivity, metal–insulator transitions, magnetism, and especially spin liquids. Molecular and crystal structures of organic conductors look rather complicated, but their electronic structures can be described simply by considering the frontier orbitals of the constituting molecules. Apart from the simple electronic structure, unconventional physical properties occur because of many-body interactions. After a brief introduction of the history of organic conductors, the chapter provides a discussion of recent topics on organic spin liquid candidates having triangular lattice.

Keywords Organic conductors · Superconducors · Mott insulators · Spin liquids · Spin frustration · Solid-state physics

15.1 Brief History of Organic Conductors

Solid materials based on organic molecules were considered as insulators until the discovery of organic semiconductors. Akamatu et al. (1954) discovered a large increase in the conductivity of perylene, an aromatic hydrocarbon, by doping the solid with iodine (Fig. 15.1). Although iodine-doped perylene was not stable, this work opened the door to the exploration of organic conductors and superconductors. To obtain stable organic conductors, the development of molecules that are stable against oxidation or reduction was required. A number of radical cation and anion molecules have been investigated to date. Among them, donor molecule tetrathiafulvalene (TTF) and accepter molecule 7,7,8,8-tetracyanoquinodimethane (TCNQ) have emerged as key molecules for the development of organic conductors. Indeed, TTF-TCNQ was the first "organic metal" reported in 1973 (Coleman et al., 1973; Ferraris et al., 1973). The segregated stacks of TTF and TCNQ molecules and charge

M. Maesato (✉)
Department of Chemistry, Graduate School of Science, Kyoto University, Kitashirakawa, Oiwake-Cho, Sakyo-ku, Kyoto 606-8502, Japan
e-mail: maesato@kuchem.kyoto-u.ac.jp

K. Nishimura et al. (eds.), *Creative Complex Systems*, Creative Economy,
https://doi.org/10.1007/978-981-16-4457-3_15

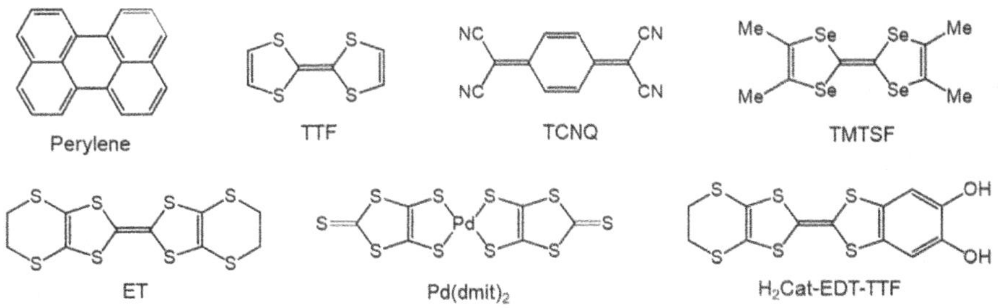

Fig. 15.1 Molecular structures

transfer between them gave rise to highly one-dimensional (1D) conducting properties. This charge-transfer salt underdoes a metal–insulator transition at 59 K due to Peierls instability (Conwell, 1988; Jérome & Schulz, 1982). The charge-density wave (CDW) ground state was confirmed experimentally.

To prevent the Peierls transition at low temperature in 1D conductors, the extension of dimensionality of electronic states was required. The radical cation salts based on tetramethyltetraselenafulvalene (TMTSF) was developed by Bechgaard et al. (1979), where the substitution of sulfur with selenium in the TTF skeleton enhanced the side-by-side interaction between donor molecules besides the π-π interactions along the direction perpendicular to the molecular plane. The quasi-1D conductors $(TMTSF)_2X$ showed metallic behavior down to about 10 K (Bechgaard et al., 1979). The nesting instability of the Fermi surface and electron–electron interaction lead to a spin-density-wave (SDW) transition below about 10 K (Bechgaard et al., 1979; Jérome, 1991; Jérome & Schulz, 1982). Notably, the first organic superconductor was discovered in $(TMTSF)_2PF_6$ by suppressing the SDW under pressure with a superconducting critical temperature T_C of about 1 K (Jérome et al., 1980).

Besides the heavy atom substitution, many chemists investigated peripheral additions of alkylchalcogeno groups to the TTF skeleton. One of the successful examples was bis-(ethylenedithio)tetrathiafulvalene (BEDT-TTF or ET) because a variety of two-dimensional (2D) conductors were developed using the ET molecule. A metallic behavior was observed in $(ET)_2ClO_4(TCE)_{0.5}$ down to 1.4 K without applying external pressure (Saito et al., 1982). $(ET)_2I_3$ exhibited a superconducting transition at 8 K under a moderate pressure (Murata et al., 1985). Using polyanion, a high superconducting transition above 10 K was found in κ-$(ET)_2Cu(NCS)_2$ (Urayama et al., 1988).

15.2 Introduction of *κ*-Type ET Salts

ET-based organic conductors have attracted much attention because of the variety of molecular arrangements, electronic states, and charge dynamics (Dressel & Tomić, 2020; Ishiguro et al., 1998; Saito & Yoshida, 2007). Owing to many chalcogen–chalcogen intermolecular interactions between neighboring ET molecules, the ET molecule tends to form 2D conducting sheets in between the insulating counter anion layers (Fig. 15.2a). These layers are stacked alternately to form layered organic conductors. Many kinds of molecular arrangements within the conducting donor layer have been reported to date. To distinguish them, each arrangement is labeled using Greek characters, such as α, β, θ, κ, etc. (Fig. 15.2b).

Although the crystal structures of organic conductors are rather complicated, their band structures are rather simple. This is because the bands derived from frontier orbitals are well separated from other bands. Actually, first-principles band structure calculations confirmed that the conduction band derived from the highest occupied molecular orbital (HOMO) of ET molecule is well separated from other bands (Nakamura et al., 2009). The semiempirical extended Hückel method combined with a tight binding approximation has been used as an efficient way of evaluating band structures of organic conductors (Mori et al., 1984). Transfer integrals t were evaluated from the intermolecular overlap integrals S of the HOMO by simple multiplication with a constant value E ($t = ES$). The calculated band structures and Fermi surfaces with a tight binding approximation reasonably explained the experimental observations in most cases.

The 2:1 salt $(ET)_2X$ composed of ET and anion X^- is essentially a 3/4-filled band system. When dimerization of ET is strong, it can be regarded effectively as a 1/2-filled system. In the case of a 3/4-filled system like θ-$(ET)_2X$, the intersite Coulomb interaction V plays an important role in determining their electronic states. The charge-ordered insulating state was systematically investigated in θ-$(ET)_2X$ (Mori et al., 1998). The frustration of charge gave rise to charge glass (Sasaki et al., 2017; Sato et al., 2017). α-$(ET)_2X$ shows a variety of electronic states such as charge-ordered, density wave, and superconducting states in addition to a metallic state. Recently, by suppressing a charge-ordered state, a Dirac fermion was discovered in α-$(ET)_2I_3$. This system has since attracted much attention as a bulk Dirac fermion system (Kajita et al., 2014).

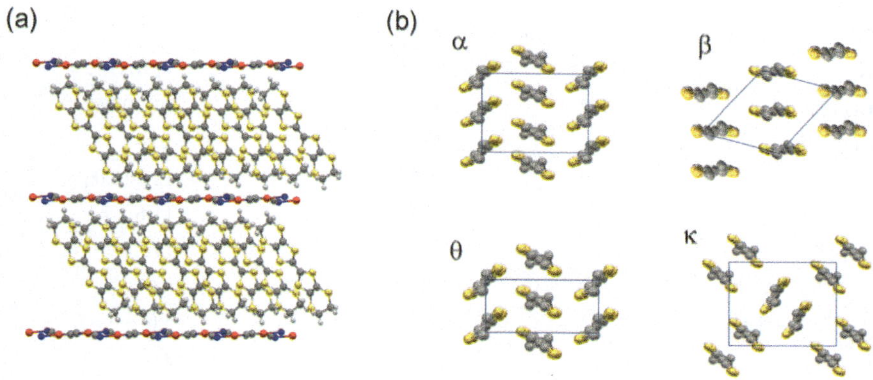

Fig. 15.2 **a** Layered structure of $(ET)_2X$ and **b** arrangement of ET molecules within the conducting plane

ET molecules form dimers in β-$(ET)_2X$ and κ-$(ET)_2X$ salts, which are regarded effectively as 1/2-filled band systems. κ-$(ET)_2X$ is one of the most studied organic conductors because of the high-T_C superconductivity, Mott transition, and spin-liquid behavior. In the κ-type arrangement, $(ET)_2{}^{\bullet+}$ dimers are arranged nearly orthogonally as shown in Figs. 15.2b and 15.3a. Since the discovery of 10 K-class high-T_C superconductor κ-$(ET)_2Cu(NCS)_2$ (Urayama et al., 1988), a series of κ-$(ET)_2X$ has been studied extensively. The κ-$(ET)_2X$ salts show a variety of ground states depending on the counter anions X such as metallic, superconducting, and antiferromagnetic insulating states (Fig. 15.3c). For example, κ-$(ET)_2Cu[N(CN)_2]Br$ shows a superconducting transition at 11.8 K (Kini et al., 1990), while κ-$(ET)_2Cu[N(CN)_2]Cl$ is an antiferromagnetic insulator with a Neel temperature T_N ~ 27 K (Miyagawa et al., 1995). Interestingly, the latter shows an insulator–superconductor transition by the application of pressure at low temperature (Table 15.1). While the band structure calculation gives 2D Fermi surfaces, some compounds are insulators, indicating

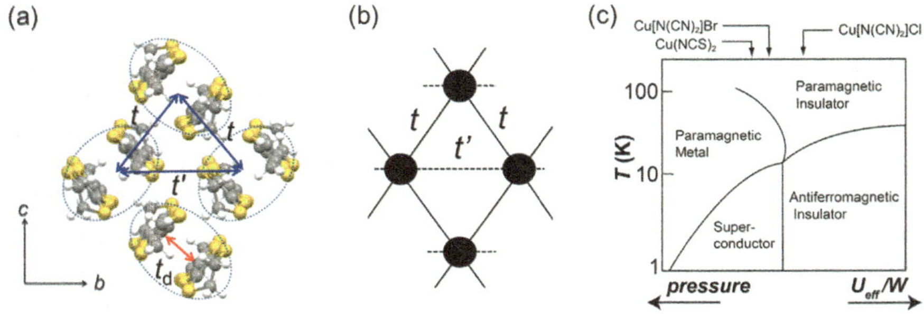

Fig. 15.3 **a** κ-Type arrangement of ET in $(ET)_2Cu_2(CN)_3$. Arrows represent interdimer transfer integrals (t and t') and intradimer transfer (t_d). **b** Triangular lattice in a dimer model. **c** Temperature–pressure (T–P) phase diagram of κ-$(ET)_2X$ salts

Table 15.1 Parameters at 100 K and maximum T_C of κ-$(ET)_2X$ salts

X	U_{eff} (eV)	W (eV)	t'/t	T_C (K)
$Cu_2(CN)_3$	0.483	0.480	1.074	3.9 (0.06)
$Cu(NCS)_2$	0.520	0.555	0.803	10.4
$Cu[N(CN)_2]Cl$	0.549	0.607	0.727	12.8 (0.03)
$Cu[N(CN)_2]Br$	0.531	0.593	0.672	11.6
$Cu(CN)[N(CN)_2]$	0.503	0.553	0.643	11.2
$Ag(CN)[N(CN)_2]$	0.482	0.544	0.615	6.6
$Ag(CN)_2 \cdot H_2O$	0.525	0.584	0.609	5.0
I_3	0.536	0.660	0.541	3.6

The on-site Coulomb energy on an ET dimer U_{eff} is estimated by the intradimer transfer energy as U_{eff} ~ 2 t_d. W is the bandwidth of the upper HOMO band. t'/t is the ratio of interdimer transfer energies based on a dimer model. The numerical values given in the parentheses are applied pressures (in GPa) for the maximum T_C.

the importance of electron correlation. The effect of on-site Coulomb interaction was examined theoretically within the Hartree–Fock approximation by Kino and Fukuyama (1995). They clarified the important role of the Coulomb interaction for the realization of insulating (antiferromagnetic) ground state and metal–insulator transition in κ-$(ET)_2X$. Based on this, it has been understood that the ratio of the effective on-site Coulomb interaction U_{eff} on a dimer to the band width U_{eff}/W dominates the electronic states in this system (Fig. 15.3c) (Kanoda,). In Fig. 15.3c, the superconductivity zone sits adjacent to the antiferromagnetic state, suggesting the importance of spin fluctuation on high-T_C superconductivity (Ishiguro et al., 1998; Kanoda, 1997a, 1997b, 2006).

Table 15.1 shows the band parameters of some κ-$(ET)_2X$ salts calculated by extended Hückel method combined with a tight binding approximation as well as U_{eff} and maximum T_C. Based on a dimer model, κ-$(ET)_2X$ salts are regarded as triangular lattice systems as shown in Fig. 15.3b. The t'/t ratio is a ratio of interdimer transfer energies, and a measure of spin frustration. Most κ-$(ET)_2X$ salts have highly anisotropic triangular lattice systems. However, t'/t is close to 1 in κ-$(ET)_2Cu_2(CN)_3$, suggesting strong spin frustration. It is noteworthy that first principles calculations tend to give smaller t'/t values (Hiramatsu et al., 2017; Kandpal et al., 2009; Koretsune & Hotta, 2014; Nakamura et al., 2009). Exotic physical properties caused by spin frustration are discussed in the following section.

15.3 Spin-Liquid Behavior and Superconductivity in κ-Type ET Salts

In 1973, Philip W. Anderson theoretically proposed a quantum spin liquid state in strongly frustrated spin systems (Anderson, 1973). As a representative of frustrated spins, he considered a triangular spin lattice with antiferromagnetic interactions. He proposed a resonating valence bond (RVB) state, which is described by a quantum mechanical superposition of singlet pairs. However, the search for spin liquids in real materials remains challenging.

κ-$(ET)_2Cu_2(CN)_3$ has attracted much attention as a strong candidate as a quantum spin liquid with triangular lattice (Shimizu et al. 2003b). It was first synthesized in 1991 (Bu et al., 1991; Geiser et al., 1991). Semiconducting behavior, hydrostatic pressure-induced insulator–metal transition (Mott transition), and superconducting transitions were reported (Geiser et al., 1991). An electron spin resonance (ESR) study indicated no magnetic transition down to 1.7 K (Komatsu et al., 1996). It is interesting that its T_C was low (T_C = 3.9 K) compared with other 10 K superconducting κ-ET salts (Komatsu et al., 1996). At that time, the so-called κ'-$(ET)_2Cu_2(CN)_3$ was also reported (Komatsu et al., , 1991, 1996). Even though the crystal structure was very similar to κ-$(ET)_2Cu_2(CN)_3$, κ'-$(ET)_2Cu_2(CN)_3$ showed a superconducting transition at ambient pressure. The ESR measurement revealed the contamination of Cu^{2+} ions in κ'-$(ET)_2Cu_2(CN)_3$. Besides carrier doping by the contamination of Cu^{2+} instead of Cu^{1+}, detailed angle-dependent Raman measurements confirmed the contamination of $N(CN)_2^-$ anion (Drozdova et al., 2001). This is because the $N(CN)_2^-$ anion was used for the synthesis of κ'-$(ET)_2Cu_2(CN)_3$. Therefore, the chemical formula of κ'-$(ET)_2Cu_2(CN)_3$ was regarded as κ-$(ET)_2Cu^{1+}{}_{(2-x-y)}Cu^{2+}{}_x(CN)_{(3-2y)}[N(CN)_2]_y$. Such a complicated situation probably hindered the detailed investigation of κ-$(ET)_2Cu_2(CN)_3$ in the early stages.

To understand the origin of the low T_C, we performed a uniaxial strain study of κ-$(ET)_2Cu_2(CN)_3$ (Shimizu et al., 2003a, 2011). It is expected that the anisotropy of interdimer interactions can be modulated by the in-plane uniaxial strain. Given that the ET dimers are tilted about ±45 degrees from the in-plane crystallographic axes in κ-ET salts, each dimer can remain equivalent under the strain. We found that κ-$(ET)_2Cu_2(CN)_3$ shows higher T_C under the in-plane uniaxial strain than that under hydrostatic pressure (Shimizu et al. 2003a, 2011). This fact motivated us to explore the magnetic ground state of the Mott insulator κ-$(ET)_2Cu_2(CN)_3$ at ambient pressure.

We measured the static magnetic susceptibility χ of κ-$(ET)_2Cu_2(CN)_3$ by a superconducting quantum interference device (SQUID). The temperature dependence of χ was well described by the $S = 1/2$ Heisenberg antiferromagnetic triangular-lattice model with an exchange interaction energy of $J/k_B \sim 250$ K (Shimizu et al. 2003b). Despite the large J, no magnetic transition was observed down to 1.9 K. Further investigation was performed by nuclear magnetic resonance (NMR) measurements. ^{1}H NMR is a highly sensitive probe of magnetic order because the ET molecule has eight hydrogens at the terminal. However, we observed neither broadening or splitting in ^{1}H NMR spectra of κ-$(ET)_2Cu_2(CN)_3$ down to 32 mK, which is about three orders of magnitude smaller than the exchange interaction ($J/k_B \sim 250$ K) of localized $S = 1/2$ spins on the dimers. The experimental results suggested a quantum spin liquid ground state because of strong spin frustration in the nearly regular triangular lattice (Shimizu et al. 2003b).

Low-temperature heat capacity measurement revealed a T-linear term of heat capacity in κ-$(ET)_2Cu_2(CN)_3$ (Yamashita et al., 2008). The T-linear term appears normally in metals (Fermi liquids) but is absent in insulators without mobile electrons. This abnormal heat capacity suggested gapless fermionic low-lying excitations of frustrated spins. The possible existence of spinon Fermi surface was discussed, but it has not been observed experimentally to date. On the other hand, however, the thermal conductivity showed an activation behavior at low temperature indicative of gapped excitation (Yamashita et al., 2009). The origin of this discrepancy remains an open question.

There are other problematic issues such as the presence of C/N disorder in the anion layer (Fig. 15.4a) and a relaxor-like dielectric response in κ-$(ET)_2Cu_2(CN)_3$. In the early stages, the crystal structure was refined with $P2_1/c$ space group, where an inversion center was assumed in the center of the C–N bond as a result of C/N disorder. However, strictly speaking, the space group is wrong and the effects of symmetry lowering on, for example, lattice vibrations have been investigated (Dressel et al.,

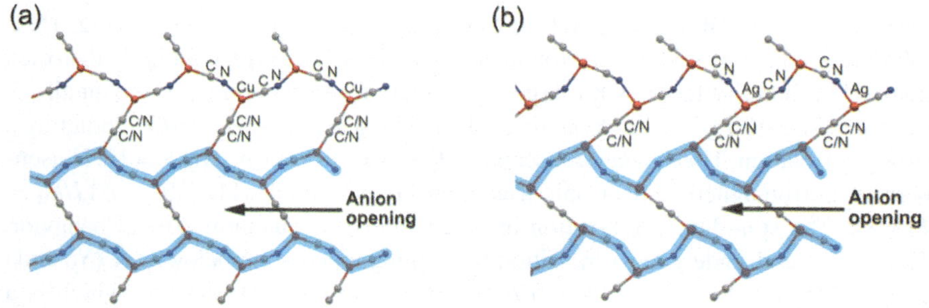

Fig. 15.4 Structure of anion layer in **a** κ-$(ET)_2Cu_2(CN)_3$ and **b** κ-$(ET)_2Ag_2(CN)_3$

2016; Foury-Leylekian et al., 2018; Ilakovac et al., 2017). An anomalous relaxor-like dielectric response was observed in κ-$(ET)_2Cu_2(CN)_3$ as with other dimer-Mott insulators (Abdel-Jawad et al., 2010). Theoretical models such as quantum-electric dipole and paired-electron crystal have been proposed based on the assumption of intradimer charge disproportionation because of intersite Coulomb interaction (Clay et al., 2012; Dayal et al., 2011; Gomi et al., 2013; Hotta, 2010, 2012; Li et al., 2010; Naka & Ishihara, 2010). However, the signature of charge disproportionation has not been observed experimentally.

There is also a puzzling anomaly at 6 K that appears in measurements such as NMR spin–lattice relaxation rate (Shimizu et al., 2003b), heat capacity (Yamashita et al., 2008), thermal conductivity (Yamashita et al., 2009), thermal expansion (Manna et al., 2010), and ultrasonic wave propagation (Poirier et al., 2014). These results suggest a second-order phase transition at 6 K. Very recently, a multifrequency ESR study reported a rapid decrease of spin susceptibility below 6 K, indicating an opening of spin gap (Miksch et al., 2021). Given that the onset of spin gap is associated with lattice softening (Poirier et al., 2014), the authors concluded that the ground state of κ-$(ET)_2Cu_2(CN)_3$ is a broken-symmetry state, namely, a valence bond solid (VBS) state (Miksch et al., 2021). There remains a question why the spin gap behavior does not appear in the static magnetic susceptibility and torque measurements. The ESR detected an additional new signal at low temperature that was attributed to unpaired defect spins. The localized unpaired spins were considered to interact with Cu^{2+} impurity spins via dipole–dipole interaction. The impurity effect is distinct from that in the 1D spin-Peierls system where a small impurity induces an antiferromagnetic order or a staggered moment around the doped site (Fukuyama et al., 1996; Regnault et al., 1995). A recent NMR study also suggests that the low-temperature NMR properties are dominated by extrinsic contribution of impurity spins in the frustrated spin systems (Pustogow et al., 2020).

It is crucial to obtain the unified phase diagram of κ-$(ET)_2X$ to fully understand the nature of strongly correlated electrons and the exotic phenomena in organic conductors. Besides the strength of the Coulomb interaction relative to bandwidth U_{eff}/W, the degree of anisotropy of lattice (t'/t) is an important parameter to determine the ground states of κ-$(ET)_2X$. Therefore, it is desirable to synthesize new κ-$(ET)_2X$ salts with different U_{eff}/W (or U/t) and t'/t toward the unified phase diagram (Fig. 15.5) (Yoshida et al., 2015). The new Mott insulator κ-$(ET)_2B(CN)_4$ has a highly distorted quasi 1D triangular lattice with large t'/t (~1.8), in sharp contrast to the small t'/t (<1) in most κ-$(ET)_2X$ salts (Yoshida et al., 2015). It also has a large U/t, indicting a strong correlation. The magnetic susceptibility was well described by $S = 1/2$ Heisenberg antiferromagnetic anisotropic triangular lattice model with $J'/J = 2$ ($J/k_B =$ 118 K). The spin–lattice relaxation in NMR indicated quantum critical behavior. This compound undergoes a transition to a spin-gapped state below 5 K (Yoshida et al., 2015). In the limit of large t'/t ($= \infty$) is a purely 1D system, which is a well-known Tomonaga-Luttinger liquid. However, the intermediate t'/t (>1) region is largely underexplored. In this sense, the result of κ-$(ET)_2B(CN)_4$ is important.

By substituting Cu with Ag, new spin liquid candidate κ-$(ET)_2Ag_2(CN)_3$ was synthesized by Hiramatsu and coworkers (Hiramatsu et al., 2017; Shimizu et al., 2016). The anion network of the insulating layer in κ-$(ET)_2Ag_2(CN)_3$ is similar to that of κ-$(ET)_2Cu_2(CN)_3$, although the shape of anion opening is different slightly as shown in Fig. 15.4. The C/N disorder still exists in κ-$(ET)_2Ag_2(CN)_3$ (Fig. 15.4b). Given that the ionic radius of Ag(I) (1.29 Å for a six-coordinate system) is larger than that of Cu(I) (0.91 Å for a six-coordinate system), lattice expansion occurs. In other words, negative chemical pressure can be applied. In fact, the critical pressure of Mott transition P_C of κ-$(ET)_2Ag_2(CN)_3$ ($P_C = 0.95$ GPa) is higher than that of κ-$(ET)_2Cu_2(CN)_3$ ($P_C = 0.36$ GPa). The critical superconducting temperature T_C

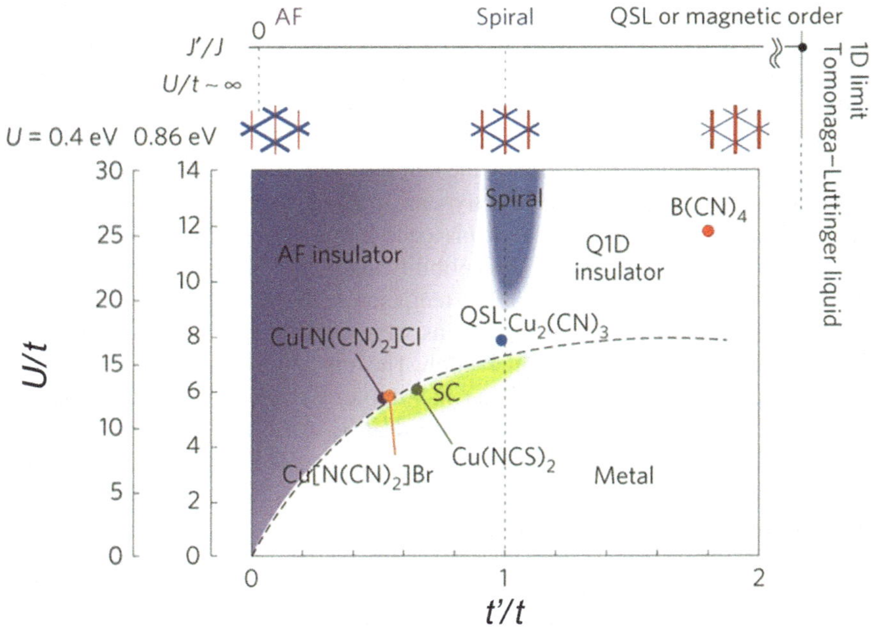

Fig. 15.5 Schematic unified phase diagram of κ-$(ET)_2X$. Reproduced from Yoshida et al., (2015), Macmillan Publishers Ltd.

= 5.2 K of κ-$(ET)_2Ag_2(CN)_3$ is slightly higher than that of κ-$(ET)_2Cu_2(CN)_3$ (T_C = 3.9 K). As a result, the overall temperature–pressure (*T*–*P*) phase diagram looks similar by considering the offset of the critical pressure (Fig. 15.6a). The exchange interaction energy (J/k_B ~ 175 K) of κ-$(ET)_2Ag_2(CN)_3$ at ambient pressure is smaller than that of κ-$(ET)_2Cu_2(CN)_3$ (J/k_B ~ 250 K), and it increased with increasing hydrostatic pressure as expected (Shimizu et al., 2016). The ^{1}H-NMR and ^{13}C-NMR measurements indicate absence of magnetic order down to 0.1 K. The low temperature *T*-linear term of heat capacity also exists in κ-$(ET)_2Ag_2(CN)_3$ (Shimizu et al., 2016). These results suggest a spin liquid ground state in κ-$(ET)_2Ag_2(CN)_3$. In the case of κ-$(ET)_2Cu_2(CN)_3$, the contamination of Cu(II) was a problem. However, Ag(I) is stable and there need be no concern about the inclusion of Ag(II) in κ-$(ET)_2Ag_2(CN)_3$. This is a big advantage of κ-$(ET)_2Ag_2(CN)_3$ for the investigation of intrinsic low-lying excitations. We also found a higher T_C under uniaxial strain than under hydrostatic pressure, which is similar to the behavior of κ-$(ET)_2Cu_2(CN)_3$ as shown in Fig. 15.6b (Tomeno et al. 2020a).

The partial substitution of Ag for Cu in κ-$(ET)_2Cu_2(CN)_3$ is also possible. High school students were involved in this study by chance, and brief results at an initial stage were published (Matsubara et al,. 2018). The systematic study of mixed crystals, κ-$(ET)_2Ag_{2x}Cu_{2(1-x)}(CN)_3$ [$0.24 < x < 0.71$], was performed afterward (Yoshida et al., 2019). The Ag and Cu atoms were distributed homogeneously in the mixed crystals. We expected ordering of Ag/Cu as well as C/N in the anion network. However, they were found to be disordered statistically in the mixed crystals. The anion structures were similar to that of κ-$(ET)_2Cu_2(CN)_3$. The t'/t value was increased with increasing the *x*. The sample with $x = 0.49$ did not show magnetic order down to 2 K in a Mott insulating state at ambient pressure, and showed a pressure-induced Mott and superconducting transitions like mother compounds (Yoshida et al., 2019).

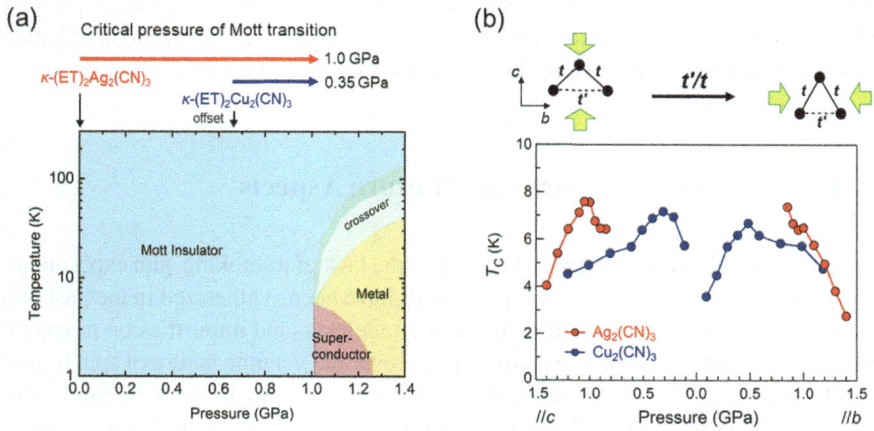

Fig. 15.6 **a** Phase diagram of κ-$(ET)_2M_2(CN)_3$ [M = Cu, Ag] under hydrostatic pressure. **b** Uniaxial strain dependence of T_C in κ-$(ET)_2M_2(CN)_3$ [M = Cu, Ag]

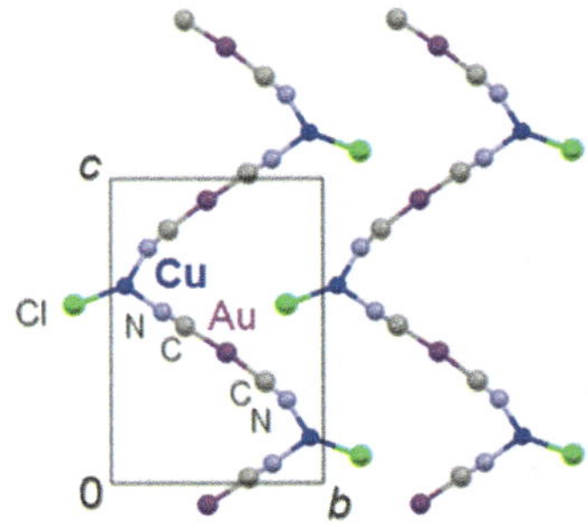

Fig. 15.7 Ordered anion structure in κ-$(ET)_2Cu[Au(CN)_2]Cl$

Spin-liquid behaviors have been reported in several organic Mott insulators since 2003. The candidates include $EtMe_3Sb[Pd(dmit)_2]_2$ (Itou et al., 2008, 2010), κ-H_3(Cat-EDT-TTF$)_2$ (Isono et al., 2014; Shimozawa et al., 2017; Yamashita et al., 2017), κ-$(ET)_2Ag_2(CN)_3$ (Hiramatsu et al., 2017; Shimizu et al., 2016), κ-$(ET)_2CuAg(CN)_3$ (Yoshida et al., 2019), and X-ray irradiated κ-$(ET)_2Cu[N(CN)_2]Cl$ (Furukawa et al., 2015). Notably, all these compounds have some structural disorders. Because the strong disorder can suppress a long-range magnetic order, the disorder-free compound is desirable to realize a genuine quantum spin liquid. In this context, we have been trying to develop a disorder-free spin liquid candidate having triangular lattice.

Recently, by using bi-metal ions, we succeeded in obtaining a triangular-lattice organic Mott insulator with a disorder-free polyanion, κ-$(ET)_2Cu[Au(CN)_2]Cl$ (Tomeno et al., 2020b). The CN groups of polyanion network $\{Cu[Au(CN)_2]Cl^-\}_\infty$ is ordered as shown in Fig. 15.7. The band structure calculation based on the extended Hückel method and tight-binding approximation demonstrated a nearly regular triangular lattice ($t'/t \sim 1.19$) in κ-$(ET)_2Cu[Au(CN)_2]Cl$. No indication of magnetic order down to 2 K and semiconducting behaviors suggest that this compound is the first ET-based quantum spin liquid candidate having a nearly regular triangular lattice with a disorder-free anion. More detailed studies on this compound are in progress.

15.4 Concluding Remarks and Future Aspects

The quantum spin liquid is elusive because of the lack of a smoking-gun experiment. However, attractive candidates for spin liquids have been synthesized in the past two decades. The extrinsic contribution of disorder, defects, and impurities on magnetic properties should be carefully examined to uncover the genuine nature of spin liquid. Superconductivity emerging from spin liquid is also attractive. Besides the bandwidth control of Mott and superconducting transitions, band-filling control using a field-effect Mott transistor was recently reported (Kawasugi et al., 2016; Yamamoto et al., 2013). The latter would provide deep insights into organic Mott insulators and spin liquid candidates. Another issue of interest is the use of spin currents in organic conductors and spin liquid. The spin polarization and possible use of spin current in κ-$(ET)_2X$ has been proposed theoretically (Naka et al., 2019, 2020). The spin current might also be used to probe quantum spin liquid (Aftergood & Takei, 2020). There remain many attractive and challenging issues in the study of organic spin liquid.

Acknowledgements The author thanks collaborators for the study of organic spin liquid, especially Y. Shimizu, Y. Yoshida, T. Hiramatsu, S. Tomeno, H. Kitagawa, and G. Saito. The author also acknowledges financial support from JSPS (JP20H02709, JP23225005, JP15205019).

References

Abdel-Jawad, M., Terasaki, I., Sasaki, T., Yoneyama, N., Kobayashi, N., Uesu, Y., & Hotta, C. (2010). Anomalous dielectric response in the dimer Mott insulator κ-(BEDT-TTF)$_2$Cu$_2$(CN)$_3$. *Physical Review B, 82*, 125119/1–5.
Aftergood, J., & Takei, S. (2020). Probing quantum spin liquids in equilibrium using the inverse spin Hall effect. *Physical Review Research, 2*, 033439/1–18.
Akamatu, H., Inokuchi, H., & Matsunaga, Y. (1954). Electrical conductivity of the perylene-bromine complex. *Nature, 173*, 168–169.
Anderson, P. W. (1973). Resonating valence bonds: a new kind of insulator? *Materials Research Bulletin, 8*, 153–160.
Bechgaard, K., Jacobsen, C., Mortensen, K., Pedersen, H., & Thorup, N. (1979). The properties of five highly conducting salts: (TMTSF)$_2$X, X = PF_6^-, AsF_6^-, SbF_6^-, BF_4^- and NO_3^-, derived from tetramethyltetraselenafulvalene (TMTSF). *Solid State Communications, 33*, 1119–1125.
Bu, X., Jensen, A. F., Allendoerfer, R., & Coppens, P. (1991). Structure and properties of a new κ-phase organic metal: (BEDT-TTF)$_2$Cu$_2$(CN)$_3$. *Solid State Communications, 79*, 1053–1057.
Clay, R. T., Dayal, S., Li, H., & Mazumdar, S. (2012). Beyond the quantum spin liquid concept in frustrated two dimensional organic superconductors. *Physica Status Solidi (b), 249*, 991–994.
Coleman, L. B., Cohen, M. J., Sandman, D. J., Yamagishi, F. G., Garito, A. F., & Heeger, A. J. (1973). Superconducting fluctuations and the Peierls instability in an organic solid. *Solid State Communications, 12*, 1125–1132.
Conwell, E. (Ed.). (1988). *Semiconductors and semimetals* (Vol. 27). New York: Academic Press.
Dayal, S., Clay, R. T., Li, H., & Mazumdar, S. (2011). Paired electron crystal: Order from frustration in the quarter-filled band. *Physical Review B, 83*, 245106/1–12.
Dressel, M., & Tomić, S. (2020). Molecular quantum materials: electronic phases and charge dynamics in two-dimensional organic solids. *Advances in Physics, 69*, 1–120.
Dressel, M., Lazić, P., Pustogow, A., Zhukova, E., Gorshunov, B., Schlueter, J.A., Milat, O., Gumhalter, B., & Tomić, S. (2016). Lattice vibrations of the charge-transfer salt κ-(BEDT-TTF)$_2$Cu$_2$(CN)$_3$: Comprehensive explanation of the electrodynamic response in a spin-liquid compound. *Physical Review B, 93*, 081201(R)/1–5.
Drozdova, O., Saito, G., Yamochi, H., Ookubo, K., Yakushi, K., Uruichi, M., & Ouahab, L. (2001). Composition and structure of the anion layer in the organic superconductor κ'-(ET)$_2$Cu$_2$(CN)$_3$: optical study. *Inorganic Chemistry, 40*, 3265–3266.
Ferraris, J., Cowan, D. O., Walatka, V., & Perlstein, J. H. (1973). Electron transfer in a new highly conducting donor-acceptor complex. *Journal of the American Chemical Society, 95*, 948–949.
Foury-Leylekian, P., Ilakovac, V., Balédent, V., Fertey, P., Arakcheeva, A., Milat, O., Petermann, D., Guillier, G., Miyagawa, K., Kanoda, K., Alemany, P., Canadell, E., Tomić, S., & Pouget, J.-P. (2018). (BEDT-TTF) $_2$Cu$_2$(CN)$_3$ Spin Liquid: Beyond the Average Structure. *Crystals, 8*, 158/1–17.

Fukuyama, H., Tanimoto, T., & Saito, M. (1996). Antiferromagnetic long range order in disordered spin-peierls systems. *Journal of the Physical Society of Japan, 65*, 1182–1185.

Furukawa, T., Miyagawa, K., Itou, T., Ito, M., Taniguchi, H., Saito, M., Iguchi, S., Sasaki, T., & Kanoda, K. (2015). Quantum spin liquid emerging from antiferromagnetic order by introducing disorder. *Physical Review Letters, 115*, 077001/1–5.

Geiser, U., Wang, H. H., Carlson, K. D., Williams, J. M., Charlier, H. A., Jr., Heindl, J. E., Yaconi, G. A., Love, B. H., Lathrop, M. W., Schirber, J. E., Overmyer, D. L., Ren, J., & Whangbo, M.-H. (1991). Superconductivity at 2.8 K and 1.5 kbar in κ-$(BEDT\text{-}TTF)_2Cu_2(CN)_3$: The first organic superconductor containing a polymeric copper cyanide anion. *Inorganic Chemistry, 30*, 2586–2588.

Gomi, H., Ikenaga, M., Hiragi, Y., Segawa, D., Takahashi, A., Inagaki, T. J., & Aihara, M. (2013). Ferroelectric states induced by dimer lattice disorder in dimer Mott insulators. *Physical Review B, 87*, 195126/1–12.

Hiramatsu, T., Yoshida, Y., Saito, G., Otsuka, A., Yamochi, H., Maesato, M., Shimizu, Y., Ito, H., Nakamura, Y., Kishida, H., Watanabe, M., & Kumai, R. (2017). Design and preparation of a quantum spin liquid candidate κ-$(ET)_2Ag_2(CN)_3$ having a nearby superconductivity. *Bulletin of the Chemical Society of Japan, 90*, 1073–1082.

Hotta, C. (2010). Quantum electric dipoles in spin-liquid dimer Mott insulator κ-$ET_2Cu_2(CN)_3$. *Physical Review B, 82*, 241104(R)/1–4.

Hotta, C. (2012). Theories on Frustrated Electrons in Two-Dimensional Organic Solids. *Crystals, 2*, 1155/1–46.

Ilakovac, V., Carniato, S., Foury-Leylekian, P., Tomić, S., Pouget, J.-P., Lazić, P., Joly, Y., Miyagawa, K., Kanoda, K., & Nicolaou, A. (2017). Resonant inelastic x-ray scattering probes the electron-phonon coupling in the spin liquid κ-$(BEDT\text{-}TTF)_2Cu_2(CN)_3$. *Physical Review B, 96*, 184303/1–9.

Ishiguro, T., Yamaji, K., & Saito, G. (1998). *Organic superconductors* (2nd ed.). Springer.

Isono, T., Kamo, H., Ueda, A., Takahashi, K., Kimata, M., Tajima, H., Tsuchiya, S., Terashima, T., Uji, S., & Mori, H. (2014). Gapless quantum spin liquid in an organic spin-1/2 triangular-lattice κ-$H_3(Cat\text{-}EDT\text{-}TTF)_2$. *Physical Review Letters, 112*, 177201/1–5.

Itou, T., Oyamada, A., Maegawa, S., & Kato, R. (2010). Instability of a quantum spin liquid in an organic triangular-lattice antiferromagnet. *Nature Physics., 6*, 673–676.

Itou, T., Oyamada, A., Maegawa, S., Tamura, M., & Kato, R. (2008). Quantum spin liquid in the spin-1/2 triangular antiferromagnet $EtMe_3Sb[Pd(dmit)_2]_2$. *Physical Review B, 77*, 104413/1–5.

Jérome, D. (1991). The physics of organic superconductors. *Science, 252*, 1509–1514.

Jérome, D., & Schulz, H. J. (1982). Organic conductors and superconductors. *Advances in Physics, 31*, 299–490.

Jérome, D., Mazaud, A., Ribault, M., & Bechgaard, K. (1980). Superconductivity in a synthetic organic conductor $(TMTSF)_2PF_6$. *Journal De Physique Lettres, 41*, L95-98.

Kajita, K., Nishio, Y., Tajima, N., Suzumura, Y., Kobayashi, A. (2014). Molecular Dirac Fermion systems—theoretical and experimental approaches. *Journal of the Physical Society of Japan, 83*, 072002/1–31.

Kandpal, H.C., Opahle, I., Zhang, Y.Z., Jeschke, H.O., Valentí, R. (2009). Revision of model parameters for κ-type charge transfer salts: An Ab Initio study. *Physical Review Letters, 103*, 067004/1–4.

Kanoda, K. (1997). Electron correlation, metal-insulator transition and superconductivity in quasi-2D organic systems, $(ET)_2X$. *Physica C, 287*, 299–302.

Kanoda, K. (1997). Recent progress in NMR studies on organic conductors. *Hyperfine Interactions, 104*, 235–249.

Kanoda, K. (2006). Metal–Insulator Transition in κ-$(ET)_2X$ and $(DCNQI)_2M$: Two contrasting manifestation of electron correlation. *Journal of the Physical Society of Japan, 75*, 051007/1–16.

Kawasugi, Y., Seki, K., Edagawa, Y., Sato, Y., Pu, J., Takenobu, T., Yunoki, S., Yamamoto, H. M., & Kato, R. (2016). Electron-hole doping asymmetry of Fermi surface reconstructed in a simple Mott insulator. *Nature Communications, 7*, 12356/1–8.

Kini, M., Geiser, U., Wang, H. H., Carlson, K. D., Williams, J. M., Kwok, W. K., Vandervoort, K. G., Thompson, J. E., Stupka, D. L., Jung, D., & Whangbo, M.-H. (1990). A new ambient-pressure organic superconductor, κ-$(ET)_2Cu[N(CN)_2]Br$, with the highest transition temperature yet observed (inductive onset $T_C = 11.6$ K, resistive onset = 12.5 K). *Inorganic Chemistry, 29*, 2555–2557.

Kino, H., & Fukuyama, H. (1995). Electronic states of conducting organic κ-$(BEDT\text{-}TTF)_2X$. *Journal of the Physical Society of Japan, 64*, 2726–2729.

Komatsu, T., Matsukawa, N., Inoue, T., & Saito, G. (1996). Realization of superconductivity at ambient pressure by band-filling control in κ-$(BEDT\text{-}TTF)_2Cu_2(CN)_3$. *Journal of the Physical Society of Japan, 65*, 1340–1354.

Komatsu, T., Nakamura, T., Matsukawa, N., Yamochi, H., & Saito, G. (1991). New ambient-pressure organic superconductors based on BEDT-TTF, Cu, $N(CN)_2$ and CN with $T_C = 10.7$ K and 3.8 K. *Solid State Communications, 80*, 843–847.

Koretsune, T., & Hotta, C. (2014). Evaluating model parameters of the κ- and β'-type Mott insulating organic solids. *Physical Review B, 89*, 045102/1–6.

Li, H., Clay, R. T., & Mazumdar, S. (2010). The paired-electron crystal in the two-dimensional frustrated quarter-filled band. *Journal of Physics: Condensed Matter, 22*, 272201/1–7.

Manna, R. S., De Souza, M., Brühl, A., Schlueter, J. A., & Lang, M. (2010). Lattice effects and entropy release at the low-temperature phase transition in the spin-liquid candidate κ-$(BEDT\text{-}TTF)_2Cu_2(CN)_3$. *Physical Review Letters, 104*, 016403/1–4.

Matsubara, E., Miyazono, H., Tomeno, S., Kimura, Y., & Maesato, M. (2018). Synthesis and physical properties of new organic conductor with triangular lattice. *ELCAS Journal, 3*, 13–15.

Miksch, B., Pustogow, A., Rahim, M. J., Bardin, A. A., Kanoda, K., Schlueter, J. A., Hübner, R., Scheffler, M., & Dressel, M. (2021). Gapped magnetic ground state in quantum spin liquid candidate κ-$(BEDT\text{-}TTF)_2Cu_2(CN)_3$. *Science, 372*, 276–279.

Miyagawa, K., Kawamoto, A., Nakazawa, Y., & Kanoda, K. (1995). Antiferromagnetic ordering and spin structure in the organic conductor, κ-$(BEDT\text{-}TTF)_2Cu[N(CN)_2]Cl$. *Physical Review Letters, 75*, 1174–1177.

Mori, H., Tanaka, S., & Mori, T. (1998). Systematic study of the electronic state in θ-type BEDT-TTF organic conductors by changing the electronic correlation. *Physical Review B, 57*, 12023/1–7.

Mori, T., Kobayashi, A., Sasaki, Y., Kobayashi, H., Saito, G., & Inokuchi, H. (1984). The Intermolecular Interaction of Tetrathiafulvalene and Bis(ethylenedithio)tetrathiafulvalene in organic metals. Calculation of orbital overlaps and models of energy-band structures. *Bulletin of the Chemical Society of Japan., 57*, 627–633.

Murata, K., Tokumoto, M., Anzai, H., Bando, H., Saito, G., Kajimura, K., & Ishiguro, T. (1985). Superconductivity with the onset at 8 K in the organic conductor β-$(BEDT\text{-}TTF)_2I_3$ under pressure. *Journal of the Physical Society of Japan, 54*, 1236–1239.

Naka, M., & Ishihara, S. (2010). Electronic ferroelectricity in a Dimer Mott insulator. *Journal of the Physical Society of Japan, 79*, 063707/1–4.

Naka, M., Hayami, S., Kusunose, H., Yanagi, Y., Motome, Y., & Seo, H. (2019). Spin current generation in organic antiferromagnets. *Nature Communications, 10*, 4305/1–8.

Naka, M., Hayami, S., Kusunose, H., Yanagi, Y., Motome, Y., & Seo, H. (2020). Anomalous Hall effect in κ-type organic antiferromagnets. *Physical Review B, 102*, 075112/1–11.

Nakamura, K., Yoshimoto, Y., Kosugi, T., Arita, R., & Imada, M. (2009). Ab initio derivation of low-energy model for κ-ET Type organic conductors. *Journal of the Physical Society of Japan, 78*, 083710/1–4.

Poirier, M., De Lafontaine, M., Miyagawa, K., Kanoda, K., & Shimizu, Y. (2014). Ultrasonic investigation of the transition at 6 K in the spin-liquid candidate κ-$(BEDT\text{-}TTF)_2Cu_2(CN)_3$. *Physical Review B, 89*, 045138/1–5.

Pustogow, A., Le, T., Wang, H. H., Luo, Y., Gati, E., Schubert, H., Lang, M., & Brown, S. E. (2020). Impurity moments conceal low-energy relaxation of quantum spin liquids. *Physical Review B, 101*, 140401(R)/1–6.

Regnault, L. P., Renard, J. P., Dhalenne, G., & Revcolevschi, A. (1995). Coexistence of dimerization and antiferromagnetism in Si-doped $CuGeO_3$. *Europhysics Letters, 32*, 579–584.

Saito, G., & Yoshida, Y. (2007). Development of conductive organic molecular assemblies: Organic metals, superconductors, and exotic functional materials. *Bulletin of the Chemical Society of Japan, 80*, 1–137.

Saito, G., Enoki, T., Toriumi, K., & Inokuchi, H. (1982). Two-dimensionality and suppression of metal-semiconductor transition in a new organic metal with alkylthio substituted TTF and perchlorate. *Solid State Communications, 42*, 557–560.

Sasaki, S., Hashimoto, K., Kobayashi, R., Itoh, K., Iguchi, S., Nishio, Y., Ikemoto, Y., Moriwaki, T., Yoneyama, N., Watanabe, M., Ueda, A., Mori, H., Kobayashi, K., Kumai, R., Murakami, Y., Müller, J., & Sasaki, T. (2017). Crystallization and vitrification of electrons in a glass-forming charge liquid. *Science, 357*, 1381–1385.

Sato, T., Miyagawa, K., & Kanoda, K. (2017). Electronic crystal growth. *Science, 357*, 1378–1381.

Shimizu, Y., Hiramatsu, T., Maesato, M., Otsuka, A., Yamochi, H., Ono, A., Itoh, M., Yoshida, M., Takigawa, M., Yoshida, Y., & Saito, G. (2016). Pressure-Tuned Exchange Coupling of a Quantum Spin Liquid in the Molecular Triangular Lattice κ-$(ET)_2Ag_2(CN)_3$. *Physical Review Letters, 117*, 107203/1–6.

Shimizu, Y., Maesato, M., & Saito, G. (2011). Uniaxial strain effects on Mott and superconducting transitions in κ-$(ET)_2Cu_2(CN)_3$. *Journal of the Physical Society of Japan, 80*, 074702/1–7.

Shimizu, Y., Maesato, M., Saito, G., Drozdova, O. O., & Ouahab, L. (2003). Transport properties of a Mott insulator κ-$(ET)_2Cu_2(CN)_3$ under the uniaxial strain. *Synthetic Metals, 133*, 225–226.

Shimizu, Y., Miyagawa, K., Kanoda, K., Maesato, M., & Saito, G. (2003). Spin liquid state in an organic Mott insulator with a triangular lattice. *Physical Review Letters, 91*, 107001/1–4.

Shimozawa, M., Hashimoto, K., Ueda, A., Suzuki, Y., Sugii, K., Yamada, S., Imai, Y., Kobayashi, R., Itoh, K., Iguchi, S., Naka, M., Ishihara, S., Mori, H., Sasaki, T., & Yamashita, M. (2017). Quantum-disordered state of magnetic and electric dipoles in an organic Mott system. *Nature Communications, 8*, 1821/1–6.

Tomeno, S., Maesato, M., Yoshida, Y., Hiramatsu, T., Saito, G., & Kitagawa, H. (2020a). Uniaxial strain induced superconductivity in quantum spin liquid κ-$(ET)_2Ag_2(CN)_3$. *Journal of the Physical Society of Japan, 89*, 054709/1–5.

Tomeno, S., Maesato, M., Yoshida, Y., Kiswandhi, A., & Kitagawa, H. (2020). Triangular-lattice organic Mott insulator with a disorder-free polyanion. *Inorganic Chemistry, 59*, 8647–8651.

Urayama, H., Yamochi, H., Saito, G., Nozawa, K., Sugano, T., Kinoshita, M., Sato, S., Oshima, K., Kawamoto, A., & Tanaka, J. (1988). A new ambient pressure organic superconductor based on BEDT-TTF with T_C higher than 10 K ($T_C = 10.4$ K). *Chemistry Letters, 17*, 55–58.

Yamamoto, H. M., Nakano, M., Suda, M., Iwasa, Y., Kawasaki, M., & Kato, R. (2013). A strained organic field-effect transistor with a gate-tunable superconducting channel. *Nature Communications, 4*, 2379/1–7.

Yamashita, M., Nakata, N., Kasahara, Y., Sasaki, T., Yoneyama, N., Kobayashi, N., Fujimoto, S., Shibauchi, T., & Matsuda, Y. (2009). Thermal-transport measurements in a quantum spin-liquid state of the frustrated triangular magnet κ-(BEDT-TTF)$_2Cu_2(CN)_3$. *Nature Physics, 5*, 44–47.

Yamashita, S., Nakazawa, Y., Oguni, M., Oshima, Y., Jojiri, H., Miyagawa, K., & Kanoda, K. (2008). Thermodynamic properties of a spin-1/2 spin-liquid state in a κ-type organic salt. *Nature Physics, 4*, 459–462.

Yamashita, S., Nakazawa, Y., Ueda, A., & Mori, H. (2017). Thermodynamics of the quantum spin liquid state of the single-component dimer Mott system κ-H_3(Cat-EDT-TTF)$_2$. *Physical Review B, 95*, 184425/1–5.

Yoshida, Y., Ito, H., Maesato, M., Shimizu, Y., Hayama, H., Hiramatsu, T., Nakamura, Y., Kishida, H., Koretsune, T., Hotta, C., & Saito, G. (2015). Spin-disordered quantum phases in a quasi-one-dimensional triangular lattice. *Nature Physics, 11*, 679–683.

Yoshida, Y., Maesato, M., Tomeno, S., Kimura, Y., Saito, G., Nakamura, Y., Kishida, H., & Kitagawa, H. (2019). Partial Substitution of Ag(I) for Cu(I) in Quantum Spin Liquid κ-$(ET)_2Cu_2(CN)_3$, Where ET Is Bis(ethylenedithio)tetrathiafulvalene. *Inorganic Chemistry, 58*, 4820–4827.

Mitsuhiko Maesato was born in Okinawa, Japan, in 1973. He graduated from the University of Tokyo (1995), and received his Ph.D. from the Department of Basic Science, the University of Tokyo (2000), under the guidance of Prof. Seiichi Kagoshima. He moved to Kyoto in 2000 to assume a position of Assistant Professor with Prof. Gunzi Saito's group at Kyoto University. He joined Prof. Hiroshi Kitagawa's group in 2009. Since 2013 he has been an Associate Professor in the Department of Chemistry, Graduate School of Science, Kyoto University. He was a visiting scientist at Rennes University, France (2007), and at the National High Magnetic Field Laboratory, USA (2011). His research interests include multifunctional materials, organic (super)conductors, spin liquids, and emergent phenomena in strongly correlated materials.

Part IV
Creative Dynamics of Complex "Living" Systems: From Molecules to Health and Disease in Life and Its Evolution

Chapter 16
Impact of Reactive Oxygen Species and G-Quadruplexes in Telomeres and Mitochondria

Madhu Malinee and Hiroshi Sugiyama

Abstract Possible formation of G-quadruplex (G4) structures in telomere and mitochondrial DNA (mtDNA) are the subjects of intense current research interest. The telomere, a complex of repetitive nucleotide sequences and binding proteins, is present at the end of chromosomes to safeguard genomic stability. G4, a noncanonical DNA structure, forms in guanine-rich nucleic acid sequences such as immunoglobulin switch regions, telomeres, promoter regions (especially of oncogenes), and 5′-untranslated regions of genes. G4 motifs have been recognized as regulatory structures and are associated with genome instability and gene expression defects in the nuclear genome. The mitochondrial genome is a circular double-stranded DNA with significant asymmetry in strand composition. Mitochondrial DNA, especially the heavy (H) strand, is rich in guanine sequences and hence shows a strong propensity to form G4 structures that have been associated with mtDNA deletion breakpoints. mtDNA deletions are notably observed in various genetic disorders, mitochondrial dysfunction, cancer, and aging. Telomerase, which maintains telomere length, shuttles dynamically between different cellular locations. Under oxidative stress, telomerase localizes to mitochondria; however, little is known about the role of the interaction between the G4 structures in mtDNA and telomerase in determining mitochondrial function and fate. In this review, we cover the recent evidences supporting the potential of G4 structures and telomerase to regulate mitochondrial function and mitochondrial fate.

Keywords Mitochondria · Telomere · Telomerase · Aging · TERT · TERC · G-Quadruplex · Energy metabolism · Respiration · Autophagy · Oxidative phosphorylation (OXPHOS) · Fatty acid oxidation

---

M. Malinee
Department of Anatomy and Developmental Biology, Graduate School of Medicine, Kyoto University, Kyoto 606-8501, Japan

M. Malinee · H. Sugiyama (✉)
Department of Chemistry, Graduate School of Science, Kyoto University, Kitashirakawa-Oiwakecho, Sakyo-Ku, Kyoto 606-8502, Japan
e-mail: hs@kuchem.kyoto-u.ac.jp

K. Nishimura et al. (eds.), *Creative Complex Systems*, Creative Economy,
https://doi.org/10.1007/978-981-16-4457-3_16

16.1 Introduction

16.1.1 Telomere

In the late nineteenth century, a German zoologist, August Weismann, proposed the concept of cellular senescence and predicted that somatic cells would have a finite lifespan. Subsequently, Hayflick and Moorhead proved experimentally that when cultured in vitro, embryo-derived fibroblasts can divide 50–100 times before reaching replicative senescence (Hayflick & Moorhead, 1961). Mounting evidence suggested that telomere shortening was the causal factor in replicative senescence (Harley et al., 1990). The telomere, a complex of repetitive nucleotide sequences and six telomere-specific binding proteins (shelterins), is present at each end of a chromosome and protects them from deterioration or fusion with neighboring chromosomes (de Lange, 2005; Greider, 1991). Telomere shortening eventually leads to dysfunctional telomeres, growth arrest, and replicative senescence. Telomerase, a unique reverse transcriptase enzyme, elongates and maintains the telomere. Telomerase consists of two parts: an RNA component TERC (antisense template for telomere synthesis) and a catalytic protein reverse transcriptase enzymatic subunit (TERT). Bodnar et al. (1998) showed that ectopic expression of TERT in retinal pigment epithelial cells and foreskin fibroblast cells leads to their immortalization. Different factors regulate the expression and function of telomerase, including epigenetic modification, transcriptional activation, repression by multiple transcription factors (TFs)/other signaling molecules, and alternative splicing of TERT mRNA.

Telomeres contain approximately 2500 tandem repeats of the repetitive sequence 5′-TTAGGG-3′, comprising a double-stranded region, and a 3′ G-rich single-stranded TTAGGG overhang. The three shelterin subunits, TRF1, TRF2, and POT1, directly recognize the TTAGGG repeats and are interconnected by three additional shelterin proteins, TIN2, TPP1, and Rap1, forming a complex that allows cells to distinguish telomeres from sites of DNA damage (de Lange, 2005). On average, the length of human telomeres declines from 11 kb at birth to 4 kb in old age. Telomerase expression is higher in germline/stem cells and cancer cells but is reduced greatly in adult somatic tissues. During cell division, the telomere shortens because of the end-replication problem. A critical length of telomere is a prerequisite to avoid the activation of DNA damage pathways. Telomerase maintains the length of the telomere and contributes to genome stability (Geserick & Blasco, 2006). Constitutively active and high expression of telomerase in cancer and stem cells allows their unlimited proliferation and explains the immortality of these cells. Marion et al. (2009) demonstrated the indispensable role of telomerase in generating iPS, by showing that iPS cannot be derived from a TERC knockout (KO) mouse but that stem cell development can be rescued by reintroducing telomerase.

16.1.2 Mitochondria

Mitochondria, the powerhouse of the cell, generate chemical energy in the form of adenosine triphosphate (ATP) to power the cell's biochemical reactions. High-energy-demanding cells such as muscle, neuron, and kidney cells contain a greater number of mitochondria than lower-energy-demanding cells like fibroblasts. Apart from producing energy, mitochondria are also known to be involved in regulating a number of signaling pathways (such as calcium homeostasis and immune signaling) and cellular processes such as cellular differentiation, thermogenesis, cell death (or apoptosis), cell cycle, and growth. As the hub of cellular metabolism, mitochondria are the site of processes including fatty acid biosynthesis and catabolism, the TCA cycle, oxidative phosphorylation (OXPHOS; i.e., electron transport chain and ATP generation), heme synthesis, the urea cycle, gluconeogenesis, ketogenesis, Fe–S cluster biosynthesis, and branched-chain amino acid synthesis. Mitochondria contain many copies of a circular double-stranded (ds) DNA (size 16.6 kb) that encodes 13 proteins required for OXPHOS function, as well as the 22 tRNAs and 2 rRNAs required for their translation; other OXPHOS proteins are encoded by the nuclear genome. Mitochondrial DNA (mtDNA) encodes 7 (of 44) proteins of complex I, 1 (of 11) protein of complex III, 3 (of 14) proteins of complex IV, and 2 (of 16) proteins of complex V subunits. mtDNA copy number varies from hundreds to millions per cell (Kang et al., 2016). mtDNA contains a 1.1 kb noncoding region (NCR), called a displacement loop, which is involved in the regulation of transcription and replication of DNA. mtDNA encodes 12S rRNA (MT-RNR1) and 16S rRNA (MT-RNR2) and also encodes the tRNA that provides RNA components for mitochondrial protein synthesis (Stewart & Chinnery, 2015).

mtDNA has several distinctive features, including a lack of histones and a diverse DNA repair mechanism. Reactive oxygen species (ROS) generation by OXPHOS complexes renders mtDNA more vulnerable to damage because mitochondrial DNA repair pathways are limited. The high mtDNA copy number is thought to be the primary defense against ROS-induced DNA damage. In addition, the constant mtDNA damage from ROS is considered to contribute to progressive age-related mitochondrial dysfunction.

16.1.3 G-Quadruplex

G-Quadruplex (G4) is a secondary DNA structure (helical in shape) that forms in guanine-rich DNA/RNA sequences. Four guanine bases from one, two, or four strands of DNA/RNA associate through Hoogsteen hydrogen bonding and form a planar guanine tetrad (Fig. 16.1). G4 is the stacked helical structure composed of two to four guanine tetrads. A univalent cation (generally potassium) at the center of each tetrad stabilizes the G4 structure. G4 can have parallel or antiparallel orientation based on the direction of the strands and can be intramolecular or intermolecular

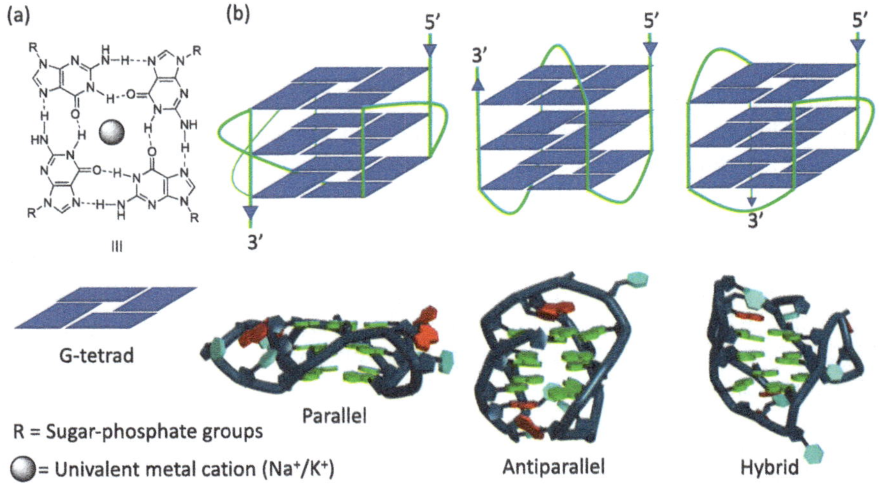

Fig. 16.1 Structures of G-quadruplex. **a** G-Tetrad, a square planar alignment of four guanine bases (connected by cyclic Hoogsteen hydrogen bonding between the N1, N2 and O6, N7 of guanine bases) with a univalent cation at the center. **b** G-Quadruplexes made of stacked G-tetrads in parallel or antiparallel orientation

based on whether the G4 is formed entirely from a single strand or from multiple strands. G4 structures are most prevalent in telomeres and regulatory regions such as the promoters of some oncogenes (β-globin, myc), and 5′-untranslated regions. G4 formation within key regulatory regions of the genome indicates the potential role of G4 as a secondary structure-based epigenetic mechanism for controlling transcription, replication, and even telomere maintenance. Vertebrate telomeres contain repeats of the sequence 5′-TTAGGG-3′ that form stable G4s in vitro as well as in vivo (Henderson et al., 1987). G-Quadruplex DNA structures have been visualized quantitatively by a structure-specific antibody, BG4 (which has a high affinity for intramolecular and intermolecular DNA G4s with K_ds of 0.5 nM and 2.0 nM, respectively) (Biffi et al., 2013). Biffi et al. (2013) have shown that G4 formation in DNA is modulated during cell cycle progression with the maximum occurring during S phase. Of note, they have also shown that not all TRF2 colocalizes with BG4 staining (36.8%), which suggests that G4 structures also form in locations other than telomeres.

Both beneficial and harmful effects of G4 have been reported. DNA repair enzymes consider linear ends of DNA as damaged and hence induce apoptotic signaling. The telomere with its G-rich sequences forms G4 structures to protect these vulnerable ends from instability.

In cancer cells, the role of G4 structures in DNA replication, telomere maintenance, and regulation of transcription is clear (Rhodes & Lipps, 2015). Putative G4-DNA forming sequences are pervasive in the genome and more than half a million G4-DNA structures have been identified by G4-sequencing (Chambers et al., 2015).

16.2 Noncanonical Roles of Telomerase

The primary role of telomerase is to maintain the length of telomeres and contribute to genome stability. However, recent researches have identified some additional noncanonical roles that include (a) improved DNA repair, (b) increased resistance to apoptosis, (c) changes in chromatin structure and gene expression, and (d) resistance to oxidative stress.

16.2.1 Role of Telomerase in Regulating Mitochondrial Function

Ahmed et al. (2008) and Gorbunova et al. (2002) have shown that under conditions of mild oxidative stress, telomerase or TERT overexpression rescues the cells from stress-induced apoptosis and necrosis. It is important to understand the biology and consequences of the telomere- and nontelomere-dependent activity of telomerase to illustrate the significance of telomerase activity against oxidative stress. The separate roles of TERT, TERC, and telomerase have been studied using mice with KO of different components of telomerase. In a very interesting study, Lee et al. (2008) showed using mTERT and mTERC KO mice that withdrawal of TERT but not TERC has a role in the survival and stress resistance of the host. They demonstrated the protective activity of TERT against staurosporine in 293T cells and against *N*-methyl-D-aspartic acid-induced mortality in TERT KO mice. TERT overexpression causes reduced cytochrome C release and cleavage of caspase 3 in 293T cells. Their work indicates that the protective effect of TERT is independent of telomerase activity (Lee et al., 2008).

16.2.2 Telomere–Mitochondrial Aging Axis

The primary hallmarks of aging are genomic instability, telomere damage, and mitochondrial dysfunction; these are interdependent and occur in the mitochondria and the nucleus. Telomere damage leads to genomic instability (Fumagalli et al., 2012; Hewitt et al., 2012) as well as mitochondrial dysfunction (Correia-Melo et al., 2016; Sahin et al., 2011). Mitochondrial dysfunction also causes telomerase attrition and aging, and is closely related to genomic instability (Liu et al., 2002).

Apart from generating energy, mitochondria also generate ROS when electrons pass from a reduced substrate to the terminal electron acceptor oxygen (Raha & Robinson, 2000; Wallace, 2005). The "free radical theory of aging" proposes that ROS causes oxidation damage in both mtDNA and nuclear DNA, which leads to the accumulation of mutations and eventually aging (Harman, 1956).

High oxidative stress causes breaks in single-stranded or dsDNA (Sohal et al., 1994). Because guanine has a lower oxidation potential compared with other nucleobases, the stacking interaction of two consecutive guanine bases would create a site in duplex DNA that has an extremely low ionization potential (Sugiyama & Saito, 1996). We provided direct evidence for a single-electron transfer from a guanine base in duplex DNA to a triplet excited photocleaving amino acid (PCA) and demonstrated that the 5′-G of the 5′-GG-3′ sequence is the most readily oxidizable site in B-form DNA (Sugiyama & Saito, 1996). Furthermore, it has been reported that guanines are oxidized to 8-oxo-7,8-dihydroguanine (8-oxoG) and 2,6-diamino-4-hydroxy-5-formamidopyrimidine (FapyG) by oxidative stress (Oikawa & Kawanishi, 1999; Wang et al. 2010b). Because telomeres are guanine-rich, they are more vulnerable to oxidative stress, which results in telomere shortening. von Zglinicki and coworkers and other research groups have shown that telomere shortening accelerates under increased oxidative stress while lower stress decreases telomere shortening (von Zglinicki, 2000). Multiple reports suggest telomere damage as an important indicator of oxidative stress accumulated in the body and have been advocated as an underlying event in many age-related disorders (Passos et al., 2007). Both oxidative stress and decreased telomerase activity have been identified as causative agents for shortened telomeres in blood cells during aging (von Zglinicki, 2000). Many studies support the connection of mitochondrial dysfunction with telomere shortening. In one such study, researchers showed that (a) MitoQ, an antioxidant that specifically targets mitochondria, reduces telomere shortening (Saretzki et al., 2003); and that (b) depolarizing mitochondria using carbonyl cyanide 4-(trifluoromethoxy)phenylhydrazone results in ROS production and telomere shortening (Liu et al., 2002). Furthermore, it has been reported that patients with primary and secondary mitochondrial dysfunctions have shorter telomeres than healthy persons (Gonzales-Ebsen et al., 2017).

Cells protect themselves from free radicals by enzymatic and nonenzymatic (i.e., vitamins) antioxidant mechanisms. The enzymatic mechanism includes superoxide dismutases (SOD), catalases, thioredoxin reductases, or glutathione peroxidases (Espinosa-Diez et al., 2015). SOD catalyzes the conversion of O_2^- to H_2O_2 and O_2. Depending upon cellular localization and the metal cofactor used for their catalytic activity, SOD enzymes have different forms (Fridovich, 1995), namely SOD1 [soluble Cu/Zn enzyme; a dimer; mainly present in cytosol (Sturtz et al., 2001)], SOD2 [located exclusively in mitochondria; tetramer; use Mn cofactor, also known as MnSOD (Karnati et al., 2013)], and SOD3 [a tetramer; secretory extracellular Cu/ZnSOD expressed in the gastrointestinal tract, blood vessels, lung, or kidneys (Qin et al., 2008)]. When H_2O_2 level increases, SOD1 moves to the nucleus and promotes the expression of oxidative resistance and repair genes by binding with DNA promoters (Tsang et al., 2014). MnSOD is the primary defense against mitochondrial oxidative stress. After release in extracellular space, SOD3 binds to the surface of endothelial cells rich in sulfated polysaccharides (e.g., heparin and heparan sulfate), which helps endothelial cells to function and protect against oxidant-mediated damage and inflammation (Kliment et al., 2008). In one interesting study, researchers showed that mice with higher MnSOD activity had longer telomeres

(Stauffer et al., 2018). It is the balance of these ROS-generating and ROS-detoxifying processes that decides whether the cell will undergo oxidative stress.

The presence of ROS is equally beneficial for the onset of specific signaling and functions in particular cells. ROS generation in the electron transport chain determines T-cell activation and effector functions (Murphy & Siegel, 2013). The killer cells of the innate immune system [e.g., macrophages, neutrophils, and natural killer (NK) cells] use ROS as a key weapon against pathogenic organisms by releasing ROS to irreversibly oxidize and damage the cellular structures of pathogens (Shekhova, 2020). ROS act as important intracellular mediators for driving the inflammatory functions of NK cells (Kim et al., 2017). Extending the positive roles of ROS, Chamoto et al. (2017) showed that mitochondrial uncouplers and ROS generators synergize with immune checkpoint (PD-1)-blockade therapy. Such findings emphasize the importance of ROS for immune responses. Furthermore, it has been shown in a *Caenorhabditis elegans* model that mitochondrial ROS sensed by the intrinsic apoptosis signaling pathway can, independent of apoptosis, protect from the consequences of mitochondrial dysfunction by triggering a unique pattern of gene expression that modulates stress sensitivity and promotes survival (Yee et al., 2014).

Mitochondrial insult by ROS-induced oxidative stress has also been reported. Oxidative stress-induced mitochondrial dysfunction and the accumulation of mutations in mtDNA have been implicated in the aging process, as postulated by Miquel et al. (1980) in the 'mitochondrial theory of aging'. Various reports suggest that the energy-generating ability of mitochondria declines while oxidative stress-induced cellular and mitochondrial damage and mutation in mtDNA increases with aging (Krishnan et al., 2007).

PGC-1 regulates mitochondrial biogenesis, which is in turn modulated by the energy sensors AMPK and SIRT1 (Scarpulla, 2011). It has been shown that short telomeres reduce mitochondrial hyperpolarization and calcium influx via DNA damage responses (DDRs). In addition, Sahin et al. (2011) showed that telomere dysfunction activates p53, which binds PGC-1α and PGC-1β promoters and in turn results in enhanced mitochondrial biosynthesis.

16.2.3 TERT and TERC Shuttling

The presence of telomerase protein TERT outside the nucleus has also been reported, which suggests a nontelomeric function. A range of evidence suggests a protective effect of mitochondrial TERT on mitochondrial functions. Miwa et al. (2016) showed that TERT expression decreases while ROS increases in mouse brain mitochondria. Furthermore, they showed that dietary restriction and rapamycin treatment lead to accumulation of mitochondrial TERT (the shuttling of TERT is dependent on Src kinase), decreased ROS production, and improved cognition (Miwa et al., 2016).

It has been shown that under oxidative damage (H_2O_2 treatment), overexpression of mitochondrion-localized TERT protects against nuclear DNA damage and apoptosis (Singhapol et al., 2013). This result is suggestive of nuclear exclusion of TERT and its shuttling to mitochondria to protect the nucleus from nuclear DNA damage and apoptosis (Fig. 16.2).

Not only proteins but also nuclear-encoded RNAs are imported to the mitochondria via a mitochondrial intermembrane space protein PNPASE that recognizes a small stem-loop structure of RNA, which is an import signal (Wang et al. 2010a). The TERC with a stem-loop structure that is imported into mitochondria is processed to a smaller product, TERC-53, which is then exported to the cytosol (Cheng et al., 2018). TERC-53 is independent of telomerase activity and is a signaling molecule for the initiation of cellular senescence and organismal aging (Zheng et al., 2019). TERC is expressed in most tissues and has broad functions. TERC knockdown induces apoptosis without causing telomere shortening or DDR (Gazzaniga & Blackburn, 2014). These findings are suggestive of different telomerase-independent behaviors

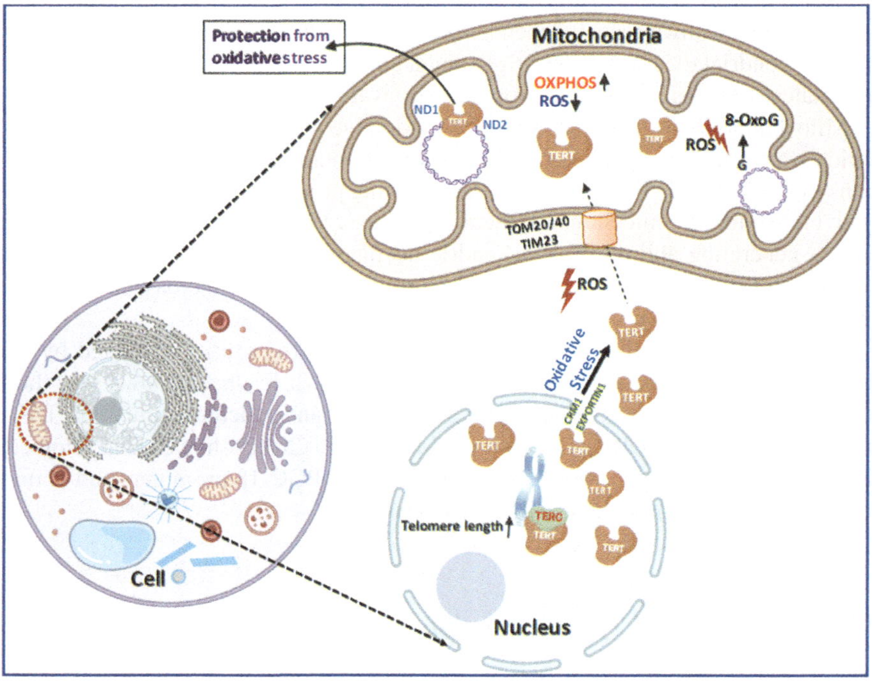

Fig. 16.2 Schematic representation of telomere-dependent/independent functions of telomerase. TERT together with the RNA component TERC makes a functional protein telomerase and maintains the length of the telomere (considered as the telomere dependent function of telomerase). Under oxidative stress, TERT is exported from the nucleus through the CRM1/exportin1 receptor complex and enters mitochondria with the aid of the TOM20/40/TIM23 complex. TERT binds to the mt-ND1/ND2 coding region on the guanine-rich heavy chain of circular double-stranded mtDNA, protecting it from ROS-induced damage and concomitantly increasing OXPHOS. Mitochondrial TERT is associated with improved OXPHOS or electron transport chain function and reduced ROS generation (considered as telomere independent function of telomerase)

or functions of TERC, depending upon its subcellular localization (i.e., nuclear, mitochondrial, or cytosolic).

16.2.4 Telomerase Involvement in Oxidative Stress in Mitochondria

Accumulated evidence suggests that under oxidative stress, telomerase can shuttle from the nucleus to the mitochondria (Ahmed et al., 2008; Santos et al., 2004). TERT overexpression counteracts the loss of mitochondrial membrane potential and the increase of ROS in these cells that occurs during withdrawal of trophic factors. Ahmed et al. (2008) have shown that overexpression of TERT in human fibroblasts protects them from both acute and chronic oxidative stress. They showed that treatment with H_2O_2 and etoposide caused less apoptosis in TERT-overexpressing fibroblast cells than in controls (Ahmed et al., 2008). Cancer cells overexpressing TERT also show shuttling of telomerase to mitochondria to evade oxidative stress-induced cellular death.

The TERT protein is unique in the sense that it contains both a nuclear/nucleolar localization signal and a nuclear export signal. Akiyama et al. (2004) reported that upon activation of lymphocytes by antigen exposure, TERT protein is transported into the nucleus. It has also been shown that telomerase is associated with protein 14-3-3, which is involved in the subcellular shuttling of proteins important for cellular functions (Seimiya et al., 2000). Sub-cellular shuttling occurs naturally within cells and is dynamically regulated by several factors, including cell-cycle phase, DNA damage, and oxidative stress.

Many explanations have been suggested for the process by which TERT protects mtDNA. Supporting one very plausible explanation, Haendeler et al. (2009) showed that the binding of TERT to mtDNA at the coding regions for ND1 and ND2 increases respiratory chain activity (importantly, that of complex I) and reduces oxidative stress-induced damage (Fig. 16.2).

Shuttling of telomerase to subcellular locations such as mitochondria might be an example of the mechanisms for regulation of telomerase catalytic activity and gene transcription. Figure 16.3 shows the localizations of telomerase and their respective functions. Improvement of mitochondrial functions and a contribution to reducing oxidative stress suggests a new function of telomerase in addition to the protection of chromosome length by telomere capping (Fig. 16.2).

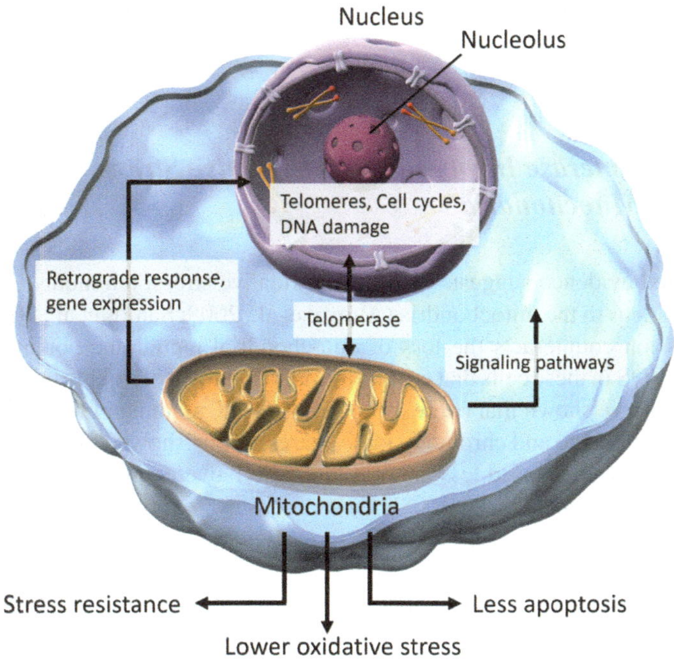

Fig. 16.3 Telomerase localization and respective functions

16.3 Implications of G-Quadruplex in Mitochondrial Genome and Gene Regulation

DNA or RNA sequences rich in guanine (G) nucleotides can adopt noncanonical conformations known as G4. Apart from their beneficial role in the telomere, some studies emphasize the involvement of G4s in the regulation of transcription. About 40% of all promoter regions contain sequences with a propensity to form G4 structures (Huppert & Balasubramanian, 2007). Evidence from multiple experimental studies, such as treatment with G4 ligand, anti-G4 antibody (scFvs), or from individuals lacking the activity of DNA helicases with G4-unwinding activity (e.g., human Werner's syndrome protein or Bloom syndrome protein) suggest the regulation of transcription by G4 (Nguyen et al., 2014; Tang et al., 2016). G4 DNA sequences are often found in oncogenes and in genes with regulatory and homeostatic functions (Huppert & Balasubramanian, 2007). Some proto-oncogene promoters (e.g., c-MYC, VEGF, c-KIT, KRAS, hTERT, and BCL2) have a strong tendency to form G4 structures (Phan et al., 2004; Sun et al., 2008). Other strategies through which G4 affects gene expression are categorized as regulation of RNA splicing, polyadenylation, stability, targeting, and translation into protein. The strong association of promoter quadruplexes with nuclease-hypersensitive sites is indicative of a direct involvement of G4 in gene regulation at the transcription level. In the nuclear genome, G4 motifs have been associated with genome instability and gene expression defects, but they are increasingly recognized as regulatory structures that control the gene expression.

16.3.1 Role of Mitochondrial G4 in Transcription/Replication Switching

Recent studies have revealed that G4 structures can form in the mtDNA. The H strand of mtDNA is G-rich and has a strong propensity to form G4 structures, and potential G4-forming sequences are associated with the origin of mtDNA deletions. However, little is known about the regulatory role of G4 structures in mitochondria. A significant asymmetry in the strand composition of mtDNA has been observed, in which the H and light (L) strands of mtDNA differ in their base composition. The H strand is rich in guanine. During mtDNA replication, selective pressure limits the cytosine content in the displaced strand, which is the reason for the strand asymmetry (Reyes et al., 1998). The longer NCR contains promoters for polycistronic transcription of H and L strands (H strand promoter, HSP; L strand promoter, LSP) as well as the origin of replication for the H strand (OH), while the origin of replication for the L strand (OL) is located outside the NCR. Furthermore, during replication and transcription, the DNA strand rich in G-quadruplex forming potential (QFP) sequences is temporarily single-stranded, suggesting an increased opportunity to form G4 structures. In addition, the high potassium concentration in the matrix provides a favorable environment for G4 formation.

For metabolically active cells, it is important to have coordinated replication and expression of the mitochondrial genome. The mtDNA transcription directed by LSP and HSP results in a genome-sized polycistronic transcript followed by polyadenylation and translation to form the mitochondrial proteins (Hallberg & Larsson, 2014). Interestingly, at the G-rich region conserved sequence block II transcription terminates prematurely because of in situ G-quadruplex formation between nascent RNA and the nontemplate strand of DNA. At the termination site, a replication primer is generated and the replication proceeds until it reaches OL in the opposite strand. Agaronyan et al. (2015) have shown that the human transcription elongation factor (TEFM) interacts with mitochondrial RNA polymerase. Furthermore, the nascent transcript formed prevents the generation of replication primers and increases transcription processivity. The increased processivity serves as a molecular switch between replication and transcription. During the early stages of embryogenesis, TEFM causes an increased transcription rate of mitochondrial genes, and as a result, mitochondrial respiration (OXPHOS) and ATP production are enhanced without the need to replicate mtDNA to fulfil the cellular demand (Agaronyan et al., 2015).

16.3.2 G-Quadruplex and Mitochondrial Gene Expression

Although a range of evidence supports the notion of G4 formation in the mtDNA of cancerous cells, it remains unclear how G4 structures help in mitochondrial nucleic acid homeostasis in noncancerous cells. In an elegant study, Falabella et al. (2019) showed a direct role for G4 perturbation in mitochondrial genome replication, transcription processivity, and respiratory function in noncancerous cells. They found that low concentrations of the G4-ligand 3,11-difluoro-6,8,13-trimethylquino[4,3,2-kl]acridinium methyl sulfate caused acute inhibition of mitochondrial transcript elongation, which led to depletion of the respiratory complex (Falabella et al., 2019). The respiratory defect was stronger when they used an mtDNA variant with increased G4 stability. They also found that the G4 ligand interfered with mtDNA synthesis (Falabella et al., 2019).

16.3.3 G-Quadruplex and Mitochondrial/Cell Fate

Cells get rid of damaged or unwanted organelles by sequestering and degrading them in a cellular process called autophagy. Autophagy is important for cellular survival, development, inflammation and immune responses, DNA repair, proteostasis, organelle quality control, and the prevention of cellular senescence and aging (Galluzzi et al., 2017). Recently, Manchon et al. (2020) showed that pharmacologically stabilizing G4-DNA with G4 ligands strongly downregulates the Atg7 gene (critical for initiating autophagy). They have shown that a putative G-quadruplex-forming sequence in the first intron of the Atg7 gene folds into a G4.

16.3.4 Hypothetical Role of G4 DNA as Oxidative Stress Sensor in Determining Cell Fate

Although G4 structures in the nuclear genome are well documented, the role of G4 in mitochondria has not been well investigated. The H chain of mitochondria contains G-rich sequences and QFPs that have a strong propensity to form G4. It is postulated that mitochondria can tolerate a high level of oxidative stress because of the high guanine content in their H chain, because guanine is easily oxidized to 8-oxoG under oxidative stress (high ROS) conditions. Fleming et al. (2017) have described the role of oxidative DNA damage in activating transcription of genes (e.g., VEGF, NTHL1) whose promoters contain QFPs with guanines. Their work also suggests the importance of 8-oxoG as an epigenetic marker for active transcription of a gene when it occurs in potential G4-forming sequences. Based on this evidence, we propose that a similar mechanism of gene regulation could occur in mtDNA under oxidative stress

(Fig. 16.2). This may explain the high tolerance of mitochondria to ROS damage. This aspect of oxidative DNA damage-induced gene activation in mtDNA should be investigated.

It is also known that telomere DNA has a G4 structure similar to that of the H chain of mtDNA, which is rich in guanine. Because guanine is easily oxidized to 8-oxoG, it may be used by the cell as a sensor. It is known that the formation of 8-oxoG occurs under oxidative stress, and that 8-oxoG is a prevalent endogenous oxidized DNA base in the genome (Fleming & Burrows, 2017). The enzyme 8-oxoG DNA glycosylase removes the oxidized 8-oxoG and generates an apurinic/apyrimidinic (AP/abasic) site that is repaired by AP endonuclease 1 (APE1) (Roychoudhury et al., 2020). The mitochondrial H chain encodes 28 important genes while the L chain encodes only 9 genes. Therefore, mitochondrial function is most likely to be impaired by selective oxidative damage to the H chain. It is thought that this causes mitophagy to control the quality of mitochondria. When oxidative and chemical stress increases at the cellular level, it is possible that this will be detected by the G4 structure at the telomere terminal in the nucleus that will send a signal to the mitochondria. As a result, cytochrome c is released and caspase is produced, which may activate the apoptotic pathway. This hypothetical role of G4 in the H chain also has evolutionary significance. In lower animals (e.g., yeast, *C. elegans*), the two strands of mtDNA show few significant differences, while in mammals, the two strands show significant asymmetry because the H strand is rich in guanine. This has evolutionary value because mammals use oxygen very efficiently. It is plausible that the H strands acquired such changes to compensate for the oxidative damage developed in response to the increasing use of oxygen. This indicates the significance of the H chain of mtDNA.

16.4 Therapeutic Approaches to Improve Mitochondrial Function

Mitochondrial diseases are characterized by dysfunctional mitochondria (i.e., dysregulated oxidative phosphorylation and other mitochondrial metabolic pathways caused by mutations in mtDNA/nuclear DNA-encoded genes for mitochondrial structural proteins) (Gorman et al., 2016). Neurological and metabolic disorders are the most common inherited diseases related to mitochondrial dysfunction. Mitochondrial diseases are reported at a frequency of 1 in 5000 adults. Around 1000 pathogenic mutations have been reported in mtDNA (Kogelnik et al., 1996). Mitochondrial diseases can occur at any age and show a broad range of clinical symptoms. They can originate in any organ or tissue. The most common syndromes associated with mitochondrial dysfunction are Leigh syndrome (also known as subacute necrotizing encephalomyelopathy) and Alpers–Huttenlocher syndrome.

To improve the quality of mitochondria, numerous preclinical approaches have been suggested. Many helicase-based and small-molecule-based therapeutic drugs are in use. We discuss the different approaches in the following sections. Both context-dependent activation and inhibition of mitochondrial functions are needed for therapeutic purposes. Donation of healthy mitochondria from a healthy woman to female patients has been proposed to prevent transmission to offspring, but this is not encouraged because of ethical concerns.

16.4.1 Gene Editing Tools for Targeting mtDNA Mutations

Mitochondrial heteroplasmy occurs when wild-type mtDNA and mutant mtDNA both reside in the mitochondria of the cell. A high level of heteroplasmy indicates higher ratios of mutant to wild-type mtDNA, and vice versa. The phenotype of a cell is decided by the relative levels of mutated versus wild-type mtDNA. High level (>50%) of mutated mtDNA results in cellular defects. The m.3243A > G mutation is the primary cause of mitochondrial encephalopathy, lactic acidosis, and stroke-like episodes (MELAS). Mitochondrially targeted gene-editing nucleases such as mito-ZFNs and mito-TALENS (Jackson et al., 2020) have been shown to effectively digest specific mtDNA mutations in vitro (Hashimoto et al., 2015; Minczuk et al., 2008) and in animal models (Reddy et al., 2015). Adeno-associated virus (AAV)-mediated gene therapy is being evaluated for the tissue-specific treatment of mitochondrial diseases by improving mitochondrial function (Di Meo et al., 2012). In one clinical trial (NCT02064569), patients were injected with GS010, a recombinant AAV vector serotype 2 (rAAV2/2) containing the wild-type ND4 gene (rAAV2/2-ND4) to treat Leber hereditary optic neuropathy.

16.4.2 Small Molecule-Based Therapeutic Approaches for Treating Mitochondrial Disorders

Most mtDNA mutations result in defects in genes associated with the respiratory complex, causing defective mitochondrial respiration. ROS are generated at a high level when the respiratory chain is impaired which subsequently damage nucleic acid, protein, lipids, and other cellular contents. Different formulations are prescribed in clinical settings for symptomatic treatment of oxidative stress caused by the high level of ROS; for example, antioxidants such as cysteine and lipoic acid. Similarly, lactic acid, the by-product of respiration, is also symptomatically treated using dichloroacetate (DCA) to prevent lactic acidosis. DCA inhibits the enzyme pyruvate dehydrogenase kinase and promotes oxidation of pyruvate. Other options to counter mitochondrial defects are: (a) increasing the concentration of a respiratory chain substrate (e.g., L-carnitine), (b) promoting electron transfer within the respiratory chain (e.g., CoQ10), or (c) biochemically bypassing specific respiratory chain complexes (e.g., succinate).

In a recent study, Kobyashi et al. (2020) identified a tryptoline derivative (tryptolinamide, TLAM) that activates mitochondrial respiration by inhibiting phosphofructokinase 1 (PFK1). Inhibition of PFK1 in turn promotes OXPHOS via enhanced AMPK-mediated fatty acid oxidation. In addition, it also leads to shunting of the carbon skeleton to the pentose phosphate pathway to bolster its antioxidative potential (Kobayashi et al., 2020). In a recent report by Bonekamp et al. (2020) of a mitochondrial DNA transcription inhibitor, IMTs selectively targeted mitochondrial RNA polymerase (POLRMT) and impaired mtDNA transcription, which is essential for biogenesis of the OXPHOS system. IMTs allosterically bind near the active center cleft of POLRMT (Bonekamp et al., 2020). Several reports suggest that the growth of cancer cells and the persistence of therapy-resistant cancer stem cells depend on OXPHOS (Bosc et al., 2017; Kuntz et al., 2017; Škrtić et al., 2011; Vasan et al., 2020). IMTs show potent effects in the treatment of cancer in preclinical mouse models (Bonekamp et al., 2020).

Many small molecule-based formulations have been prescribed to improve mitochondrial dysfunction (Avula et al., 2014; Nicolson, 2014; Nightingale et al., 2016; Wang et al., 2016). Table 16.1 lists the chemicals used to improve mitochondrial functions in preclinical settings.

Mitochondrial dysfunction leads to reduced mitochondrial respiration. This reduced mitochondrial respiration is the cause of many illnesses, even reduced immune responses against foreign pathogens or tumors in the host. A nuclear cofactor, peroxisome proliferator-activated receptor (PPAR) γ coactivator 1 α/β (PGC-1α/β), is the master regulator of mitochondrial biogenesis and respiration and has been considered as a potential treatment for mitochondrial diseases. PGC-1 signaling enhances the gene expression associated with oxidative phosphorylation, fatty acid oxidation, and improves the health of mitochondria. Bezafibrate is a pan-PPAR agonist that promotes the PGC-1/PPAR axis and enhances mitochondrial biogenesis, oxidative phosphorylation, fatty acid oxidation, and anti-apoptotic gene expression (Chowdhury et al., 2018; Kanabus et al., 2014). It has been shown that combining bezafibrate treatment with checkpoint-blockade therapy enhances antitumor immunity (Chowdhury et al., 2018; Kumar et al., 2020). There is mitochondrial dysregulation in Alzheimer disease, where enhancing mitochondrial respiration benefits the patients (Cenini & Voos, 2019).

Resveratrol (RSV) is a natural polyphenol and potent activator of SIRT1 that in turn activates PGC-1α by deacetylation (Lagouge et al., 2006; Price et al., 2012). PGC-1 activation leads to enhanced mitochondrial biogenesis and function and has been reported to restore regular function in human fibroblasts with inborn errors of fatty acid β-oxidation (in mitochondria) (Bastin et al., 2011; Kanabus et al., 2014). Low-dose RSV enhances cell-mediated immune response by promoting Th1 cytokine production (Malaguarnera, 2019). Reduced longevity has been correlated with diminished mitochondrial OXPHOS and aerobic capacity. Treating mice with RSV extended their lifespan by increasing their aerobic capacity (Lagouge et al., 2006).

Table 16.1 List of chemicals to improve mitochondrial function

Chemical/drug	Target	Function	Reference (clinical trial, if any)
Tryptolinamide (TLAM)	Inhibit phosphofructokinase 1 (PFK1)	Inhibition of PFK1 leads to enhanced OXPHOS	Kobayashi et al. (2020)
Idebenone	Synthetic analogue of coenzyme Q10, antioxidant	Improves MELAS syndromes	NCT00887562
Pyruvate	NAD donor	MELAS and MELA syndromes	JMA-IIA00093
MTP-131	Cardiolipin stabilizing peptide	Mitochondrial myopathy	NCT02367014
Cysteamine bitartrate delayed-release	Cystine-depleting agent	Childhood mitochondrial diseases	NCT02023866 and NCT02473445
MitoQ	Antioxidant	Act as antioxidant and removes the ROS produced in the mitochondria	Saretzki et al. (2003)
EPI-743	NADP dehydrogenase modulator	Children with mitochondrial or metabolic diseases (Leigh syndrome)	NCT01642056
5-ALA and SFC	NADP dehydrogenase modulator	Cranial nerve symptoms associated mitochondrial diseases	JMA-IIA00200
RTA 408	Activation of Nrf2 and inhibition of Nf-κB	Mitochondrial myopathy	NCT02255422
Bezafibrate (Bz)	Pan-PPAR agonist	Bz enhances mitochondrial biogenesis and FAO. FDA approved Bz as a drug to treat hyperlipidemia. Bz treatment synergize with PD-1 blockade therapy and enhances antitumor immunity	Chowdhury et al. (2018), Kumar et al. (2020), Kumar (2019) NCT02398201
Resveratrol	SIRT1 activator	SIRT1 activate PGC-1 by deacetylation, which leads to enhanced mitochondrial biogenesis and OXPHOS	Bastin et al. (2011), Kanabus et al. (2014), Lagouge et al. (2006), Price et al. (2012)

(continued)

Table 16.1 (continued)

Chemical/drug	Target	Function	Reference (clinical trial, if any)
L-Arginine	Nitric oxide donor for endothelial dysfunction	MELAS	JMA-IIA00023; JMA-IIA00025
Metformin	AMPK activator	Metformin activates the AMPK pathway, which leads to enhanced OXPHOS. This causes enhanced memory formation in T cells	Eikawa et al. (2015)
GW501516	PGC-1α/PPARα and δ activator	GW501516 boosts FAO in T cells that results in enhanced persistence of effector T cells	Saibil et al. (2019)

PPAR, peroxisome proliferator-activated receptor; Nrf2, nuclear factor-erythroid 2 related factor 2; Nf-κB, nuclear factor kappa B; OXPHOS, oxidative phosphorylation; FAO, fatty acid oxidation; MELAS, mitochondrial encephalopathy, lactic acidosis, and stroke-like episodes

16.4.3 Pyrrole–Imidazole Polyamide (PIP)-Based Therapeutic Approaches to Improve Mitochondrial Function

Small molecules based therapeutic drugs gain much attention because of their beneficial roles compared with large molecules (e.g., antibody-based treatment). Large molecules such as antibodies are applicable therapeutically only against cell-surface proteins. The quality, as well as the quantity, of mitochondria can be improved using small-molecule-based drugs or formulations. A recent approach has been based on the synthetic ligand pyrrole-imidazole polyamide (PIP). PIP, which comprises *N*-methylpyrrole units (*P*y) and *N*-methylimidazole units (*I*m), selectively recognizes Watson–Crick base pairs located within the DNA minor groove. Their properties of being unaffected by nucleases, sequence specificity, smooth permeability to the cell/nuclear membrane, binding affinities similar to those of natural TFs, and their stability inside cells make PIPs a better candidate for therapeutic drugs than other nucleic acid-based drugs. We can harness the benefit of PIP sequence selectivity for activating or inhibiting a gene of interest by supplementing PIPs with epigenetic modulators (HAT or HDAC modulators) (Malinee et al., 2020). In one such approach, we have developed a PIP-based epigenetic ligand for activating PGC-1 in immune effector T cells. We have shown that mitochondrial functions in the immune effector T cells can be improved using PIP-based designer DNA ligands (unpublished work). The conjugated PIP enhances the mitochondrial activation in CD8 T cells in vitro as well as in vivo. Combining the conjugated PIP with immune checkpoint-blockade

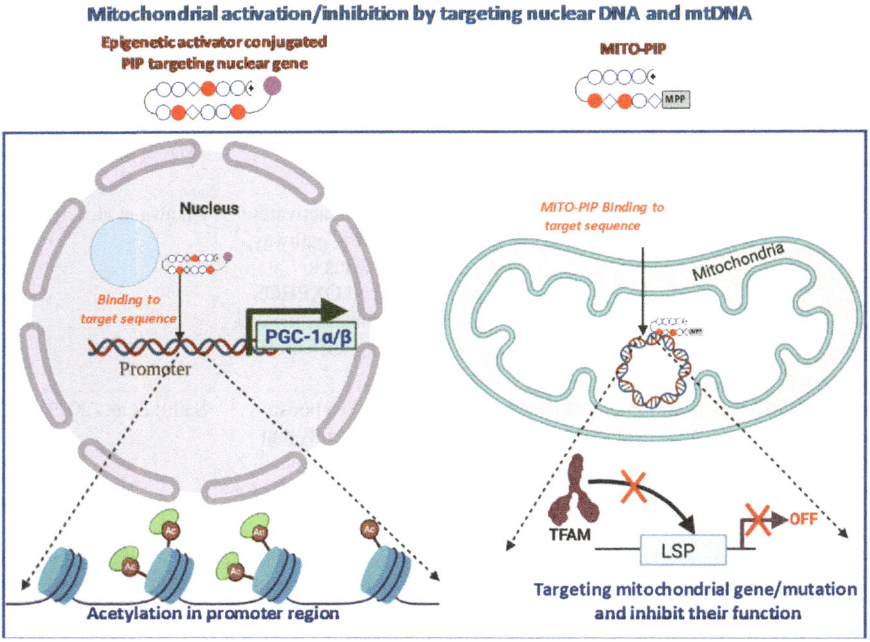

Fig. 16.4 Mitochondrial activation/inhibition by PIPs. Epigenetic activator-conjugated PIP targets the promoter region of PGC-1, which causes acetylation, opening of chromatin, and starts the transcription. To target mtDNA genes, we developed MITO-PIP (PIP conjugated to a mitochondrial-penetrating peptide), which restricts the binding of TFAM transcription factor to LSP and stops the transcription of downstream genes

therapy enhanced antitumor immunity and improved the survival of a tumor-bearing host (unpublished work) (Fig. 16.4).

Using target-specific PIP, we can target not only activation but also nuclear/mitochondrial gene aberrations. Mitochondrial dysfunction is associated with mutations or high heteroplasmy in mitochondria. Mitochondrial membranes are not easy to cross for many molecules. In a novel strategy, our group reported a new type of mitochondria-specific synthetic ligand, termed MITO-PIPs, by conjugating a mitochondria-penetrating peptide with a PIP (Hidaka et al., 2017). We have shown the specific localization of MITO-PIPs inside mitochondria in HeLa cells and that they recognized the target DNA in a sequence-specific manner. The tunability of the properties of PIPs suggests the potential of these MITO-PIPs as potent modulators of not only mitochondrial gene transcription, but also its DNA mutations (Fig. 16.4).

16.5 Conclusion and Future Perspectives

Mitochondria, the powerhouses of the cell, not only generate energy but also decide cell fate and perform various metabolic functions and regulate many signaling pathways. The H chain of mtDNA is highly enriched for guanine content and QFP sequences and shows a strong propensity to form G4 structures. A high mtDNA copy number is thought to be the primary defense against ROS-induced DNA damage. Mitochondrial dysfunction and telomere shortening are the hallmarks of the aging process. There is a strong interdependency between mitochondria, telomeres, and the different units of telomerase (the RNA component TERC and protein component TERT). Mitochondrial dysfunction causes telomere shortening while telomere damage causes mitochondrial reprograming. Under oxidative stress, TERT moves to mitochondria and binds to the coding region of ND1 and ND2 genes on the guanine-rich H chain of mtDNA. The complex regulation network means that telomeres, the nuclear genome, and mitochondria are interdependent and coregulated by multiple localizations of multifunctional proteins and RNAs. Recent studies indicate a direct regulatory role for G4 perturbation in mitochondrial genome replication, transcription processivity, and respiratory function in cells. Based on evidence from the nuclear telomeric G4 structure, it is postulated that mitochondria can tolerate a high level of oxidative stress because of their high G content that can easily be oxidized to 8-oxoG under oxidative stress conditions. This aspect should be investigated in detail to clearly understand its mechanism. Detailed knowledge of these processes may pave the way for the development of new therapeutic strategies that can be of use in mitochondrial medicine.

Competing Financial Interest The authors declare no competing financial interests.

References

Agaronyan, K., Morozov, Y. I., Anikin, M., & Temiakov, D. (2015). Mitochondrial biology. Replication-transcription switch in human mitochondria . *Science, 347*(6221), 548–551. https://doi.org/10.1126/science.aaa0986

Ahmed, S., Passos, J. F., Birket, M. J., Beckmann, T., Brings, S., Peters, H., Birch-Machin, M. A., von Zglinicki, T., & Saretzki, G. (2008). Telomerase does not counteract telomere shortening but protects mitochondrial function under oxidative stress. *Journal of Cell Science, 121*(Pt 7), 1046–1053. https://doi.org/10.1242/jcs.019372

Akiyama, M., Yamada, O., Hideshima, T., Yanagisawa, T., Yokoi, K., Fujisawa, K., Eto, Y., Yamada, H., & Anderson, K. C. (2004). TNFα induces rapid activation and nuclear translocation of telomerase in human lymphocytes. *Biochemical and Biophysical Research Communications, 316*(2), 528–532. https://doi.org/10.1016/j.bbrc.2004.02.080

Avula, S., Parikh, S., Demarest, S., Kurz, J., & Gropman, A. (2014). Treatment of mitochondrial disorders. *Current Treatment Options in Neurology, 16*(6), 292–292. https://doi.org/10.1007/s11940-014-0292-7

Bastin, J., Lopes-Costa, A., & Djouadi, F. (2011). Exposure to resveratrol triggers pharmacological correction of fatty acid utilization in human fatty acid oxidation-deficient fibroblasts. *Human Molecular Genetics, 20*(10), 2048–2057. https://doi.org/10.1093/hmg/ddr089

Biffi, G., Tannahill, D., McCafferty, J., & Balasubramanian, S. (2013). Quantitative visualization of DNA G-quadruplex structures in human cells. *Nature Chemistry, 5*(3), 182–186. https://doi.org/10.1038/nchem.1548

Bodnar, A. G., Ouellette, M., Frolkis, M., Holt, S. E., Chiu, C. P., Morin, G. B., Harley, C. B., Shay, J. W., Lichtsteiner, S., & Wright, W. E. (1998). Extension of life-span by introduction of telomerase into normal human cells. *Science, 279*(5349), 349–352. https://doi.org/10.1126/science.279.5349.349

Bonekamp, N. A., Peter, B., Hillen, H. S., Felser, A., Bergbrede, T., Choidas, A., Horn, M., Unger, A., Di Lucrezia, R., Atanassov, I., & Li, X. (2020). Small-molecule inhibitors of human mitochondrial DNA transcription. *Nature, 588*(7839), 712–716. https://doi.org/10.1038/s41586-020-03048-z

Bosc, C., Selak, M. A., & Sarry, J.-E. (2017). Resistance Is futile: Targeting mitochondrial energetics and metabolism to overcome drug resistance in cancer treatment. *Cell Metabolism, 26*(5), 705–707. https://doi.org/10.1016/j.cmet.2017.10.013

Cenini, G., & Voos, W. (2019). Mitochondria as potential targets in alzheimer disease therapy: An update. *Frontiers in Pharmacology, 10*(902). https://doi.org/10.3389/fphar.2019.00902

Chambers, V. S., Marsico, G., Boutell, J. M., Di Antonio, M., Smith, G. P., & Balasubramanian, S. (2015). High-throughput sequencing of DNA G-quadruplex structures in the human genome. *Nature Biotechnology, 33*(8), 877–881. https://doi.org/10.1038/nbt.3295

Chamoto, K., Chowdhury, P. S., Kumar, A., Sonomura, K., Matsuda, F., Fagarasan, S., & Honjo, T. (2017). Mitochondrial activation chemicals synergize with surface receptor PD-1 blockade for T cell-dependent antitumor activity. *Proceedings of the National Academy of Sciences of the United States of America, 114*(5), E761–E770. https://doi.org/10.1073/pnas.1620433114

Cheng, Y., Liu, P., Zheng, Q., Gao, G., Yuan, J., Wang, P., Huang, J., Xie, L., Lu, X., Tong, T., & Chen, J. (2018). Mitochondrial Trafficking and Processing of Telomerase RNA TERC. *Cell Reports, 24*(10), 2589–2595. https://doi.org/10.1016/j.celrep.2018.08.003

Chowdhury, P. S., Chamoto, K., Kumar, A., & Honjo, T. (2018). PPAR-induced fatty acid oxidation in T cells increases the number of tumor-reactive CD8(+) T cells and facilitates anti-PD-1 therapy. *Cancer Immunology Research, 6*(11), 1375–1387. https://doi.org/10.1158/2326-6066.CIR-18-0095

Correia-Melo, C., Marques, F. D., Anderson, R., Hewitt, G., Hewitt, R., Cole, J., Carroll, B. M., Miwa, S., Birch, J., Merz, A., & Rushton, M. D. (2016). Mitochondria are required for pro-ageing features of the senescent phenotype. *The EMBO Journal, 35*(7), 724–742. https://doi.org/10.15252/embj.201592862

de Lange, T. (2005). Shelterin: The protein complex that shapes and safeguards human telomeres. *Genes & Development, 19*(18), 2100–2110. https://doi.org/10.1101/gad.1346005

Di Meo, I., Auricchio, A., Lamperti, C., Burlina, A., Viscomi, C., & Zeviani, M. (2012). Effective AAV-mediated gene therapy in a mouse model of ethylmalonic encephalopathy. *EMBO Molecular Medicine, 4*(9), 1008–1014. https://doi.org/10.1002/emmm.201201433

Eikawa, S., Nishida, M., Mizukami, S., Yamazaki, C., Nakayama, E., & Udono, H. (2015). Immune-mediated antitumor effect by type 2 diabetes drug, metformin. *Proceedings of the National Academy of Sciences of the United States of America, 112*(6), 1809–1814. https://doi.org/10.1073/pnas.1417636112

Espinosa-Diez, C., Miguel, V., Mennerich, D., Kietzmann, T., Sanchez-Perez, P., Cadenas, S., & Lamas, S. (2015). Antioxidant responses and cellular adjustments to oxidative stress. *Redox Biology, 6*, 183–197. https://doi.org/10.1016/j.redox.2015.07.008

Falabella, M., Kolesar, J. E., Wallace, C., De Jesus, D., Sun, L., Taguchi, Y. V., Wang, C., Wang, T., Xiang, I. M., Alder, J. K., & Maheshan, R. (2019). G-quadruplex dynamics contribute to regulation of mitochondrial gene expression. *Scientific Reports, 9*(1), 5605. https://doi.org/10.1038/s41598-019-41464-y

Fleming, A. M., & Burrows, C. J. (2017). 8-Oxo-7,8-dihydro-2′-deoxyguanosine and abasic site tandem lesions are oxidation prone yielding hydantoin products that strongly destabilize duplex DNA. *Organic & Biomolecular Chemistry, 15*(39), 8341–8353. https://doi.org/10.1039/C7OB02096A

Fleming, A. M., Ding, Y., & Burrows, C. J. (2017). Oxidative DNA damage is epigenetic by regulating gene transcription via base excision repair. *Proceedings of the National Academy of Sciences of the United States of America, 114*(10), 2604–2609. https://doi.org/10.1073/pnas.1619809114

Fridovich, I. (1995). Superoxide radical and superoxide dismutases. *Annual Review of Biochemistry, 64*, 97–112. https://doi.org/10.1146/annurev.bi.64.070195.000525

Fumagalli, M., Rossiello, F., Clerici, M., Barozzi, S., Cittaro, D., Kaplunov, J. M., Bucci, G., Dobreva, M., Matti, V., Beausejour, C. M., & Herbig, U. (2012). Telomeric DNA damage is irreparable and causes persistent DNA-damage-response activation. *Nature Cell Biology, 14*(4), 355–365. https://doi.org/10.1038/ncb2466

Galluzzi, L., Baehrecke, E. H., Ballabio, A., Boya, P., Bravo-San Pedro, J. M., Cecconi, F., Choi, A. M., Chu, C. T., Codogno, P., Colombo, M. I., & Cuervo, A. M. (2017). Molecular definitions of autophagy and related processes. *The EMBO Journal, 36*(13), 1811–1836. https://doi.org/10.15252/embj.201796697

Gazzaniga, F. S., & Blackburn, E. H. (2014). An antiapoptotic role for telomerase RNA in human immune cells independent of telomere integrity or telomerase enzymatic activity. *Blood, 124*(25), 3675–3684. https://doi.org/10.1182/blood-2014-06-582254

Geserick, C., & Blasco, M. A. (2006). Novel roles for telomerase in aging. *Mechanisms of Ageing and Development, 127*(6), 579–583. https://doi.org/10.1016/j.mad.2006.01.017

Gonzales-Ebsen, A. C., Gregersen, N., & Olsen, R. K. (2017). Linking telomere loss and mitochondrial dysfunction in chronic disease. *Front Bioscience (landmark Ed), 22*, 117–127. https://doi.org/10.2741/4475

Gorbunova, V., Seluanov, A., & Pereira-Smith, O. M. (2002). Expression of human telomerase (hTERT) does not prevent stress-induced senescence in normal human fibroblasts but protects the cells from stress-induced apoptosis and necrosis. *Journal of Biological Chemistry, 277*(41), 38540–38549. https://doi.org/10.1074/jbc.M202671200

Gorman, G. S., Chinnery, P. F., DiMauro, S., Hirano, M., Koga, Y., McFarland, R., Suomalainen, A., Thorburn, D. R., Zeviani, M., & Turnbull, D. M. (2016). Mitochondrial diseases. *Nature Reviews Disease Primers, 2*, 16080. https://doi.org/10.1038/nrdp.2016.80

Greider, C. W. (1991). Telomeres. *Current Opinion in Cell Biology, 3*(3), 444–451. https://doi.org/10.1016/0955-0674(91)90072-7

Haendeler, J., Dröse, S., Büchner, N., Jakob, S., Altschmied, J., Goy, C., Spyridopoulos, I., Zeiher, A. M., Brandt, U., & Dimmeler, S. (2009). Mitochondrial telomerase reverse transcriptase binds to and protects mitochondrial DNA and function from damage. *Arteriosclerosis, Thrombosis, and Vascular Biology, 29*(6), 929–935. https://doi.org/10.1161/ATVBAHA.109.185546

Hallberg, B. M., & Larsson, N. G. (2014). Making proteins in the powerhouse. *Cell Metabolism, 20*(2), 226–240. https://doi.org/10.1016/j.cmet.2014.07.001

Harley, C. B., Futcher, A. B., & Greider, C. W. (1990). Telomeres shorten during ageing of human fibroblasts. *Nature, 345*(6274), 458–460. https://doi.org/10.1038/345458a0

Harman, D. (1956). Aging: A theory based on free radical and radiation chemistry. *Journal of Gerontology, 11*(3), 298–300. https://doi.org/10.1093/geronj/11.3.298

Hashimoto, M., Bacman, S. R., Peralta, S., Falk, M. J., Chomyn, A., Chan, D. C., Williams, S. L., & Moraes, C. T. (2015). MitoTALEN: A general approach to reduce mutant mtDNA loads and restore oxidative phosphorylation function in mitochondrial diseases. *Molecular Therapy, 23*(10), 1592–1599. https://doi.org/10.1038/mt.2015.126

Hayflick, L., & Moorhead, P. S. (1961). The serial cultivation of human diploid cell strains. *Experimental Cell Research, 25*(3), 585–621. https://doi.org/10.1016/0014-4827(61)90192-6

Henderson, E., Hardin, C. C., Walk, S. K., Tinoco, I., Jr., & Blackburn, E. H. (1987). Telomeric DNA oligonucleotides form novel intramolecular structures containing guanine-guanine base pairs. *Cell, 51*(6), 899–908. https://doi.org/10.1016/0092-8674(87)90577-0

Hewitt, G., Jurk, D., Marques, F. D., Correia-Melo, C., Hardy, T., Gackowska, A., Anderson, R., Taschuk, M., Mann, J., & Passos, J. F. (2012). Telomeres are favoured targets of a persistent DNA damage response in ageing and stress-induced senescence. *Nature Communications, 3*, 708. https://doi.org/10.1038/ncomms1708

Hidaka, T., Pandian, G. N., Taniguchi, J., Nobeyama, T., Hashiya, K., Bando, T., & Sugiyama, H. (2017). Creation of a synthetic ligand for mitochondrial DNA sequence recognition and promoter-specific transcription suppression. *Journal of the American Chemical Society, 139*(25), 8444–8447. https://doi.org/10.1021/jacs.7b05230

Huppert, J. L., & Balasubramanian, S. (2007). G-quadruplexes in promoters throughout the human genome. *Nucleic Acids Research, 35*(2), 406–413. https://doi.org/10.1093/nar/gkl1057

Jackson, C. B., Turnbull, D. M., Minczuk, M., & Gammage, P. A. (2020). Therapeutic manipulation of mtDNA heteroplasmy: A shifting perspective. *Trends in Molecular Medicine, 26*(7), 698–709. https://doi.org/10.1016/j.molmed.2020.02.006

Kanabus, M., Heales, S. J., & Rahman, S. (2014). Development of pharmacological strategies for mitochondrial disorders. *British Journal of Pharmacology, 171*(8), 1798–1817. https://doi.org/10.1111/bph.12456

Kang, E., Wu, J., Gutierrez, N. M., Koski, A., Tippner-Hedges, R., Agaronyan, K., Platero-Luengo, A., Martinez-Redondo, P., Ma, H., Lee, Y., & Hayama, T. (2016). Mitochondrial replacement in human oocytes carrying pathogenic mitochondrial DNA mutations. *Nature, 540*(7632), 270–275. https://doi.org/10.1038/nature20592

Karnati, S., Luers, G., Pfreimer, S., & Baumgart-Vogt, E. (2013). Mammalian SOD2 is exclusively located in mitochondria and not present in peroxisomes. *Histochemistry and Cell Biology, 140*(2), 105–117. https://doi.org/10.1007/s00418-013-1099-4

Kim, Y.-H., Kumar, A., Chang, C.-H., & Pyaram, K. (2017). Reactive oxygen species regulate the inflammatory function of NKT cells through promyelocytic leukemia zinc finger. *The Journal of Immunology, 199*(10), 3478. https://doi.org/10.4049/jimmunol.1700567

Kliment, C. R., Tobolewski, J. M., Manni, M. L., Tan, R. J., Enghild, J., & Oury, T. D. (2008). Extracellular superoxide dismutase protects against matrix degradation of heparan sulfate in the lung. *Antioxidants & Redox Signaling, 10*(2), 261–268. https://doi.org/10.1089/ars.2007.1906

Kobayashi, H., Hatakeyama, H., Nishimura, H., Yokota, M., Suzuki, S., Tomabechi, Y., Shirouzu, M., Osada, H., Mimaki, M., Goto, Y. I., & Yoshida, M. (2020). Chemical reversal of abnormalities in cells carrying mitochondrial DNA mutations. *Nature Chemical Biology*. https://doi.org/10.1038/s41589-020-00676-4

Kogelnik, A. M., Lott, M. T., Brown, M. D., Navathe, S. B., & Wallace, D. C. (1996). MITOMAP: A human mitochondrial genome database. *Nucleic Acids Research, 24*(1), 177–179. https://doi.org/10.1093/nar/24.1.177

Krishnan, K. J., Greaves, L. C., Reeve, A. K., & Turnbull, D. (2007). The ageing mitochondrial genome. *Nucleic Acids Research, 35*(22), 7399–7405. https://doi.org/10.1093/nar/gkm635

Kumar A. (2019). Mitochondrial manipulation improves anti-tumor immunity. *Acta Scientific Cancer Biology, 3*(12), 28–30.

Kumar, A., Chamoto, K., Chowdhury, P. S., & Honjo, T. (2020). Tumors attenuating the mitochondrial activity in T cells escape from PD-1 blockade therapy. *eLife, 9*. https://doi.org/10.7554/eLife.52330

Kuntz, E. M., Baquero, P., Michie, A. M., Dunn, K., Tardito, S., Holyoake, T. L., Helgason, G. V., & Gottlieb, E. (2017). Targeting mitochondrial oxidative phosphorylation eradicates therapy-resistant chronic myeloid leukemia stem cells. *Nature Medicine, 23*(10), 1234–1240. https://doi.org/10.1038/nm.4399

Lagouge, M., Argmann, C., Gerhart-Hines, Z., Meziane, H., Lerin, C., Daussin, F., Messadeq, N., Milne, J., Lambert, P., Elliott, P., & Geny, B. (2006). Resveratrol improves mitochondrial function and protects against metabolic disease by activating SIRT1 and PGC-1α. *Cell, 127*(6), 1109–1122. https://doi.org/10.1016/j.cell.2006.11.013

Lee, J., Sung, Y. H., Cheong, C., Choi, Y. S., Jeon, H. K., Sun, W., Hahn, W. C., Ishikawa, F., & Lee, H. W. (2008). TERT promotes cellular and organismal survival independently of telomerase activity. *Oncogene, 27*(26), 3754–3760. https://doi.org/10.1038/sj.onc.1211037

Liu, L., Trimarchi, J. R., Smith, P. J., & Keefe, D. L. (2002). Mitochondrial dysfunction leads to telomere attrition and genomic instability. *Aging Cell, 1*(1), 40–46. https://doi.org/10.1046/j.1474-9728.2002.00004.x

Malaguarnera, L. (2019). Influence of resveratrol on the immune response. *Nutrients, 11*(5), 946. https://doi.org/10.3390/nu11050946

Malinee, M., Kumar, A., Hidaka, T., Horie, M., Hasegawa, K., Pandian, G. N., & Sugiyama, H. (2020). Targeted suppression of metastasis regulatory transcription factor SOX_2 in various cancer cell lines using a sequence-specific designer pyrrole-imidazole polyamide. *Bioorganic & Medicinal Chemistry, 28*(3), 115248. https://doi.org/10.1016/j.bmc.2019.115248

Marion, R. M., Strati, K., Li, H., Tejera, A., Schoeftner, S., Ortega, S., Serrano, M., & Blasco, M. A. (2009). Telomeres acquire embryonic stem cell characteristics in induced pluripotent stem cells. *Cell Stem Cell, 4*(2), 141–154. https://doi.org/10.1016/j.stem.2008.12.010

Moruno-Manchon, J. F., Lejault, P., Wang, Y., McCauley, B., Honarpisheh, P., Scheihing, D. A. M., Singh, S., Dang, W., Kim, N., Urayama, A., & Zhu, L. (2020). Small-molecule G-quadruplex stabilizers reveal a novel pathway of autophagy regulation in neurons. *eLife*, 9, e52283. https://doi.org/10.7554/eLife.52283

Minczuk, M., Papworth, M. A., Miller, J. C., Murphy, M. P., & Klug, A. (2008). Development of a single-chain, quasi-dimeric zinc-finger nuclease for the selective degradation of mutated human mitochondrial DNA. *Nucleic Acids Research, 36*(12), 3926–3938. https://doi.org/10.1093/nar/gkn313

Miquel, J., Economos, A. C., Fleming, J., & Johnson, J. E., Jr. (1980). Mitochondrial role in cell aging. *Experimental Gerontology, 15*(6), 575–591. https://doi.org/10.1016/0531-5565(80)90010-8

Miwa, S., Czapiewski, R., Wan, T., Bell, A., Hill, K. N., von Zglinicki, T., & Saretzki, G. (2016). Decreased mTOR signalling reduces mitochondrial ROS in brain via accumulation of the telomerase protein TERT within mitochondria. *Aging (albany NY), 8*(10), 2551–2567. https://doi.org/10.18632/aging.101089

Murphy, M. P., & Siegel, R. M. (2013). Mitochondrial ROS fire up T cell activation. *Immunity, 38*(2), 201–202. https://doi.org/10.1016/j.immuni.2013.02.005

Nguyen, G. H., Tang, W., Robles, A. I., Beyer, R. P., Gray, L. T., Welsh, J. A., Schetter, A. J., Kumamoto, K., Wang, X. W., Hickson, I. D., & Maizels, N. (2014). Regulation of gene expression by the BLM helicase correlates with the presence of G-quadruplex DNA motifs. *Proceedings of the National Academy of Sciences of the United States of America, 111*(27), 9905–9910. https://doi.org/10.1073/pnas.1404807111

Nicolson, G. L. (2014). Mitochondrial dysfunction and chronic disease: Treatment with natural supplements. *Integrative Medicine (Encinitas, California), 13*(4), 35–43.

Nightingale, H., Pfeffer, G., Bargiela, D., Horvath, R., & Chinnery, P. F. (2016). Emerging therapies for mitochondrial disorders. *Brain, 139*(6), 1633–1648. https://doi.org/10.1093/brain/aww081

Oikawa, S., & Kawanishi, S. (1999). Site-specific DNA damage at GGG sequence by oxidative stress may accelerate telomere shortening. *FEBS Letters, 453*(3), 365–368. https://doi.org/10.1016/s0014-5793(99)00748-6

Passos, J. F., Saretzki, G., & von Zglinicki, T. (2007). DNA damage in telomeres and mitochondria during cellular senescence: Is there a connection? *Nucleic Acids Research, 35*(22), 7505–7513. https://doi.org/10.1093/nar/gkm893

Phan, A. T., Modi, Y. S., & Patel, D. J. (2004). Propeller-type parallel-stranded G-quadruplexes in the human c-myc promoter. *Journal of the American Chemical Society, 126*(28), 8710–8716. https://doi.org/10.1021/ja048805k

Price, N. L., Gomes, A. P., Ling, A. J., Duarte, F. V., Martin-Montalvo, A., North, B. J., Agarwal, B., Ye, L., Ramadori, G., Teodoro, J. S., & Hubbard, B. P. (2012). SIRT1 is required for AMPK activation and the beneficial effects of resveratrol on mitochondrial function. *Cell Metabolism, 15*(5), 675–690. https://doi.org/10.1016/j.cmet.2012.04.003

Qin, Z., Reszka, K. J., Fukai, T., & Weintraub, N. L. (2008). Extracellular superoxide dismutase (ecSOD) in vascular biology: An update on exogenous gene transfer and endogenous regulators of ecSOD. *Translational Research, 151*(2), 68–78. https://doi.org/10.1016/j.trsl.2007.10.003

Raha, S., & Robinson, B. H. (2000). Mitochondria, oxygen free radicals, disease and ageing. *Trends in Biochemical Sciences, 25*(10), 502–508. https://doi.org/10.1016/s0968-0004(00)01674-1

Reddy, P., Ocampo, A., Suzuki, K., Luo, J., Bacman, S. R., Williams, S. L., Sugawara, A., Okamura, D., Tsunekawa, Y., Wu, J., & Lam, D. (2015). Selective elimination of mitochondrial mutations in the germline by genome editing. *Cell, 161*(3), 459–469. https://doi.org/10.1016/j.cell.2015.03.051

Reyes, A., Gissi, C., Pesole, G., & Saccone, C. (1998). Asymmetrical directional mutation pressure in the mitochondrial genome of mammals. *Molecular Biology and Evolution, 15*(8), 957–966. https://doi.org/10.1093/oxfordjournals.molbev.a026011

Rhodes, D., & Lipps, H. J. (2015). G-quadruplexes and their regulatory roles in biology. *Nucleic Acids Research, 43*(18), 8627–8637. https://doi.org/10.1093/nar/gkv862

Roychoudhury, S., Pramanik, S., Harris, H. L., Tarpley, M., Sarkar, A., Spagnol, G., Sorgen, P. L., Chowdhury, D., Band, V., Klinkebiel, D., & Bhakat, K. K. (2020). Endogenous oxidized DNA bases and APE1 regulate the formation of G-quadruplex structures in the genome. *Proceedings of the National Academy of Sciences, 117*(21), 11409. https://doi.org/10.1073/pnas.1912355117

Sahin, E., Colla, S., Liesa, M., Moslehi, J., Müller, F. L., Guo, M., Cooper, M., Kotton, D., Fabian, A. J., Walkey, C., & Maser, R. S. (2011). Telomere dysfunction induces metabolic and mitochondrial compromise. *Nature, 470*(7334), 359–365. https://doi.org/10.1038/nature09787

Saibil, S. D., Paul, M. S., Laister, R. C., Garcia-Batres, C. R., Israni-Winger, K., Elford, A. R., Grimshaw, N., Robert-Tissot, C., Roy, D. G., Jones, R. G., & Nguyen, L.T. (2019). Activation of peroxisome proliferator-activated receptors alpha and delta synergizes with inflammatory signals to enhance adoptive cell therapy. *Cancer Research, 79*(3), 445–451. https://doi.org/10.1158/0008-5472.Can-17-3053

Santos, J. H., Meyer, J. N., Skorvaga, M., Annab, L. A., & Van Houten, B. (2004). Mitochondrial hTERT exacerbates free-radical-mediated mtDNA damage. *Aging Cell, 3*(6), 399–411. https://doi.org/10.1111/j.1474-9728.2004.00124.x

Saretzki, G., Murphy, M. P., & von Zglinicki, T. (2003). MitoQ counteracts telomere shortening and elongates lifespan of fibroblasts under mild oxidative stress. *Aging Cell, 2*(2), 141–143. https://doi.org/10.1046/j.1474-9728.2003.00040.x

Scarpulla, R. C. (2011). Metabolic control of mitochondrial biogenesis through the PGC-1 family regulatory network. *Biochimica Et Biophysica Acta, 1813*(7), 1269–1278. https://doi.org/10.1016/j.bbamcr.2010.09.019

Seimiya, H., Sawada, H., Muramatsu, Y., Shimizu, M., Ohko, K., Yamane, K., & Tsuruo, T. (2000). Involvement of 14–3-3 proteins in nuclear localization of telomerase. *EMBO Journal, 19*(11), 2652–2661. https://doi.org/10.1093/emboj/19.11.2652

Shekhova, E. (2020). Mitochondrial reactive oxygen species as major effectors of antimicrobial immunity. *PLOS Pathogens, 16*(5), e1008470. https://doi.org/10.1371/journal.ppat.1008470

Singhapol, C., Pal, D., Czapiewski, R., Porika, M., Nelson, G., & Saretzki, G. C. (2013). Mitochondrial telomerase protects cancer cells from nuclear DNA damage and apoptosis. *PLoS ONE, 8*(1), e52989. https://doi.org/10.1371/journal.pone.0052989

Škrtić, M., Sriskanthadevan, S., Jhas, B., Gebbia, M., Wang, X., Wang, Z., Hurren, R., Jitkova, Y., Gronda, M., Maclean, N., & Lai, C. K. (2011). Inhibition of mitochondrial translation as a therapeutic strategy for human acute *Myeloid leukemia. Cancer Cell, 20*(5), 674–688. https://doi.org/10.1016/j.ccr.2011.10.015

Sohal, R. S., Ku, H. H., Agarwal, S., Forster, M. J., & Lal, H. (1994). Oxidative damage, mitochondrial oxidant generation and antioxidant defenses during aging and in response to food restriction in the mouse. *Mechanisms of Ageing and Development, 74*(1–2), 121–133. https://doi.org/10.1016/0047-6374(94)90104-x

Stauffer, J., Panda, B., & Ilmonen, P. (2018). Telomere length, sibling competition and development of antioxidant defense in wild house mice. *Mechanisms of Ageing and Development, 169*, 45–52. https://doi.org/10.1016/j.mad.2017.10.002

Stewart, J. B., & Chinnery, P. F. (2015). The dynamics of mitochondrial DNA heteroplasmy: Implications for human health and disease. *Nature Reviews Genetics, 16*(9), 530–542. https://doi.org/10.1038/nrg3966

Sturtz, L. A., Diekert, K., Jensen, L. T., Lill, R., & Culotta, V. C. (2001). A fraction of yeast Cu,Zn-superoxide dismutase and its metallochaperone, CCS, localize to the intermembrane space of mitochondria. A physiological role for SOD1 in guarding against mitochondrial oxidative damage. *Journal of Biological Chemistry, 276*(41), 38084–38089. https://doi.org/10.1074/jbc.M105296200

Sugiyama, H., & Saito, I. (1996). Theoretical studies of GG-specific photocleavage of DNA via electron transfer: Significant lowering of ionization potential and 5'-localization of HOMO of stacked GG bases in B-form DNA. *Journal of the American Chemical Society, 118*(30), 7063–7068. https://doi.org/10.1021/ja9609821

Sun, D., Liu, W. J., Guo, K., Rusche, J. J., Ebbinghaus, S., Gokhale, V., & Hurley, L. H. (2008). The proximal promoter region of the human vascular endothelial growth factor gene has a G-quadruplex structure that can be targeted by G-quadruplex-interactive agents. *Molecular Cancer Therapeutics, 7*(4), 880–889. https://doi.org/10.1158/1535-7163.MCT-07-2119

Tang, W., Robles, A. I., Beyer, R. P., Gray, L. T., Nguyen, G. H., Oshima, J., Maizels, N., Harris, C. C., & Monnat, R. J. Jr. (2016). The Werner syndrome RECQ helicase targets G4 DNA in human cells to modulate transcription. *Human Molecular Genetics, 25*(10), 2060–2069. https://doi.org/10.1093/hmg/ddw079

Tsang, C. K., Liu, Y., Thomas, J., Zhang, Y., & Zheng, X. F. (2014). Superoxide dismutase 1 acts as a nuclear transcription factor to regulate oxidative stress resistance. *Nature Communications, 5*, 3446. https://doi.org/10.1038/ncomms4446

Vasan, K., Werner, M., & Chandel, N. S. (2020). Mitochondrial metabolism as a target for cancer therapy. *Cell Metabolism, 32*(3), 341–352. https://doi.org/10.1016/j.cmet.2020.06.019

von Zglinicki, T. (2000). Role of oxidative stress in telomere length regulation and replicative senescence. *Annals of the New York Academy of Sciences, 908*, 99–110. https://doi.org/10.1111/j.1749-6632.2000.tb06639.x

Wallace, D. C. (2005). A mitochondrial paradigm of metabolic and degenerative diseases, aging, and cancer: A dawn for evolutionary medicine. *Annual Review of Genetics, 39*, 359–407. https://doi.org/10.1146/annurev.genet.39.110304.095751

Wang, G., Chen, H. W., Oktay, Y., Zhang, J., Allen, E. L., Smith, G. M., Fan, K. C., Hong, J. S., French, S. W., McCaffery, J. M., & Lightowlers, R. N. (2010a). PNPASE regulates RNA import into mitochondria. *Cell, 142*(3), 456–467. https://doi.org/10.1016/j.cell.2010.06.035

Wang, Z., Rhee, D. B., Lu, J., Bohr, C. T., Zhou, F., Vallabhaneni, H., de Souza-Pinto, N. C., & Liu, Y. (2010b). Characterization of oxidative guanine damage and repair in mammalian telomeres. *PLoS Genetics, 6*(5), e1000951–e1000951. https://doi.org/10.1371/journal.pgen.1000951

Wang, W., Karamanlidis, G., & Tian, R. (2016). Novel targets for mitochondrial medicine. *Science Translational Medicine, 8*(326), 326rv323–326rv323. https://doi.org/10.1126/scitranslmed.aac7410

Yee, C., Yang, W., & Hekimi, S. (2014). The intrinsic apoptosis pathway mediates the pro-longevity response to mitochondrial ROS in C. elegans. *Cell, 157*(4), 897–909. https://doi.org/10.1016/j.cell.2014.02.055

Zheng, Q., Liu, P., Gao, G., Yuan, J., Wang, P., Huang, J., Xie, L., Lu, X., Di, F., Tong, T., & Chen, J. (2019). Mitochondrion-processed TERC regulates senescence without affecting telomerase activities. *Protein & Cell, 10*(9), 631–648. https://doi.org/10.1007/s13238-019-0612-5

Madhu Malinee is an Indian origin research scholar, currently working under the supervision of Prof. Hiroshi Sugiyama. She joined the doctoral course in 2016 at the Graduate School of Medicine, Kyoto University. Earlier, she received her master's degree in biochemistry in 2013 from Hyderabad Central University, Hyderabad. Her research interest is cancer immunotherapy, and she is currently working on employing epigenetic modulation to improve the efficacy of immune checkpoint blockade-based cancer immunotherapy.

Hiroshi Sugiyama is professor in the Department of Chemistry, Graduate School of Science, Kyoto University, teaching chemical biology and bioorganic chemistry of nucleic acids since 2003. He received his Ph.D. in 1984 from the Department of Synthetic Chemistry, Graduate School of Engineering at Kyoto University. After postdoctoral studies at the University of Virginia with Sidney M. Hecht, he returned to Kyoto University in 1986 as an assistant professor and became an associate professor in 1993. In 1996, he joined the Institute of Biomaterials and Bioengineering at Tokyo Medical and Dental University.

Chapter 17
Evolution, Motor of the Changing Biosphere

Johann Hohenegger

Abstract The biosphere, representing the entity of organisms, has changed since it started 3,490 million years before present. Biodiversity increased steadily in the Phanerozoic starting at 541 million years ago, and was interrupted by five mass extinctions. The motor of this increase is organismic evolution caused by changing environmental conditions. Microevolution explains changes in the species' life span and diversification by speciation. Macroevolution is responsible for extreme changes during speciation leading to higher taxonomic units (genera and families). Macroevolution is the main motor for rapid recovering of environments after mass extinction.

Keywords Biosphere · Biodiversity evolution · Microevolution · Macroevolution · Life span

17.1 The Changing Biosphere

> Die Pflanze, welche ihre Wurzeln Nahrung suchend in den Boden senkt und gleichzeitig sich athmend in die Luft erhebt, ist ein gutes Bild der Stellung organischen Lebens in der Region der Wechselwirkung der oberen Sphären und der Lithosphäre, und es lässt sich auf der Oberfläche des Festen eine selbständige Biosphäre unterscheiden.
>
> Eduard Suess
>
> Cited from: Die Entstehung der Alpen, p. 157
>
> Braumüller Verlag, 1875

The atmosphere, lithosphere, hydrosphere, and biosphere denominate the layers of the Earth's surface. The biosphere indicates the totality of organisms settling in the other spheres. First described by de Lamarck (1802) and coined by Suess (1875), Teilhard de Chardin (1957) understood the biosphere in the strict sense mentioned above. The definition provided by Vernadsky (1926) includes the relationships to

J. Hohenegger (✉)
Department of Palaeontology, University of Vienna, Geozentrum, Althanstrasse 14, 1090 Vienna, Austria
e-mail: johann.hohenegger@univie.ac.at

K. Nishimura et al. (eds.), *Creative Complex Systems*, Creative Economy,
https://doi.org/10.1007/978-981-16-4457-3_17

other spheres, which is matched by the ecosphere (Cole, 1958) and its totality of ecosystems (for detailed discussion of the biosphere problem, see Hugget, 1999).

The biosphere interacts with the other spheres, and changes in the latter are responded to by the former. For example, long-term changes in the atmosphere and hydrosphere by oscillations in the Earth's movement have led to strong reactions in the biosphere. Different period lengths and phases in *eccentricity* (the deviations from a circular movement around the sun with period lengths of 100 and 400 ky; ky = 1,000 years), *tilting* of the earth axis with a period length of 41 ky, and *precession* (i.e., orientation changes of the rotating earth axis with periods of 19 and 21 ky) caused differences in insolation depending on latitudes. Interferences in orbital cycles are thus responsible for the development of hothouse or icehouse conditions over long-lasting periods (Kerr, 1987). Tectonics are responsible for the main changes in the lithosphere, ranging from earthquakes and volcanic eruptions to plate tectonics (van der Pluijm & Marshak, 2004).

Changes in the biosphere on its part affect the other spheres. For instance, photosynthesis and vegetation cover are responsible for the production of oxygen and climate change, respectively. Today's global warming is partially caused by the strong reduction of tropical rainforests that hinders fixation of CO_2 by the plants (Simmons, 2018).

Regarding earth history, the biosphere started about 3,490 my (my = million years) ago with prokaryotic organisms belonging to Eubacteria and Archaea living in the oceans. Cyanobacteria initiated the production of oxygen by photosynthesis. At the beginning of the Proterozoic Eon about 2,420 my ago, the incidence of eukaryotic unicellular algae led to a strong increase in oxygen, which effectively changed the composition of the atmosphere. After a long-lasting period of stability with reducing deep oceans, the development of various "snowball earth" conditions between 850 and 635 my ago (Gradstein et al., 2012) led to the transformation from unicellular to multicellular organisms.

Organisms became abundant in Phanerozoic time, thereby making the estimation of biodiversity, reduced to organisms with fossilized parts, possible (Fig. 17.1). This time segment can be divided into three groups of stratigraphic periods distinguished by strong changes in their organismic composition. The first group of periods (Cambrian, Ordovician) starting at 541 my ago was limited by a glaciation event at 444 my where the decreasing Cambrian fauna became rare and the Paleozoic fauna became dominant. At the beginning of the following group of stratigraphic periods (Silurian to Permian) the first herbaceous plants occupied the land, followed by invertebrate animals. This extended the biosphere to the atmosphere. At around 400 my ago, the first trees and terrestrial vertebrates appeared. Soils as the result of plant roots released nutrients into the ocean leading to planktonic algal blooms. "Marine snow"

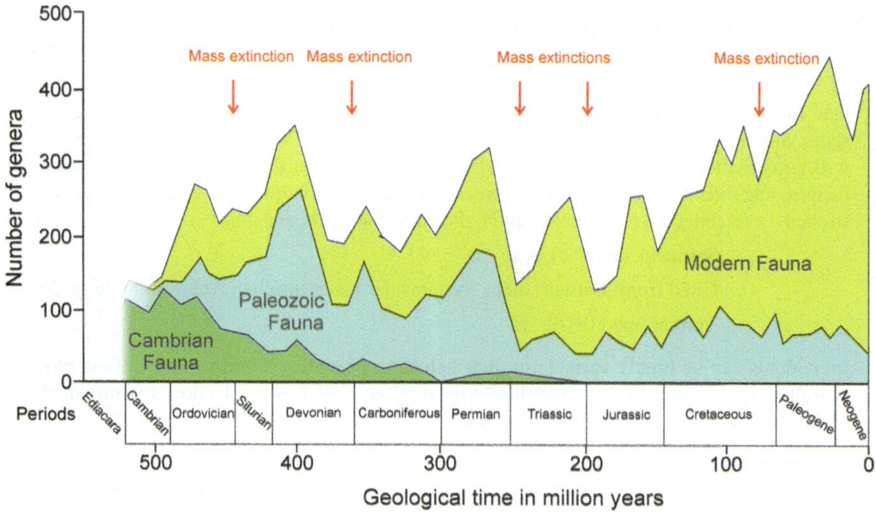

Fig. 17.1 Marine biodiversity during the Phanerozoic (after Alroy, 2010)

covered the sea floor and reduced oxygen in the oceans. This led to a weak extinction event of the marine biosphere around 375 my ago. This second group of periods was limited at 252 my ago with the strongest extinction event caused by the most extreme volcanisms in earth history. Only 25% of terrestrial animals survived and the vegetation cover was strongly reduced. Much more dramatic was this turnover in the sea where only 5% of eukaryotic species survived (Benton, 2005).

"Modern" eukaryotic species became dominant in the following stratigraphic periods from the Triassic to Recent. At the end of the Triassic at 201.3 my ago, 80% of species were lost and the few survivors of the Paleozoic biosphere became extinct. This again was caused by an extreme volcanism accompanying plate tectonics, where the unitary Pangea continent began to disrupt and started to open the Atlantic Ocean (Blackburn et al., 2013). After this event, further development of the biosphere was interrupted at 66 my ago (Cretaceous/Paleogene boundary) by a meteoric impact, where 76% of species were lost (Schulte et al., 2010). Finally, farming and stock-breeding by humans starting around 9,500 BC in the Middle East and around 6,000 BC in East Asia and Central America changed the biosphere, becoming extreme in modern times. This could be the beginning of the sixth mass extinction event, in this case caused by humans (Ceballos et al., 2015).

The reaction mechanisms of the biosphere to environmental influences differing in intensities and with short to longer duration will be described in the following.

17.2 Organisms, Populations, and Species

> Den Inbegriff einer gewissen Anzahl zusammengesetzter, miteinander verknüpfter. gegenseitig aufeinander wirkender, und durch ihr gemeinsames Wirken einander thätig und wirksam erhaltener Apparate, welche gerade durch ihre gegenseitige harmonische Zusammensetzung einem allgemein großen Zweck, nämlich der Erhaltung des Ganzen, entgegenstreben, bezeichnet die Naturkunde durch die Benennung Organismus.
>
> Samuel Christian Lucae
>
> Cited from: Entwurf eines Systems der medicinischen Anthropologie, p. 26
>
> Varrentrapp (1816)

> Individuals can be highly variable in their behavior, physiology, and genetics; these have profound implications for the dynamics of populations, which are collections of individuals of the same species linked by reproduction.
>
> Markus P. Eichhorn
>
> Cited from: Natural Systems, p. 3
>
> Wiley Blackwell, 2016

Within the hierarchical system of biological organization topped by the biosphere, individuals defined as spatiotemporal units possessing their own history and location can be found twice; first as organisms at the unicellular and multicellular level, and second at the population level. Individuality is also discussed for the species category, which represents a group of populations based on different properties (Mayden, 1997). These differences in species concepts led to the discussion about individuality of the species, which is regarded as the unit in biological evolution (Hull, 1976, 1978; Kitcher, 1984; Splitter, 1988).

At the organism level, either as a single eukaryotic cell or a multicellular organism, all start with a single cell as the result of asexual or sexual reproduction (with the exception of budding and dividing; e.g., in colonial organisms, but colonies must be regarded as a single organism). The following growth to adult individuals, called ontogeny or development, is complex in multicellular organisms producing tissues and organs. Because every cell of an organism possesses identical DNA, cell differentiation is controlled during development by regulatory genes acting at different times and gene locations. Development is governed by a semi-hierarchical gene regulatory network (GRN), where the normal gene expression from high-ranking regulatory genes to batteries of lower-level genes can be responded to by feedback from low-level genes modulating the expression of regulatory genes (Peter & Davidson, 2015; Rebeiz et al., 2015).

Phenotypic variability is not only caused by differences in genotypes, the genetic makeup of an individual, but is also influenced by epigenetics acting during development. Epigenetics is thus responsible for phenotypic variability in asexually reproducing organisms (clones), which possess identical genotypes. Environmental factors (e.g., salinity, diet, pH, humidity, temperature, photoperiod, seasonality, population density, or the presence of predators) induce epigenetic mechanisms such as DNA methylation, histone modification, and chromatin structure that have the ability to program and alter gene expression. Heritability of epigenetics is only transgenerational, but can lead to epigenetic mutations in the germ lines (Skinner, 2015).

The level above organisms showing individuality are populations, groups of (organismic) individuals interconnected by asexual or sexual reproduction and sharing the same region. Each individual within a population represents a genotype consisting of alleles, the variants of a gene. Therefore, populations can always be characterized by frequency distributions of genotypes. In diploid organisms, when only two alleles A and a can be found at a single gene location, the three genotypes *homozygote* (*AA, aa*) and *heterozygote* (*Aa*) are present. With given proportion p for allele A and $q = 1 - p$ for allele a, the frequency distribution of genotypes follows the Hardy–Weinberg law by

$$p^2 AA + 2pq\,Aa + q^2 aa$$

This works only in populations with *panmixia*, where all individuals are potential mating partners. Consistency of frequency distributions in following generations marks stability, while statistically significant deviations signalize factors disturbing the Hardy–Weinberg equilibrium.

The genetic base of continuous variation in a meristic character is caused by interaction of several genes (polygene). Frequency distributions of quantitative traits can be modeled by bell-shaped normal distributions with the mean μ and the standard deviation σ . The positions of genotypes are then determined by $Aa = \mu$, $AA = \mu + \sigma$ and $aa = \mu - \sigma$. Introducing a dominance factor d signalizes the distribution shift to the dominant allele by $\mu \pm d$ (Hartl & Clark, 2007). Because frequency distributions are based on phenotype reactions to the genotypes, phenotypic variances can be partitioned into genotypic and environmental variances. The latter shows the impact of epigenetic factors acting during development.

Several populations covering a region can be linked by the dispersal of individuals leading to gene flow (Eichhorn, 2016). These metapopulations are found in organisms where the mobility of reproducing individuals or the dispersal of germ cells is weak. Metapopulations are characteristic for terrestrial organisms, especially plants, and for benthic organisms in shallow marine environments (e.g., macroalgae, stony corals, bivalves).

Changes in frequency distributions of genotypes in a sequence of generations depend on population size, life span, and reproductive strategies of individuals. While reproduction at the end of an individual's life (semelparity, monocarpy) separates the filial generations, reproduction taking place at several times during life (iteroparity, polycarpy) leads to time-overlapping filial generations.

Selection is the driving force for changes in frequency distributions of homogeneous populations without connections to other populations. Individuals differ in fitness, which is the contribution to the next generation in relation to the average of the population. Genotypes that are preferred in reproduction by higher fitness lead to frequency distribution shifts towards the preferred genotypes. In *stabilizing selection*, genotypes near the population mean μ are preferred by eliminating both extremes (Fig. 17.2). *Disruptive selection*, in contrast, prefers both extremes $\mu \pm d$ leading

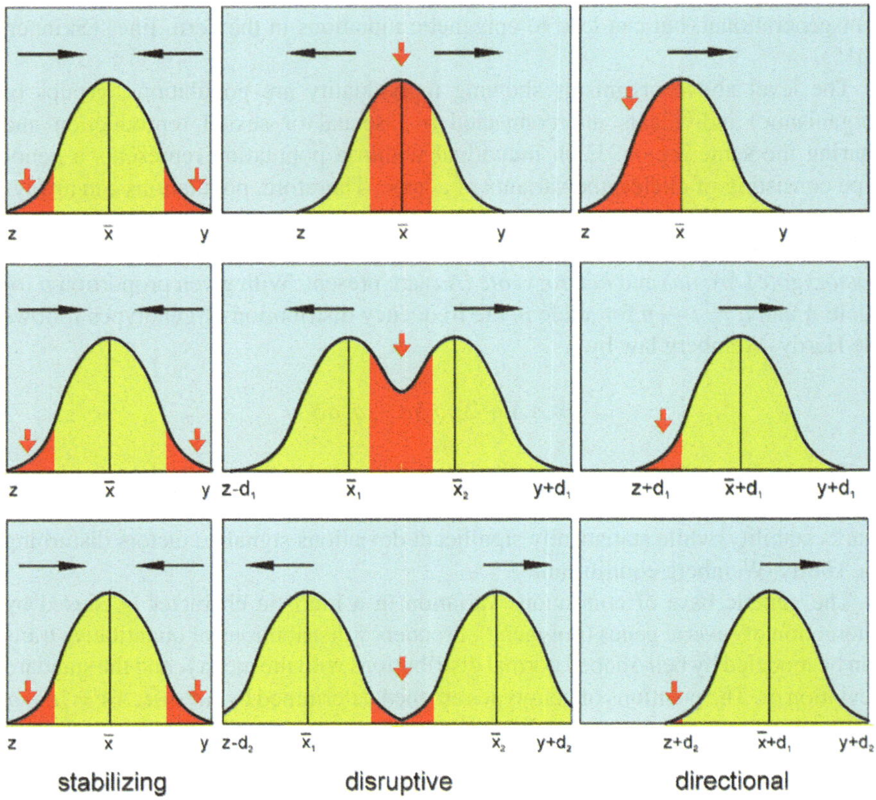

Fig. 17.2 Different forms of selection (after Solbrig & Solbrig, 1979)

to a partition of populations in following generations. The most common selection mode caused by changing environmental conditions is *directional selection*, where one of the extremes $\mu + d$ or $\mu - d$ is preferred (Fig. 17.2).

Populations can be characterized by mean fitness. Within a stable environment, proportions of genotypes possess different fitness in the population. This can be depicted in an adaptive landscape with valleys and peaks caused by isolines of fitness (Wright, 1967). Peaks in these landscapes show better adaptations, and populations with low mean fitness put into this landscape (by mutation or migration) try to reach the closest peak by directional selection. After attaining an adaptive peak, stabilizing selection starts. Continuous or abrupt environmental changes effectuate synchronous transformations of the adaptive landscape, whereby the intensity of abrupt changes determines the population's survival.

Metapopulations are internally connected in different ways. Four general relations are used to define a species (Pigliucci & Kaplan, 2006):

1. Phylogenetic (molecular-genetic and/or morphological-monophyletic) distance
2. Reproductive (genetic) isolation

3. Ecological/functional differentiation
4. Phenetic (morphological) differentiation.

Regarding the importance of the environment on both the development of phenotypes and restricting genotype frequencies by selection, the *Ecological Species Concept* (van Valen, 1976) seems to be the most appropriate for combining populations to a unit. Within the numerous species concepts, the ecological species belongs to "explanatory concepts" (Hohenegger, 2012) or "species conceptualization" concepts (de Queiroz, 2007). Phenotypic frequency distributions describe variations caused by genotypes together with environmental influences. They can thus be used to find ecological species checking significant differences between populations within metapopulations. Using MANOVA (Multiple Analysis of Variance; e.g., see Zar, 2010), species defined by phenetic homogeneities are easy to recognize and can preferably be used in biodiversity analyses (Fig. 17.3). They belong to "operational concepts" (Hohenegger, 2012) allowing "species delimitations" (de Queiroz, 2007).

17.3 Microevolution

> The production of form from formlessness in the egg-derived individual, the multiplication of parts and the orderly creation of diversity among them, is an actual evolution, of which anyone may ascertain the facts, but of which no one has dissipated the mystery in any significant measure. This microevolution forms an integral part of the grand evolution problem and lies at the base of it, so that we shall have to understand the minor process before we can thoroughly comprehend the more general one.
>
> Robert Greenleave Leavitt
>
> Cited from: Botanical Gazette, vol. 47, p. 30
>
> University Chicago Press (1909)

Like ontogeny in organisms, microevolution exanimates changes in species during longer (geological) time intervals. The number of generations within these time intervals starting at the speciation event, the birth of a species, and ending by extinction or splitting up into daughter species, depends on the life span of individuals. While mutation in combination with directional selection are the single mechanisms for the evolution of a species consisting of a homogeneous population, interconnections between subpopulations of an inhomogeneous species cause additional evolutionary mechanisms.

Mutation changes the stock of genotypes by introducing new alleles depending on mutation rates. Under stable environmental conditions, the extremes in the new genotypes were eliminated by stabilizing selection, while less extreme new genotypes will be preserved. This explains the random shift of species distributions through time.

Genetic Drift is a stochastic model explaining the random enrichment of genotypes in filial generations, where the following rule is valid: "the smaller population size

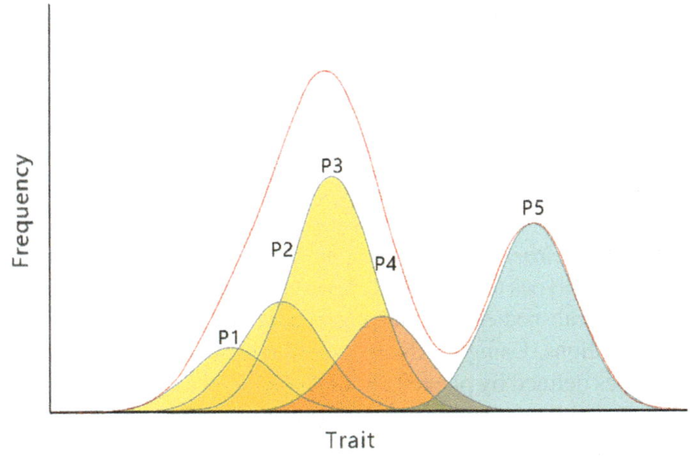

Mean
19.5

Standard deviation
11.1

Probability (Homogeneity)
0.005

Probability of concordance

	P1	P2	P3	P4
P2	0.563			
P3	0.328	0.577		
P4	0.083	0.231	0.587	
P5	**4.5E-3**	**0.004**	**0.043**	0.076

Fig. 17.3 Determination of ecological species by frequency distributions of phenotypes using a theoretical metapopulation composed of five populations with identical standard deviations ($\sigma = 5$) and different means ($\mu_1 = 5, \mu_2 = 10, \mu_3 = 15, \mu_4 = 20, \mu_5 = 35$). The red line marks the summary distribution, the general mean and standard deviation are gained by analysis of variance (ANOVA), where homogeneity is checked by Fisher's F-test. Multiple comparisons between single populations using the T-test are given in the probability matrix of pairwise comparisons. While populations 1 to 4 are homogeneous representing an (ecological) species, population 5 is significantly different and can be regarded as another species (after Hohenegger, 2013)

the higher the enrichment of a randomly chosen genotype." The *Founder Effect*, where new genotypes becoming rapidly dominant in small populations is based on genetic drift.

Gene Flow is the interconnection of populations exchanging genotypes by migration. Size differences of populations determine the strength of flow directions from larger to smaller populations, and geographic distances between populations are also responsible for gene flow intensities.

Similar to species with homogeneous frequency distributions, where in stable environmental conditions stasis or random walks are explained by mutation and stabilizing selection, stasis and random walks in species based on several populations are more complicated combining the factors mutation, genetic drift, gene flow and stabilizing selection. Differing intensities of these interfering factors can result in clear trends even though the environment remains stable. Otherwise, continuously changing environmental conditions are responded to in species with homogenous frequency distributions by directional selection leading to lineages. The complex interaction of evolutionary factors in species built on several populations explains environmentally induced lineages by replacement between directional change, random walks, and stasis (Hunt, 2007; Hunt et al., 2015).

Speciation, the birth of a species, is always caused putting the complete or parts of species populations into new environmental conditions where they have to adapt. These adaptive zones can be stable or changing through time, sometimes with instantaneous transitions or suddenly finished by a quite different zone. Speciation thus depends on the onset of a second stable or changing adaptive zone that parallels the former zone.

The various forms of speciation can be divided into "*split-off*" and "*split-up*" speciation (Hohenegger, 2013). Split-off speciation can be found in homogeneous as well as inhomogeneous species, while split-up speciation is restricted to species built up by several populations.

a. Split-Off Speciation

Gradualistic speciation (Fig. 17.4). The adaptive zone occupied by the mother species changes continuously. Over time, a new, continuously changing adaptive zone opens, which steadily transforms from the initial adaptive zone and becomes contemporaneous with it.

Punctuated equilibrium (Fig. 17.5). This model is similar to gradualistic speciation based on two compact adaptive zones, where the second zone occurs over a time and becomes contemporaneous with the first zone. In contrast to the gradualistic model, both adaptive zones are stable. The transition from the first zone to the second is short (punctual) in the geological sense.

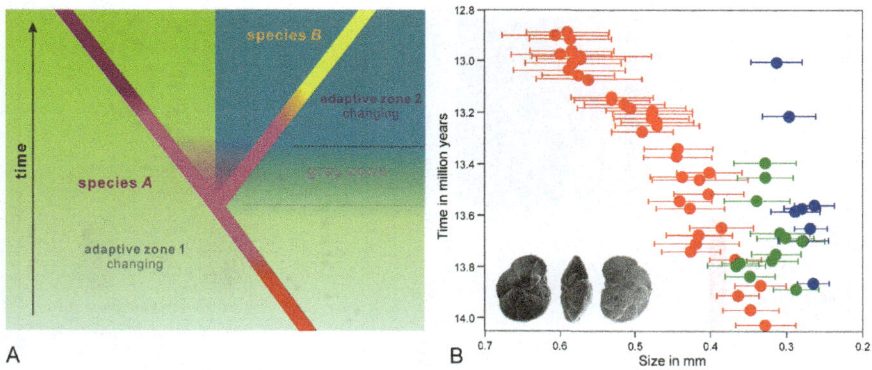

Fig. 17.4 Gradualistic speciation: **a** Theoretical model (Hohenegger, 2013). **b** Example based on size changes of the plankton foraminifer *Fohsella fohsi*. Three cores from different regions marked by colors (after Si et al., 2018)

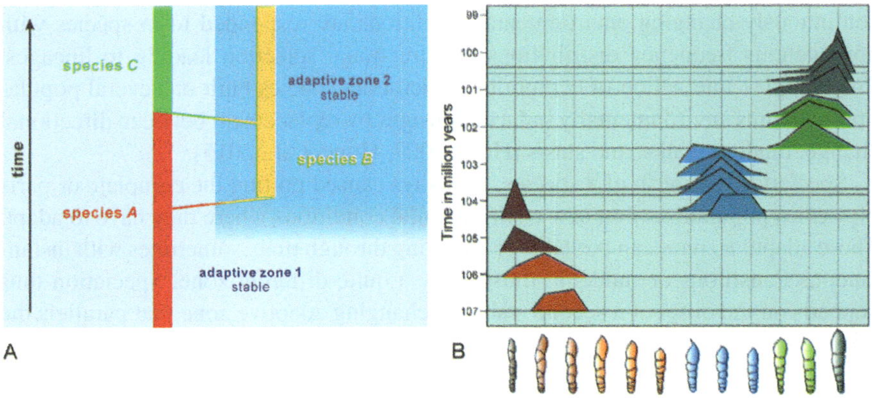

Fig. 17.5 Punctuated equilibrium speciation: **a** Theoretical model (Hohenegger, 2013). **b** Example based on the benthic foraminifer *Pleurostomella* from the section Zeltberg, NW Germany (after Bettenstaedt & Spiegler, 1982)

Steady quantum speciation (Fig. 17.6). This speciation type derives from the punctuated equilibrium model. The difference from the former model is the contemporaneous closing of the first zone and opening of the second zone, both being stable.

Gradual quantum speciation (Fig. 17.7). The difference between this model and the former is the gradual change in both adaptive zones. Again, the mother species becomes extinct at the border between the two differing adaptive zones, while the daughter species adapts to the new zone, following its gradual change by continuously changing genotype compositions (evolutionary line). Quantum speciation must not be confused with an instantaneous change in transformation rates within the lifetime

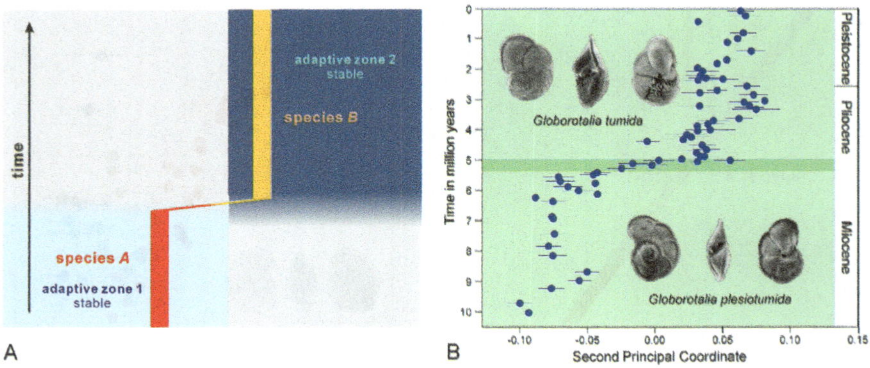

Fig. 17.6 Steady quantum speciation: **a** Theoretical model (Hohenegger, 2013). **b** Example based on form factors (eigenshape analysis) of the plankton foraminifers *Globorotalia plesiotumida* and *G. tumida*; mean diameter 0.6 mm (after Malmgren et al., 1983)

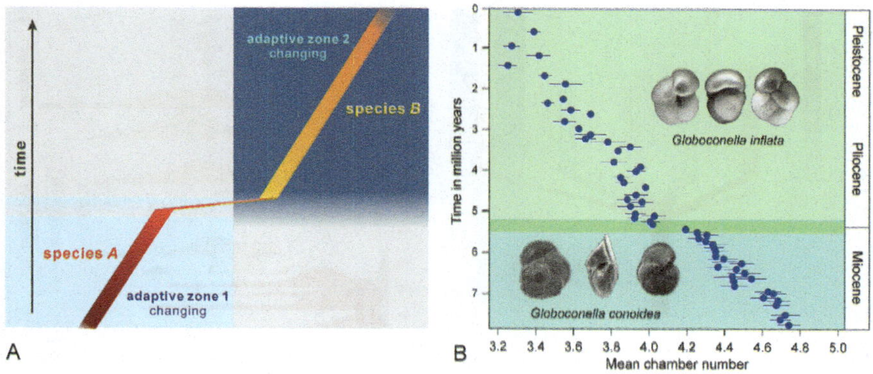

Fig. 17.7 Gradual quantum speciation: **a** Theoretical model (Hohenegger, 2013). **b** Example based on the plankton foraminifers *Globoconella conoidea* with keeled periphery and *G. inflata* with rounded periphery, both sharply differentiated at the Miocene/Pliocene boundary; mean diameter 0.6 mm (after Malmgren & Kennett, 1981)

of a species. In contrast to quantum speciation, genotype frequencies are the same before and after starting a different transformation rate.

Split-off speciation types are the commonly used speciation models often checked with fossil marine invertebrates. Low abundance and many gaps in the sedimentary record entertained doubts of the models (Hunt, 2007; Hunt et al., 2015). The stratigraphical record of marine microfossils, especially the abundant foraminifera with sizes <1 mm and possessing a high fossilization potential, is continuous without sedimentation gaps at depths of more than 200 m (bathyal and abyssal). Non-interrupted sequences of so-called "fossil populations" can be found there, which embarrass many generations. Therefore, fossil populations act against random walking in succeeding generations. The number of generations depends on the sedimentation rate, reproduction type (semelparity in foraminifera) and life span. Within undisturbed sediments of 5 cm thickness from the upper bathyal (200–500 m) with a sedimentation rate of 0.1 mm per year and a life span of 0.5 years for benthic foraminifera, the number of generations is 1000. In plankton foraminifera deposited at a depth of 1500 m, the short life span of 1 month and low sedimentation rate of 0.01 mm per year raises the number of generations in 5 cm of sediment to 60,000 (Hohenegger, 2007). Thus, random walking is totally eliminated in these lineages.

b Split-Up Speciation

Reticulate speciation (Fig. 17.8). This type of speciation starts with a single species adapted to a stable or changing adaptive zone. Subsequently, several adaptive zones open by slow transformation. The vague differentiation between zones in this transformation period leads to the development of various evolving genotype groups. Hybridization between subgroups is possible when barriers, mostly caused by geographic separation, break. This leads to a reticulate pattern of subgroups within the transition zone, complicating the identification of limits between subgroups in space

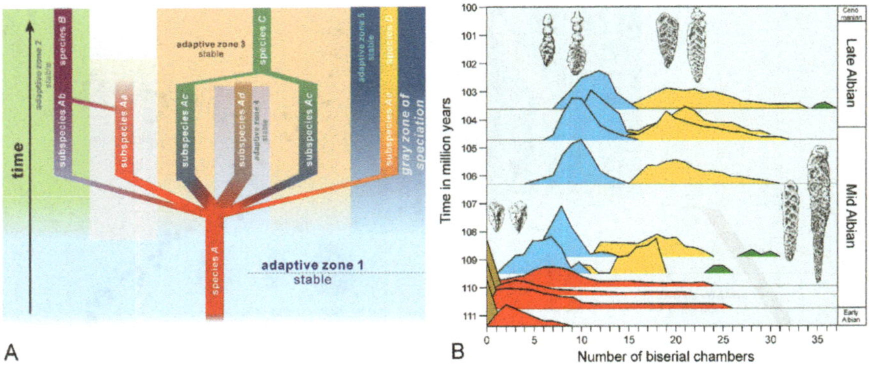

Fig. 17.8 Reticulate speciation: **a** Theoretical model (Hohenegger, 2013). **b** Example based on biserial chamber numbers of the benthic foraminifers *Gaudryina* (light brown) and *Spiroplectinata*. After reticulate speciation at the base of the Middle Albian (red color), three different lines with parallelism in gradual changes occur (after Grabert, 1959)

and time. Not until the adaptive zones are clearly separated without contemporary transition possibilities do the genotype groups adapted to these zones become species. Within the gray zone of speciation, these groups could be treated as subspecies.

Reticulate speciation was often neglected in the evolution models of marine invertebrates. First introduced for stony corals by Veron (1995), this is the "normal" evolution mechanism for terrestrial plants (Linder & Rieseberg, 2004), terrestrial animals (Sylvester-Bradley, 1977), and shallow marine invertebrates, where locations are often strongly separated. Therefore, reticulate speciation seems to be the main evolutionary model for these groups.

17.4 Macroevolution

> The classical view of macroevolution held by Darwin and the majority of paleontologists up to the present day is that species in the course of their gradual evolution in time change to such a degree that they will become different genera, or taxa of still higher rank, and acquire in the process all the adaptations and specializations of the world of organic diversity.
>
> Ernst Mayr
>
> Cited from: Evolution, vol. 36, p. 1126
>
> Wiley (1982)

Macroevolution is broadly defined as evolution above the species level (Hautmann, 2020). Different factors lead to novelties in traits creating new clades at the genus and higher taxonomic levels. Anisotropic variation caused by not-random mutations can act during epigenetically influenced development. Changes in nodes of the gene regulatory network lead to novel circuits affecting the network gene periphery, while modularity in development controlled by pleiotropic genes (i.e., one gene affects several traits) can also be changed by mutations within these gene complexes (Jablonski, 2017a, 2017b).

Especially after larger extinction events, a broad spectrum of adaptive zones opens for the few survivors. The following speciations using intensified macroevolutionary mechanisms allowed rapid adaptations to different zones becoming occupied by species groups (genus category) with strong phenetic differences. After a longer geological period of stability, the constant "normal" speciation rates could become weak (aging) leading again to intensified macroevolutionary mechanisms and the occurrence of genera strongly deviating from the normal "bauplan". This sequence is documented in the phylogeny of ammonites, fossil cephalopods with planispirally coiled shells starting in the Devonian (Fig. 17.1). The rapid rise of genera after the Permian/Triassic mass extinction ended at the Triassic/Jurassic boundary with a number of suddenly appearing aberrant (heteromorphic) genera deviating from the planispiral "bauplan." Only two genera with planispirally coiled shells survived the Triassic/Jurassic mass extinction. Stability in the Jurassic and Cretaceous led to abundant genera acting as stratigraphic index fossils. In the Late Cretaceous, just before the extinction of ammonites at the Cretaceous/Paleogene boundary, heteromorphic ammonites appeared again.

17.5 Conclusion

Organization of the biosphere is driven by evolution acting at the species level. Stable environments regulate interaction between species causing stability in communities, the next higher level of organismic organization. Long-lasting or abrupt changes in abiotic environmental factors lead to species reactions by evolution. These disturb and rearrange interactions between species in communities. Rearrangements can be continuous in constantly changing environments, but are intensified by abrupt environmental changes. Changes of community structure not only affect the relations between species, but also influence the physical and chemical factors of the abiotic environment. Thus, environmental changes are not only driven by external acting abiotic forces like irradiation, but are also influenced by biotic factors originating from communities (e.g., supply of oxygen by the presence of photosynthetic organisms). The biological motor of changes in circuits between communities and environment is evolution, influencing the biosphere through variable intensities.

References

Alroy, J. (2010). The shifting balance of diversity among major marine animal groups. *Science, 329*, 1191–1194.

Benton, M. J. (2005). *When life nearly died: The greatest mass extinction of all time*. Thames & Hudson.

Bettenstaedt, F., & Spiegler, D. (1982). *Pleurostomella* (Foram.) in der Unterkreide Nordwestdeutschlands. *Geologisches Jahrbuch, A, 65*, 445–479.

Blackburn, T. J., Olsen, P. E., Bowring, S. A., McLean, N. M., Kent, D. V., Puffer, J., McHone, G., Rasbury, T., & Et-Touhami, M. (2013). Zircon U-Pb geochronology links the end-triassic extinction with the Central Atlantic Magmatic Province. *Science, 340*(6135), 941–945.

Ceballos, G., Ehrlich, P. R., Barnosky, A. D., García, A., Pringle, R. M., & Palmer, T. M. (2015). Accelerated modern human–induced species losses: Entering the sixth mass extinction. *Science Advances,1*, e1400253.

Cole, L. C. (1958). The ecosphere. *Scientific American, 198*, 83–96.

de Lamarck, J. P. (1802). *Hydrogéologie*. Agasse.

de Queiroz, K. (2007). Species concepts and species delimitation. *Systematic Biology, 56*, 879–886.

Eichhorn, M. P. (2016). *Natural systems. The organization of life*. Wiley Blackwell.

Grabert, B. (1959). Phylogenetische Untersuchungen an *Gaudryina* und *Spiroplectinata* (Foram.), besonders aus dem nordwestdeutschen Apt und Alb. *Abhhandlungen Der Senckenbergischen Naturforschenden Gesellschaft, 498*, 1–71.

Gradstein, F. M., Ogg, J. G., Schmitz, J. G., & Ogg, G. M. (2012). *The geologic time scale 2012*. Elsevier.

Greenleave Leavitt, R. (1909). A vegetative mutant, and the principle of homeosis in plants. *Botanical Gazette, 47*, 30–68.

Hartl, D. L., & Clark, A. G. (2007). *Principles of population genetics* (4th ed.). Sinauer.

Hautmann, M. (2020). What is macroevolution? *Palaeontology, 63*, 1–11.

Hohenegger, J. (2007). Populationsgenetik und Mikropaläontologie – Möglichkeiten zur Überprüfung unterschiedlicher Modelle der Artbildung. *Denisia, 20*, 59–74.

Hohenegger, J. (2012). Transferability of genomes to the next generation: The fundamental criterion for the biological species. *Zootaxa, 3572*, 11–17.

Hohenegger, J. (2013) Species as the basic units in evolution and biodiversity: Recognition of species in the recent and geological past as exemplified by larger foraminifera. *Gondwana Research, 25*, 707–728.

Hugget, R. J. (1999). Ecosphere, biosphere, or Gaia? What to call the global ecosystem. *Global Ecology and Biogeography, 8*, 425–431.

Hull, D. L. (1976). Are species really individuals? *Systematic Zoology, 25*, 174–191.

Hull, D. L. (1978). A matter of individuality. *Philosophy of Science, 45*, 335–360.

Hunt, G. (2007). The relative importance of directional change, random walks, and stasis in the evolution of fossil lineages. *PNAS, 104*, 18404–18408.

Hunt, G., Hopkins, M. J., & Lidgard, S. (2015). Simple versus complex models of trait evolution and stasis as a response to environmental change. *PNAS, 112*, 4885–4890.

Jablonski, D. (2017a). Approaches to macroevolution: 2. Sorting of variation, some overarching issues, and general conclusions. *Evolutionary Biology, 44*, 451–473.

Jablonski, D. (2017b). Approaches to macroevolution: 1. General concepts and origin of variation. *Evolutionary Biology, 44*, 424–450.

Kerr, R. A. (1987). Milankovitch climate cycles through the ages: Earth's orbital variations that bring on ice ages have been modulating climate for hundreds of millions of years. *Science, 235*(4792), 973–974.

Kitcher, P. (1984). Species. *Philosophy of Science, 51*, 308–333.

Linder, R. C., & Rieseberg, L. H. (2004). Reconstruction pattern of reticulate evolution in plants. *American Journal of Botany, 91*, 1700–1708.

Lucae, S. Ch. (1816) *Entwurf eines Systems der medicinischen Anthropologie. Zum Gebrauche beim Studium der Natur- und Heilkunde des menschlichen Organismus. Erster Band. Geschichte des vegetativen Lebens im Individuum*. Varrentrapp.

Malmgren, B. A., Berggren, W. A., & Lohmann G. P. (1983). Evidence of punctuated gradualism in the late Neogene *Globorotalia tumida* lineage of planktonic foraminifera. *Paleobiology, 9*, 377–389.

Malmgren, B. A., & Kennett, J. P. (1981). Phyletic gradualism in a Late Cenozoic planktonic foraminiferal lineage; DSDP Site 284, southwest Pacific. *Paleobiology, 74*, 230–240.
Mayden, R. L. (1997). A hierarchy of species concepts: the denouement in the saga of the species problem. In M. F. Claridge, H. A. Dawah, & M. R. Wilson (Eds.), *Species: The units of biodiversity* (pp. 381–424). Chapman and Hall.
Mayr, E. (1982). Speciation and macroevolution. *Evolution, 36*, 1119–1132.
Peter, I. S., & Davidson, E. H. (2015). *Genomic control process: Development and evolution.* Academic Press.
Pigliucci, M., & Kaplan, J. (2006). *Making sense of evolution. The conceptual foundations of evolutionary biology.* The University of Chicago Press.
Rebeiz, M., Patel, N. H., & Hinman, V. F. (2015). Unraveling the tangled skein: The evolution of transcriptional regulatory networks in development. *Annual Review of Genomics and Human Genetics, 16*, 103–131.
Schulte, P., Alegret, L., Arenillas, I., Arz, J. A., Barton, P. J., Bown, P. R., Bralower, T. J., Christeson, G. L., Claeys, P., Cockell, C. S., Collins, G. S., Deutsch, A., Goldin, T. J., Goto, K., Grajales-Nishimura, J. M., Grieve, R. A. F., Gulick, S. P. S., Johnson, K. R., Kiessling, W. et al. (2010). The Chicxulub asteroid impact and mass extinction at the cretaceous-paleogene boundary. *Science, 327*(5970), 1214–1218.
Si, W., Berggren, W. A., & Aubry, M. P. (2018). Mosaic evolution in the middle Miocene planktonic foraminifera *Fohsella* lineage. *Paleobiology, 44*(2), 263–272.
Simmons, D. (2018). Loss of tropical forests makes climate change worse. Yale Climate Connections (online).
Skinner, M. K. (2015). Environmental epigenetics and a unified theory of the molecular aspects of evolution: A neo-Lamarckian concept that facilitates neo-Darwinian evolution. *Genome Biology and Evolution, 7*(5), 1296–1302.
Solbrig, O. T., & Solbrig, D. J. (1979). *The introduction to population biology and evolution.* Addison-Wesley Publishing Company.
Splitter, L. J. (1988). Species and identity. *Philosophy of Science, 55*, 323–348.
Suess, E. (1875). *Die Entstehung der Alpen.* Braumüller.
Sylvester-Bradley, P. C. (1977). Biostratigraphical tests of evolutionary theory. In E. G. Kaufmann & J. E. Hazel (Eds.), *Concepts and methods in biostratigraphy* (pp. 41–63). Dowden, Hutchinson and Ross.
Teilhard de Chardin. (1957). *La Vision du Passé*. Éditions du Seuil.
van der Pluijm, B. A., & Marshak, S. (2004). *Earth structure—An introduction to structural geology and tectonics* (2nd ed.). W.W. Norton.
van Valen, L. (1976). Ecological species, multispecies, and oaks. *Taxon, 25*, 233–239.
Vernadsky, V. I. (1926). *Biosfera.* Nauchnoe khimiko-technicheskoye izdatel'stvo.
Veron, J. E. N. (1995) *Corals in space and time. The biogeography and evolution of the Scleratinia.* Comstock/Cornell.
Wright, S. (1967). Surfaces" of selective value. *PNAS, 58*, 165–172.
Zar, J. H. (2010). *Biostatistical analysis* (5th ed.). Pearson.

Johann Hohenegger is a retired professor at the Department of Palaeontology, Vienna University, Austria, where he also obtained his undergraduate and postgraduate degrees. His main interests with more than 150 publications are focused on population dynamics, taphonomy and carbonate production of larger foraminifera, recognition of species in the present and past, modeling speciation events, morphometrics, morphogenetic programs and their phylogenetic implications in comparison to molecular genetic trees, morphocoenoclines along environmental gradients and their importance to decipher paleoenvironmental conditions, integrated stratigraphy and paleoenvironmental analysis in the Neogene of the Paratethys (Central Europe) concentrating on astronomic cycles, and, finally, spatial distributions on micro- and macroscale. Hohenegger started his scientific work on Late Paleozoic to Early Jurassic foraminifera. He spent long periods in Japan working at the Sesoko Marine Laboratory of Ryukyu University, at the Kagoshima University Museum, and at the Kyoto University Museum. At Vienna University, he led basic courses for earth science students in geostatistics, biology, micropaleontology, and biostratigraphy.

Chapter 18
New Horizons in Brain Science

Toshihiko Aso and Hidenao Fukuyama

Abstract As a complex system, the brain is considered to ultimately defy understanding through a reductionist approach. However, at present, scientists are busy addressing technical problems before considering such a theoretical dichotomy. Advances in neuroimaging methods have allowed observation of living brains at work, but many issues remain to be overcome; some fundamental, and some arising through poor utilization. For example, functional MRI suffers signal contamination by chaotic behavior of physiological confound. Nevertheless, functional neuroimaging is finally about to secure its position as a noninvasive method of physiological measurement that can be used to investigate in great detail, thus meeting the huge expectations from society.

Keywords MRI · Brain · Positron emission tomography · Transcranial magnetic stimulation

18.1 Rise of Functional Neuroimaging

The past three decades have seen explosive growth in the scientific application of neuroimaging, boosted by expectations of breakthroughs in brain research. One of the earliest compelling achievements was a positron emission tomography (PET) study that showed recruitment of the primary visual cortex when people suffering from early-onset blindness read using braille (Sadato et al., 1996). Owing to the lesion studies of the nineteenth century, the occipital lobe was already known to be involved in visual processing and perception (Glickstein, 1988). Using PET, the authors of this work investigated a group of blind people who had little visual experience. They observed activation—inferred from an increase in blood flow—in the

T. Aso
Laboratory for Brain Connectomics Imaging, RIKEN Center for Biosystems Dynamics Research, 2-3 Minamijmachi, Minatojima, Chuou-ku, Kobe 650-0047, Japan

H. Fukuyama (✉)
Center for the Promotion of Interdisciplinary Education and Research, Kyoto University, Kyoto, Japan
e-mail: fukuyama@kuhp.kyoto-u.ac.jp

K. Nishimura et al. (eds.), *Creative Complex Systems*, Creative Economy,
https://doi.org/10.1007/978-981-16-4457-3_18

medial parts of the occipital lobe, which corresponds to the primary visual cortices. In marked contrast, when sighted individuals were tested for comparison, tactile perception tasks actually suppressed occipital activity, probably reflecting reduced visual attention when focusing on finger sensations. The results suggested a differential functional development of the occipital region caused by a lack of visual input from the retina. Such an environmental effect on cortical development had been previously shown by famous experiments in newborn cats, in which restricted visual input (e.g., showing only vertical lines) modified the ability to perceive visual features (Blakemore & Cooper, 1970). In cases of early blindness, the primary visual area appears to serve as a spatial sketchpad that mediates some process for letter-by-letter reading or whole-word pattern recognition of braille codes.

This finding alone was captivating enough to draw people's attention, but the story did not end there. The same group published a follow-up report using transcranial magnetic stimulation (TMS) to prove the functional relevance of the occipital activity (Cohen et al., 1997). They successfully demonstrated a temporal disruption in the ability to read braille by applying TMS to the occipital region, which arrested ongoing local information processing. Later, in 1998, another group confirmed the differential recruitment of visual brain regions in those who were blind or sighted, although the interpretations were not fully consistent (Büchel et al., 1998). Finally, in 2000, one clinical case report described a blind woman who lost the ability to read braille after a brain infarction (Hamilton et al., 2000). The woman was a fluent braille reader, having obtained a college degree, held a job that involved actively communicating in braille. Importantly, the bilateral infarction, presumably a rare case of vessel-to-vessel embolism, involved the early visual cortices and the braille alexia was the only symptom to persist after the acute period.

This series of findings not only strongly substantiated the idea that cortical plasticity occurs across different sensory modalities, but also provided very strong empirical support for the neuroimaging technique. A fundamental difference between classic lesion studies and functional neuroimaging is that in the latter, not all observed signal changes are considered imperative for task execution (Bauer et al., 2014). While the repeated coincidence of occipital lesions in visually incapacitated soldiers can collectively prove the region as indispensable for vision, functional activation can be a mere accompanying phenomenon (e.g., an auditory response to one's own speech). Therefore, cases in which the activation of interest was proven relevant have been generally rare. These sets of reports have supported the validity of functional neuroimaging based on neurovascular coupling, such as the water PET, which is an indirect measure of local neuronal activity.

18.2 Neurovascular Coupling: An Unexpected Convenience

Since the initial experiences with PET, magnetic resonance imaging has emerged as the dominant neuroradiology technique because its safe and noninvasive nature enhances the feasibility of safe human study. In addition, the high temporal resolution of functional MRI (fMRI), at one frame per 2–3 s, allows observation of the dynamic vascular response. Thirty years ago, however, neurovascular coupling was not what scientists expected from fMRI; instead, they dreamed of a more direct observation of neuronal activity (Bandettini, 2020). In an early paper by Seiji Ogawa, who was the first to present a model of fMRI signal dynamics for detecting neurovascular coupling, deoxygenated hemoglobin (deoxy-Hb) was assumed to be a marker for local oxygen consumption (Ogawa et al., 1990). If we could detect an acute increase in local deoxy-Hb, indeed, it would have allowed a direct measurement of neural metabolism and activity. Unfortunately, this was not the case because neurovascular coupling leads to an excessive supply of fresh blood, which dramatically dilutes the concentration of local deoxy-Hb. This results in a signal change with a polarity opposite to what is expected from oxygen consumption, and on a far larger scale. In fact, the paradoxical increase of oxygen, which disproportionately surpasses the demand, remains a mystery of neurovascular coupling (Bandettini, 2020; Leithner & Royl, 2014).

After determining the behavior of MR signals in response to local neural activity (Logothetis et al., 2001), the surrogate marker for neural activity was supposed to be fast dilation of the vessels, triggered by neuronal activity and mediated by astrocytes or signaling agents such as nitrogen monoxide. Scientists then asked for a dynamic physiological model, in addition to Ogawa's MR signal model, to explain the observed temporal profile of the signal change. The balloon model based on the expansion of the venules was the most popular in the early days, and was later modified to incorporate the observation of arteries and even capillaries dilating after local neuronal activation (Kim & Ress, 2016). In any case, the main source of fMRI signal change remains the dilution of deoxy-Hb by the fresh blood supply. Importantly, neurovascular coupling itself is still a matter of vigorous debate (Grutzendler & Nedergaard, 2019). Some studies have followed on from Logothetis's work with the aim to directly validate correspondence between neural activity and fMRI signal in various parts of the brain (Ojemann et al., 2013). We take an in-depth look at neurovascular coupling later in this chapter.

18.3 Rise of the Global Signal Problem

In 1996, Aguirre et al. (1998) published an article entitled "The Inferential Impact of Global Covariates in Functional Neuroimaging Analyses." At that time, no consensus had been reached on how to deal with the global signal (mean or sum of the whole brain signal), which, believe it or not, remains the case today. Briefly, the issue concerns the source of the global signal fluctuation and its correlation with structures across the brain. If it is of purely non-neuronal origin, removing it by linear regression or whole-brain scaling should be justified. Indeed, this sort of treatment is popular in neuroradiological methods, such as SPECT and PET, under the assumption that global cerebral blood flow is roughly stable. However, the signal change in fMRI is not as simple as that in imaging techniques based on radiation count. According to the models, it is determined by a combination of blood flow and blood volume that are mutually dependent.

Then, a question arises: if there was such an uncertainty as to whether a signal component was signal or noise, why did a considerable proportion of scientists choose to remove the global fluctuation? The answer is that they found the effect "favorable"; more confined activations and even deactivations appeared occasionally for the convenience of interpretation. Back then, to our relief, many scientists seemed to heed the warnings given by Aguirre and others, and those treatments became less common in fMRI research.

18.4 Default Mode and Network Problems

As a preface to the upcoming confusion, in 2001, an influential group in the PET field published a paper introducing the concept of the brain's "default mode" (Raichle et al., 2001). This was motivated by their previous findings that a set of brain regions was regularly "deactivated" by cognitive tasks (Shulman et al., 1997). This behavior indicated that the set of regions was more active during rest. This new concept strongly supported their next step, which proposed a cortical "network" by treating correlation maps as measures of continuous, fluctuating activity (Fox et al., 2005). Although Biswal's pioneering work had been published 10 years earlier (Biswal et al., 1995), the popularity of resting-state fMRI (rs-fMRI) did not begin to rise until this point.

As a result, two major problems emerged. The first was that some scientists immediately responded by refusing to grant such a privileged status to resting conditions (Morcom & Fletcher, 2007). Although the state of mind during rest is difficult to specify, quietly lying down without movement should still recruit a range of brain activity, including information processing and concentration. The notion that the default mode network (DMN) is regularly deactivated by tasks is also misleading in itself; some cognitive tasks do activate the DMN, they were just not popular at that time. Cognitive components such as self-reflection or perspective taking, which are evoked when reading stories, recruit the DMN with high reproducibility (Mano et al., 2009). According to a large review paper from 1997, only one then-popular task was considered to be of this sort: episodic memory retrieval (Cabeza & Nyberg, 1997). So, what happens during the rest period between periods of a task that activates the DMN? Reports show a very different "deactivation" during narrative reading, which proves that non-DMN areas are also active during rest (Aso & Fukuyama, 2015). Such observations at least partly invalidate the original concept of the default mode (Harrison et al., 2011).

The term "deactivation" is also confusing when it implies a mechanism different from simple termination of activation. DMN deactivation reportedly reflects the current task load but is also nonspecific to the task itself (Čeko et al., 2015). If this is the case, how can we make unambiguous inference when comparing two task conditions, both of which include DMN-deactivating components like working memory? Does the difference reflect a qualitative difference in tasks or the strength of deactivation? If we are to adopt cognitive subtraction as a method, we should not have a baseline in an absolute sense, and everything we see in fMRI must be accounted for by activations alone (Morcom & Fletcher, 2007).

18.5 Gray Matter in rs-fMRI

The second problem was the validity of rs-fMRI as an actual functional MRI. It would indeed be fascinating if we could measure the fluctuating neural activity in the brain that is not possible to capture with tasks. Accordingly, below is the hierarchy of implicit assumptions that seems popular among those who use rs-fMRI.

- fMRI signals reflect neural activity in the same way during tasks and rest.
- Regions presenting synchronized activity have a mutual connection.
- Regularly observed functional connectivity should be considered a network.
- Networks can be considered a new unit of functional segregation (modules) in the brain.
- rs-fMRI can provide biomarkers by somehow summarizing the brain's condition.

The second item is already troublesome because any unconnected neuronal units will act synchronously if they have common inputs, as pointed out half a century ago (Perkel et al., 1967). The third through fifth items occasionally create *petitio principii*, or circular logic underlying rsfMRI studies. However, without doubt, the first item of this list is far more critical than the others, and every effort must be made to ensure its validity. Contrary to this, the seminal work by Fox et al. (2005) re-introduced the fundamental and persistent problem of global signal regression (GSR). Their preprocessing steps included removal of nuisance parameters by multiple regression, which means that the unexplained residual variance of the fitting is used as the cleaned data. In addition to the global mean, these parameters included head motion and fMRI signal averaged over the cerebral white matter and cerebrospinal fluid. Hence, it was a modified GSR, but like the original, it lacked a sound basis. Indeed, the authors did not provide an explanation for this treatment (Fox et al., 2005).

GSR thus became a part of the de facto standard, but only for rs-fMRI, not for fMRI recorded during explicit tasks. Unlike task fMRI, in which non-neuronal fluctuation is separable by trial averaging or multiple regression, source separation of rs-fMRI is innately blind. Surprisingly, studies that try to support the reliability of rs-fMRI tend to rely on trial averaging, which fails to appeal to those who are serious about using this tool (Bruyns-Haylett et al., 2013; Ojemann et al., 2013; Ossandón et al., 2011). Animal experiments have shown that signals without averaging are not reliable because of vascular fluctuation that occurs with or without neural activity (Cardoso et al., 2012; Winder et al. 2017). Furthermore, using fMRI requires that care be taken regarding additional noisy components, which are absent when using invasive techniques that can be used in animals (Das et al., 2021; Murphy et al., 2013). Today, this scenario remains a focus of controversy (Aquino et al., 2020).

Why, then, were those poorly supported fMRI data treatments readily accepted? One answer is that the fMRI user population was trained to do so. In the early days, attractive neuroimaging results such as the one regarding reading braille encouraged the use of fMRI. GSR was accepted because the results were interpretable. This time, it was the tempting theories, as listed above, about the concept of DMN. Accepting that the method is valid because the results are beautiful is similar to being persuaded that the ends justify the means. An early promise of rs-fMRI was being able to diagnose Alzheimer's disease through DMN "connectivity" (Buckner & Vincent, 2007; Greicius et al., 2004). Although this claim attracted a great deal of attention, it has yet to be fulfilled. Before achieving reliable diagnoses, reproducibility of the rs-fMRI studies must be secured first (Teipel et al., 2017; King et al., 2019; He et al. 2020). In view of this situation, David Van Essen proposed that the rs-fMRI technique must be established based on some grounded truth concerning connectivity, such as animal tracer studies (Van Essen et al., 2019). Lack of a gold standard to perfect this method is indeed critical and needs to be addressed to fulfill the promise, or at least to see what the method can do in finding useful biomarkers.

18.6 Current Status of Neurovascular and Neuro-BOLD Coupling

In the current result-replication crisis in science, reproducibility of fMRI studies is a major topic (Horster et al., 2020; Kampa et al., 2020; Nee, 2019; Renvall et al., 2014). Although the explosive increase of rs-fMRI studies should have accelerated this concern, the situation itself is perfectly sound and constructive. Another favorable outcome of the popularity of rs-fMRI is that it encouraged studies that focus on fMRI artifacts, although the number of publications is still very limited compared with those just using rs-fMRI. Formulation of the respiration-related signal components is clearly a major advance in this regard (Birn et al., 2006; Weiss et al., 2020). Because validating neuronal activity itself is generally difficult, we can only employ the second-best approach to reliable fMRI: find and eradicate noise components.

Before going into the details of how to clean up fMRI noise, let us note that neurovascular coupling itself remains an enigma, contrary to the belief of most fMRI users. Blood flow regulation has always been an important topic in clinical neurology, a field in which researchers have recently become interested in the "neurovascular unit" (Iadecola, 2017). While many textbooks mention the role of astrocytes in dilating arterioles within the brain, the details are far from resolved. For example, in 2016, reports claimed that the focus of adjustment is capillary pericytes, not the arterioles (Biesecker et al., 2016; Mishra et al., 2016). Since then, an opinion paper has claimed that those observations in vitro, or in brain preparations, may be incorrect (Grutzendler & Nedergaard, 2019). There is even a theory that red blood cells regulate the microcirculation via changes in plasticity that are linked to oxygenation level (Wei et al., 2016). Thus, building a theoretical model to account for the

BOLD signal response already seems challenging, considering that the underlying physiology is yet to be settled.

Several options exist for the find-and-eradicate approach that is designed to achieve better neuro-BOLD coupling in practice. Retrospective correction using simultaneous physiological recordings of multiple cardiorespiratory parameters is at present the least controversial strategy (Glover et al., 2000). Regrettably, this method is not popular because its implementation workload in both data acquisition and analysis is too high. Being easy to implement was indeed an obvious reason for the popularity of rs-fMRI over fMRI with tasks. The most popular and accepted data-driven approach is independent component analysis (ICA)-based denoising. This technique covers head motion and scanner noise, as well as physiological artifacts such as cardiac pulsation and cerebrospinal fluid movement, which have spatial structure (Griffanti et al., 2017; Salimi-Khorshidi et al., 2014). However, it might remain insufficient for perfusion-related, physiological noise (Glasser et al., 2018). In the following sections, recent advances regarding this global noise structure, potentially involving a renewal of the fMRI signal models, are introduced.

18.7 Perfusion-Related Structure in fMRI Data

Whether it is considered a target signal or noise, an fMRI signal change can be explained by either neurovascular coupling or autoregulation as long as the participant is lying very still in the scanner. A local blood flow increase over 30% creates a 0.5–2% signal change with around a 5-s delay (Buxton, 2013). This is the target phenomenon of fMRI as a tool for functional brain mapping. At the same time, autoregulatory changes in vessel diameter can be found in a wide range of arteries, including the middle cerebral or internal carotid artery (Hoiland et al., 2016). Based on this phenomenon, a line of clinical studies of cerebrovascular disease have attempted to detect perfusion deficit using the BOLD fMRI signal (Murphy et al., 2011). Response to gas or respiratory challenges is degraded in vessels that have already dilated to maximum because of poor oxygen supply. Importantly, neurovascular coupling and autoregulation have no clear distinction because both involve multiple overlapping pathways (Willie et al., 2014). Their contributions to fMRI signal changes are also uniformly postulated to reflect increased cerebral blood flow (CBF), cerebral blood volume (CBV), and eventual dilution of deoxy-Hb, a natural contrast agent that suppresses the local signal (Kim & Ress, 2016).

As mentioned above, several studies have revealed various systemic components in BOLD signal fluctuations. Physiological parameters, such as cardiac pulsation (Chang et al., 2009), blood pressure, and end-tidal carbon dioxide (CO_2) (Murphy et al., 2013; Wise et al., 2004), are considered sources of signal artifacts. However, in rs-fMRI, discrimination between neuronal and non-neuronal components is a major challenge because of a lack of validation techniques that have the spatial and temporal precision comparable with those found in fMRI (Winder et al. 2017). Another source of difficulty is the intercorrelation across many neural and non-neural parameters that fall within this low-frequency range upon which the original method for rs-fMRI focuses (Murphy et al., 2013).

Scientists using near-infrared spectroscopy (NIRS), which is also used as a brain-mapping method based on plasma hemoglobin, had also been struggling with slow baseline fluctuations. Spontaneous low-frequency oscillations (sLFOs) are a major confound in NIRS measurement because they are also detectable from non-brain structures, which means that they are of systemic origin. Optically detected sLFOs were found in both oxy- and deoxy-Hb that possessed an interesting phase difference that was shown to be emphasized in brain tissue (Rayshubskiy et al., 2014; Tgavalekos et al., 2016). Similar to the case with fMRI, these slow fluctuations might encompass multiple sources of physiological artifacts (Tong et al., 2017; Zhu et al., 2015). An sLFO found in systemic blood pressure, known as the Traube–Hering–Mayer wave, seems to originate from the autonomic loop (Guyton & Harris, 1951; Julien, 2006). Associated sLFOs in blood flow and velocity have also been found (Fagrell et al., 1977; Killip, 1962), whose traces were then confirmed within the brain via transcranial Doppler ultrasonography (Giller et al., 1999; Obrig et al., 2000). Even electroencephalographic recordings have been reported to detect sLFOs, for which arterial vasomotion was suggested as the origin (Nikulin et al., 2014).

In 2010, a group from Harvard Medical School reported a significant correlation between NIRS sLFOs and simultaneously obtained fMRI data (Tong and Frederick 2010). The most intriguing finding was the variation in sLFO phase across brain structures: the earlier phase was observed in the upstream arteries, while the late phase was seen in the venous sinuses. Because Hb-sLFOs are postulated to be of systemic origin (Katura et al., 2006; Sassaroli et al., 2012; Tian et al., 2011), finding a correlation between the global fMRI signal and extracerebral signals was not surprising. However, the constant time shift across body/brain structures was unexpected (Anderson et al., 2011; Tong et al., 2012). The similarity between the low-frequency phase map and perfusion MRI (using contrast agents) was a milestone in this direction, because it is proof that sLFO waveforms move with the blood. Mapping of the phase within the brain parenchyma is now called lag mapping (Fig. 18.1).

Finding that the perfusion time lag is embedded in the BOLD signal was a surprise (Tong et al., 2017). Based on this finding, a number of clinical studies have been conducted to establish a biomarker for cerebrovascular disorders that relate to the perfusion delay (Amemiya et al., 2013; Christen et al., 2015; Khalil et al., 2020; Lv et al., 2013; Ni et al., 2017; Nishida et al., 2018). Moreover, normal gross vascular anatomy was consistently detected in the unaffected hemisphere in these studies, replicating the results from respiratory challenges (Blockley et al., 2011; Chang & Glover, 2009), and suggesting an equivalence between the sLFO and manipulated circulatory turbulences. In the venous side of the vasculature, recent studies of a large group of healthy participants revealed changes in venous drainage patterns with normal aging (Aso et al., 2020; Satow et al., 2017). In parallel, apart from patient data, the resilience of the perfusion lag map against the fMRI task condition confirmed its non-neuronal origin (Aso et al., 2017). A body of evidence thus empirically supports the biological significance of the low-frequency lag structure and its underlying

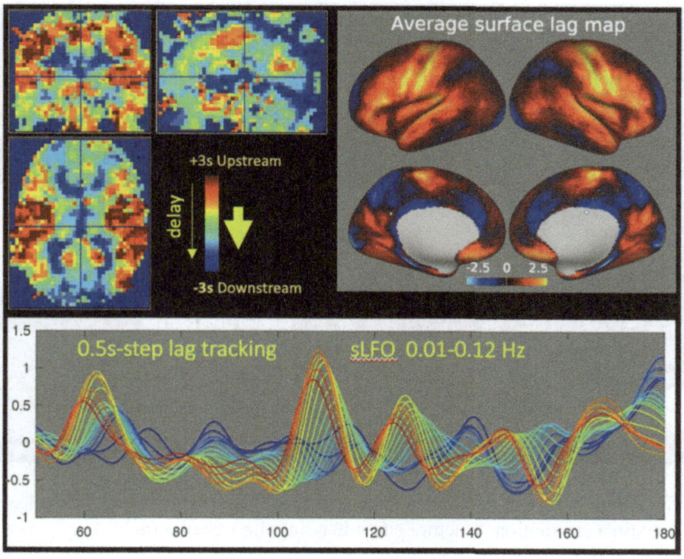

Fig. 18.1 Perfusion lag structure in the whole-brain BOLD fMRI signal. In the bottom panel, traces in warm colors are the averaged timeseries for the voxels of the same color in the top-left panel

principles, in the same way that early fMRI supported the neuronal origin of the BOLD signal.

Once we accept that the sLFO is of systemic origin and not a product of local neurovascular coupling, it should be somehow eliminated from the fMRI signal. Mathematically, this is easy using regression because we have complete knowledge of its time course and phase variation across voxels (Byrge & Kennedy, 2018; Erdoğan et al., 2016; Frederick et al., 2012). Of note, the title of this pioneering work ("Correcting for Blood Arrival Time in Global Mean Regression Enhances Functional Connectivity Analysis of Resting State fMRI-BOLD Signals") is misleading because the regression would not correct the blood arrival time; it is removing the whole signal component that encompasses the perfusion timing information. In actuality, it is highly uncertain whether some effect of blood arrival timing, particularly an effect on neurovascular coupling, is corrected by the treatment (Chen et al., 2020). Removal of this spatiotemporal lag structure in the signal can also be useful in task fMRI, as long as slow components (<0.1 Hz) are not introduced into the study design (Aso et al., 2019). ICA-based noise removal is being updated to remove the global noise structure whose relationship with this lag structure remains to be resolved (Glasser et al., 2018). There is also a concern that full removal of the lag structure might harm the neurovascular coupling-component because it also creates minor lag structures (Aso et al., 2017).

The implication of this lag structure on the long-standing GSR problem is simple. While conventional GSR removes a considerable amount of variance that can be attributed to sLFO, for rs-fMRI, it leaves a troublesome correlation structure in the

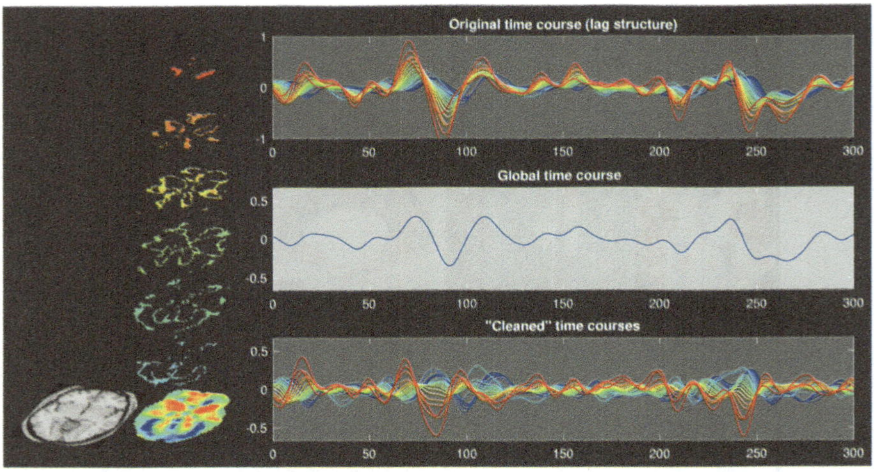

Fig. 18.2 Effect of global signal regression on the perfusion lag structure. Regression leaves residual signals that contain a correlation structure encompassing the vasculature

relatively higher frequency ranges (Fig. 18.2). In the aforementioned work, removal of the lag structure is described as "dynamic GSR," assuming that the major portion of GSR is sLFO, which might not always be the case (Erdoğan et al., 2016). In the following section, we investigate how the discovery of lag structure impacted theories regarding fMRI signals beyond the GSR problem.

18.8 Dissociation of Neurovascular Coupling from Neuro-BOLD Coupling

As mentioned in the previous section, BOLD fMRI signal changes have been explained solely by local dilution of deoxy-Hb by fresh arterial blood, triggered by increases in CBF and CBV. According to this model, researchers have tried to model sLFO and its lag structure via vascular mechanisms (Tong et al., 2019). For example, the Mayer wave was rejected as the main source because arterial pressure should affect disparate brain regions almost simultaneously, instead of creating a lag structure. Vasomotion, an active diameter change in the precapillary vessels, was considered a much better candidate, as was plasma CO_2, which can be its driving force. However, vasomotion is a phenomenon specifically found in the arteries, and its presumed role is in facilitating microcirculation (Kapela et al., 2012). Can vasomotion fully account for the time lag between the venous structures and the capillary bed, both of which are non-arterial vessels that are consistently detected by lag mapping? In addition, is the Mayer wave completely unrelated despite its tight relationship with autonomic loops?

Every numerical model of BOLD fMRI signaling has a constant term for the magnetic susceptibility that determines the baseline. It is a product of the scanner's magnetic field strength, the difference in susceptibility between oxy- and deoxy-Hb, the hematocrit (density of the red blood cells), and the fraction of oxygen extraction (An & Lin, 2002; Yablonskiy & Haacke, 1994). The last two items determine the amount of deoxy-Hb, but under the assumption of a constant inflow-oxygen saturation of 100%. This is obviously a simplification, and the question is whether the assumption is totally safe and sound. Briefly, is deoxy-Hb inflow stable at zero?

Ventilated with room air, animals cannot actually exhibit 100% oxygen saturation in reality; change in SaO_2 ranges from 94 to 98% in the artery (Collins et al., 2015; Intaglietta et al., 1996). In the presence of deoxy-Hb in the blood, a small fluctuation of hematocrit alone can directly affect the fMRI signal (Dent et al., 2016). Axial variation of hematocrit in the vessels has been theoretically suggested in relation to both neurovascular coupling (Chen & Pike, 2009; Kleinschmidt et al., 1996; Siegel et al., 2003) and sLFOs (Fagrell et al., 1980; Mayhew et al., 1996). Vasomotion drives local velocity fluctuations, termed "flowmotion" (Intaglietta, 1990), which reportedly accompany local hematocrit fluctuation (Fagrell et al., 1980; Hudetz et al., 1999). In addition, the respiration-related BOLD signal component is thought to be mediated by blood CO_2 levels and pH, which can shift the oxygen dissociation curve (Birn et al., 2006; Chang & Glover, 2009). These parameters are considered to fluctuate in the blood as part of the autonomic loop, and as far as it is random, such a signal component would have escaped detection in studies of neurovascular coupling that use trial averaging.

A signal origin intrinsic to flowing blood might thus explain the constant phase difference across signals that come from different body/brain parts, like a waveform traveling in the vessel (Tong et al., 2012, 2017). Variations in the base susceptibility had once been considered to cause intersession variabilities (Cohen et al., 2002; Tuunanen & Kauppinen, 2006), but not for within-session fluctuations (Thomas et al., 2000). Although the assumption of constant base deoxy-Hb concentration may be sufficient for modeling signals with a vascular response (Ogawa et al., 1998), other effects might not be negligible in non-neuronal fluctuations. According to our calculations, fluctuation in oxygen saturation between 97 and 98% can create a signal change of approximately 1% (Aso et al., 2019), which is sufficient to affect any sort of fMRI (Fig. 18.3). Thus, vascular responses seem unable to fully account for neuro-BOLD coupling. Even the autoregulatory response to respiratory challenges might contain this component via fluctuation of blood gas content.

18.9 Complex System Within a Complex System

Today, real-time fMRI denoising is required in neurofeedback study settings (Weiss et al., 2020). Neurofeedback is an approach to disease treatment via fMRI in which participants adjust their state of mind while monitoring their own brain activity. The

Fig. 18.3 Inhomogeneity in the blood deoxy-Hb, such as variation in oxygen saturation, can create very similar timeseries that differ gradually in phase. Different parts of the vascular tree structure possess different lag times, thereby enabling mapping of temporal distance from artery (A) or vein (V)

technical challenges are very similar to those of rs-fMRI because time-locked analyses are not possible. However, real-time removal of sLFO ("find and eradicate" approach) might be far more difficult than offline data cleaning such as lag-structure removal. The behavior of the cardiopulmonary system, which is governed by autonomic loops, is itself a chaotic complex system (Tsuda et al., 1992). In addition, sympathetic arousal, which depends on the state of mind (i.e., brain activity), is known to affect heart rate and blood pressure. Moreover, the human brain receives 14% of the total cardiac output, which means that cerebral vascular resistance might modify systemic circulation.

Furthermore, the total circulation time for the human body is reportedly between 10 and 20 s (Sutton et al., 1950), which is an average value across various path lengths that the blood can take. This duration of time is very close to the wavelength of the sLFO, which means that a fraction of the blood is re-entering the brain and contributing to the oscillation more than once within a single phase. Modelling of the sLFO is thus complicated and ultimately requires brain activity itself in the equation.

18.10 Future Directions

With or without cognitive tasks, removal of the non-neuronal, global correlation structures will be a minimum requirement for fMRI studies. Scientists are becoming aware that physiological artifacts create spurious correlations in raw fMRI data, which greatly affect how results are interpreted (Chen et al., 2020; Das et al., 2021). One easy step would be to limit the neuronal activity to within a higher temporal frequency range by using an event-related design instead of a boxcar or blocked design. Because the troublesome physiological noise occurs predominantly in the lower range, recent protocols tend to preserve higher frequency ranges by abandoning lowpass filtering (Smith et al., 2013).

Departure from the resting state may be another fundamental option. Recently, a study proposed to gather as much information as possible from the participant to attenuate the randomness of the resting state (Gonzalez-Castillo et al., 2021). A standardized "movie fMRI" has been proposed, obviously as an alternative to rs-fMRI, on which development of clinical biomarkers are to be based (Eickhoff et al., 2020). This trend can be interpreted as a reaction to the disappointing outcome of the craze for rs-fMRI, driven by the promise of early diagnoses of Alzheimer's disease. At this point, how "brain networks" are biologically relevant and informative remains unclear. The classical analysis approaches are hence again becoming obsolete for highly empirical reasons. If functional connectivity based on rs-fMRI is going to endure as a research topic, it must be rebuilt on a firm basis such as axonal connections probed by tracer studies (Van Essen et al., 2019).

What will be the next major direction of research? Key to maximizing fMRI reliability seems to be the volume of data (Nee, 2019). Combined with the newly developed data-cleaning methods, thanks to the noise issues in rs-fMRI, we might now be able to proceed to fMRI studies of individuality (Dubois & Adolphs, 2016), rather than making inferences for humans in general. Characteristics that make us human mostly reside in the inter-individual differences of our brains. Once fMRI secures its position as a tool that can differentiate individual brains, it can help resolve clinical or even social problems. Combination with comparative neuroscience, namely cross-species studies with primates, would help to answer the most primordial questions of brain science. Taking advantage of the spatial resolution of ultra-high magnetic fields at 7 T, separation of the activities in different cortical layers are now under exploration (Kashyap et al., 2018).

Too much was expected of fMRI when it was first being applied to neuroscience and the method was still in its infancy. As brain science has continued to garner broad interest, findings based on this premature methodology have damaged the reputation of the tool (Resnick, 2016; Turner et al., 2018). It is now time to banish "classical" approaches and the resulting non-replicable studies to establish efficacy of neuroimaging as *in vivo* brain slicing.

References

Aguirre, G. K., Zarahn, E., & D'Esposito, M. (1998). The inferential impact of global signal covariates in functional neuroimaging analyses. *NeuroImage, 8*, 302–306. https://doi.org/10.1006/nimg.1998.0367

Amemiya, S., Kunimatsu, A., Saito, N., & Ohtomo, K. (2013). Cerebral hemodynamic impairment: Assessment with resting-state functional MR imaging. *Radiology, 270*, 1–8. https://doi.org/10.1148/radiol.13130982

An, H., & Lin, W. (2002). Cerebral oxygen extraction fraction and cerebral venous blood volume measurements using MRI: Effects of magnetic field variation. *Magnetic Resonance in Medicine, 47*, 958–966. https://doi.org/10.1002/mrm.10148

Anderson, J. S., Druzgal, T. J., Lopez-Larson, M., et al. (2011). Network anticorrelations, global regression, and phase-shifted soft tissue correction. *Human Brain Mapping, 32*, 919–934. https://doi.org/10.1002/hbm.21079

Aquino, K. M., Fulcher, B. D., Parkes, L., et al. (2020). Identifying and removing widespread signal deflections from fMRI data: Rethinking the global signal regression problem. *NeuroImage, 212*, 116614. https://doi.org/10.1016/j.neuroimage.2020.116614

Aso, T., & Fukuyama, H. (2015). Functional heterogeneity in the default mode network edges. *Brain Connectivity, 5*, 203–213. https://doi.org/10.1089/brain.2014.0256

Aso, T., Jiang, G., Urayama, S., & Fukuyama, H. (2017). A resilient, non-neuronal source of the spatiotemporal lag structure detected by BOLD signal-based blood flow tracking. *Frontiers in Neuroscience, 11*, 256. https://doi.org/10.3389/fnins.2017.00256

Aso, T., Sugihara, G., Murai, T., et al. (2020). A venous mechanism of ventriculomegaly shared between traumatic brain injury and normal ageing. *Brain, 143*, 1843–1856. https://doi.org/10.1093/brain/awaa125

Aso, T., Urayama, S., Fukuyama, H., & Murai, T. (2019). Axial variation of deoxyhemoglobin density as a source of the low-frequency time lag structure in blood oxygenation level-dependent signals. *PLOS ONE, 14*, e0222787. https://doi.org/10.1371/journal.pone.0222787

Bandettini, P. A. (2020). *FMRI*. MIT Press.

Bauer, P. R., Reitsma, J. B., Houweling, B. M., et al. (2014). Can fMRI safely replace the Wada test for preoperative assessment of language lateralisation? A meta-analysis and systematic review. *Journal of Neurology, Neurosurgery and Psychiatry, 85*, 581–588. https://doi.org/10.1136/jnnp-2013-305659

Biesecker, K. R., Srienc, A. I., Shimoda, A. M., et al. (2016). Glial cell calcium signaling mediates capillary regulation of blood flow in the retina. *Journal of Neuroscience, 36*, 9435–9445. https://doi.org/10.1523/JNEUROSCI.1782-16.2016

Birn, R. M., Diamond, J. B., Smith, M. A., & Bandettini, P. A. (2006). Separating respiratory-variation-related fluctuations from neuronal-activity-related fluctuations in fMRI. *NeuroImage, 31*, 1536–1548. https://doi.org/10.1016/j.neuroimage.2006.02.048

Biswal, B., Yetkin, F. Z., Haughton, V. M., & Hyde, J. S. (1995). Functional connectivity in the motor cortex of resting human brain using echo-planar MRI. *Magnetic Resonance in Medicine, 34*, 537–541. https://doi.org/10.1002/mrm.1910340409

Blakemore, C., & Cooper, G. F. (1970). Development of the brain depends on the visual environment. *Nature, 228*, 477–478. https://doi.org/10.1038/228477a0

Blockley, N. P., Driver, I. D., Francis, S. T., et al. (2011). An improved method for acquiring cerebrovascular reactivity maps. *Magnetic Resonance in Medicine, 65*, 1278–1286. https://doi.org/10.1002/mrm.22719

Bruyns-Haylett, M., Harris, S., Boorman, L., et al. (2013). The resting-state neurovascular coupling relationship: Rapid changes in spontaneous neural activity in the somatosensory cortex are associated with haemodynamic fluctuations that resemble stimulus-evoked haemodynamics. *European Journal of Neuroscience, 38*, 2902–2916. https://doi.org/10.1111/ejn.12295

Büchel, C., Price, C., Frackowiak, R. S., & Friston, K. (1998). Different activation patterns in the visual cortex of late and congenitally blind subjects. *Brain, 121*, 409–419. https://doi.org/10.1093/brain/121.3.409

Buckner, R. L., & Vincent, J. L. (2007). Unrest at rest: Default activity and spontaneous network correlations. *NeuroImage, 37*, 1091–1096. https://doi.org/10.1016/j.neuroimage.2007.01.010

Buxton, R. B. (2013). The physics of functional magnetic resonance imaging (fMRI). *Reports on Progress in Physics, 76*, 096601. https://doi.org/10.1088/0034-4885/76/9/096601

Byrge, L., & Kennedy, D. P. (2018). Identifying and characterizing systematic temporally-lagged BOLD artifacts. *NeuroImage, 171*, 376–392. https://doi.org/10.1016/j.neuroimage.2017.12.082

Cabeza, R., & Nyberg, L. (1997). Imaging cognition: An empirical review of PET studies with normal subjects. *Journal of Cognitive Neuroscience, 9*, 1–26. https://doi.org/10.1162/jocn.1997.9.1.1

Cardoso, M. M. B., Sirotin, Y. B., Lima, B., et al. (2012). The neuroimaging signal is a linear sum of neurally distinct stimulus- and task-related components. *Nature Neuroscience, 15*, 1298–1306. https://doi.org/10.1038/nn.3170

Čeko, M., Gracely, J. L., Fitzcharles, M.-A., et al. (2015). Is a responsive default mode network required for successful working memory task performance? *Journal of Neuroscience, 35*, 11595–11605. https://doi.org/10.1523/JNEUROSCI.0264-15.2015

Chang, C., Cunningham, J. P., & Glover, G. H. (2009). Influence of heart rate on the BOLD signal: The cardiac response function. *NeuroImage, 44*, 857–869. https://doi.org/10.1016/j.neuroimage.2008.09.029

Chang, C., & Glover, G. H. (2009). Relationship between respiration, end-tidal CO_2, and BOLD signals in resting-state fMRI. *NeuroImage, 47*, 1381–1393. https://doi.org/10.1016/j.neuroimage.2009.04.048

Chen, J. E., Lewis, L. D., Chang, C., et al. (2020). Resting-state "physiological networks." *NeuroImage, 213*, 116707. https://doi.org/10.1016/j.neuroimage.2020.116707

Chen, J. J., & Pike, G. B. (2009). BOLD-specific cerebral blood volume and blood flow changes during neuronal activation in humans. *NMR in Biomedicine, 22*, 1054–1062. https://doi.org/10.1002/nbm.1411

Christen, T., Jahanian, H., Ni, W. W., et al. (2015). Noncontrast mapping of arterial delay and functional connectivity using resting-state functional MRI: A study in Moyamoya patients. *Journal of Magnetic Resonance Imaging, 41*, 424–430. https://doi.org/10.1002/jmri.24558

Cohen, E. R., Ugurbil, K., & Kim, S.-G. (2002). Effect of basal conditions on the magnitude and dynamics of the blood oxygenation level-dependent fMRI response. *Journal of Cerebral Blood Flow and Metabolism, 22*, 1042–1053. https://doi.org/10.1097/00004647-200209000-00002

Cohen, L. G., Celnik, P., Pascual-Leone, A., et al. (1997). Functional relevance of cross-modal plasticity in blind humans. *Nature, 389*, 180–183. https://doi.org/10.1038/38278

Collins, J.-A., Rudenski, A., Gibson, J., et al. (2015). Relating oxygen partial pressure, saturation and content: The haemoglobin-oxygen dissociation curve. *Breathe (sheffield, England), 11*, 194–201. https://doi.org/10.1183/20734735.001415

Das, A., Murphy, K., & Drew, P. J. (2021). Rude mechanicals in brain haemodynamics: Non-neural actors that influence blood flow. *Philosophical Transactions of the Royal Society B: Biological Sciences, 376*, 20190635. https://doi.org/10.1098/rstb.2019.0635

Dent, P., Deng, B., Goodisman, J., et al. (2016) Noninvasive in vivo plasma volume and hematocrit in humans: observing long-term baseline behavior to establish homeostasis for intravascular volume and composition. In *Biophotonics: Photonic solutions for better health care V* (p. 98871S). International Society for Optics and Photonics.

Dubois, J., & Adolphs, R. (2016). Building a science of individual differences from fMRI. *Trends in Cognitive Sciences, 20*, 425–443. https://doi.org/10.1016/j.tics.2016.03.014

Eickhoff, S. B., Milham, M., & Vanderwal, T. (2020). Towards clinical applications of movie fMRI. *NeuroImage, 217*, 116860. https://doi.org/10.1016/j.neuroimage.2020.116860

Erdoğan, S. B., Tong, Y., Hocke, L. M., et al. (2016) Correcting for blood arrival time in global mean regression enhances functional connectivity analysis of resting state fMRI-BOLD signals. *Frontiers in Human Neuroscience, 10*. https://doi.org/10.3389/fnhum.2016.00311

Fagrell, B., Fronek, A., & Intaglietta, M. (1977). A microscope-television system for studying flow velocity in human skin capillaries. *American Journal of Physiology, 233*, H318–H321.

Fagrell, B., Intaglietta, M., & Ostergren, J. (1980). Relative hematocrit in human skin capillaries and its relation to capillary blood flow velocity. *Microvascular Research, 20*, 327–335. https://doi.org/10.1016/0026-2862(80)90033-3

Fox, M. D., Snyder, A. Z., Vincent, J. L., et al. (2005). The human brain is intrinsically organized into dynamic, anticorrelated functional networks. *Proceedings of the National Academy of Sciences, 102*, 9673–9678. https://doi.org/10.1073/pnas.0504136102

Frederick, B. deB., Nickerson, L. D., & Tong, Y. (2012). Physiological denoising of BOLD fMRI data using Regressor Interpolation at Progressive Time Delays (RIPTiDe) processing of concurrent fMRI and near-infrared spectroscopy (NIRS). *NeuroImage, 60*, 1913–1923. https://doi.org/10.1016/j.neuroimage.2012.01.140

Giller, C. A., Hatab, M. R., & Giller, A. M. (1999). Oscillations in cerebral blood flow detected with a transcranial Doppler index. *Journal of Cerebral Blood Flow and Metabolism, 19*, 452–459. https://doi.org/10.1097/00004647-199904000-00011

Glasser, M. F., Coalson, T. S., Bijsterbosch, J. D., et al. (2018). Using temporal ICA to selectively remove global noise while preserving global signal in functional MRI data. *NeuroImage, 181*, 692–717. https://doi.org/10.1016/j.neuroimage.2018.04.076

Glickstein, M. (1988). The discovery of the visual cortex. *Scientific American, 259*, 118–127. https://doi.org/10.1038/scientificamerican0988-118

Glover, G. H., Li, T.-Q., & Ress, D. (2000). Image-based method for retrospective correction of physiological motion effects in fMRI: RETROICOR. *Magnetic Resonance in Medicine, 44*, 162–167. https://doi.org/10.1002/1522-2594(200007)44:1%3c162::AID-MRM23%3e3.0.CO;2-E

Gonzalez-Castillo, J., Kam, J. W. Y., Hoy, C. W., & Bandettini, P. A. (2021). How to interpret resting-state fMRI: Ask your participants. *Journal of Neuroscience, 41*, 1130–1141. https://doi.org/10.1523/JNEUROSCI.1786-20.2020

Greicius, M. D., Srivastava, G., Reiss, A. L., & Menon, V. (2004). Default-mode network activity distinguishes Alzheimer's disease from healthy aging: Evidence from functional MRI. *Proceedings of the National Academy of Sciences, 101*, 4637–4642. https://doi.org/10.1073/pnas.0308627101

Griffanti, L., Douaud, G., Bijsterbosch, J., et al. (2017). Hand classification of fMRI ICA noise components. *NeuroImage, 154*, 188–205. https://doi.org/10.1016/j.neuroimage.2016.12.036

Grutzendler, J., & Nedergaard, M. (2019). Cellular control of brain capillary blood flow. In vivo imaging veritas. *Trends in Neurosciences, 42*, 528–536. https://doi.org/10.1016/j.tins.2019.05.009

Guyton, A. C., & Harris, J. W. (1951). Pressoreceptor-autonomic oscillation; a probable cause of vasomotor waves. *American Journal of Physiology, 165*, 158–166. https://doi.org/10.1152/ajplegacy.1951.165.1.158

Hamilton, R., Keenan, J. P., Catala, M., & Pascual-Leone, A. (2000). Alexia for Braille following bilateral occipital stroke in an early blind woman. *NeuroReport, 11*, 237–240. https://doi.org/10.1097/00001756-200002070-00003

Harrison, B. J., Pujol, J., Contreras-Rodríguez, O., et al. (2011). Task-induced deactivation from rest extends beyond the default mode brain network. *PLOS ONE, 6*, e22964. https://doi.org/10.1371/journal.pone.0022964

He, Y., Byrge, L., & Kennedy, D. P. (2020). Nonreplication of functional connectivity differences in autism spectrum disorder across multiple sites and denoising strategies. *Human Brain Mapping*. https://doi.org/10.1002/hbm.24879

Hoiland, R. L., Tymko, M. M., Bain, A. R., et al. (2016). Carbon dioxide-mediated vasomotion of extra-cranial cerebral arteries in humans: A role for prostaglandins? *Journal of Physiology, 594*, 3463–3481. https://doi.org/10.1113/JP272012

Horster, I., Nickel, K., Holovics, L., et al. (2020). A Neglected topic in neuroscience: Replicability of fMRI results with specific reference to ANOREXIA NERVOSA. *Front Psychiatry, 11*. https://doi.org/10.3389/fpsyt.2020.00777

Hudetz, A. G., Wood, J. D., Biswal, B.B., et al. (1999). Effect of hemodilution on RBC velocity, supply rate, and hematocrit in the cerebral capillary network. *Journal of Applied Physiology (Bethesda, Md : 1985), 87*, 505–509.

Iadecola, C. (2017). The neurovascular unit coming of age: A journey through neurovascular coupling in health and disease. *Neuron, 96*, 17–42. https://doi.org/10.1016/j.neuron.2017.07.030

Intaglietta, M. (1990). Vasomotion and flowmotion: Physiological mechanisms and clinical evidence. *Vascular Medicine Review, 1*, 101–112.

Intaglietta, M., Johnson, P. C., & Winslow, R. M. (1996). Microvascular and tissue oxygen distribution. *Cardiovascular Research, 32*, 632–43. 0008636396001101 [pii].

Julien, C. (2006). The enigma of Mayer waves: Facts and models. *Cardiovascular Research, 70*, 12–21. https://doi.org/10.1016/j.cardiores.2005.11.008

Kampa, M., Schick, A., Sebastian, A., et al. (2020). Replication of fMRI group activations in the neuroimaging battery for the Mainz Resilience Project (MARP). *NeuroImage, 204*, 116223. https://doi.org/10.1016/j.neuroimage.2019.116223

Kapela, A., Parikh, J., & Tsoukias, N. M. (2012). Multiple factors influence calcium synchronization in arterial vasomotion. *Biophysical Journal, 102*, 211–220. https://doi.org/10.1016/j.bpj.2011.12.032

Kashyap, S., Ivanov, D., Havlicek, M., et al. (2018). Resolving laminar activation in human V1 using ultra-high spatial resolution fMRI at 7T. *Scientific Reports, 8*, 17063. https://doi.org/10.1038/s41598-018-35333-3

Katura, T., Tanaka, N., Obata, A., et al. (2006). Quantitative evaluation of interrelations between spontaneous low-frequency oscillations in cerebral hemodynamics and systemic cardiovascular dynamics. *NeuroImage, 31*, 1592–1600. https://doi.org/10.1016/j.neuroimage.2006.02.010

Khalil, A. A., Villringer, K., Filleböck, V., et al. (2020). Non-invasive monitoring of longitudinal changes in cerebral hemodynamics in acute ischemic stroke using BOLD signal delay. *Journal of Cerebral Blood Flow and Metabolism, 31*. https://doi.org/10.1177/0271678X18803951

Killip, T. (1962). Oscillation of blood flow and vascular resistance during Mayer waves. *Circulation Research, 11*, 987–993. https://doi.org/10.1161/01.RES.11.6.987

Kim, J. H., & Ress, D. (2016). Arterial impulse model for the BOLD response to brief neural activation. *NeuroImage, 124*, 394–408. https://doi.org/10.1016/j.neuroimage.2015.08.068

King, J. B., Prigge, M. B. D., King, C. K., et al. (2019). Generalizability and reproducibility of functional connectivity in autism. *Molecular Autism, 10*, 27. https://doi.org/10.1186/s13229-019-0273-5

Kleinschmidt, A., Obrig, H., Requardt, M., et al. (1996). Simultaneous recording of cerebral blood oxygenation changes during human brain activation by magnetic resonance imaging and near-infrared spectroscopy. *Journal of Cerebral Blood Flow and Metabolism, 16*, 817–826. https://doi.org/10.1097/00004647-199609000-00006

Leithner, C., & Royl, G. (2014). The oxygen paradox of neurovascular coupling. *Journal of Cerebral Blood Flow and Metabolism, 34*, 19–29. https://doi.org/10.1038/jcbfm.2013.181

Logothetis, N. K., Pauls, J., Augath, M., et al. (2001). Neurophysiological investigation of the basis of the fMRI signal. *Nature, 412*, 150–157. https://doi.org/10.1038/35084005

Lv, Y., Margulies, D. S., Cameron Craddock, R., et al. (2013). Identifying the perfusion deficit in acute stroke with resting-state functional magnetic resonance imaging. *Annals of Neurology, 73*, 136–140. https://doi.org/10.1002/ana.23763

Mano, Y., Harada, T., Sugiura, M., et al. (2009). Perspective-taking as part of narrative comprehension: A functional MRI study. *Neuropsychologia, 47*, 813–824. https://doi.org/10.1016/j.neuropsychologia.2008.12.011

Mayhew, J. E. W., Askew, S., Zheng, Y., et al. (1996). Cerebral vasomotion: A 0.1-Hz oscillation in reflected light imaging of neural activity. *NeuroImage, 4*, 183–193. https://doi.org/10.1006/nimg.1996.0069

Mishra, A., Reynolds, J. P., Chen, Y., et al. (2016). Astrocytes mediate neurovascular signaling to capillary pericytes but not to arterioles. *Nature Neuroscience, 19*, 1619–1627. https://doi.org/10.1038/nn.4428

Morcom, A. M., & Fletcher, P. C. (2007). Does the brain have a baseline? Why we should be resisting a rest. *NeuroImage, 37*, 1073–1082. https://doi.org/10.1016/j.neuroimage.2006.09.013

Murphy, K., Birn, R. M., & Bandettini, P. A. (2013). Resting-state fMRI confounds and cleanup. *NeuroImage, 80*, 349–359. https://doi.org/10.1016/j.neuroimage.2013.04.001

Murphy, K., Harris, A. D., & Wise, R. G. (2011). Robustly measuring vascular reactivity differences with breath-hold: Normalising stimulus-evoked and resting state BOLD fMRI data. *NeuroImage, 54*, 369–379. https://doi.org/10.1016/j.neuroimage.2010.07.059

Nee, D. E. (2019). fMRI replicability depends upon sufficient individual-level data. *Communications Biology, 2*, 1–4. https://doi.org/10.1038/s42003-019-0378-6

Ni, L., Li, J., Li, W., et al. (2017). The value of resting-state functional MRI in subacute ischemic stroke: Comparison with dynamic susceptibility contrast-enhanced perfusion MRI. *Scientific Reports, 7*, 41586. https://doi.org/10.1038/srep41586

Nikulin, V. V., Fedele, T., Mehnert, J., et al. (2014). Monochromatic Ultra-Slow (~0.1Hz) Oscillations in the human electroencephalogram and their relation to hemodynamics. *NeuroImage, 97*, 71–80. https://doi.org/10.1016/j.neuroimage.2014.04.008

Nishida, S., Aso, T., Takaya, S., et al. (2018). Resting-state functional magnetic resonance imaging identifies cerebrovascular reactivity impairment in patients with arterial occlusive diseases: A pilot study. *Neurosurgery*. https://doi.org/10.1093/neuros/nyy434

Obrig, H., Neufang, M., Wenzel, R., et al. (2000). Spontaneous low frequency oscillations of cerebral hemodynamics and metabolism in human adults. *NeuroImage, 12*, 623–639. https://doi.org/10.1006/nimg.2000.0657

Ogawa, S., Lee, T.-M., Nayak, A. S., & Glynn, P. (1990). Oxygenation-sensitive contrast in magnetic resonance image of rodent brain at high magnetic fields. *Magnetic Resonance in Medicine, 14*, 68–78. https://doi.org/10.1002/mrm.1910140108

Ogawa, S., Menon, R. S., Kim, S. G., & Ugurbil, K. (1998). On the characteristics of functional magnetic resonance imaging of the brain. *Annual Review of Biophysics and Biomolecular Structure, 27*, 447–474. https://doi.org/10.1146/annurev.biophys.27.1.447

Ojemann, G.A., Ojemann, J., & Ramsey, N.F. (2013). Relation between functional magnetic resonance imaging (fMRI) and single neuron, local field potential (LFP) and electrocorticography (ECoG) activity in human cortex. *Frontiers in Human Neuroscience, 7*. https://doi.org/10.3389/fnhum.2013.00034

Ossandón, T., Jerbi, K., Vidal, J. R., et al. (2011). Transient suppression of broadband gamma power in the default-mode network is correlated with task complexity and subject performance. *Journal of Neuroscience, 31*, 14521–14530. https://doi.org/10.1523/JNEUROSCI.2483-11.2011

Perkel, D. H., Gerstein, G. L., & Moore, G. P. (1967). Neuronal Spike trains and stochastic point processes: II. Simultaneous spike trains. *Biophysical Journal, 7*, 419–440. https://doi.org/10.1016/S0006-3495(67)86597-4

Raichle, M. E., MacLeod, A. M., Snyder, A. Z., et al. (2001). A default mode of brain function. *Proceedings of the National Academy of Sciences, 98*, 676–682. https://doi.org/10.1073/pnas.98.2.676

Rayshubskiy, A., Wojtasiewicz, T. J., Mikell, C. B., et al. (2014). Direct, intraoperative observation of ~0.1Hz hemodynamic oscillations in awake human cortex: Implications for fMRI. *NeuroImage, 87*, 323–331. https://doi.org/10.1016/j.neuroimage.2013.10.044

Renvall, V., Nangini, C., & Hari, R. (2014). All that glitters is not BOLD: Inconsistencies in functional MRI. *Scientific Reports, 4*, 3920. https://doi.org/10.1038/srep03920

Resnick, B. (2016). There's a lot of junk fMRI research out there. Here's what top neuroscientists want you to know. In: Vox. https://www.vox.com/2016/9/8/12189784/fmri-studies-explained. Accessed February 24, 2021.

Sadato, N., Pascual-Leone, A., Grafman, J., et al. (1996). Activation of the primary visual cortex by Braille reading in blind subjects. *Nature, 380*, 526–528. https://doi.org/10.1038/380526a0

Salimi-Khorshidi, G., Douaud, G., Beckmann, C. F., et al. (2014). Automatic denoising of functional MRI data: Combining independent component analysis and hierarchical fusion of classifiers. *NeuroImage, 90*, 449–468. https://doi.org/10.1016/j.neuroimage.2013.11.046

Sassaroli, A., Pierro, M., Bergethon, P. R., & Fantini, S. (2012). Low-frequency spontaneous oscillations of cerebral hemodynamics investigated with near-infrared spectroscopy: A review. *IEEE Journal on Selected Topics in Quantum Electronics, 18*, 1478–1492. https://doi.org/10.1109/JSTQE.2012.2183581

Satow, T., Aso, T., Nishida, S., et al. (2017). Alteration of venous drainage route in idiopathic normal pressure hydrocephalus and normal aging. *Frontiers in Aging Neuroscience, 9*. https://doi.org/10.3389/fnagi.2017.00387

Shulman, G. L., Fiez, J. A., Corbetta, M., et al. (1997). Common blood flow changes across visual tasks: II. Decreases in cerebral cortex. *Journal of Cognitive Neuroscience, 9*, 648–663. https://doi.org/10.1162/jocn.1997.9.5.648

Siegel, A. M., Culver, J. P., Mandeville, J. B., & Boas, D. A. (2003). Temporal comparison of functional brain imaging with diffuse optical tomography and fMRI during rat forepaw stimulation. *Physics in Medicine and Biology, 48*, 1391–1403.

Smith, S. M., Beckmann, C. F., Andersson, J., et al. (2013). Resting-state fMRI in the human connectome project. *NeuroImage, 80*, 144–168. https://doi.org/10.1016/j.neuroimage.2013.05.039

Sutton, G. C., Karnell, J., & Nylin, G. (1950). Studies on the rapidity of complete blood circulation in man. *American Heart Journal, 39*, 741–748. https://doi.org/10.1016/0002-8703(50)90134-7

Teipel, S. J., Wohlert, A., Metzger, C., et al. (2017). Multicenter stability of resting state fMRI in the detection of Alzheimer's disease and amnestic MCI. *NeuroImage: Clinical, 14*, 183–194. https://doi.org/10.1016/j.nicl.2017.01.018

Tgavalekos, K. T., Kainerstorfer, J. M., Sassaroli, A., & Fantini, S. (2016). Blood-pressure-induced oscillations of deoxy- and oxyhemoglobin concentrations are in-phase in the healthy breast and out-of-phase in the healthy brain. *Journal of Biomedical Optics, 21*, 101410. https://doi.org/10.1117/1.JBO.21.10.101410

Thomas, D. L., Lythgoe, M. F., Pell, G. S., et al. (2000). The measurement of diffusion and perfusion in biological systems using magnetic resonance imaging. *Physics in Medicine and Biology, 45*, R97-138. https://doi.org/10.1088/0031-9155/45/8/201

Tian, F., Niu, H., Khan, B., et al. (2011). Enhanced functional brain imaging by using adaptive filtering and a depth compensation algorithm in diffuse optical tomography. *IEEE Transactions on Medical Imaging, 30*, 1239–1251. https://doi.org/10.1109/TMI.2011.2111459

Tong, Y., Frederick, B, deB. (2010). Time lag dependent multimodal processing of concurrent fMRI and near-infrared spectroscopy (NIRS) data suggests a global circulatory origin for low-frequency oscillation signals in human brain. *NeuroImage, 53*, 553–564. https://doi.org/10.1016/j.neuroimage.2010.06.049

Tong, Y., Hocke, L. M., & Frederick, B. B. (2019). Low frequency systemic hemodynamic "noise" in resting state BOLD fMRI: Characteristics, causes, implications, mitigation strategies, and applications. *Frontiers in Neuroscience, 13*, 787. https://doi.org/10.3389/fnins.2019.00787

Tong, Y., Hocke, L. M., Licata, S. C., & Frederick, B. deB. (2012). Low-frequency oscillations measured in the periphery with near-infrared spectroscopy are strongly correlated with blood oxygen level-dependent functional magnetic resonance imaging signals. *Journal of Biomedical Optics, 17*, 106004. https://doi.org/10.1117/1.JBO.17.10.106004

Tong, Y., Lindsey, K. P., Hocke, L. M., et al. (2017). Perfusion information extracted from resting state functional magnetic resonance imaging. *Journal of Cerebral Blood Flow and Metabolism, 37*, 564–576. https://doi.org/10.1177/0271678X16631755

Tsuda, I., Tahara, T., & Iwanaga, H. (1992). Chaotic pulsation in human capillary vessels and its dependence on mental and physical conditions. *International Journal of Bifurcation and Chaos, 02*, 313–324. https://doi.org/10.1142/S0218127492000318

Turner, B. O., Paul, E. J., Miller, M. B., & Barbey, A. K. (2018). Small sample sizes reduce the replicability of task-based fMRI studies. *Communications Biology, 1*, 1–10. https://doi.org/10.1038/s42003-018-0073-z

Tuunanen, P. I., & Kauppinen, R. A. (2006). Effects of oxygen saturation on BOLD and arterial spin labelling perfusion fMRI signals studied in a motor activation task. *NeuroImage, 30*, 102–109. https://doi.org/10.1016/j.neuroimage.2005.09.021

Van Essen, D. C., Donahue, C. J., Coalson, T. S., et al. (2019). Cerebral cortical folding, parcellation, and connectivity in humans, nonhuman primates, and mice. *Proceedings of the National Academy of Sciences, 116*, 26173–26180. https://doi.org/10.1073/pnas.1902299116

Wei, H. S., Kang, H., Rasheed, I.-Y.D., et al. (2016). Erythrocytes are oxygen-sensing regulators of the cerebral microcirculation. *Neuron, 91*, 851–862. https://doi.org/10.1016/j.neuron.2016.07.016

Weiss, F., Zamoscik, V., Schmidt, S. N. L., et al. (2020). Just a very expensive breathing training? Risk of respiratory artefacts in functional connectivity-based real-time fMRI neurofeedback. *NeuroImage*, 116580. https://doi.org/10.1016/j.neuroimage.2020.116580

Willie, C. K., Tzeng, Y.-C., Fisher, J. A., & Ainslie, P. N. (2014). Integrative regulation of human brain blood flow. *Journal of Physiology, 592*, 841–859. https://doi.org/10.1113/jphysiol.2013.268953

Winder, A. T., Echagarruga, C., Zhang, Q., & Drew, P. J. (2017). Weak correlations between hemodynamic signals and ongoing neural activity during the resting state. *Nature Neuroscience, 20*, 1761–1769. https://doi.org/10.1038/s41593-017-0007-y

Wise, R. G., Ide, K., Poulin, M. J., & Tracey, I. (2004). Resting fluctuations in arterial carbon dioxide induce significant low frequency variations in BOLD signal. *NeuroImage, 21*, 1652–1664. https://doi.org/10.1016/j.neuroimage.2003.11.025

Yablonskiy, D. A., & Haacke, E. M. (1994). Theory of NMR signal behavior in magnetically inhomogeneous tissues: The static dephasing regime. *Magnetic Resonance in Medicine, 32*, 749–763.

Zhu, D. C., Tarumi, T., Khan, M. A., & Zhang, R. (2015). Vascular coupling in resting-state fMRI: Evidence from multiple modalities. *Journal of Cerebral Blood Flow and Metabolism, 35*, 1910–1920. https://doi.org/10.1038/jcbfm.2015.166

Toshihiko Aso was born and raised in Kyoto, Japan. He is a Deputy Team Leader of the Laboratory for Brain Connectomics Imaging, RIKEN Center for Biosystems Dynamics Research. He is also a part-time Lecturer at the Human Brain Research Center, Graduate School of Medicine, Kyoto University. After training in neurophysiology and clinical neurology, he received a BS and Ph.D. in Medicine at Kyoto University. He worked as a Research Fellow at NeuroSpin Centre CEA Saclay in France (2007–2008). For the ultimate goal of achieving reliable fMRI as a tool for neuroscience, his research has been focused on functional MRI methodology including diffusion-weighted fMRI and physiological denoising of BOLD fMRI.

Hidenao Fukuyama, born in Fukui Prefecture, Japan, is currently the Coordinator of the Graduate School of Medicine Educational Program at Kyoto University (KU). After serving as a Lecturer at the Department of Neurology (KU) until June 1995, and Associate Professor at the Department of Brain Pathophysiology (KU), he was appointed as Professor of Functional Brain Imaging (KU) in January 2001. He has been the Director of the Human Brain Research Center, Kyoto University Graduate School of Medicine since 2003. His main works have been directed to the imaging of neurological disorders including cerebrovascular diseases and dementia such as Alzheimer's disease. Recent MRI research in collaboration with Denis Le Bihan (Neurospin, Paris) is part of a leading project in a new field of functional MRI using diffusion MRI.

Chapter 19
Evolutionary Perspective on Suffering: Murase's Self–nonself Circulation Theory of Life Applied to PRISM (Pictorial Representation of Illness and Self Measure)

Stefan Büchi and Masatoshi Murase

Abstract Based on Murase's seminal theory of living creatures, simple and stable rules govern the dynamic of all life on every level from microscopic to macroscopic systems. The basic dynamic process is a "Self–Nonself circulation", where the emergence of new paradigms occurs in science through five circumscribed steps. Such steps are deduced from evolutionary processes. This chapter considers how this theory could be applied to the development and therapeutic use of PRISM (Pictorial Representation of Illness and Self Measure) and then evaluates how a new visual measure to assess suffering could be obtained utilizing the assessment of the patient's (Self) relationship with his/her illness (Nonself). Finally a novel understanding of suffering from an evolutionary perspective is presented.

Keywords Self–nonself circulation · PRISM (Pictorial Representation of Illness and Self Measure) · Therapeutic interaction · Evolutionary theory of suffering

---

S. Büchi (✉)
Clinic for Psychotherapy and Psychosomatics "Hohenegg", Meilen, Switzerland
e-mail: stefan.buechi@hohenegg.ch; buechi@prism-coop.ch

Privatklinik Hohenegg, 8706 Meilen Schweiz, 555 Zurich, Switzerland

M. Murase
Yukawa Institute for Theoretical Physics (YITP), Kyoto University, Kitashirakawa, Kyoto 606-8502, Japan

S. Büchi · M. Murase
International Research Unit of Advanced Future Studies (IRU-AFS), Kyoto University, Kyoto, Japan

K. Nishimura et al. (eds.), *Creative Complex Systems*, Creative Economy,
https://doi.org/10.1007/978-981-16-4457-3_19

19.1 Murase's Self–nonself Circulation Theory of Life

(a) Introduction of Self–nonself Circulation Theory

Murase (2000) proposed the Self-Nonself Circulation Theory of Life to explain diverse life phenomena on the basis of dynamical processes in contrast with the traditional Western thinking of element-reductionism. The most important emphasis was made on process-reductionism in this theory because diverse life phenomena have provided us with universal dynamics regardless of space and time scale differences. It would be possible to understand the universal principles behind the diverse phenomena from the microscopic molecular level to the macroscopic behavioral or social levels. Although the Self-Nonself Circulation Theory of Life has been described in detail (Murase, 2018), the essential features of this model are shown in Fig. 19.1.

The self indeed possesses nonself because of the continuous exchange of materials, information, energy, and so on. This kind of dynamical behavior is observed not only at the level of individual cells, but also at the level of cognitive level and much higher levels such as social groups. Murase (1996) explained clearly that aging occurs due to the process when the internalized nonself materials start to accumulate, although self-nonself circulation is required for living organism adaptation, learning, and even evolution. Interestingly, this kind of paradox occurs everywhere within us, between us, and even beyond us. This chapter describes the collaboration of the authors in their efforts to understand human suffering.

(b) NECTE Theory

Murase (2018) first introduced the NECTE Theory to understand evolutionary processes and learning on the assumption that there must be scale-invariant principles behind the diverse living phenomena as previously suggested (Murase, 1996).

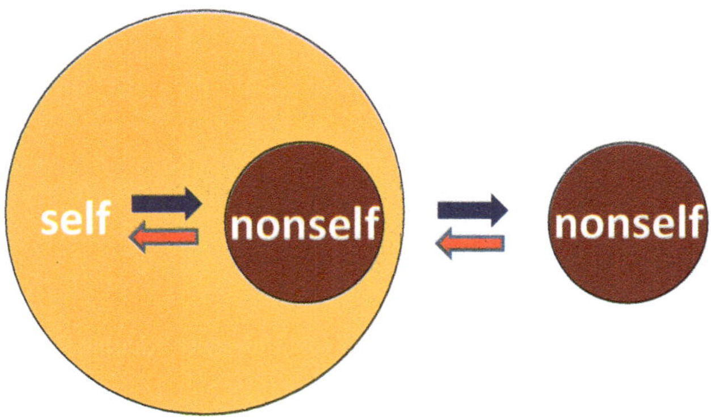

Fig. 19.1 Dynamics of self–nonself circulation theory

NECTE Theory

1. Negation
2. Expansion
3. Convergence
4. Transference
5. Emergence.

The theory is based on five steps with the first step of Negation representing denial of the present situation. In the mathematical sense, it is relating to the set Not-A in contrast to Set A. The second step of Expansion to induction is to get the general scheme behind the diverse events. The third step of Convergence is the reverse process of expansion, which means that the deduction process to understand specific events is based on general rules. The fourth step of Transference is often referred to as Abduction, which can completely connect different events and/or processes. The final fifth step of Emergence represents the appearance of new unexpected dynamics and/or systems through these previous four steps.

Murase (2018) discussed that NECTE Theory applies not only to evolutionary dynamics but also to learning processes and everyday life phenomena at all scales.

19.2 PRISM (Pictorial Representation of Illness and Self Measure)

(a) Introduction of PRISM

We report here a simple and novel method of eliciting personally salient information that is easy to apply. Pictorial Representation of Illness and Self Measure (PRISM) (Büchi et al., 1998) was originally conceived as a simple assessment of how people coped with serious illness. However, initial pilot work demonstrated that PRISM was not measuring coping. Nevertheless, the pilot work also indicated that individuals consistently responded to PRISM. Quantitative and qualitative data demonstrated that, in its initial form, PRISM behaved exactly as expected as a measure of suffering (Büchi & Sensky, 1999; Büchi et al., 2002). According to Cassell's seminal definition (Cassell, 1982), suffering occurs when the individual perceives a serious threat to his or her personhood, defined by things or qualities uniquely important to him or her. Among people who have serious illnesses, suffering is expected to be influenced more by the individual's appraisal of the threats to personhood than by the seriousness of the illness as defined by healthcare specialists.

In the standard PRISM task, based on the original task to measure suffering (Büchi et al., 2002) (see Fig. 19.2), informants were shown a white A4-size metal board (297 × 210 mm), with a fixed, yellow disk (7 cm in diameter) at the bottom right-hand corner. Informants were instructed: "Please imagine that this white sheet is your life at the moment, and the yellow disk [point to the disk] is your 'Self'.' Informants

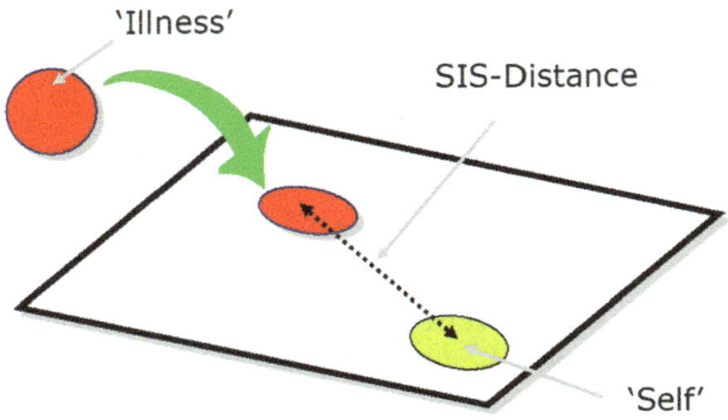

Fig. 19.2 PRISM instrument

were then handed a red disk (5 cm in diameter) and told: "Now imagine this red disk is your illness. Where would you put the red Illness disk in your life at the moment?" The main quantitative measure derived from PRISM is the distance between the centers of the Self and Illness disks (termed the Self-Illness Separation, or SIS).

In all the studies carried out to date using PRISM, the vast majority of informants have understood these simple instructions and placed the Illness disk on the PRISM board. As will be noted below, the resultant Self-Illness Separations were not random but behaved as expected as a measure of suffering (the smaller SIS, the greater the person's suffering). Having placed the Illness disk on the PRISM board, informants were asked why they placed their Illness in the position they chose. In most instances, a brief and succinct explanation was offered, which could be written down or audiotaped.

Some of the decisions regarding the design of the PRISM task were arbitrary, such as the colors of the Self and Illness disks, and their diameters. The decision to place the Self disk in the corner of one's life, rather than at its center, was a purely pragmatic one to allow the greatest range of distance between Illness and Self.

Qualitative data from the initial studies indicated that despite the simple instructions for PRISM, informants appeared to understand the task consistently. In general, the closer the Illness disk was placed to the Self disk, the more difficult informants perceived their symptoms were to control, and the more their illness intruded in their lives (Büchi et al., 2002). The same themes have emerged in later qualitative studies (Wittmann et al., 2009).

Quantitatively, the most consistent significant correlations of Self-Illness Separation (SIS) have been negative with depression (Büchi et al., 1998, 2002; Gielissen et al., 2013; Rumpf et al., 2004; Streffer et al., 2009) and with pain (Büchi et al., 1998, 2002; Streffer et al., 2009), particularly in the appraisal of pain (Kassardjian et al., 2008). Studies have also reported significant negative correlations between SIS and disease-specific quality-of-life measures (Büchi et al., 2002; Meyer et al., 2014; Mühleisen et al., 2009; Tõndury et al., 2011).

(b) **PRISM as a Measure of Suffering**

There is no gold standard for a measure of suffering against which PRISM can be validated. However, a recent systematic review (Krikorian et al., 2013) identified ten measures of suffering for use in palliative care, and PRISM was scored as superior to other measures for its overall psychometric properties. The reviewers also noted PRISM to be the most conceptually coherent instrument to measure suffering.

A consistent finding from published studies, as noted above, is that SIS, the quantitative measure derived from PRISM, behaves as would be predicted by Cassell's conceptualization of suffering (Cassell, 1982). Summarizing the results presented above, SIS has shown statistically significant correlations with other measures of the experience of illness, but the strength of these correlations has varied according to diagnosis and between samples with the same diagnosis. In essence, this is exactly what would be expected of a measure of suffering, as defined by Cassell (2004) as a perceived severe threat to an individual's Personhood (Büchi et al., 2002). Thus, for example, fatigue is likely to represent a more severe and salient threat to Personhood in cancer survivors who experience severe fatigue, than among people with rheumatoid arthritis, where pain is likely to have a greater influence. In conditions where symptoms or handicaps constitute a major threat to Personhood, SIS would be expected to correlate with disease-specific symptom inventories, or possibly with more generic measures such as the SF-36.

How a person appraises the threat from an illness to his or her Personhood is complex and involves factors beyond the burden of symptoms or the extent of disabilities. This is illustrated by the inverse correlation of suffering, measured by SIS, and sense of coherence as a measure of personal resilience against stress (Büchi et al., 1998, 2002). There is considerable evidence that, independent of illness severity, a high sense of coherence is associated with less depression (Büchi et al. 1998) and better health outcomes (Eriksson & Lindström, 2006). An effect of age and/or time since diagnosis might also be apparent in some samples. While some studies have reported no relationship between SIS and age (Reinhardt et al., 2006; Rumpf et al., 2004), others have reported correlations with age (Büchi et al., 2002) or time since diagnosis (Krikorian et al., 2013). With some conditions, longer time since diagnosis might lead a person to reappraise his or her Personhood—essentially reviewing what is important in one's life—hence offering scope to also alter suffering (Sensky, 2010). Aging could lead a person to relinquish previously valued aspects of Personhood, this again working to reduce suffering. However, if someone has already experienced a "narrowing" of Personhood through persistent symptoms and/or disability, the threat of losing even more of one's Personhood could lead to greater suffering.

19.3 Understanding Suffering in Medicine

(a) Paradox Behind Development of Medicine

"The goal of medicine is to heal and, where this is no longer possible, to alleviate suffering" (Cassell, 1982, 2004). In the future, medicine will be even more concerned than before with the dimension of alleviating suffering. The morbidity structure in developed industrial societies is changing, with a significant increase in chronic diseases. Since the early 1960s, epidemiological statistics have shown that the formerly predominant infectious diseases have lost their significance and that the chronic diseases of the cardiovascular system and the musculoskeletal system, as well as cancer and mental disorders, are the most common health impairments. The prevalence and incidence of chronic diseases will continue to increase in the future. This will be helped by demographic developments with an increasing proportion of elderly people, which will lead to an increase in the incidence of chronic degenerative diseases.

Paradoxically, progress in acute medicine will also lead to an increase in chronic disabilities: in many cases survival is possible, but often disabilities remain. Living with chronic diseases and the many functional limitations and psychosocial burdens that result from them has long since become the number one health problem in industrial societies (Gerdes & Weiss, 2000). If healing is no longer possible for an increasing number of patients, the dimension of suffering, or rather the therapy of suffering, would have to be given a new central place in medicine.

Suffering is a deeply human dimension, with which medicine was intensively concerned until the time of Descartes. Since the sixteenth century, when it was split into body-oriented medicine and a spirit-oriented religion, scientific medicine has turned away from this topic. The last 20 years has brought about a paradigm shift in clinical medicine and research towards a more individual-oriented medicine. In this context, quality of life has become a central theme in medicine. In terms of content, this interest meant an upgrading of the patient's subjective experience. The individuality of the patient was approached by considering the patient's life processes and the impairments caused by the illness. Dealing with the causes and treatment of suffering, a dimension that is inseparably linked to the experience of the person in a holistic physical–psychological–spiritual sense represents an attempt to contribute to overcoming the body–soul division in medicine.

(b) What is Suffering?

"All is suffering" was once said by the German philosopher Arthur Schopenhauer. This quotation indicates the complexity and comprehensive nature of this basic category of human existence. Suffering cannot be reduced to a phenomenon caused solely by pain, illness, or loss. Even the old German word stem for suffering (*lidan*) includes meanings such as "to go away, to drive, to travel, to move to a foreign country." This points to the processual character of suffering, which describes a movement in the context of meaning.

Suffering plays a central role in religions. In Christianity, the problem of the suffering of innocent people in the face of an almighty and benevolent God led to the problem of theodicy: how can God allow even the righteous to suffer misfortune? In the biblical figure of Job, this argument is shown in an exemplary way. From it also developed the idea of the *oeconomia salutis*, the economy of salvation, according to which in a mixed calculation extending to the hereafter suffering is ultimately set off. The test character of suffering is of central importance, as numerous martyr stories impressively point out. In Buddhism, suffering is of central importance. In the fundamental teaching of the Four Noble Truths, the explanations of the recognition, origin, and cessation of suffering form the core statement (Hebblethwaite, 1979).

In Western philosophy, the relationship to suffering is more ambivalent. There are probably philosophers like Schopenhauer, quoted above, who gave suffering a central meaning and regard compassion—the ability to recognize the suffering of others and accordingly contribute to the alleviation of suffering—as a central ethical category. Nevertheless, suffering appears rather as something that cannot be completely eradicated but should nevertheless be minimized. Utilitarianism, which sees the reduction of suffering and an increase of happiness as ethical guiding maxims, is an example of this concept.

In a separate empirical study (Wittmann et al., 2009), we asked patients with a chronic illness about their ideas of suffering. Physical impairments such as pain, fatigue, or limited mobility were mentioned as causes of suffering. Almost as often, impairments in areas of central individual importance such as professional plans or motherhood were mentioned. Social isolation was cited as the third causal category of suffering. Another noteworthy finding was that three quarters of the interviewees wanted to withdraw and be alone during phases of suffering.

(c) Pain and Suffering

Although often equated, pain and suffering can be distinguished. For suffering, the meaning of pain is crucial. For example, extremely severe pain during childbirth is not synonymous with pronounced suffering, because this pain does not have the meaning of a future loss of the person's integrity. However, previously bearable back pain, for example, can suddenly increase acutely and lead to severe suffering if osseous metastases are found as the cause. The extent of suffering depends on the patient's assessment of how well he or she can control pain and understand its cause. Only where the quantity of pain is experienced as overwhelming is pain, regardless of its significance, also suffering. In situations where the pain is perceived as a threat to one's existence or integrity, suffering arises. Circumstances that favorably influence the assessment of a pain problem (e.g., the diagnosis of a benign lung tumor in the case of suspected bronchus carcinoma) lead—with unchanged physical symptoms—to a reduction in suffering.

Pain can become suffering if it is experienced as pointless and useless. Until the development of anesthesia in the nineteenth century, medicine had only limited possibilities for analgesia. Therefore, the pressure to put suffering caused by pain and illness into a context of meaning was greater than today. The well-known Swiss

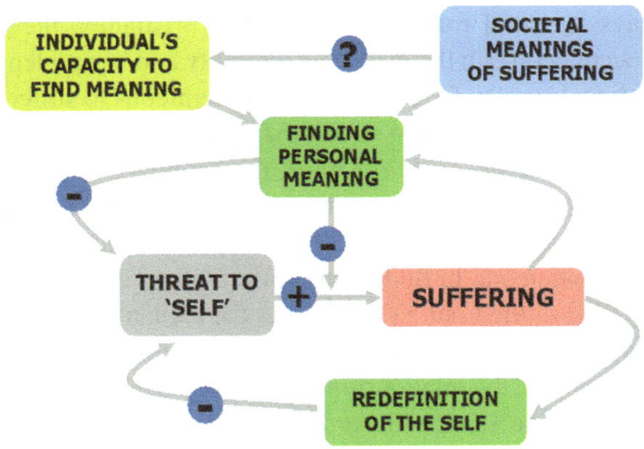

Fig. 19.3 Model of suffering

philosopher Hans Saner formulated the hypothesis that in our current "analgesic" culture (Stulz & Kaegi, 2006), which is aversively oriented towards pain, the extent of suffering caused by the remaining pain that cannot be eliminated therapeutically could be greater than the suffering caused by the untreated pain of earlier centuries, which was placed in a context of meaning. The connections between suffering, the threatened integrity of the person, and the influence of cultural sense responses are shown in Fig. 19.3.

(d) Suffering and Narration

Humans differ from other living beings by the meaning of language. Humans can create a sense of self through the use of stories. The processes of becoming ill, being ill, getting healthy, and dealing with illness can also be understood as individual stories. In the medical context of a hospital or a medical practice, there is hardly any room for these stories. An analysis of the communication in the medical context shows that in most cases a medical definition claim prevails, which attempts to classify illness in a given diagnostic context, and the patient's history becomes the medical history. The medical sociologist Arthur Frank (2013) described three different forms of patient narratives, which he distinguished based on content and form of experience.

1. *Healing Narrative*

The content of the story is simple: the narrator was sick and got well again. As the circumstances involved in overcoming the illness become more dramatic, the story becomes stronger. Often, medicine plays an important role through medication or surgical interventions. It is the narrative that is understood and loved by the medical profession, and it is the central myth of hospital medicine and medical education.

2. *Chaos Narrative*

The patient experiences a complete overtaxing by the disease and is in a state of being ill rather than having the illness. This state is accompanied by speechlessness and lack of communication of the affected person. This experience is suffered in extreme pain, but also in a state of shock, such as after notification of a carcinoma diagnosis. The medical system often ignores such situations.

3. *Search Narrative*

Chronic and incurable diseases, and disorders with an unpredictable course, change the self-image and life perspectives of the affected person in the long term. This constellation challenges the individual to adapt to the existing life plan and self-image. Central to this is the search for a new identity, or a new self-image.

(e) **Suffering and Meaning**

The central problem in dealing with the disease is often the question of meaning. Viktor Frankl, a world-famous Austrian-Jewish psychiatrist who survived the Holocaust, developed logotherapy based on his shocking experiences. This school of therapy gives central importance to the individual process of finding meaning. Frankl was convinced that "suffering stops being suffering when it makes sense" (Frankl & Weigel, 1978).

Empirical research has shown that the question of meaning in illness or loss has two independent dimensions (Davis et al., 1998). The first question that most sufferers ask is "Why me?" This is the causal question of understanding the loss of health in a larger context. Later on, an intensive process of wringing a positive meaning out of the new situation becomes apparent in many of those affected and they also define it as an opportunity for something new. The new question is "What for?" Clinical studies have shown that the why-me question correlates with symptoms of anxiety and depression; the more individuals deal with this question, the more these psychopathological symptoms are found. It is noteworthy that the why-me question did not correlate with psychopathological symptoms. In people who were more concerned with the what-for question, stronger maturation processes were found, and these will be discussed in more detail in the next section.

(f) **Suffering and Maturation**

For a few years, psychosomatic research has been dealing with the question of whether there are connections between suffering and maturation. Maturation is defined and recorded as "subjective experience of positive change in the concept of self and the world, deepened connection with other people, positive search for meaning as well as distinct philosophical reflection and deepening of spirituality." In a study at the University Hospital of Zurich, 54 parents were examined 2–6 years after the death of a prematurely born child who died at birth (24–26 weeks of pregnancy) through questionnaires and interviews (Büchi et al., 2007). The grief over the death of the child was still very present in 80% of the mothers and 60% of the fathers. At the same time, the majority of the parents interviewed stated that they had undergone major processes of change as a result of losing their child. Positive maturation processes were more frequently observed in mothers than in fathers. For example, almost 80% of the mothers and a good 40% of the fathers stated that they had set completely new priorities in life since the death of the child. Surprisingly, maturation was greatest among those parents who experienced the strongest feelings of grief 2–6 years after the death of their child.

Other empirical studies also confirmed the links between traumatic experiences and self-assessed maturation: the more severe the trauma caused by an accident, the occurrence of a serious illness or the loss of a loved one, the greater the maturation experienced in the years following the trauma (Schnyder et al., 2003). In a group of young female patients suffering from the autoimmune disease systemic lupus erythematosus (SLE), interviews, PRISM, and questionnaires were used to investigate the interrelationships between suffering and maturation processes (Wittmann et al., 2009). Of central interest was the importance of the individual ability to find meaning, as measured by individual stress resistance, for suffering and maturation processes. Based on the parameters of "individual stress resistance," "subjective suffering pressure," and "maturation triggered by illness," three groups of patients were defined who did not differ in terms of physical impairment. In the first group of patients who showed a low-stress resistance, the disease was experienced as overwhelming. The suffering was very great, and pronounced psychological symptoms of anxiety and depression were present. In these patients, only a few maturation processes were found; life was perceived as destroyed and broken. In contrast, the second group of patients were particularly resistant to stress. They had a low level of suffering and few psychological symptoms. As in the first group, these patients also showed few positive changes in their maturation; life as a whole had changed very little. In the third group with medium stress resistance, very interesting correlations emerged: SLE disease occupied a relatively central place in life and the patients also showed clear psychological symptoms and significant physical impairments caused by the disease. At the same time, however, these patients described distinct processes of maturation and change, which they experienced as very positive. New priorities in life were repeatedly reported, and the quality of personal relationships was significantly improved. In some patients, the disease triggered a new form of spirituality and religious searching.

What can be derived from these studies? Adaptation processes in chronic diseases are strongly influenced by individual resistance. Diseases that are experienced as overstraining lead to pronounced suffering and little maturation. Particularly resistant patients can live with the disease without undergoing major changes and therefore mature little. In contrast, those patients who are very burdened by the disease, but not completely overstrained, experience a change in their life perspective, which is experienced as personal maturation. The suffering process thus triggers personal growth and maturation in this group.

With the PRISM method, the patient can visually represent how he/she perceives his/her life situation from within. This view from the inside out leads to surprising insights into the human experience of suffering: a group of these people evaluates their suffering experience positively, as a process of maturation. To date, psychosomatic research has focused on the group of particularly stress-resistant and symptom-free patients. Because of the new findings on post-traumatic maturation, attention is now being focused for the first time on those people who are maturing in suffering. They claim to live more in connection with other people and in greater solidarity. They experience their lives as meaningful in an unusual way. From this point of view, the question of the nature of health arises anew: is health only freedom from symptoms, or can it also mean maturing in suffering?

The present research results question this one-sidedly functional image of man: experiences of limited functional efficiency and the associated symptoms of suffering can fundamentally change a person's self-image as well as their relationship to their environment and lead to deepened compassion and an intensive search for meaning.

What does this maturation aspect of suffering mean to the treatment of suffering people? To alleviate suffering is the noblest and most important task of therapeutic action. The prerequisite for a healing approach to suffering is to accept it as an unavoidable human condition. Our interest is in the knowledge of the growth processes triggered by suffering and how it promotes respect for what is happening, as well as the processes of change in the suffering person.

19.4 Discussions: Application of the Self–nonself-Circulation-Theory on PRISM

(a) Therapeutic Use of PRISM

A basic understanding of Murase's model is that in all evolutionary processes, there are similar Self–Nonself interactions that create life. In humans, an essential part of evolution includes learning processes, so similar rules are expected for biological development. In psychological therapeutic processes, specific learning processes are fostered. We discuss these interactions from an evolutionary perspective.

PRISM was developed to ameliorate communication between doctors and patients and is, therefore, an instrument to foster the personal learning processes of patients. Longstanding discussion about PRISM visualizing the Self–Nonself activity within a patient but also "between" therapist and patients provided new insights on the basic functions of PRISM. This process was reinforced by a practical roleplay constellation with Prof. Murase.

As a result of the discussion, the dynamics of the doctor–patient use of PRISM are shown in Fig. 19.4.

The interaction of therapist and patients is interlinked and synchronized by using PRISM. The dynamics of therapists and patients can be conceptualized similarly to strict biological evolutionary processes by using the model of Murase into five distinct NECTE steps. In summary, the use of PRISM as a haptic instrument implies the

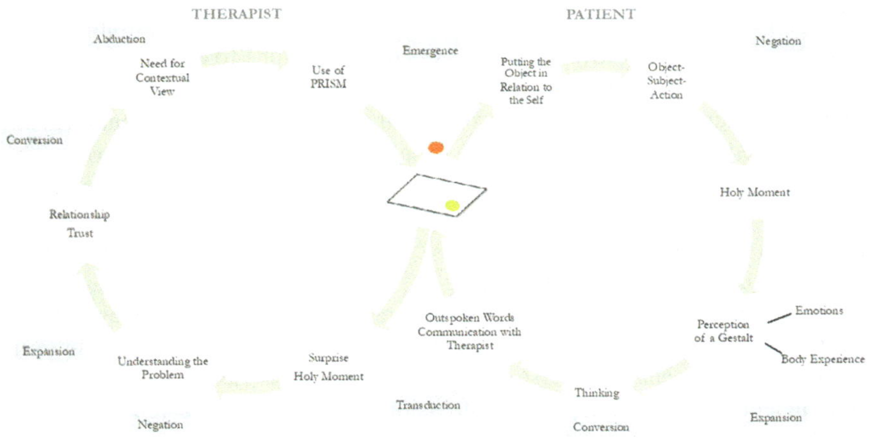

Fig. 19.4 Dynamic perspective of the use of PRISM including NECTE steps

construction of new meaning structures in human learning constellations including the patient as well as the therapist:

1. Patient level: the perspective of the subject–object relation within the patient.
2. Therapist level: the perspective of the interaction between patient and the therapist.

1. ***Patient Perspective***

The use of PRISM within a therapeutic process drastically changes the perception of the patient's view of his/her problem. The focus within the normal framework of psychotherapy is on understanding the problems and trying to find meaningful solutions to resolve them. Therefore, the process is object oriented, or, in other words, problem oriented. When PRISM is introduced, the focus is shifted from the content of the problem to the relation of the patient in dealing with the problem. The patient is exposed to a new experience of relationship with the problem and is forced to be active in dealing with the problem by grasping the disk and putting it in his/her life.

The right side of Fig. 19.3 describes the subject–object dynamic within Murase's framework of NECTE:

1. Negation

The patient is confronted with a new perspective on his/her situation: the problem disk is placed within the patient's life and this is perceived emotionally and physically as a new Gestalt.

2. Expansion

The patient perceives the disk in his/her life and in relation to self as a Gestalt constituting emotions, physical perception, and nonvalidating thoughts.

3. Convergence

The patient puts the experience into words.

4. Transference

The patient is motivated to think about his/her relation to the problem at different times, indicating a dynamic relation, rather than a static subject–object relation.

5. Emergence

The patient posits the disk expressing the dynamic dealing with his/her problem at different times.

This patient dynamic is linked with the second level: the learning process of the therapist.

2. ***Therapist Perspective***

When using PRISM, the therapist is actively guiding the learning process of the patient. Meanwhile, the patient is exploring his/her relationship with the "problem," and the therapist is reflecting on his/her understanding of the patient. PRISM interlinks the learning processes of patient and therapist.

An analytical time sequence of the therapist applying the NECTE-framework includes the following steps:

1. Negation

When the patient applies PRISM and puts the disk on his/her life, the therapist is confronted with new and often surprising information.

2. Expansion

The therapist perceives the patient in watching the PRISM-result and looks for body signals of effects.

3. Conversion

The therapist asks the patient about the meaning of the PRISM result and tries to integrate his/her former understanding of the patient by asking questions.

4. Transference

The therapist induces the patient to think about his/her problem not in a static way but in a dynamic way. The therapist asks for differences in time and activities that affect the relationship with the problem.

5. Emergence

The patient starts to actively use PRISM for other problems.

In clinical practice, the dynamic learning circle of the patient and the learning dynamic of the therapist are strongly interlinked by the use of PRISM. Therapists guide patients only with words, but are not allowed to interfere actively with patients' reasoning processes. Therapists are learning from the patients by observing their actions, and listen to the words they mention when performing PRISM. So a circular learning process is guaranteed—patients are confronted with new insights when performing PRISM and doctors learn from their patients at the same time—both the patients and their therapist develop.

(b) Theory of Suffering

Our research on suffering was deeply discussed and analyzed from an evolutionary perspective. Individual learning can be understood in terms of Piaget's concept of assimilation (adaptation of new information within the old schemas) and accommodation (new information induces the construction of new schemas) (Piaget & Inhelder, 1948). From this perspective, suffering is a comprehensive, subjectively averse, and nonvoluntary accommodation process concerning the metacognition of the meaning of life as well as the concepts about oneself:

Suffering is a painful learning process to adjust core concepts about oneself and the functioning of life.

Suffering and health are supported by the same overarching Self–Nonself circulation process. Mental disorders are states in which the Nonself is dominating, absorbing, and overwhelming, taking control of the Self. A healthy state is where the Nonself is in a dynamic and fluid exchange with the Self.

Recovery from illness is therefore reached when the state of extreme internalization of the Nonself is not exchanged by complete separation from the Nonself, but a proper and adapted inclusion of the Nonself problem. This crucial core process of therapy is called acceptance or "active resignation." This implies finding a balance between the extremes of being only passively overwhelmed as a "victim" of the Nonself problem on one hand, and on the other to heroically fight the Nonself problem to extricate it from the Self. Acceptance, therefore, includes an active exposition in dealing with a problem, but at the same time keeping in touch with other important aspects of life and aspects of one's Self.

In conclusion, suffering and recovery or mental illness and mental health are based on one single Self–Nonself circulation process. Through the use of PRISM, the relationship with the Nonself problem can be visualized and perceived by the patient, as well as the therapist. This learning process is interlinked, so we could talk of co-learning or co-evolution because both patients and therapists are evolving their minds. When this insight is established, it is important to enlarge the basic principles of healthy Self–Nonself circulation in the normal struggles of daily life.

References

Büchi, S., et al. (1998). Graphic representation of illness: A novel method of measuring patients' perceptions of the impact of illness. *Psychotherapy and Psychosomatics, 67*(4–5), 222–225.

Büchi, S., et al. (2002). Preliminary validation of PRISM (Pictorial Representation of Illness and Self Measure)—a brief method to assess suffering. *Psychotherapy and Psychosomatics, 71*(6), 333–341.

Büchi, S., et al. (2007). Grief and post-traumatic growth in parents 2–6 years after the death of their extremely premature baby. *Psychotherapy and Psychosomatics, 76*(2), 106–114.

Büchi, S., & Sensky, T. (1999). PRISM: Pictorial Representation of Illness and Self Measure. A brief nonverbal measure of illness impact and therapeutic aid in psychosomatic medicine. *Psychosomatics, 40*(4), 314–320.

Cassell, E. J. (1982). The nature of suffering and the goals of medicine. *The New England Journal of Medicine, 306*(11), 639–645.

Cassell, E. J. (2004). *The nature of suffering and the goals of medicine.* Oxford University Press.

Davis, C. G., et al. (1998). Making sense of loss and benefiting from the experience: Two construals of meaning. *Journal of Personality and Social Psychology, 75*(2), 561.

Eriksson, M., & Lindström, B. (2006). Antonovsky's sense of coherence scale and the relation with health: A systematic review. *Journal of Epidemiology & Community Health, 60*(5), 376–381.

Frank, A. W. (2013). *The wounded storyteller: Body, illness, and ethics.* University of Chicago Press.

Frankl, V. E., & Weigel, H. (1978). Trotzdem Ja zum Leben sagen: ein Psychologe erlebt das Konzentrationslager, Kösel.

Gerdes, N., & Weiss, J. (2000). Zur Theorie der Rehabilitation. In. 1. Bengel & U. Koch (Hrsgj), Grundlagen der Rehabilitationswissenschofien: Themen, Strategien und Methoden der Rehabilitarionsforschnng (S. 41–6S). Springer.

Gielissen, M. F. M., et al. (2013). Pictorial Representation of Self and Illness Measure (PRISM): A graphic instrument to assess suffering in fatigued cancer survivors. *Psychological Assessment, 25*(2), 658–663.

Hebblethwaite, B. (1979). Evil, Suffering and Religion.

Kassardjian, C. D., et al. (2008). Validating PRISM (Pictorial Representation of Illness and Self Measure) as a measure of suffering in chronic non-cancer pain patients. *The Journal of Pain: Official Journal of the American Pain Society, 9*(12), 1135–1143.

Krikorian, A., et al. (2013). Suffering assessment: A review of available instruments for use in palliative care. *Journal of Palliative Medicine, 16*(2), 130–142.

Meyer, M., et al. (2014). Disentangling tinnitus distress and tinnitus presence by means of EEG power analysis. *Neural Plasticity, 2014*, 468546–468546.

Mühleisen, B., et al. (2009). Pictorial Representation of Illness and Self Measure (PRISM): A novel visual instrument to measure quality of life in dermatological inpatients. *Archives of Dermatology, 145*(7), 774–780.

Murase, M. (1996). Alzheimer's disease as subcellular 'cancer' the scale-invariant principles underlying the mechanisms of aging. *Progress of Theoretical Physics, 95*(1), 1–36.

Murase, M. (2000). *The origin and evolution of life by means of endo-exo circulation.* Kyoto University Press.

Murase, M. (2018). *A self-similar dynamic systems perspective of "living" nature: The self-nonself circulation principle beyond complexity* (pp. 257–283). The Kyoto Manifesto for Global Economics: Springer.

Piaget, J., & Inhelder, B. (1948). La représentation de l'espace chez l'enfant.

Reinhardt, S., et al. (2006). Performance of the pictorial representation of illness and self measure in individuals with alcohol dependence, alcohol abuse or at-risk drinking. *Psychotherapy and Psychosomatics, 75*(4), 249–256.

Rumpf, H.-J., et al. (2004). A self-administered version of a brief measure of suffering: First aspects of validity. *Psychotherapy and Psychosomatics, 73*(1), 53–56.

Schnyder, U., et al. (2003). Does patient cognition predict time off from work after life-threatening accidents? *American Journal of Psychiatry, 160*(11), 2025–2031.
Sensky, T. (2010). Suffering. *International Journal of Integrated Care, 10*(Suppl), e024–e024.
Streffer, M.-L., et al. (2009). PRISM (pictorial representation of illness and self measure): A novel visual instrument to assess pain and suffering in orofacial pain patients. *Journal of Orofacial Pain, 23*(2), 140–146.
Stulz, P., & Kaegi, D. (2006). *Philosophie und Medizin*, Chronos.
Töndury, B., et al. (2011). The Pictorial Representation of Illness and Self Measure (PRISM) instrument reveals a high burden of suffering in patients with chronic urticaria. *Journal of Investigational Allergology & Clinical Immunology, 21*(2), 93–100.
Wittmann, L., et al. (2009). Suffering and posttraumatic growth in women with systemic lupus erythematosus (SLE): A qualitative/quantitative case study. *Psychosomatics, 50*(4), 362–374.

Stefan Büchi received his MD degree at the University of Zurich. Between 1990 and 1995 he trained as a clinical psychiatrist in Switzerland and at the West Middlesex University Hospital in London. Until 2008, he was head of the Consultation–Liaison Service at the University Hospital in Zurich. Since 2013, he has been the medical director of the Clinic for Psychotherapy and Psychosomatics "Hohenegg" near Zurich. His research interests include quality of life, coping with chronic illness, and psycho-oncology. Together with Tom Sensky he developed PRISM (Pictorial Representation of Illness and Self Measure), the best established tool in medicine to assess suffering.

Masatoshi Murase received his Ph.D. degree from The University of Tokyo in 1987. Since 1992, he has been an associate professor at the Yukawa Institute for Theoretical Physics, Kyoto University. In 1987 and 1988, he was a visiting scientist at the Duke University Medical Center, USA. Since 2010, he has been a member of the Cooperation Promotion Committee of the International Research Unit of Integrated Complex System Science, Kyoto University. Between 2015 and 2020, he was a director of the Research Promotion Strategy Office of the International Research Unit of Advanced Future Studies, Kyoto University.

Part V
Integrated Complex Science: Theory and Its Application in Transdisciplinary Studies

Chapter 20
Machine Learning for Metabolic Identification

Dai Hai Nguyen, Canh Hao Nguyen, and Hiroshi Mamitsuka

Abstract Metabolic identification is an essential part of metabolomics to understand biochemical characteristics of metabolites, which are small molecules that play important functions in biological systems. However, this field remains challenging with many unknown metabolites in existence. Mass spectrometry (MS) is a common technology that deals with such small molecules. Over recent decades, many methods have been proposed for MS-based metabolite identification, but machine learning has been a key process in recent progress in metabolite identification. This chapter provides a survey on computational methods for metabolic identification with the focus on *machine learning*, with a discussion on potential improvements for this task.

Keywords Machine learning · Metabolic identification · Mass spectrometry (MS) · Electron ionization (EI) · Electrospray ionization (ESI)

20.1 Introduction

Metabolomics involves the study of many metabolites, which are small molecules present in biological systems. They play roles in many important functions such as energy transport, building blocks of cells, and so on (Wishart, 2007). Identification of metabolites or understanding their biochemical characteristics is an essential and

D. H. Nguyen
Graduate School of Frontier Sciences, The University of Tokyo, 5-1-5 Kashiwa-no-ha, Kashiwa 277-8561, Chiba, Japan
e-mail: hai@k.u-tokyo.ac.jp

D. H. Nguyen · C. H. Nguyen · H. Mamitsuka (✉)
Bioinformatics Center, Institute for Chemical Research, Kyoto University, Uji 611-0011, Japan
e-mail: mami@kuicr.kyoto-u.ac.jp

C. H. Nguyen
e-mail: canhhao@kuicr.kyoto-u.ac.jp

H. Mamitsuka
Department of Computer Science, Aalto University, 02150 Espoo, Finland

K. Nishimura et al. (eds.), *Creative Complex Systems*, Creative Economy,
https://doi.org/10.1007/978-981-16-4457-3_20

significant part of metabolomics to enlarge the knowledge of biological systems. Metabolomics is also key to the development of many application domains such as biomedicine, biotechnology, or pharmaceuticals. However, metabolite identification remains a challenging task in metabolomics with a huge number of potentially interesting but unknown metabolites in existence.

Mass spectrometry (MS) is a common technique in analytical chemistry (De Hoffmann et al., 1997; Gross, 2006; McLafferty et al., 1993) used to measure the mass-to-charge ratio (*m/z*) of one or more molecules in a chemical sample. The output of a mass spectrometer, given a sample, is a mass spectrum, which is simply represented by a graph with *m/z* on the *x*-axis and the relative abundance of ions on the *y*-axis. Another way to represent a mass spectrum is as a list of peaks, each of which is defined by its *m/z* and its relative abundance. An illustration of a mass spectrum is shown in Fig. 20.1. A mass spectrometer consists of at least three components: ionization source, mass analyzer, and a detector (De Hoffmann et al., 1997). The ionization source is the component by which input molecules become charged ions. Two common forms of ionization are electron ionization (EI) and electrospray ionization (ESI). The mass analyzer is the component to physically separate ions according to *m/z*. The common types of mass analyzer are quadrupole, time-of-flight, and orbitrap devices. Once the ions have been separated, the detector is responsible for detecting

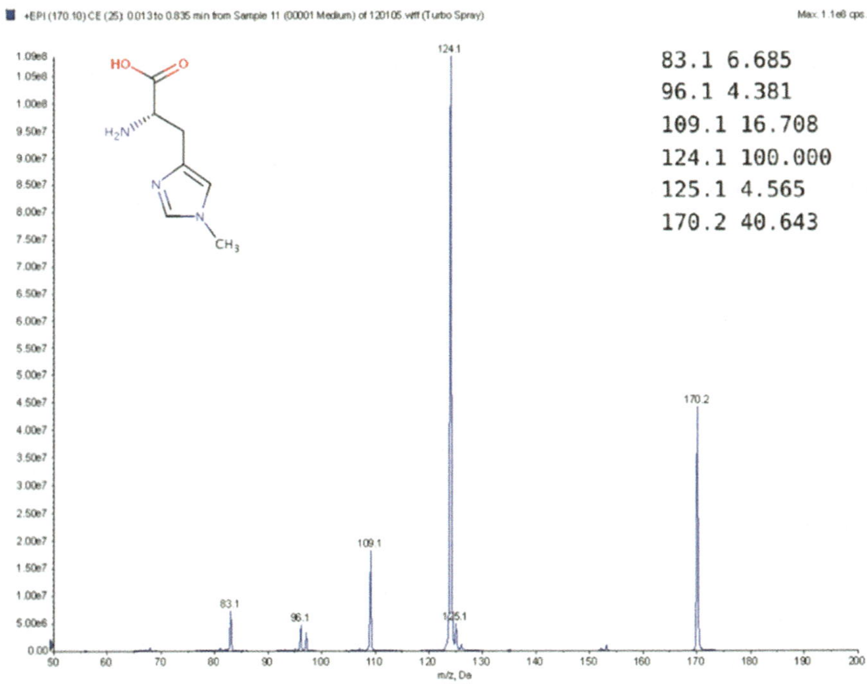

Fig. 20.1 Example MS from Human Metabolome Database (Wishart, 2007) for 1-methylhistidine (HMBD00001) with the corresponding chemical structure (top left) and list of peaks (top right)

and quantifying these ions. The mass spectrum contains peaks corresponding to the masses and relative abundance of the charged fragments as well as precursor ions. Because these values provide masses of some of the substructures, they can be used to elucidate the structure of the measured molecule. In practice, MS/MS, also known as tandem MS, is often preferred, and has been versatile and powerful for many applications. It consists of two mass analyzers coupled with collision-induced dissociation (CID). Ions are separated in the first mass analyzer (MS1), then enter a collision or fragmentation cell where fragmentation occurs and product ions are generated. The product ions are separated in the second mass analyzer (MS2) and then detected, eventually providing the MS/MS or tandem mass spectrum. Multistage MS allows further fragmentation of the product ions, providing ways to link these product ions to their precursor ions, thus offering more information about the fragmentation process (Nguyen et al., 2019a).

Metabolic identification from (tandem) mass spectra is an important step for further biochemical interpretation of metabolomics samples. In practice, this process is presumed to be challenging and is a time-consuming task in metabolomics experiments. Different from peptides and proteins where the fragmentation is generally simple because of repetition of the chemical structures, fragmentation of metabolites under varying conditions is a more complicated stochastic process. Thus, the interpretation of mass spectra can be cumbersome and requires expert knowledge. MS-based metabolite identification can be regarded as a retrieval task; that is, given a query spectrum of an unknown molecule, the goal is to find molecules (usually from a given reference database) that have similar spectra. It is straight-forward to directly compare the query against reference spectra in a reference spectra database (also known as a spectra library). The candidate molecules from the spectra library are ranked based on the similarities between their reference spectra and the query, and the best matched molecules are returned. However, spectra libraries often contain spectra of a small fraction of possible molecules, which can lead to unreliable results if the molecule corresponding to the query spectrum is not in the spectra library. Therefore, to mitigate issues of insufficient database breadth, alternative computational approaches have been devised (Nguyen et al., 2019a).

A number of computational methods and software tools have been developed to deal with the task of metabolite identification. We systematically organize them into four groups based on their methodologies described in Nguyen et al. (2019a): (1) mass spectra library; (2) in silico fragmentation; (3) fragmentation trees, and (4) machine learning (see Fig. 20.2 for details). We briefly describe the approaches as follows: (1) with a given query spectrum of an unknown molecule, the mass spectra library is used to compare the query against spectra in the library. (2) In silico fragmentation-based methods attempt to generate simulated spectra from chemical structures of reference compounds in the structural databases, which leverage a huge number of chemical structures of known molecules, and then compare the query against the simulated spectra. (3) Fragmentation tree-based approaches take advantage of relations of peaks in spectra, represented by fragmentation trees, which can be directly predicted from spectra by combinatorial optimizations and used to cluster molecules into categories. (4) Machine learning (ML) approaches are used to learn and predict

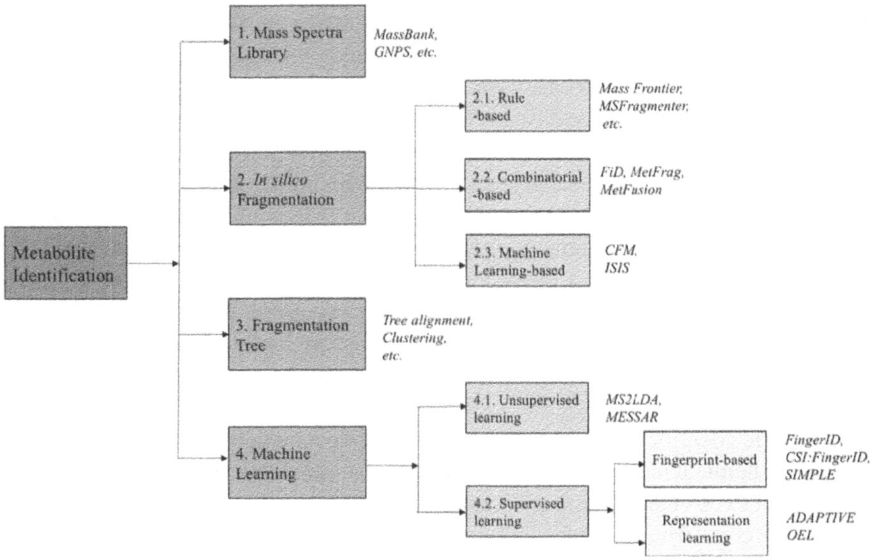

Fig. 20.2 Overview of approaches for metabolic identification. Adapted from Nguyen et al. (2019a)

intermediate representations between spectra and molecular structures and then use such representations for matching and retrieval. In this chapter, we focus on the use of in silico fragmentation (2) and ML (4) for metabolite identification.

20.2 In Silico Fragmentation Tools to Aid Metabolic Identification

Given the lack of MS/MS data of compounds in mass spectra libraries, the ability to identify unknown molecules by searching in mass spectra libraries is limited. Therefore, the use of software tools for predicting fragments and their abundance from the molecular structures of compounds can fill the gap between spectra and structural databases. This strategy has been successfully applied in protein studies to construct databases containing data on trypsin-associated cleavage and MS/MS of peptides, such as MASCOT (Perkins et al., 1999) and SEQUEST (Eng et al., 1994). However, in comparison with the prediction of fragmentation mechanisms for peptides and protein, which are simple because of the repetition in their structures, the fragmentation of product ions of metabolites in the tandem mass spectrometer is a much more complicated stochastic process and depends on a variety of factors. These factors may include: the detailed three-dimensional structure of the metabolite, the amount of energy required to break particular bond types to give the product ion, the probabilities of different dissociation reactions, which can be considered as a function of the applied collision energy and the pressure in the collision chamber, and so on. Nowadays, many in silico fragmentation software tools are available and are widely used to elucidate MS/MS when the sizes of spectra libraries are limited. In this section, we survey different computational tools and methods that use various algorithms for in silico fragmentation. The algorithms differ in the way they deploy different strategies to generate in silico fragments from the molecular *structure/graph* of the candidate compounds. We divide them into three subgroups (Nguyen et al., 2019a): rule-based, combinatorial-based, and ML-based fragmentation tools (see Fig. 20.2).

20.2.1 Rule-Based Methods

Rule-based in silico fragmentation tools are used to predict/generate theoretical spectra from molecular structures/graphs of compounds in the database using a set of chemical rules. This set of rules is a collection of general and heuristic rules of fragmentation processes extracted from datasets of elucidated MS/MS in the literature. The predicted spectra of candidate compounds from the database are then compared with the query (Hill et al., 2008; Kumari et al., 2011).

A typical commercial software tool, Mass Frontier (Mistrik, 2004), developed by HighChem, can generate fragments according to known general rules, or to specific rule libraries. The libraries can be defined by users or provided by HighChem or as a combination of both. ACD/MS Fragmenter (http://www.acdlabs.com), another commercial tool, also uses a comparable set of chemical rules to generate fragments. MOLGEN-MSF (Schymanski et al., 2009), developed by the University of Bayreuth, uses general fragmentation rules and is able to accept additional rules as an optional input file when calculating fragments. In addition, noncommercial rule-based software tools, like MASSIS (Chen et al., 2003) and MASSIMO (Gasteiger et al., 1992), use different approaches. In particular, structure-specific cleavage rules contained in MASSIS are divided into 26 different molecular classes. An input molecule is classified into one or some of these classes and the corresponding fragmentation rules are applied to obtain a set of fragments. MSSIMO uses a small set of general fragmentation reactions parameterized with reaction probabilities drawn from a collection of determined fragmentations.

In practice, these rule-based methods are not widely used because of several drawbacks: (1) the fragmentation process can vary significantly with small changes in the structure of the molecule. Thus, a fragmentation rule collected from a known fragmentation of a molecule may not properly apply to another, despite highly similar molecular structures. (2) It has been experimentally demonstrated that a single set of general rules is insufficient to identify some observed fragments with a reasonably high accuracy. Even though specific rules are constantly added to rule databases, it is not necessary to apply them to a new undiscovered compound in many cases. (3) The product ions of generated spectra have the same intensities because the bond cleavage rates are ignored. In reality, different molecules can generate the same product ions and the relative intensities can play an important role in discriminating these molecules.

20.2.2 Combinatorial-Based Methods

Different from the above software tools, which are mainly based on fragmentation rule databases, combinatorial-based methods are used to generate a graph of substructures from the chemical structure of a candidate compound in the database (see Fig. 20.3), and then find the most likely subset of substructures or so-called fragmentation trees that best matches the query by solving optimization problems. An advantage offered by this approach is in situations where MS/MS of compounds with less known fragmentation rules are queried. Some typical methods are reviewed in this subsection. In general, methods belonging to this subsection differ in the way they find the fragmentation tree that best matches the query spectra to produce similarity scores.

FiD (Fragment iDentificator, Heinonen et al., 2008) performs a search over all possible fragmentation paths and outputs a ranked list of alternative structures. In particular, given a graph structure of a precursor ion and its MS/MS, FiD first generates all possible connected subgraphs by a depth-first graph traversal, then computes the masses of product ions corresponding to the generated subgraphs to match with observed peak masses in the spectrum. A list of candidate fragments is then obtained, each of which is assigned a cost, namely, the standard bond energy required to cleave

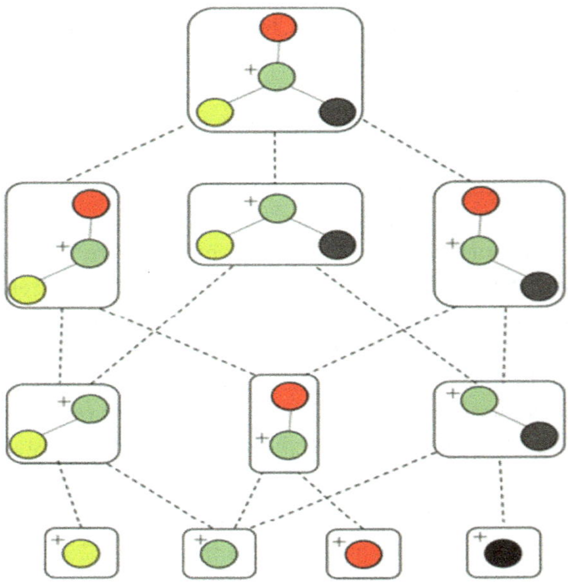

Fig. 20.3 An illustration of generating all connected subgraphs of the precursor graph. Adapted from Nguyen et al. (2019a)

bonds from the precursor ion. Obviously, the candidate fragments with smaller costs are preferred. Finally, a combinatorial optimization method, such as mixed-integer linear programing (MILP) is used to assign candidate fragments to measured peaks with minimal cost. Their experimental results showed that the product ions predicted by FiD are more consistent with the manual identification produced by domain experts than those of the rule-based fragment identification tools mentioned in the previous section. However, the main drawback of FiD is the computational cost because of: (1) rapid increase in the number of connected subgraphs; (2) the computational complexity of MILP to explain peaks with most likely candidate fragments. Thus, FiD can be applied only to small molecules.

Another combinatorial-based method is MetFrag (Wolf et al., 2010), which uses heuristic strategies such as a breadth-first search algorithm with a maximum tree depth parameter or removing duplicated subgraphs, to narrow down the search space of candidate fragments, overcoming the computational issue of FiD, which employs depth-first graph traversal to generate subgraphs. Thus, it is much faster than FiD and can be applied to a full structure database to search for the compound that best explains the spectrum. MetFrag uses bond dissociation energies for the cost of cleaving bonds. The candidate fragments are then used to rank the candidate molecules in the database without finding the most likely fragments corresponding to the spectrum. In a similar fashion, Ridder et al. (2012) introduced MAGMA, an extended version to multistage spectral trees MS^n. Different from MetFrag, when a substructure is considered to explain an MS^2 product ion, which is the precursor ion of MS^3, in addition to its substructure score, the resulting MS^3 is also taken into account. This spectrum is temporarily annotated with only a subset of the substructure, similar to the MS^2 level fragmentation spectrum. Then, the substructure scores obtained at the third level are added to the score at the second level, and this total score ranks the substructure candidates for MS/MS peaks and its fragmentation spectrum. This procedure is applied recursively to handle MS^n with any level.

Gerlich and Neumann (2013) presented a system called MetFusion that combines the results from MassBank (search in spectra database) and MetFrag (in silico fragmentation). This combination aims to take advantage of complementary approaches to improve the compound identification; that is, the vast coverage of the structural databases queried by MetFrag and reliable matching results achieved by searching in spectra libraries if similar spectra are available. Experimental results showed that combination of an in silico fragmentation-based method with curated reference measurements can improve compound identification and achieve the best of two approaches. More details about this method and results are described by Gerlich and Neumann (2013).

A drawback of this approach is that the above methods are mainly based on a bond disconnection-based approach to generate fragments from molecules, for example, standard bond energy and bond dissociation energy used by FiD and MetFrag, respectively. However, these are solely approximate estimates and bond dissociation energies are much more complicated in reality. These limitations have been tackled with some methods based on learning models, which are presented the following subsections.

20.2.3 Machine Learning-Based Methods

In addition to the above approaches to generate in silico fragmentation from the graph structure of compounds, a few works have been proposed that use machine learning models to learn the fragmentation process from training data. These methods have shown great promise in generating in silico spectra for structural identification. To avoid confusion with the content in Sect. 20.3, we clarify here that machine learning methods are used to learn and predict the presence of certain fragments (e.g., whether a bond between two atoms is broken) to generate in silico spectra from molecular structures. In a different sense, methods in Sect. 20.3 are to learn and perform classification or clustering from spectra (see Fig. 20.4 for illustration).

The previously mentioned methods to generate in silico fragments from molecular structures of compounds are based on chemical reaction equations or approximate bond strength. None of them have shown sufficient accuracy in generating in silico spectra to enable automated and correct identification of metabolites. To overcome this difficulty, Kangas et al. (2012) presented a method, named ISIS, using machine learning to generate *in silico* MS/MS spectra for lipids solely from chemical structures of compounds without fragmentation rules and without the need to define bond dissociation energy. The main concept is that for every bond in the molecular structure, one artificial neural network (ANN) is designed to predict bond cleavage energy from which bond cleavage rates can be calculated to determine the relative intensities; another is to predict which side of the bond is charged and captured by the detector in the mass spectrometer. These ANNs are iterated over all bonds in a

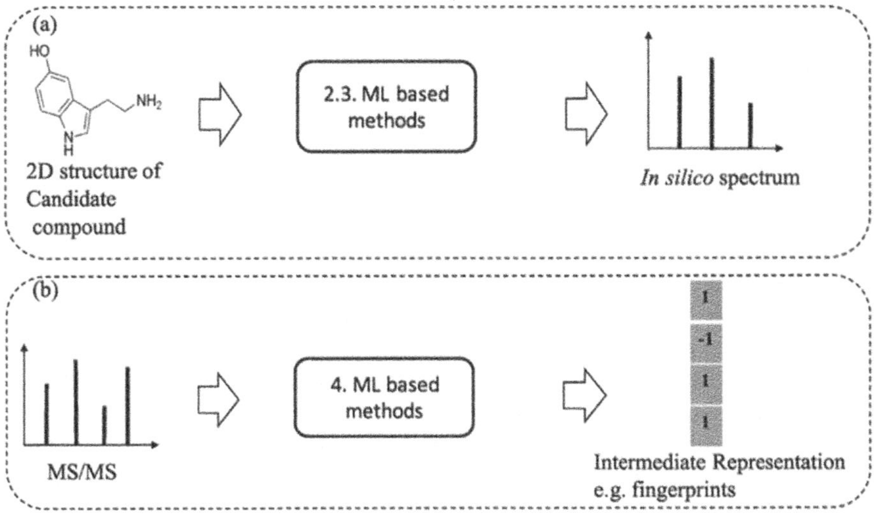

Fig. 20.4 Clarification of the difference between ML-based methods for learning and predicting in silico spectra from 2D structures of compounds **a** and ML-based methods for learning and predicting substructures or chemical properties from MS/MS **b**. Adapted from Nguyen et al. (2019a)

molecule to find bond cleavage energies and charged ions. For the learning process, the weights of the former ANN are trained by genetic algorithm (GA) to better predict the bond cleavage energies that produce ions and their corresponding intensities in the in silico spectra. The objective of GA is to have the in silico spectra match those in the experimental spectra using a Pearson R^2 correlation. The latter ANN is trained by a backpropagation algorithm in which the labels can be found by comparing the fragment masses with the experimental spectra.

Allen et al. (2015) proposed a probabilistic generative model called Competitive fragmentation mode for the fragmentation process. They assumed that each peak in the spectrum is generated by a fixed length sequence of random fragment states. It consists of two models: transition model defines the probability of each fragment leads to another at one step in the process and an observation model to map the final intermediate fragment state to the given peak. The parameter estimation for the transition and observation models is performed by an Expectation Maximization-like algorithm. The trained CFM can be used to predict peaks in the spectrum and for metabolite identification. The results showed that CFM obtained significantly better ranking for the correct candidate than MetFrag and FingerID (Heinonen et al., 2012). However, like the above methods, this method is limited to small molecules because of the combinatorial enumeration of fragmentation possibilities. It is also noteworthy that while ISIS is based on supervised machine learning, CFM is based on unsupervised learning to predict spectra.

20.3 Machine Learning for Metabolic Identification

A number of computational methods or tools have been introduced to deal with the task of metabolic identification. Remarkably, ML is the key to recent developments in this field. Besides identifying molecular compounds by searching spectra and structural databases, a number of methods have been proposed to predict predefined substructures or more generally chemical properties (Brouard et al., 2016; Dührkop et al., 2015; Heinonen et al., 2012; Nguyen et al., 2018, 2019b). Another direction is to automatically discover substructures directly from a set of MS/MS, from which we can identify the candidate compounds from the database based on their substructures (Mrzic et al., 2017; Van Der Hooft et al., 2016). This section is devoted to covering ML frameworks for these purposes, which can be divided into two groups: supervised learning for predicting substructures, and unsupervised learning for annotating substructures. Furthermore, the difference between the two subgroups can be intuitively illustrated as in Fig. 20.5.

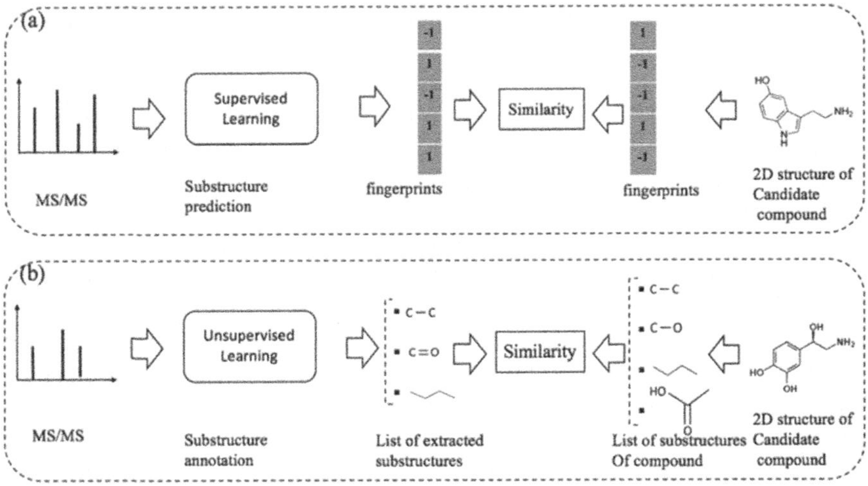

Fig. 20.5 Clarification of the difference between supervised and unsupervised learning for metabolic identification: **a** substructure prediction using supervised learning to map a given MS/MS to an intermediate representation (e.g., fingerprints), which is subsequently used to retrieve candidate metabolites in the database; **b** substructure annotation using unsupervised learning to extract biochemically relevant substructures with certain confidence from the given spectrum. Then, the similarity between the MS/MS and a chemical structure of a metabolite is estimated according to their common substructures. Note that the output of supervised learning (e.g., fingerprints) may indicate the presence/absence of all "predefined" substructures whereas the output of unsupervised learning may be a list of substructures frequently occurring in the database. Adapted from Nguyen et al. (2019a)

20.3.1 Supervised Learning for Predicting Substructures

The task of supervised learning for metabolic identification is that, given a set of MS/MS and corresponding molecular structures of known molecules, supervised learning methods are to learn a mapping function from MS/MS data to the molecular structure of a molecule. However, this task is technically challenging because both input and output spaces (spectra and molecular structures, respectively) are highly structured objects. Instead of directly learning a mapping from spectra to molecules, the approach of learning through intermediate representations between spectra and molecules has been used in many systems. The intermediate representations can be manually designed fingerprints of molecules or representations directly learned from the data. We go into details of such intermediate representations in the following subsections.

20.3.1.1 Fingerprint-Based Methods

A molecular fingerprint is a feature vector that is used to encode the structure of a molecule. In general, the values of this vector are often binary, indicating the presence or absence of a certain substructure or chemical property. The methods based on fingerprints are often carried out in two steps as illustrated in Fig. 20.6. The first step is to predict a fingerprint with supervised ML, which is regarded as a collection of binary classification tasks where each corresponds to a part in the fingerprint. The second step then uses the predicted fingerprint to query the database with techniques in ranking/information retrieval.

Kernel methods have been shown to be effective in dealing with the first step of predicting fingerprints from MS/MS. A notable method is FingerID (Heinonen et al., 2012), which used a support vector machine (SVM, Burges, 1998) with kernels defined on MS/MS to predict fingerprints. The kernels for pairs of MS/MS were defined, including an integral mass kernel and a probability product kernel (PPK, Jebara et al., 2004). It is noteworthy that the above kernels are mainly based on the information from individual peaks present in the spectra but ignores their interactions. Such information has proven to be useful in predicting fingerprints.

CSI:FingerID (Dührkop et al., 2015), an extended version of FingerID, jointly takes MS/MS and corresponding fragmentation trees (FTs, Böcker & Rasche, 2008)

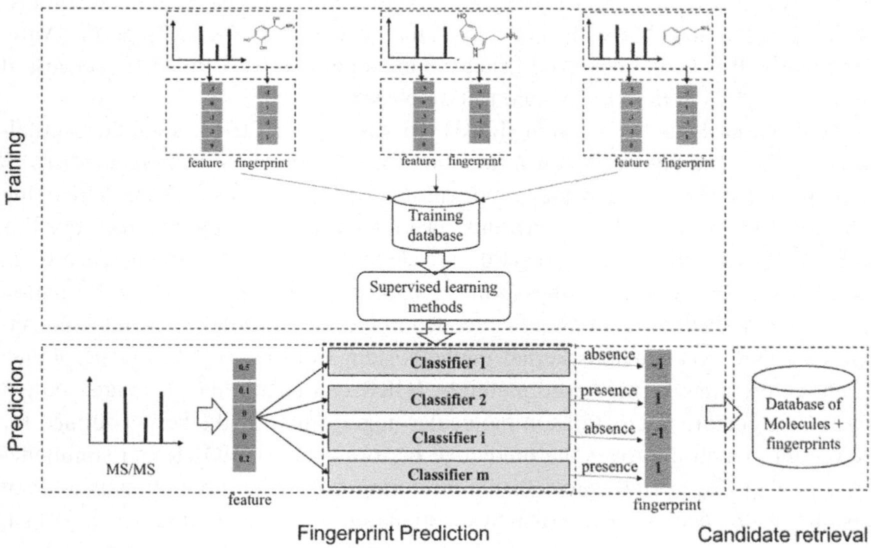

Fig. 20.6 Identification of unknown metabolites based on molecular fingerprint vectors. There are two main stages: (1) fingerprint prediction: learning a mapping from a molecule to the corresponding binary molecular fingerprint vector by classification methods, given a set of MS/MS spectra and fingerprints; (2) candidate retrieval: using the predicted fingerprints to retrieve candidate molecules from the databases of known metabolites. Adapted from Nguyen et al. (2019a)

as input to improve the predictive performance. FTs play an important role in interpreting the structure of molecules because it is usually assumed that MS/MS alone is insufficient to describe the fragmentation process. It is noteworthy that FTs are constructed from spectra and can be used to provide prior knowledge about the structure of compounds (e.g., dependencies between peaks in the spectra), which was ignored in FingerID. To incorporate MS/MS and FTs, kernels for FTs have to be defined, which range from simple ones for vertices, including node binary (NB), node intensity (NI); for edges, including loss binary (LB), loss count (LC), and loss intensity (LI), to complicated ones like common path counting (CPC), common subtree counting (CSC), and so on (see Dührkop et al., 2015 for more details). Subsequently, multiple kernel learning (MKL, Gönen & Alpaydın, 2011) is used to combine these kernels with ones defined for MS/MS using one of the following methods: centered alignment (ALIGNF), quadratic combination, and l_p-norm regularized combination. The combined kernel is then used in learning the final model for fingerprint prediction. CSI:FingerID presented improved scores against other benchmark tools but has the current limitation of processing MS/MS one at a time because of the need of computationally heavy conversion of spectra into FTs. For a similar purpose of incorporating peak dependencies into learning for fingerprint prediction, Nguyen et al. (2018) designed a kernel for peak interactions and combined this kernel with other kernels defined for individual peaks through MKL. The combined kernels were then combined with SVM to predict fingerprints. Experimental results showed that the micro-average accuracy and F1 score of the combined kernels were higher than PPK, while being comparable with the state-of-the-art kernels for CSI:FingerID. More importantly, this kernel achieved faster computation than CSI:FingerID because it does not use FTs in the training and testing phases.

Input output kernel regression (IOKR, Brouard et al., 2016) is another kernel-based method and has been shown to outperform the previous methods in terms of both predictive performance and computational speed. It learns a function mapping from MS/MS to molecular fingerprints (or molecular structures directly). For this purpose, IOKR defines two kernels to encode similarities in the input space (e.g., MS/MS) and output space (fingerprint or molecular structures). The information about input and output is implicitly encoded in these two kernels. Then, the advantage of IOKR over previous kernel methods stem from the two following points: (i) unlike previous kernel based methods, IOKR can handle the structured output space (e.g., feature interactions in molecular fingerprints) by the kernel defined for the output, which improves the predictive performance; (ii) IOKR can simultaneously predict fingerprints rather than considering fingerprint prediction as a set of separate tasks, leading to an efficient computation in the prediction stage. In fact, one can take structures of molecules into account by using graph kernels such as path, shortest-path, or graphlet kernels. It has been shown that kernels defined on molecular fingerprints obtained the best performance (Brouard et al., 2016).

Kernel-based methods have difficulty in dealing with sparse data and lack of interpretation. That is, each bit in a fingerprint represents a predetermined chemical property or substructure and its presence is often decided by a few peaks in the MS/MS. Furthermore, the number of training data is small, while sparse learning models have not been considered yet. In addition, sparse learning models are advantageous in that their results are easily interpretable. Nguyen et al. (2018) proposed a sparse, interpretable model, called SIMPLE, to incorporate peak interactions explicitly and to have a high interpretability. The model is as follows: given a MS/MS, represented by a feature $\boldsymbol{x} = [\boldsymbol{x}_1, \boldsymbol{x}_1, \ldots, \boldsymbol{x}_d]$, for one particular bit in the fingerprint, they formulated the model for individual peaks and interactions as follows:

$$\begin{aligned} f(x; w, W) &= b + \sum_{i=1}^{n} w_i x_i + \sum_{i=1}^{n} \sum_{j=1}^{n} W_{ij} x_i x_j \\ &= b + w^T + x^T W x \end{aligned}$$

where $b \in \mathbb{R}$, $w \in \mathbb{R}^d$, and $W \in \mathbb{R}^{d \times d}$. The prediction function consists of a bias b and two terms: the main effect term parameterized by the weight vector w, and the interaction term parameterized by the weight matrix W. Their roles are different, as illustrated in Fig. 20.7. While the former captures information about individual peaks, the latter captures information about peak interactions. To interprete the weights learned by the model, they imposed $l1$-norm (Tibshirani, 1996) and nuclear norm (Srebro & Shraibman, 2005) regularizations on the main effect and interaction terms to induce sparsity in w and lowrankness in W after training. The training stage is performed by minimizing a convex objective function, guaranteeing that the obtained solution is globally optimal. The incorporation of peak interactions in the prediction model $f(x; w, W)$ were found to significantly improve the prediction accuracy of not using interactions, resulting in comparable performance with the current top methods. Furthermore, SIMPLE could show interpretability of results (see Nguyen et al., 2018 for more details).

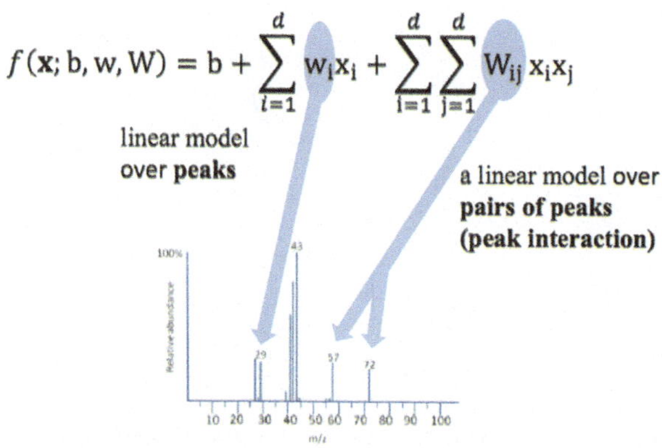

Fig. 20.7 Illustration of the predictive model of SIMPLE: the weight vector w of the main effect term captures information about the individual peaks, while interaction weight matrix W of the interaction term captures information about the peak interactions

20.3.1.2 Learning Intermediate Representations for Molecules from Spectra

Using molecular fingerprints as representations for molecules has been shown to be effective in metabolic identification. However, they have some drawbacks: (i) molecular fingerprints should be large in size to encode all possible substructures and chemical properties related to molecules, causing slow prediction in the candidate retrieval step; (ii) molecular fingerprints are not necessarily specific to any task nor data, and therefore redundant in the sense that they might contain much information irrelevant to the task and data, resulting in limited predictive performance.

Motivated by these drawbacks of molecular fingerprints, Nguyen et al. (2019b) proposed ADAPTIVE, which allows to the generation of representations for molecules using their molecular structures. The representations are learned from pairs of spectra and corresponding molecular structures, and thus are specific to the data and the task of metabolite identification. In technical detail, ADAPTIVE has two subtasks in the learning step: (1) learning a mapping $\hat{\phi}$ from molecular structure space $\mathcal{Y}$ to representation space $\mathcal{F}_d$, and (2) learning another mapping $\hat{h}$ from spectra space $\mathcal{X}$ to the representations obtained from step (1), as illustrated in Fig. 20.8a. In subtask (1), the mapping function $\hat{\phi}$ is parameterized by a model, named a message passing neural network (MPNN, Gilmer et al., 2017), for mapping a molecular structure of the molecule to a representation vector. MPNN is a framework, which can take graphs of arbitrary sizes and structures as inputs, to generate their representations at different levels (i.e., nodes, subgraphs, and the whole graph) with regard to a number of learning tasks including supervised, unsupervised, and semi-supervised learning (see Duvenaud et al., 2015; Gilmer et al., 2017; Nguyen et al., 2017 for more details). A key advantage is that MPNN can learn representations specific to the given task

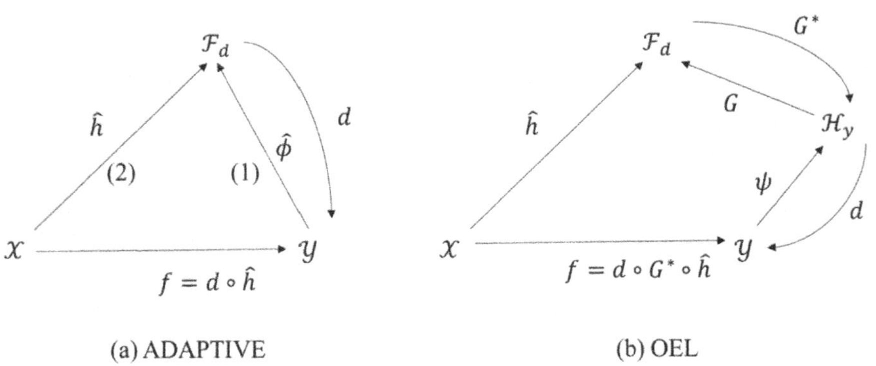

Fig. 20.8 Schematic illustration of ADAPTIVE and OEL for learning representations for molecules from the given set of spectra and molecular structures. **a** ADAPTIVE has two functions to learn: (1) function $\hat{\phi}$ (parameterized by a graph neural network) mapping from molecular structures to representations; (2) function $\hat{h}$ regressing representations from given input spectra. **b** OEL jointly learns two functions: (1) G (and G^*) mapping from molecular structures in the Hilbert space to representations; (2) function $\hat{h}$ regressing representations given input spectra

from the given data. The parameters of the MPNN are trained so that the correlation between the spectra and vectors mapped from molecular structures is maximized. The correlation is estimated by the Hilbert–Schmidt Independence Criterion (HSIC, Gretton et al., 2005). For subtask (2), IOKR is used to learn the mapping $\hat{h}$ from spectra to representations. Empirical validation of ADAPTIVE with a benchmark dataset showed the advantages of ADAPTIVE over existing methods, including the state-of-the-art IOKR, both in predictive performance and computation time for the prediction step.

In a somewhat similar fashion, Output Embedding Learning (OEL, Brogat-Motte et al., 2020) is a kernel-based framework that jointly learns a finite embedding (representation) of the outputs (chemical structures of molecules) and the regression of the representations given inputs (spectra) using prior information about the structure of outputs and unlabeled output data. Formally, given a collection of pairs of spectra and structures $\{(x_i, y_i), i = 1, \ldots, n\}$ and a collection of unlabeled molecular structures $\{y_j, j = 1, \ldots, m\}$, the optimization problem is as follows:

$$\min_{h,G} \frac{\gamma}{n} \sum_{i=1}^{n} h(x_i) - G\psi(y_i)^2_{\mathcal{F}_d} + \frac{1-\gamma}{n} \sum_{j=1}^{m} \psi(y_i) - G^*G\psi(y_i)^2_{\mathcal{F}_d}$$

where $\psi : \mathcal{Y} \to \mathcal{H}_y$ is a function that maps structured objects (molecular structures) into a Hilbert space $\mathcal{H}_y$; G is an operator from the Hilbert space $\mathcal{H}_y$ to the space of representations $\mathcal{F}_d$ and orthogonal ($GG^* = Id_d$); γ is a hyperparameter controlling the tradeoff between two terms in the objective function. It is noted that the first term corresponds to the regression of the embedding of chemical structures given spectra. The second term corresponds to learning a one-layer linear auto-encoder (parameterized by G^*G) whose inputs and outputs belong to the Hilbert space $\mathcal{H}_y$. However, the hidden layer is trained in the supervised manner by having a constraint with the regression part. The learning scheme of OEL is illustrated in Fig. 20.8b.

While both methods above have the same purpose of learning the representations for molecular structures of molecules through the available data, the key difference lies in the way they learn the representations. While the former relies on the graph neural network to parameterize the model to generate representations for chemical structures, the latter adopts kernels to encode the information about structures in an implicit way. Both methods have been shown to outperform the fingerprint-based methods for the metabolite identification tasks in terms of predictive performance and speed. The improvements suggest that learning representations for molecular structures with the available information of spectra is beneficial for the metabolite identification task in overcoming the limitations of manually designed feature vectors for molecules such as molecular fingerprints. However, a drawback of ADAPTIVE and OEL would be interpretability, because structural information of molecular structures is implicitly encoded by complicated nonlinear neural networks and kernels in ADAPTIVE and OEL, respectively, and cannot be made explicit easily. In metabolic identification, it would be desirable to connect the set of peaks to the corresponding substructures or chemical properties of molecules (see SIMPLE, Nguyen et al., 2018). Developing a model with the advantages of interpretability and high predictive performance would be interesting future work.

20.3.2 Unsupervised Learning for Substructure Annotation

Unlike supervised learning methods that aim to learn the mapping function from input to output, the purpose of unsupervised learning methods is to learn the underlying structures from a set of inputs without outputs being specified. In the context of MS/MS of metabolites, the inputs are only a collection of MS/MS. Metabolites may have common substructures, yielding similar product ions in their MS/MS. Many substructures among them contain information pertaining to the biochemical processes present. Therefore, extraction of such biochemically relevant substructures allows metabolites to be grouped based on their shared substructures regardless of classical spectral similarity. This approach can also be used to improve the accuracy of metabolic identification (Nguyen et al., 2019b).

One of the typical software tools for chemical substructure exploration is MS2Analyzer (Ma et al., 2014), which is a library-independent tool. It exploits the potential structure information contained in MS/MS. It was developed to elucidate substructures of small molecules from accurate MS/MS. The main function of this tool is to search mass spectral features including neutral loss, precursor and fragment ion masses, and mass differences in a large number of spectra. Through the combination of searching results and substructures/compound class relationship knowledge, compounds can be identified. However, MS2Analyzer can find all molecules sharing a specific set of mass spectral features provided by users, and sample-specific features are likely to be ignored. Another technique called molecular networking (Wang et al., 2016; Watrous et al., 2012; Yang et al., 2013) groups parent ions (i.e., MS1 peaks) based on their MS2 spectral similarity (e.g., cosine score) such that metabolites that are structurally annotated in a cluster can be used to annotate their neighbors. However, a drawback of molecular networks is that only MS1 peaks with high similarity are grouped and spectral features specifying the clusters must be manually extracted. Thus, it may fail to cluster molecules sharing small substructures with low MS2 spectral similarity.

MS2LDA, presented in Van Der Hooft et al. (2016), is a software tool offering benefits of both above methods, while overcoming their disadvantages. It can automatically extract relevant substructures in molecules based on their co-occurrence of mass fragments and neutral losses, and cluster the molecules accordingly. Based on the assumption that each observed MS/MS is composed of one or more substructures, MS2LDA adopts Latent Dirichlet Allocation (LDA, Blei et al., 2003), which was initially developed for text mining for extracting such substructures. LDA is a Bayesian version of probabilistic latent semantic analysis. In standard setting for text mining, each of D documents is modeled as a discrete distribution over T latent

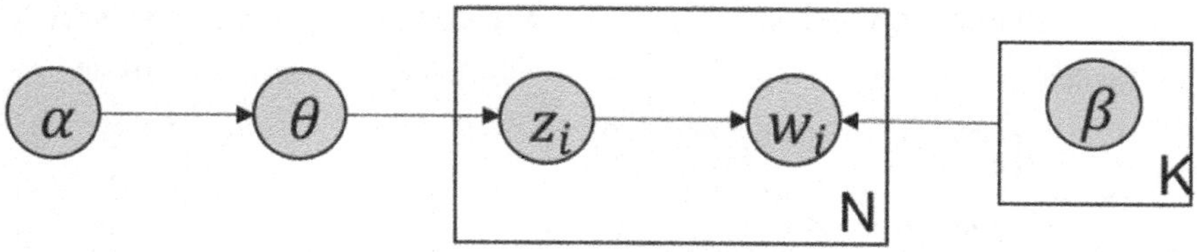

Fig. 20.9 Simplified graphical representation of LDA. Adapted from Nguyen et al. (2019a)

topics, each of which corresponds to a discrete distribution over a vocabulary of V words. For a document d, the distribution over topics, denoted by θ_d, is drawn from a Dirichlet distribution $Dir(\alpha)$, and for each topic t, the distribution over words, denoted by ϕ_t, is drawn from a Dirichlet distribution $Dir(\beta)$. A generative process in LDA is defined on document d as follows (note that the index d for document d is omitted for simplification):

(i) Choose $\theta \sim Dir(\alpha)$.
(ii) For each word w_i in document d:

a .Choose a topic $z_i \sim Multinomial(\theta)$.
b .Choose a word $w_i \sim Multinomial\left(\phi_{z_i}\right)$.

where latent variable z_i is a topic assignment for the ith word w_i in the document d. The parameters to be learned include α and β. The graphical representation of this process is illustrated in Fig. 20.9.

The correspondence between text documents and fragmentation spectra can be observed from a machine-learning perspective. LDA decomposes a document into topics based on the co-occurring words, while MS2LDA decomposes MS/MS into patterns of co-occurring fragments and losses. Learning LDA (MS2LDA) is to extract these topics [patterns or so-called (Mass2) Motifs] as illustrated in Fig. 20.10. For reference, either collapsed Gibb sampling (Griffiths & Steyvers, 2004) or Variational Bayes (Blei et al., 2003) can be used to assign topics (Mass2Motifs) to words (peaks). This step applied to MS/MS is called substructure annotation. By MS2LDA, each metabolite can be explained by one or more Mass2Motifs by which we can partly identify unknown metabolites via their spectra. Also, it can be used to quickly classify metabolites into functional classes without knowing the complete structures.

A drawback of the MS2LDA is that the extracted motifs still need to be structurally annotated based on expert knowledge, which is a complex and time-consuming process. To overcome this issue, Mrzic et al. (2017) introduced a method, named MESSAR, for automated substructure recommendation from mass spectra, motivated by frequent set mining (Goethals, 2005). Similar to MS2LDA, this method is also capable of extracting recurring patterns from a given set of spectra. Briefly, molecular substructures are first generated from molecular structures/graphs of metabolites in a given database, which consists of both MS/MS and corresponding molecular structures of known metabolites. Then, they are associated with fragment ions (i.e., peaks)

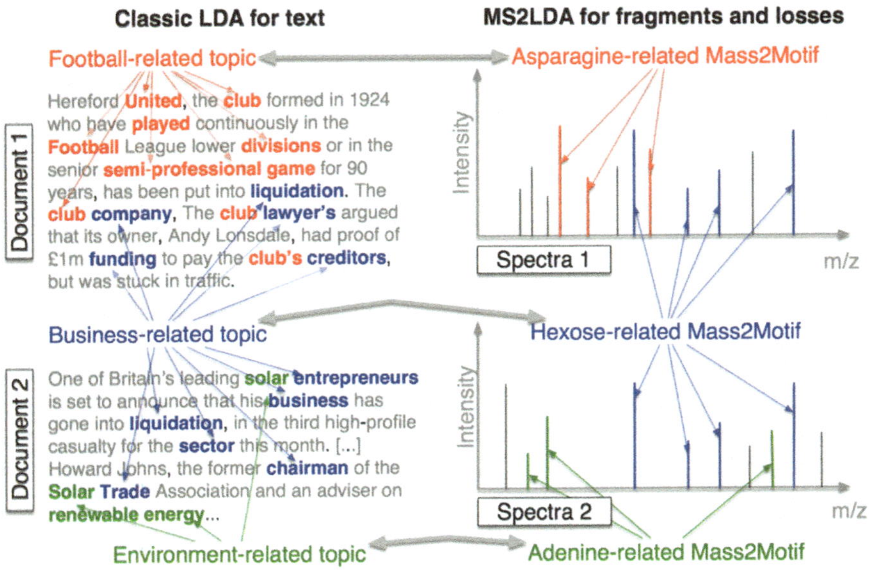

Fig. 20.10 Correspondence between LDA for text and MS2LDA for mass spectra: LDA finds topics based on the co-occurrence of words while MS2LDA finds substructures based on the co-occurrence of mass fragments and neutral losses. Adapted from Van Der Hooft et al. (2016)

and mass differences between peaks to construct a single dataset in a transactional format. Subsequently, frequent set mining techniques are applied to this set to extract rules of the following format: peaks p (or mass difference md) can be associated with substructure s with support f and confidence c. Such rules can be used to annotate substructures with calculated scores of support and confidence for mass spectra, in which the given peaks and mass differences are observed. Moreover, the recommended substructures can also be used for ranking candidate metabolites retrieved from a database by the similarity between recommended substructures and candidate molecular structures. Metabolites with a high number of substructures with high confidence are assigned higher ranks.

It is noteworthy that the aims of the aforementioned methods are similar, that is, substructure annotation. While MS2LDA needs a set of unlabeled MS/MS for learning without prior information about the molecular structures, MESSAR utilizes both experimental spectra and the corresponding structures, thus providing an automated substructure recommendation as opposed to expert-driven substructure annotation by MS2LDA. To end this section, we give a brief comparison of methods in both supervised and unsupervised approaches for substructure prediction and substructure annotation, respectively, in Table 20.1.

Table 20.1 Comparison of main representative methods for supervised and unsupervised learning approaches. The performance of supervised methods is evaluated by the accuracy of the returned list of candidates, whereas that of unsupervised methods is evaluated by their capability of substructure annotation

Approach	Method	Information type for learning	Performance	Training cost	Prediction cost
Supervised	FingerID (Heinonen et al., 2012)	Spectra	High	Low	Low
	CSI:FingerID (Dührkop et al., 2015)	Spectra + trees	High	High	High
	SIMPLE (Nguyen et al., 2018)	Spectra	High	Low	Low
	IOKR (Brouard et al., 2016)	Spectra + trees	High	Medium	Medium
	OEL (Brogat-Motte et al., 2020)	Spectra + trees	High	Medium	Medium
	ADAPTIVE (Nguyen et al. 2019b)	Spectra + trees	High	High	Low
Unsupervised	MS2Analysis (Ma et al., 2014)	User-specific features	Low	N/A	N/A
	MolecularNetwork (Wang et al., 2016)	Spectra	Low	N/A	N/A
	MS2LDA (Van Der Hooft et al., 2016)	Spectra	High (expert-driven)	N/A	N/A
	MESSAR (Mrzic et al., 2017)	Spectra + molecular graph	High (automation)	N/A	N/A

20.4 Conclusion

Metabolic identification is an essential part of metabolomics to enlarge the knowledge of biological systems. However, this remains a challenging task with a huge number of potentially interesting but unknown metabolites in existence. This chapter reviews the computational techniques and software tools used to deal with the task of metabolic identification with the focus on machine learning-based methods used in in silico fragmentation and machine learning approaches, which are the key to recent progress in metabolite identification.

It is suggested in this chapter that statistical machine learning-based methods should be a reasonable choice for the task of metabolice identification. When the volume of spectra and molecular structure data is increasing over time, the ability of machine learning algorithms to learn and predict relationships inherent in the data will be more enhanced. For example, IOKR, ADAPTIVE, and OEL are currently the best ML-based tools/methods for learning useful representations for molecules and achieving the best performance in the metabolite identification task. In addition, we also emphasize that the combination of different approaches should also be considered, by which we can take advantages of them for significant improvement. For example, IOKR and CSI:FingerID use *machine learning* and *fragmentation trees to incorporate structure information from trees and ML for prediction.* Another is MetFusion, which combines the results from MassBank (mass spectra library) and MetFrag (in silico fragmentation) to take advantage of complementary approaches.

Acknowledgements D. H. N. has been supported in part by an Otsuka Toshimi scholarship and a JSPS Research Fellowship for Young Scientists (DC2) with KAKENHI [grant number 19J14714]. C.H.N. has been supported in part by MEXT KAKENHI [grant number 18K11434]. H.M. has been supported in part by JST ACCEL [grant number JPMJAC1503], MEXT KAKENHI [grant numbers 16H02868, 19H04169], FiDiPro by Tekes (currently Business Finland), and the AIPSE program of the Academy of Finland.

References

Allen, F., Greiner, R., & Wishart, D. (2015). Competitive fragmentation modeling of ESI-MS/MS spectra for putative metabolite identification. *Metabolomics, 11*(1), 98–110.

Blei, D. M., Ng, A. Y., & Jordan, M. I. (2003). Latent dirichlet allocation. *Journal of Machine Learning Research, 3*(Jan), 993–1022.

Böcker, S., & Rasche, F. (2008). Towards de novo identification of metabolites by analyzing tandem mass spectra. *Bioinformatics, 24*(16), i49–i55.

Brogat-Motte, L., Rudi, A., Brouard, C., Rousu, J., & d'Alché-Buc, F. (2020). Learning Output Embeddings in Structured Prediction. *arXiv preprint* arXiv:2007.14703.

Brouard, C., Shen, H., Dührkop, K., d'Alché-Buc, F., Böcker, S., & Rousu, J. (2016). Fast metabolite identification with input output kernel regression. *Bioinformatics, 32*(12), i28–i36.

Burges, C. J. (1998). A tutorial on support vector machines for pattern recognition. *Data Mining and Knowledge Discovery, 2*(2), 121–167.

Chen, H., Fan, B., Xia, H., Petitjean, M., Yuan, S., Panaye, A., & Doucet, J. P. (2003). MASSIS: a mass spectrum simulation system. 1. Principle and method. *European Journal of Mass Spectrometry, 9*(3), 175–186.

De Hoffmann, E., Charette, J., & Stroobant, V. (1997). *Mass spectrometry: Principles and applications*.

Dührkop, K., Shen, H., Meusel, M., Rousu, J., & Böcker, S. (2015). Searching molecular structure databases with tandem mass spectra using CSI: FingerID. *Proceedings of the National Academy of Sciences, 112*(41), 12580–12585.

Duvenaud, D. K., Maclaurin, D., Iparraguirre, J., Bombarell, R., Hirzel, T., Aspuru-Guzik, A., & Adams, R. P. (2015). Convolutional networks on graphs for learning molecular fingerprints. In *Advances in neural information processing systems* (pp. 2224–2232).

Eng, J. K., McCormack, A. L., & Yates, J. R. (1994). An approach to correlate tandem mass spectral data of peptides with amino acid sequences in a protein database. *Journal of the American Society for Mass Spectrometry, 5*(11), 976–989.

Gasteiger, J., Hanebeck, W., & Schulz, K. P. (1992). Prediction of mass spectra from structural information. *Journal of Chemical Information and Computer Sciences, 32*(4), 264–271.

Gerlich, M., & Neumann, S. (2013). MetFusion: Integration of compound identification strategies. *Journal of Mass Spectrometry, 48*(3), 291–298.

Gilmer, J. et al. (2017). Neural message passing for quantum chemistry. In Precup, D. & Teh, Y. W. (Eds.), *Proceedings of the 34th International Conference on Machine Learning, Volume 70 of Proceedings of Machine Learning Research.* International Convention Centre, PMLR, Sydney, Australia, pp. 1263–1272.

Goethals, B. (2005). Frequent set mining. In *Data mining and knowledge discovery handbook* (pp. 377–397). Springer.

Gönen, M., & Alpaydın, E. (2011). Multiple kernel learning algorithms. *The Journal of Machine Learning Research, 12*, 2211–2268.

Gretton, A., Bousquet, O., Smola, A., & Schölkopf, B. (2005, October). Measuring statistical dependence with Hilbert-Schmidt norms. In *International conference on algorithmic learning theory* (pp. 63–77). Springer.

Griffiths, T. L., & Steyvers, M. (2004). Finding scientific topics. *Proceedings of the National Academy of Sciences, 101*(suppl 1), 5228–5235.

Gross, J. H. (2006). *Mass spectrometry: A textbook.* Springer Science & Business Media.

Heinonen, M., Rantanen, A., Mielikäinen, T., Kokkonen, J., Kiuru, J., Ketola, R. A., & Rousu, J. (2008). FiD: A software for ab initio structural identification of product ions from tandem mass spectrometric data. *Rapid Communications in Mass Spectrometry: an International Journal Devoted to the Rapid Dissemination of up-to-the-Minute Research in Mass Spectrometry, 22*(19), 3043–3052.

Heinonen, M., Shen, H., Zamboni, N., & Rousu, J. (2012). Metabolite identification and molecular fingerprint prediction through machine learning. *Bioinformatics, 28*(18), 2333–2341.

Hill, D. W., Kertesz, T. M., Fontaine, D., Friedman, R., & Grant, D. F. (2008). Mass spectral metabonomics beyond elemental formula: chemical database querying by matching experimental with computational fragmentation spectra. *Analytical Chemistry, 80*(14), 5574–5582.

Jebara, T., Kondor, R., & Howard, A. (2004). Probability product kernels. *Journal of Machine Learning Research, 5*(Jul), 819–844.

Kangas, L. J., Metz, T. O., Isaac, G., Schrom, B. T., Ginovska-Pangovska, B., Wang, L., Tan, L., Lewis, R. R., & Miller, J. H. (2012). In silico identification software (ISIS): A machine learning approach to tandem mass spectral identification of lipids. *Bioinformatics, 28*(13), 1705–1713.

Kumari, S., Stevens, D., Kind, T., Denkert, C., & Fiehn, O. (2011). Applying in-silico retention index and mass spectra matching for identification of unknown metabolites in accurate mass GC-TOF mass spectrometry. *Analytical Chemistry, 83*(15), 5895–5902.

Ma, Y., Kind, T., Yang, D., Leon, C., & Fiehn, O. (2014). MS2Analyzer: A software for small molecule substructure annotations from accurate tandem mass spectra. *Analytical Chemistry, 86*(21), 10724–10731.

McLafferty, F. W., Tureček, F., & Turecek, F. (1993). *Interpretation of mass spectra.* University science books.

Mistrik, R. (2004). A new concept for the interpretation of mass spectra based on a combination of a fragmentation mechanism database and a computer expert system. *Advances in Mass Spectrometry, Elsevier, Amsterdam, 16*, 821.

Mrzic, A., Meysman, P., Bittremieux, W., & Laukens, K. (2017). Automated recommendation of metabolite substructures from mass spectra using frequent pattern mining. *bioRxiv*, 134189.

Nguyen, D. H., Nguyen, C. H., & Mamitsuka, H. (2018). SIMPLE: Sparse Interaction Model over Peaks of moLEcules for fast, interpretable metabolite identification from tandem mass spectra. *Bioinformatics, 34*(13), i323–i332.

Nguyen, D. H., Nguyen, C. H., & Mamitsuka, H. (2019a). Recent advances and prospects of computational methods for metabolite identification: A review with emphasis on machine learning approaches. *Briefings in Bioinformatics, 20*(6), 2028–2043.

Nguyen, D. H., Nguyen, C. H., & Mamitsuka, H. (2019b). ADAPTIVE: LeArning DAta-dePendenT, concIse molecular VEctors for fast, accurate metabolite identification from tandem mass spectra. *Bioinformatics, 35*(14), i164–i172.

Nguyen, H., Maeda, S. I., & Oono, K. (2017). Semi-supervised learning of hierarchical representations of molecules using neural message passing. *arXiv preprint* arXiv:1711.10168.

Perkins, D. N., Pappin, D. J., Creasy, D. M., & Cottrell, J. S. (1999). Probability-based protein identification by searching sequence databases using mass spectrometry data. *ELECTROPHORESIS: An International Journal, 20*(18), 3551–3567.

Ridder, L., van der Hooft, J. J., Verhoeven, S., de Vos, R. C., van Schaik, R., & Vervoort, J. (2012). Substructure-based annotation of high-resolution multistage MSn spectral trees. *Rapid Communications in Mass Spectrometry, 26*(20), 2461–2471.

Schymanski, E. L., Meringer, M., & Brack, W. (2009). Matching structures to mass spectra using fragmentation patterns: Are the results as good as they look? *Analytical Chemistry, 81*(9), 3608–3617.

Srebro, N., & Shraibman, A. (2005, June). Rank, trace-norm and max-norm. In *International Conference on Computational Learning Theory* (pp. 545–560). Springer.

Tibshirani, R. (1996). Regression shrinkage and selection via the lasso. *Journal of the Royal Statistical Society: Series B (methodological), 58*(1), 267–288.

Van Der Hooft, J. J. J., Wandy, J., Barrett, M. P., Burgess, K. E., & Rogers, S. (2016). Topic modeling for untargeted substructure exploration in metabolomics. *Proceedings of the National Academy of Sciences, 113*(48), 13738–13743.

Wang, M., Carver, J. J., Phelan, V. V., Sanchez, L. M., Garg, N., Peng, Y., Nguyen, D. D., Watrous, J., Kapono, C. A., Luzzatto-Knaan, T., & Porto, C. (2016). Sharing and community curation of mass spectrometry data with Global Natural Products Social Molecular Networking. *Nature Biotechnology, 34*(8), 828–837.

Watrous, J., Roach, P., Alexandrov, T., Heath, B. S., Yang, J. Y., Kersten, R. D., van der Voort, M., Pogliano, K., Gross, H., Raaijmakers, J. M., Moore, B. S., Laskin, J., Bandeina, N., & Dorrestein, P. C., & (2012) Mass spectral molecular networking of living microbial colonies. *Proceedings of the National Academy of Sciences 109*(26), E1743–E1752.

Wishart, D. S. (2007). Current progress in computational metabolomics. *Briefings in Bioinformatics, 8*(5), 279–293.

Wolf, S., Schmidt, S., Müller-Hannemann, M., & Neumann, S. (2010). In silico fragmentation for computer assisted identification of metabolite mass spectra. *BMC Bioinformatics, 11*(1), 148.

Yang, J. Y., Sanchez, L. M., Rath, C. M., Liu, X., Boudreau, P. D., Bruns, N., Glukhov, E., Wodtke, A., De Felicio, R., Fenner, A., & Wong, W. R. (2013). Molecular networking as a dereplication strategy. *Journal of Natural Products, 76*(9), 1686–1699.

Dai Hai Nguyen received his B.S. degree in Computer Science from Hanoi University of Science and Technology, Hanoi, Vietnam, and his Ph.D. degree from Kyoto University, Japan. He has been working on machine learning and its applications. His current research interests are in machine learning for biological data.

Canh Hao Nguyen received his B.S. degree in Computer Science from the University of New South Wales, Australia, and M.S. and Ph.D. degrees from JAIST, Japan. He has been working in machine learning and bioinformatics. His current interests are in machine learning for graph data, and sparse modeling with applications in biological network analysis.

Hiroshi Mamitsuka received his B.S. degree in biophysics and biochemistry (1998), his M.E. degree in information engineering (1991), and his Ph.D. degree in information sciences (1999) from the University of Tokyo. He has worked on research in machine learning, data mining, and bioinformatics. His current research interests include mining from graphs and networks in biology and chemistry.

Chapter 21
Ignorance, Creation, Destruction

Sander van der Leeuw and Masatoshi Murase

Abstract This chapter presents a relatively simple but effective model of the co-evolution of human societies and their environments in terms of the growth of their information-processing capacity. The discussion then turns to society dynamics in terms of the process of categorization, emphasizing the role of "noise"—the information discarded in any information processing system—in the development of the path-dependent trajectories that characterize all societies. It then contrasts some of the dominant characteristics of the Western approach to science with those that developed in the Far East, and concludes with a plea to integrate the two approaches to improve our understanding of the complex phenomena facing our societies.

Keywords Information processing · Categorization · Endo–exo dynamic model · Path dependency of science · Western and Eastern scientific approaches

21.1 Introduction

The study of invention and innovation is a long-standing, widely researched field. Most of the literature is dedicated to the process of innovation, which is here understood as the spread of inventions through society (e.g., Rossi, 2002). A much less voluminous part of the literature is dedicated to the study of the process of invention (e.g., Arthur, 2009; Simondon, 1958). For a long time, study of that process has been considered the domain of creativity, and therefore out of bounds of science. Scientists have often been content to consider it as a black box—while studying the conditions

S. van der Leeuw (✉)
School of Complex Adaptive Systems, College of Global Futures, Arizona State University, PO Box 872701, 1031 S. Palm Walk, Tempe, AZ 85287, USA
e-mail: vanderle@asu.edu

Center for Social Dynamics and Complexity (CSDC), Arizona State University, PO Box 872701, 1031 S Palm Walk, Tempe, AZ 85287, USA

M. Murase
Yukawa Institute for Theoretical Physics, Kyoto University, Kitashirakawa, Kyoto 606-8502, Japan
e-mail: murase@yukawa.kyoto-u.ac.jp

K. Nishimura et al. (eds.), *Creative Complex Systems*, Creative Economy,
https://doi.org/10.1007/978-981-16-4457-3_21

under which it could be stimulated (e.g., Martins & Terblanche, 2003), and the effect of invention and innovation on the economy (e.g., Florida, 2019; Romer, 1990). One important reason for that has been the fact that most science undertook the ex-post study of the origins of the present, but not the ex-ante emergence of novelty. That has changed with the emergence of the Complex Systems approach (e.g., Mitchell, 2009).

Many people, among them many politicians and business leaders, think that we can innovate ourselves out of trouble and thereby attain sustainability. However, 250 years of unbridled innovation in every conceivable domain, material, or function has placed us *in* the predicament we find ourselves in. Hence, to innovate ourselves out of trouble, we cannot continue to use the same scattershot approach to invention and innovation. We need to favor innovation in directions that contribute to sustainability. Unfortunately, our current socio-economic-environmental system leaves the choice of effort to invent and innovate entirely dominated by financial gain. Worse, we have very little idea of how invention actually happens, and how we might therefore direct it towards results that bring us closer to sustainability.

This chapter is about consequences of the information processing dynamics underpinning both the invention and the destruction of knowledge. Both occur at the interface of the human mind and the outside world. Contrary to many discussions of invention and innovation, the main emphasis of this contribution is on the role of ignorance, how ignorance is transformed into knowledge, and how it may subsequently be destroyed or replaced.

Much research is currently happening around the cognitive interface of humans with their environment. It involves functional Magnetic Resonance Imaging (e. g., Jezzard et al., 2001) and a wide range of other experimental techniques to link cognitive performance to the dynamics of the human brain (e.g., Raz, 2000), as well as experiments to look at particular biases in the process of knowledge acquisition (Stammers, 2019). In a development originally initiated by Kahnemann and his colleagues (Kahnemann, 2011; Kahnemann et al., 1986), slowly but surely, these avenues of research, combined with anthropological and ethological observations, behavioral (Thaler, 2016) and narrative (Shiller, 2019) economics are converging to unearth the detailed dynamics of human cognition (Beckert, 2016; Iriki, 2019).

However, our purpose in this chapter is different: to present a relatively simple, functionally adequate model of the role of cognition dynamics in creating and destroying human knowledge, both individual and collective. A model that is not focused on the intricacies of the cognitive process, but that can help us understand the role of invention in societies, economies, and politics.

21.2 A Model of Knowledge Acquisition

The uninterrupted process of human individual and collective learning that has transformed human societies from small bands roaming the Earth to huge societies involving millions if not billions of people is here considered as the result of a positive feedback loop that creates shared order out of experiences of the—seemingly chaotic—world, by isolating patterns, characterizing them in terms of a limited number of dimensions, and using those patterns in the form of knowledge. It may be summarized in the following feedback loop (van der Leeuw, 2007):

> Problem-solving structures knowledge ⇒ more knowledge increases the information processing capacity ⇒ that in turn allows the cognition of new problems ⇒ solving them creates new knowledge …

Problems that exceed an individual's capacity have an impact on the structure of the groups involved, leading to a second feedback loop of the following form:

> Increasing knowledge creation involves more and more people in processing information ⇒ increasing the size of the group involved and its degree of aggregation ⇒ creates more problems (including societal ones) ⇒ increases need for problem-solving …

This process results in the definition of ever-growing numbers of cognitive (knowledge) dimensions. As more cognitive dimensions are distinguished, more and more complex challenges can be tackled, and the more quickly knowledge enabling information processing is accumulated. This in turn enables a concomitant increase in matter, energy, and information flows through a community, enabling it to grow in size. Hence, we conceive here of human societies as "dissipative flow structures" in the sense of Prigogine and Nicolis (1980)—dynamic structures dependent for their existence on flows of energy, matter, and information, which, due to their information processing, dissipate entropy or "mixedupness" and create clarity. They also align the understanding, know-how, and skills of the individuals in the group involved, including their technology, social institutions, and environments, thereby generating group cohesion.

In this chapter, we place these dynamics in their context by considering *the information that is not processed*, the potential meaning that is left unstructured, and does not enhance the society's information processing capability. By extricating from a set of observations those that "make sense" in the framework one is applying, one is also discarding observations that could make sense in another framework that one is at that moment not using, unaware of, or that remains to be developed. *"Noise" is potential meaning except for the fact that nobody has yet defined the cognitive structure that would enable it to be transformed into "signal(s)"*. What is it that leaves that information unprocessed? What is the role of those unprocessed observations? What is their relationship to the observations that were processed? How do they contribute to beliefs, rituals, and other forms of information processed, whose justification can be found in the symbolic realms?

21.2.1 Information Processing as Niche Construction

One can conceive of the interaction between a society and an environment as a case of resonance and niche construction (Laubichler & Renn, 2015; Odling-Smee et al., 2013) between the society's information-processing apparatus, including mental and material tools such as ideas, institutions and artefacts, and the environment with which the society interacts. External observations are interpreted when resonating with existing interpretative schemes (knowledge) and provide information to those schemes. In the action of the society upon its environment, the resultant changes in information processing in turn affect the aspects of the environment to which they relate, causing changes. Thus, as a result of their interaction, the society and its environment concurrently co-evolve to shape both the information processing of the society and the environment with which it interacts.

There is, however, an important difference related to the direction of the interactions between the society and its environment. The categories that societies derive from their observations selectively reduce the total, almost infinite, complexity of the dynamics observed. This creates signals, and augments the number of cognitive dimensions involved in information processing. However, the number of such dimensions is always very small compared with the high dimensionality of the environment. The conception the society has of the environment is therefore partial, but when societies act upon their environment, these partial conceptions are confronted with the much more complex and noisy, mostly unknown, dynamics in the environment. This means that society's actions have many effects that could not have been anticipated. The interaction thus triggers novel challenges to the information-processing apparatus in the form of questions or problems. Dealing with these further enhances the processing capacity involved, which will then resonate with thus far unobserved phenomena, so that the cycle starts anew. As a result, one cannot predict the outcome of society's interactions with the environment, at least at the systemic level and over the longer term. *We must conclude that as a consequence of the difference in dimensionality, any society's action upon the environment always has unanticipated consequences and is subject to "ontological uncertainty"* (Lane & Maxfield, 1997, 2005). It follows that (1) although the evolution of a society's knowledge structure *drives* the trajectory of human–environment interaction, it only partly *directs* it, and (2) information processing is part of a continuous feedback loop, and *there is no inherent limit* to the process. Invention is an inherent part of information processing, transforming unknown aspects of the environment into known ones.

21.2.2 Categorization

The core dynamic occurring when there is resonance between an existing cognitive system and novel observations is *categorization*. That process requires pattern recognition, a comparison between similarities and dissimilarities among the phenomena observed. It proceeds in two steps. In the first phase, the category to be created is the subject and the phenomena studied to do so are the referent. According to Tversky and Gati (1978),[1] this comparison is done with a bias in favor of similarity, a focus on what might unite the observed phenomena into a category. However, when the relevant category is firmly established, the process is inverted: categories become the referents and the phenomena become the subjects, biasing the comparisons towards dissimilarity and determining which phenomena, after all, do not belong in the tentative categories established in the initial comparison.

Rescher (1964) described this process in terms that derive from a very different train of thought than the work of Tversky and Gati. First, he argues that one considers an *extrinsically defined* sample of phenomena, delimited in space and time, and looks for similarities between them. These similarities describe them as *potential* members of a hypothetical category, yet do not identify other potential members of that category that may be found outside the sample. Nor can one be sure whether other characteristics will need to be used to create a meaningful category. That category is therefore an open one, a *group.* Only after one has constructed a meaningful framework by which one can include both the relevant members of the initial group and other similar members outside that group, and exclude all the phenomena that do not fit, one can transform the category from an open group into an *intrinsically defined and closed class* by choosing a number of characteristics that apply to all actual and prospective members, and only to them, and use these to transform the hypothetical category into a firm, defined one. In the first part of this feedback loop, the reasoning is inductive (from particular phenomena to more general ideas) while in the second part it is deductive (from the ideas to the particular phenomena to which they apply (van der Leeuw, 2019).

In other words, when faced with unknown phenomena, one first selects members of one or more *potential* categories by identifying dimensions of those phenomena that emphasize their similarity. Until one has defined the phenomena that do not belong in these categories, these categories are *open-ended groups*, and they are an intermediate step between noise and signal. One then whittles away at both the phenomena and the interpretations, selecting dimensions to be retained as signals in the information processing framework, while discounting others as "noise." Once one has thus defined a set of *closed classes*, those co-determine further investigations on different phenomena, so that a partly path-dependent knowledge structure is built up. The choices made in both phases of this process (which aspects of the phenomena to look at or not, and which dimensions to include in the definition or not) are fundamental for the eventual outcome of the information-processing structure adopted by a society, as they separate signals from noise. They thus partly determine the future of invention in the society.

[1] For a more extensive summary, see van der Leeuw (2019).

21.2.3 The Evolution of Information Processing

A human information processing system is not defined only by the knowledge it contains or what it is able to do, but also by the domain outside that sphere that the society has never asked any questions about, never thought about, and of which it is thus completely unaware. In this chapter, we call that domain the "*unknown unknown*," as neither signals nor noise have been defined. Another intermediate domain—of which the society is aware but has not managed sufficient coherent information processing to define either groups or classes—is here called the "*known unknown*." It is a domain that the society is aware of, but has not (yet) been able to fully understand. In it, we often find beliefs or myths that are only very loosely related to observations, such as the Ancient Greeks' ideas about the god Helios (the sun) that regulated day and night, or about the mythical ferry master Charon who brought people into the underworld after death. Over time, a growing density of observations will lead to a denser connection between observations and ideas, generating increasingly effective signals, categorizations, and theories, such as by exchanging the geocentric conception of the universe for the heliocentric one, which each have their own noise.

Thus, the trajectory of a society's information processing can be described as reaching from the unknown unknown, via the known unknown, to the sphere where effective structuration of the knowledge framework is beginning to distinguish signals and noise. Initially, that framework is still flexible due to a number of open (at least in part extrinsically defined) categories, but it then evolves into a sphere where all categories are intrinsically defined as valid signals, closed so that the structure is fixed and any new information can only be interpreted in terms of the existing framework. In that phase, only known, fixed sets of observations are taken into account.

21.2.4 The Present is Constructed by Interaction Between Past and Future

Human perception thus iteratively relates established information (existing knowledge) and new observations, in an interaction between perception from an a posteriori perspective (concerning the known past) that limits variation, and perception from an a priori point of view (concerning the unknown future), which opens opportunity for variation. The former is focused on the intrinsically defined, traditional, "closed" categories that shaped the society's existing worldview. The latter consists of extrinsically defined categories about which there are hypotheses under consideration. They are "open" categories, focused on emergence, on possibilities and probabilities, and on novelty. Thus, changes are built upon the established framework. Exactly how that happens, which exact choices are made, shapes the resulting (path-dependent) information processing framework.

All this is occurring in the interaction between the society and its external environment. As that interaction proceeds in certain domains, information processing is structured, enriched, and refined, which changes the appreciation of the outside world, but also the method by which information is organized. In the domains where this happens, we see perceptual revolutions; for example, the abandonment of Newton's conception of physics in favor of those of Einstein and Heisenberg. In that process, importantly, a perception of (an aspect of) the outside world is transformed from one that reckons with the unexpected, with the existence of unknown dynamics, into one that is much more highly (or completely) predictable because the society has developed information processing that takes more dimensions of variability into account. This fundamentally changes what are considered signals, what is deemed noise, and the structure of perception, conception, and decision-making.

21.2.5 Knowledge Destruction and Replacement

We have argued that the feedback loop that drives information processing does not in itself have any inherent limit. Out of the infinity of observable phenomena, one can in theory always define more or different signals and discard more or different noise. Yet, much knowledge has in different moments in the past been lost, and at other moments it has been replaced by other information. In the simplest of terms, one could explain this concept by referring to the limitations of human memory. However, since the invention of narratives and writing, and more recently since the invention of electronic information processing, this may be true for individuals, but is not true for societies—they do not physically forget unless they are destroyed.

In changing their information-processing mechanisms, societies dynamically change their emphasis on what they perceive, how they categorize and construct narratives, and how they use information processing in their interaction with their environment. Because that is a dynamic process, categories constituting signals, and narratives that combine them in specific ways need to be activated regularly to remain part of the individual's or the society's information processing, and thus to be available when called upon. Once they are no longer regularly referred to because they have been replaced by other signals, they slide away into the background and can eventually no longer be activated, having been reduced to noise.

The interaction between signals and noise thus defines the trajectory adopted by a society's information processing, and the collective choices made by the society in this process define the path-dependent construction of the society's information processing.

21.2.6 Path Dependency

Recall that the evolution of our society's cognitive structure *drove* the trajectory of human–environment interaction, but only partly *directed* it. In our Western societies' trajectory, a number of systemic choices were made, such as adopting the models of Copernicus, Newton, Einstein, and Heisenberg regarding certain physical phenomena. Over the long term, these choices have engendered the *path-dependent trajectory* of our Western thinking, which has resulted in our current scientific perspective. However, that trajectory is not the only conceivable one.

To illustrate that, we will next contrast our Western scientific approach with some aspects of East Asian thinking and point to both differences and similarities. As Henrich (2020) argues convincingly, these two traditions are very different indeed. The Western tradition developed, beginning in the sixteenth century, with very particular traits that differ from other traditions worldwide. For example, it decontextualizes phenomena and decomposes complex ones in the expectation that recomposing them after analysis will help understanding; it is also focused on entities and results rather than processes and relationships, and it is reductionist and thinks in terms of oppositions. Other characteristics emerge in the next section.

21.3 Some Characteristics of Western Scientific Information Processing

Taleb (2007) argues that in Europe and later in some of the parts of the world colonized by European powers, slowly but surely a perception of an unpredictable world was replaced by a perception that is simplified, controlled, and fixed. This perception no longer takes the unforeseen sufficiently into account and is therefore further and further removed from the actual dynamics that are occurring around us. In the language we have developed above, closed categories have become dominant in our thinking, and noise is no longer considered as the domain in which unexpected phenomena can emerge. Indeed, the scientific revolution of the twentieth century led to the dream that ultimately the sphere of the known unknown could be described by laws based on a limited number of dimensions, and would be controllable (Kuhn, 1962). But in the twenty-first century, it is becoming increasingly obvious that in many fields, such a dream is not the reality. The current Covid-19 epidemic overtaking the world is a case in point.

It is therefore worthwhile to consider some of the implications of Taleb's (2007) idea. He applies the *ex ante* perspective that is inherent in the Complex Adaptive Systems (CAS) approach, focusing on identifying the *emergence* of phenomena by looking at system trajectories from the past to the future, in contrast with most of science's *ex post* perspective that looks from the present to the past in an attempt to identify the *origins* of the present. In that CAS perspective, most, if not all, *emergent phenomena are unanticipated.* The world is thus subject to *ontological uncertainty*

(Lane & Maxfield, 2005), even though the common scientific perspective on the phenomena it has investigated claims that those phenomena are fully understood, and thus predictable.

Western science has in effect progressively exchanged high-dimensional, partly understood perspectives formulated in open categories for low-dimensional ones, formulated in closed categories. In the process, because Western science abhors ambiguity, numerous dimensions are relegated from (part understood, but perceived) "fuzzy signals" to (ununderstood) "noise." This fundamentally changed our appreciation of the uncertainties and risks involved in our interaction with the environment, resulting in a *false sense of control* over those interactions and, over time, an accumulation of unintended and unanticipated consequences of our interactions with the environment. Hence our current sustainability conundrum!

Increasingly, Western science has not only projected an image of understanding and controlling (part of) the world around us, but also fragmented our society's perspective on the environment by decomposing, under the impact of Empiricism, complex phenomena into constituent, presumably simpler ones. This was done in the hope that bringing those, once dealt with, together again would create an understanding of the complex phenomena. In that process, many part-understood dimensions of the "real" world have been ignored, and its immense complexity oversimplified. We are left with a deep disciplinary understanding of phenomena that have been taken out of context, and can therefore not be re-integrated into an understanding of the complexity of the real world because the context of these known phenomena has been defined as noise.

Moreover, from its roots in Aristotelian thinking, the Western scientific approach has focused on stasis and closed categories, disregarding Heraclitus of Ephesus' vision of a world that is in perpetual motion, and that therefore has to be described in open categories. In the Heraclean vision, open categorization leaves room for change and ambiguity, which in our current Western approach are insufficiently appreciated. Because the world is presumed to be continually changing, at any point in time the context of observed phenomena is essential to correct understanding, which is less the case when phenomena are assumed to be unchanging. Until recently, as long as socio-environmental change was slow, this static perspective was sufficient to deal with most challenges, but with the current acceleration of change it is no longer adequate.

Since the foundation of the Royal Academy in the 1660s, the way to be accepted by peers as a scientist required that one could demonstrate or prove one's hypotheses about the natural world. As such demonstrations or proofs could only be engineered for the present or the past, the career structure of scientists focused on explaining the *relationship between present and past*, to the detriment of that between the present and the future (for which one could not prove or demonstrate hypotheses). Thus, much science today is dedicated to the *ex-post* exploration of the origins of the present in the past, rather than the *ex-ante* exploration of the emergence of the present and future from the past. With the increasing emphasis on numerical scientific thinking, *multivariate frequency statistics* came to be a major tool to consider the outcomes of the dynamics of the past by reducing the dimensionality of phenomena. This contributed to a systematic emphasis on the "norm" rather than the diversity.

These intellectual choices in the past have shaped a worldview, and an understanding of the dynamic context in which our society functions that is reductionist, fragmented, focused on stasis and entities rather than change and processes, biased towards averages and inept at anticipating the unexpected. Over time, our societies have thus gained control over a wide range of phenomena and anchored that control in language, institutions, and technologies. But in doing so, they have created a situation in which the accumulation of (within their worldview) unperceivable consequences of the dynamics driving them threatens the whole house of cards. It has been responsible for many blind spots in our perception, the simplification of what we perceive, and thus for the inadequacies in our information processing system that prevent us from dealing with many of the realities of the context in which we find ourselves. Taleb's main message is that by thus normalizing, our biased and fragmented Western perspective has created a false sense of control over the dynamics of which our societies are part. While that inadequate perspective only touched the Western part of the world, as it did for most of recent human history, one could hope that other developments elsewhere on Earth would ensure an overall balance. However, because globalization continues to push this perspective onto many cultures that hold profoundly different worldviews, such a balance is no longer assured.

We must conclude that our aim must be to learn from the past about the present, ***for the future***, and should place more emphasis on finding approaches that can help us understand the future. A fundamental step towards that goal is to include the process of learning about phenomena in our considerations, thus linking ontologies and epistemologies in our attempts at understanding our world. Bayesian statistics do offer some opportunities for this, as does the thinking of a philosopher like Belnap (2007).

We must emphasize the importance of cultural diversity if our planet is to regain a diversity of options for its societies to implement their futures. The biological and ecological science communities have—rightly—insisted on the importance of biodiversity, and have even been able to push that issue to the level of the United Nations in the IPBES (Intergovernmental Science-Policy Platform on Biodiversity and Ecosystem Services) convention. From a complex systems perspective, in which we need to take different futures into account, it is just as important that we *promote cultural diversity*. Thus, when one approach to the survival of humanity fails, others might be in a better position. That is how the Earth has survived thus far in a habitable state—when various socio-environmental dynamics led to societal collapse through their unintended consequences, in other locations other societies could hope to survive because they had different knowledge systems.

Ultimately, therefore, globalization is as big a danger to sustainability as environmental change or biodiversity loss. The fact that what is a signal in one culture is not in another (or in any other, often), so that *open or closed categories in one knowledge system are part of the unknown unknown, or maybe the known unknown, in other societies' knowledge systems* is the only guarantee we have of long-term sustainability for human life on this planet.

Returning to the history and present state of the Western knowledge system, we need to accept that its intellectual tradition, and in particular its science, certainly since the Enlightenment, has played a major role in driving our societies to the point where their sustainability is in question. To find the deep roots of the current situation, we may need to go back to the time of Aristotle and Heraclitus or to the plague epidemics of the fourteenth century, which introduced a linear temporal perspective on life and death and the separation between changing and nonchanging phenomena that later led to the distinction between "nature" and "nurture," between humanity and its natural environment.

We now turn to Eastern approaches to achieve two goals. On the one hand, to point to some of the conceptual differences and similarities between the two traditions, which, after all, were confronted with the same observations, and, on the other hand, to show how very different intellectual traditions approach some of the issues we have also identified in the Western tradition. That then leads us to the question whether it would be useful to "hybridize" the two traditions, redefining what in the combined worldviews are signals and noise, and thus opening the way for different kinds of signals and a different path-dependent development of our understanding of the world.

21.4 Troublesome Dichotomies Never Worry Eastern Philosophers

Schrödinger (1958) clearly mentioned "*our present way of thinking does need to be amended, perhaps by a bit of blood-transfusion from Eastern thought.*" In the Western intellectual-scientific tradition, Schrödinger is an exception, referring to the relative isolation of the Western perspective at the time, and arguing for the utility of adopting some of the ideas of the East Asian perspective that go back to Confucius and the Daoists.

The entity-focused, dichotomous perspective typical of Western science assumes that opposites are mutually exclusive and even contradictory, while the complementarity perspective typical of Eastern philosophy suggests that opposites are not mutually exclusive, but merely complementary to one another. That vision is rooted in the idea that opposites are only different aspects of the same wholeness, just like the two sides of the same coin.[2]

A strict Western logic, based on the dichotomy perspective would always distinguish opposites into two groups, and would thus have to face the challenge summarized by the following story (Nicolescu, 2002). "A man desperately wants to separate the two ends of a stick. He cuts the stick and then sees that instead of having two

---

[2] This kind of Eastern thinking is very similar to Niels Bohr's complementarity principle in modern physics (Bohr, 1985). It has led to revolutionary breakthroughs in the field of particle physics that show how moving beyond the limits of Western scientific perspective has helped enlighten major issues concerning nature (Atmanspacher & Dalenoort, 1994).

separated ends, he has now two sticks, both of which have two ends of their own! He continues to cut these sticks, and ultimately realizes that he ends up with an infinite number of sticks, each of which still has two ends. It is impossible to separate the two ends".

The troublesome Western dichotomies of opposition would not worry a Zen master, for the simple reason that he would not believe that dichotomous oppositions exist at all. Such oppositions as "nature and nurture", "inside and outside" are both separate from, as well as merged into one another. In his idea, separation and merging are one and the same action, applied to the originally undivided "something" (Izutsu, 1973).[3] From the Zen point of view, the problem of "inside and outside," is only a pseudo-problem, no matter in what form it may be raised. This is because, if it is seen with the eyes of an enlightened (Eastern) person, the inside and outside are not two regions to be distinguished from one another. *The distinction has no reality: it is nothing but a thought-construct peculiar to the discriminating activity of the mind* (Izutsu, 1975).

The complementarity principle at the root of Eastern religious and related thought thus offers us a way to overcome dichotomies inherent in Western concepts such as "nature" and "nurture." To do so one may adopt the self–nonself circulation theory (Murase, 2000).

21.5 The Self–Nonself (or Endo–Exo) Circulation Theory

Instead of thinking of "reality" as an existing object or condition, as we would in the Western approach with its emphasis on stasis and entities, we could consider reality as a complex of dynamic processes involving disintegration and integration from times past into the future. In the terms used in the first part of this chapter, this reflects exactly the signal processing of the human mind, which takes signals in, transforms them, and then sends them back out to confront the environment. As a result, what we in the West think of as "reality" is, in the Eastern approach, not fixed, but undergoes both epistemological and ontological changes at any instance of signal processing.

This self–nonself (or endo–exo) circulation theory (Murase, 2000) suggests that we should not only consider the dichotomy of interior and exterior, but also the dichotomy of one and many in terms of relationships, and phenomena linked rather than divided by process.

The reason for introducing both these dichotomies is as follows. The former involves the emergence of the closed endo-system (or the self), which holds an entity, organism, or individual's identity through interaction between the self and its

[3] Surprisingly similar methodologies were not only proposed by the great Western poet Johann Wolfgang von Goethe (Bortoft, 1996), but also by the Western particle physicist Geoffrey Chew (Nicolescu, 2017) and by the Western art educator, Edwards (1986). The same has occurred in the medical sciences (Doidge, 2016) and the social sciences (Scharmer, 2016).

exterior environment, and eventually shapes the information processing apparatus in its interior (as in the mind). However, wholeness does not exist as a simple collection of many entities but involves the dialectical dynamics between an entity and a whole that consists of many entities. Translated in Western terms, one could see this as the relationship between subject and context, in which the former is unidimensional and the latter multidimensional. To use another image, one could see this as the relationship, in the complex systems approach, between individual entities at the micro level and the pattern that their interaction creates at the macro level, which in turn impacts on the behavior of the individual entities at the micro level. This idea was promoted in the social sciences by Bourdieu (1977) and became a standard theme in the Complex Adaptive Systems movement (Fig. 21.2). Note that there is a similarity between Figs. 21.1 and 21.2. When we look at the bottom panel of Fig. 21.1, we can realize the diagram shown in Fig. 21.2. In effect, Fig. 21.2 is a two-dimensional description of the dynamic process described in three dimensions in Fig. 21.1.

Figure 21.1 illustrates how a closed self (endo-system) establishes identity in contrast with the open environment (exo-world). While evolving through processes of variation and selection triggered by interaction with its environment, the system maintains its identity by means of the boundaries that isolate its internal components from the exo-world. By supposing that the closed self is subject to integration and disintegration cycles, we could imagine how complex structures grow out of rather simple ones, in interaction with the external world, but separated from that world

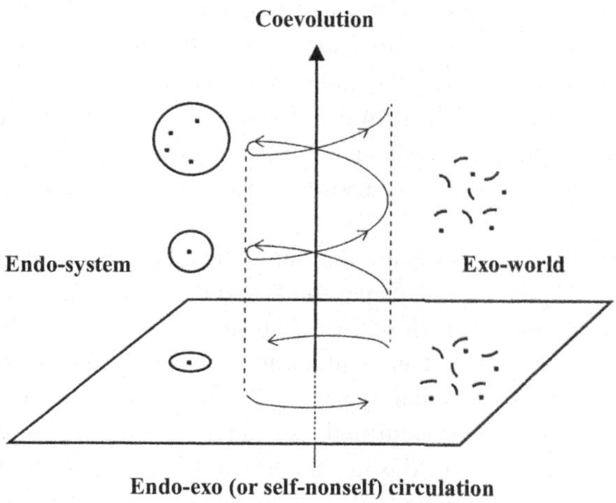

Fig. 21.1 Co-evolution of a closed self (or endo-system) and an exo-world (or environment) by means of endo–exo (or self–nonself) circulation. Note that there are empty regions besides filled ones within the closed endo-system (on the left, in the circles), which in turn play an important role for the further evolution (after Murase, 2000, p. 371)

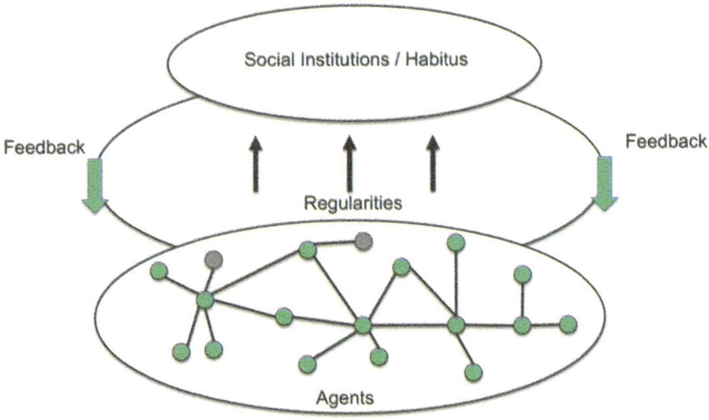

Fig. 21.2 Interaction between individuals and an overarching whole, as viewed in the CAS approach (after van der Leeuw 2019a)

by a semi-permeable boundary. During such integration and collapse cycles, boundaries are selectively crossed in one direction, at certain times allowing entry of new elements from the exo-world into an endo-world with open categories before the latter is resealed when closed categories become dominant again. With each successive cycle, the composition of the closed endo-system will change, evolving in a spiraling fashion. Murase described this interactive process between an endo-system and the exo-world as a self–nonself circulation theory for the origin and evolution of life (Murase, 2000, 2008, 2011; see also van der Leeuw, 2019b).

The theory of self–nonself (or endo–exo) circulation suggests that nature and nurture should be taken together and cannot be considered in isolation. Each is tied to the other, inseparable within a dynamic circulatory process. This endless circulation between nature (self or endo-) and nurture (nonself or exo-) triggers the emergence of new orders or disorders whenever the surrounding environment changes, or vice-versa.

It is very important to be aware that there must be other possibilities for further "co-evolution." Generally, when we think about co-evolution, we are tempted to consider it as the mutual evolution of both the endo-system and exo-world. This is, of course, the case, but it is insufficient from the point of view of wholeness that is central to the East Asian approach. Another possibility is that co-evolution occurs between the endo-system and its contents! The co-evolution between the endo-system and the exo-world is just the mirror image of the co-evolution between the endo-system and its contents. We should realize that there are three different levels: the endo-system, its interior contents, and its exterior environments. All are multidimensional. If there are nested hierarchical endo-systems (Murase and Murase, 2021), we can think about the interaction between the interior and exterior of each endo-system. This possibility leads to Iriki's (2019) conception of human cognition as an interactive process occurring between two niches—one internal and one external. Each niche constructs its identity between itself and its contents.

In that process, there will inevitably be noise as potential, but discarded, information—empty areas alongside the areas populated with known phenomena in the closed endo-system. Such empty areas will be dismissed as noise—as potential, but discarded, information. But they will also play an important role in the further evolution of the (path-dependent) system (Murase, 2011, 2018), enabling "creativity" by filling gaps in the endo-system.

Prigogine (1977) observed that chemical, biochemical, and sociobiological transitions occur because of "order through fluctuation." In such transitions, it is not the macroscopic averages that play transformative roles, but the internal amplification and breakthrough of fluctuations that started at the almost invisible micro level (Jantsch 1980). This is the basis for the spread of particular "imagined futures" in societies, which manifest themselves as new ideas (Beckert, 2016).

All self-organizing evolutionary systems contain spatial voids and/or temporal nothingness. In physics, for example, if a spatially distributed system is homogeneous and consists only of active elements, it shows spatio-temporal chaos but when spatial voids are introduced which make the system inhomogeneous, then it shows ordered dynamics (Murase, 1992). Phase transitions are thus caused by the presence of spatial voids. Likewise, small empty regions in social systems are very important in maintaining resilience against sudden changes of the external environment, because such empty regions may absorb stresses caused by external and internal stimuli (Servigne & Stevens, 2015). Finally, in the case of temporal dynamics such as nerve excitation, there is always a refractory period (when nothing happens even in the presence of stimuli) just after the excitation period. Such "silent" periods enhance the coherent temporal dynamics typical of living phenomena (Murase, 1992). Nothingness or emptiness are merely the opposite sides of the same coin. In the next section, we consider the implications of the introduction of emptiness, which was conceived by Indian philosophers.

21.6 Discovery of Nothingness Leading to Infinity

The Indian religious traditions accept the concept of nonbeing on an equal footing with that of being. Like other Eastern religions, Indian culture considered nothing as a state from which one might have come and to which one might return without beginning or ending (Barrow, 2001). This implies ontological equality (Tanabe, 2018): all beings, whether alive or not, are equal at the level of existence. Based on this ontological equality, Indian logic developed the idea of the tetralemma. The four logical possibilities of the tetralemma are as follows: (1) A (affirmation); (2) not-A (negation); (3) both A and not-A (both); and (4) neither A nor not-A (neither). Of these, the fourth case implies nothingness (neither held nor not held).

In this domain, the Greek tradition upon which Western science was built was in complete contrast to the Eastern one. The Greeks considered logic as a central importance of human thinking, and thus they believed that empty space could not exist and that no one could think about empty space. Accordingly, the Indian introduction of the symbol zero in the Western counting system owes much to the Indian accommodation of a variety of concepts relating to nothingness and emptiness.

In India, the zero could thus be introduced without deforming any other philosophical framework, while in Greece it could not. The appreciation of the presence of nothingness in India enabled the modern method of expressing every possible number by using a set of ten symbols, each of which has a position value and an absolute value. The Indian counting system must be the most successful intellectual innovation in Western mathematics, but could not have happened without Eastern philosophers' ideas about nothingness. This paradox is important in the following discussions.

21.7 How Can We Imagine Something that Thought Cannot Think?

Kierkegaard mentioned "The supreme paradox of all thought is the attempt to discover something that thought cannot think" (Edwards, 1986). Whereas this is a paradox within the framework of Western thinking, that is not always the case in the thinking typical of the Zen masters. Getting into meditation, for example, has often been understood in the West as the development of special skills, but this completely misunderstands the Zen perspective. The point is that the usual way of Western thinking would focus on learning additional skills, but not on unlearning habitual attitudes. As clearly mentioned by Varela et al. (1991), "This unlearning may take training and effort, but it is a different sense of effort from the acquiring of something new." We have to challenge the dichotomy between learning and unlearning again, because they are merely the opposite sides of the same coin. The idea of cumulative learning without unlearning is at the root of the Western idea of continual progress, whereas the Indian conception is best represented by the term "creative destruction" (Schumpeter, 1942), representing a dynamic unity between signals and noise.

21.8 Towards a New Synthesis

Most of us think in one of two patterns—Western or Eastern (Gieser, 2005; Jung, 1933; Nisbett, 2003). Both are illustrated in Fig. 21.3. The Western mindset thinks in terms of a linear progression, starting at ① and working through ② to ③ to ④ and, finally, to ⑤. In that approach, it will always be important to ask: "What came first—the chicken or the egg?" and to answer firmly yes or no, to establish cause, and, most of all, to trace the effect that must always follow—never precede—a given cause. This way of thinking is bound to view conflicting results as mutually exclusive or even contradictory. It is also at the root of the Western concept of "progress", which is one of the major contributors to the current sustainability conundrum.

By contrast, the Eastern mindset is able to think in simultaneities, allowing for the coexistence of ①, ②, ③, ④, and ⑤ at any one time. This mindset views results and ideas that are conflicting in the Western approach as complementary rather than oppositional.

Nisbett (2003) has emphasized that both Western and Eastern thinking have an important role to play in our science and our lives. Complementarity of each of these approaches is required for an improved understanding of the complex problems in today's world. In other words, as Murase and others discuss elsewhere in this volume, we must combine intellectually logical thinking with nonlogical intuition, or even with an evolutionary perspective responsible for *novelty* or *creativity*. Whereas the former is reductionist and promotes clarity and definition, the latter is the opposite—it is highly dimensional and is thus able to accommodate categories that partly fit multiple, part-perceived perspectives. The combination is needed to understand and survive in our *unpredictable* world. It is the essence of the Complex Adaptive Systems approach and enables us to deal with Taleb's (2007) "Black Swans."

21.9 Ignorance is Necessary for Acquiring Knowledge

Both ways of thinking, as depicted in Figs. 21.1, 21.2, and 21.3, are analogous to each other in functionality, although the spatial and temporal dimensions are quite different. One of the essential differences is between linear chains and looped configurations. For the evolution of life, interaction between the two has been shown to be essential. At the level of molecular evolution, for example, DNA-strands

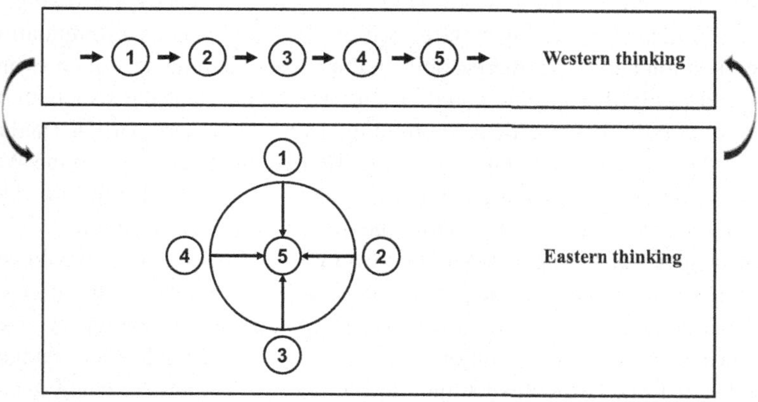

Fig. 21.3 Western (linear) and Eastern (multidimensional, contextual) ways of thinking (after Murase and Murase, 2020)

containing genetic information as well as proteins change their configuration cyclically between chains and looped configurations. Viral infection occurs through such cyclical dynamics between hosts and pathogens.

Likewise, scientific revolutions in the form of paradigm shifts always occur when we consider not only the logical linear way of thinking, but also the nonlogical nonlinear (intuitive) way of thinking. It is well known that beginner's luck often occurs because a beginner does not have a complete (linear) knowledge about the problem, but instead has an incomplete understanding. That lack of knowledge (or nothingness) leads the beginner to rely on intuition to make a breakthrough.

This principle is expressed more formally by Heisenberg in his work on the uncertainty principle (1925) in modern physics. That principle implies that observing a particle is also perturbing it, and that we cannot avoid an *essential uncertainty* with regard to the state of the particle. It follows that we cannot obtain certain kinds of information on a phenomenon without losing some other information. We must consider using the co-evolution of both Western and Eastern ways of thinking to be able to take care of such unavoidable ignorance or essential uncertainty in our reasoning. These ways of thinking should not be separated but should be considered as different sides of the same coin.

Likewise, construction and destruction, constructive destruction, and destructive construction are also opposite sides of the same coin. They can never be separated from each other! Taking information processing as the pillar of the argument as we do in this chapter reflects the fact that it is both subject and object, and that "reality" must be considered the interaction between nothingness and wholeness, the dynamic circular process between subject and object that is fundamental to our learning of phenomena and processes. We become aware of, and instantiate, phenomena by learning about them.

To show the difference between the two approaches, we could take this a step further and imagine, from the Eastern perspective, a situation in which a "co-evolution" of both the micro world (Fig. 21.1) and the macro world (Fig. 21.3) occurs—a mutual co-evolution of the conditions for simultaneous differentiation and complexification along the microscopic and macroscopic branches of co-evolution. That would imply that no one could any longer assume that the structures in the universe were built up one-sidedly, bottom up, from elementary particles and atoms to stars, galaxies, and so on (Jantsch, 1980). That would be contrary to the Western idea of (neo-)Darwinian evolution, which is usually attributed to a build-up of higher life through micro evolution, neglecting the impact of macro evolution.

Such an approach taking account of co-evolution of both branches would provide us with radical new insights, such as so-called self-transcendence (the reaching out beyond the boundaries of one's own existence); that is, the evolution of evolutionary processes or meta-evolution (Jantsch, 1980) or second-order change (van der Leeuw, 2019a). Through such meta-evolution or self-transcendental dynamics, a novel mental system may emerge within an individual person. As indicated by Murase (2018, Fig. 13, p. 278), the individual person carries the additional dimensions in the form of the macro world within himself. Societies could then enter into a new phase such as co-evolution with themselves (Jantsch, 1980).

This leads to the reversal of the time series typical of Western thinking, because humans can imagine and plan. The plan can turn into reality through complex evolutionary dynamics, meaning that mental anticipation can pull the future into the present. The direction of causality is reversed, which evokes the deductive process. As time proceeds in inductive ways, we have an example of circular dynamics between present (or past) and future, as in the "Future Design" approach that is being developed in Japan by Saijo (2020).

Next, if microscopic co-evolution occurs as in Fig. 21.1 and macroscopic co-evolution appears as in Fig. 21.3, then further co-evolution appears through the interplay between micro and macro co-evolution; for example, between humans and themselves. Finally, at this point, we may imagine that a new kind of co-evolution dynamic could emerge between the present (or the past) and the future.

In general, as argued as part of the Panarchy perspective on multilevel dynamic evolution (Gunderson & Holling, 2002), the collective mind would try to dampen the fluctuation to maintain the old structure of the world or the traditional mindset on how we see the world. However, when a person enters into such complex evolutionary dynamics, there is a hope that the person could drastically transform the traditional mindset towards a creative mode to realize an alternative reality behind the visible world. This could exemplify a future perspective on order through fluctuations. We *are* evolving through evolutionary learning and unlearning dynamics. It is now time not only to appreciate the obtained results, but also to invent novel, radical methodologies for speculating about multiple futures, thus inventing our own future!

References

Arthur, W. B. (2009). *The nature of technology*. Free Press.

Atmanspacher, H., & Dalenoort, G. J. (Eds.) (1994). *Inside versus outside: Endo- and exo-concepts of observation and knowledge in physics, philosophy and cognitive science*. Springer-Verlag.

Barrow, J. D. (2001). *The book of nothing: Vacuums, voids, and the latest ideas about the origins of the universe*. CPI COX & Wyman Ltd.

Beckert, J. (2016). *Imagined futures: Fictional expectations and capitalist dynamics*. Harvard University Press.

Belnap, N. (2007). From Newtonian determinism to branching space-time indeterminism. In T. Müller & A. Newen (Eds.), *Logik, Begriffe, Prinzipien des Handelns* (Logic, Concepts, Principles of Action), (pp. 13–31). Mentis Verlag.

Bohr, N. (1985). Light and life. In A. P. French & P. J. Kennedy (Eds.), *Niels Bohr* (pp. 311–319). Harvard University Press.

Bortoft, H. (1996). *The wholeness of nature: Goethe's way toward a science of conscious participation in nature*. Lindisfarne Press.

Bourdieu, P. (1977). *Outline of a theory of practice*. Cambridge University Press.

Doidge, N. (2016). *The brain's way of healing: Stories of remarkable recoveries and discoveries*. Penguin Books Ltd.

Edwards, B. (1986). *Drawing on the artist within*. Simon & Schuster, Inc.

Florida, R. (2019). *The rise of the creative class*. Basic Books.

Gieser, S. (2005). *The innermost kernel: Depth psychology and quantum physics. Wolfgang Pauli's dialogue with C. G. Jung*. Springer-Verlag.
Gunderson, L. H., & Holling, C. S. (2002). *Panarchy: Understanding transformations in human and natural systems*. Island Press.
Heisenberg, W. (1925). Über quantentheoretishe Umdeutung kinematisher und mechanischer Beziehungen. [English translation: Quantum-Theoretical Re-interpretation of Kinematic and Mechanical Relations. In B. L. van der Waerden, (Ed.), *Sources of quantum mechanics* (1968) Dover Publications.
Henrich, J. (2020). *The WEIRDest people in the world: How the west became psychologically peculiar and particularly prosperous*. Farrar Straus and Giroux.
Iriki, A. (2019). The brain in the ecosystem: Cognition, culture, and the environment. Video of *Kavli Plenary Address at International Convention of Psychological Science*, March 2019. https://www.youtube.com/watch?v=XdX0xkPxS0c
Izutsu, T. (1973). The concept and reality of existence. *Philosophy East and West, 23*(1), 258–259.
Izutsu, T. (1975). The interior and exterior in Zen Buddhism. In A. Portmann & R. Ritsena (Eds.), *Correspondence of man and world*, Eranos Yearbook (Vol. 42, pp. 581–618). Brill.
Jantsch, E. (1980). The self-organizing universe: Scientific and human implications of the emerging paradigm of evolution. Pergamon Press.
Jezzard, P., Matthews, P. M., & Smith, S. M. (Eds.). (2001). *Functional magnetic resonance imaging: An introduction to methods*. Oxford University Press.
Jung, C. G. (1933). *Modern man in search of a soul* (translated by W. S. Dell and C. F. Baynes), Kegan Paul, Trench, Trubner & Co.
Kahnemann, D. (2011). *Thinking fast and slow*. Farrar Strauss & Giroux.
Kahnemann, D., Slovic, P., & Tversky, A. (Eds.). (1986). *Judgment under uncertainty: Heuristics and biases*. Cambridge University Press.
Kuhn, T. (1962). *The structure of scientific revolutions*. University of Chicago Press.
Lane, D. A., & Maxfield, R. (1997). Foresight, Complexity and Strategy. In W. B. Arthur, S. Durlauf, D. Lane (Eds.), *Economy as a complex evolving system II* (pp. 169–198). Westview Press.
Lane, D. A., & Maxfield, R. (2005). Ontological uncertainty and innovation. *Journal of Evolutionary Economics, 15*, 3–50.
Laubichler, M. D., & Renn, J. (2015). Extended evolution: A conceptual framework for integrating regulatory networks and niche construction. *Journal of Experimental Zoology Part B: Molecular and Developmental Evolution, 324*(B), 565–577.
Martins, E. C., & Terblanche, F. (2003). Building organisational culture that stimulates creativity and innovation. *European Journal of Innovation Management, 6*(1), 64–74.
Mitchell, M. (2009). *Complexity: A guided tour*. Oxford University Press.
Murase, M. (1992). *The dynamics of cellular motility*. Wiley.
Murase, M. (2000). *Life as history: The construction of self-nonself circulation theory* (in Japanese), pp. 369–376 (in English) Kyoto University Press. https://repository.kulib.kyoto-u.ac.jp/dspace/bitstream/2433/49765/1/Murase2000b2.pdf
Murase, M. (2008). Environmental pollution and health: An interdisciplinary study of the bioeffects of electromagnetic fields. *SANSAI, an Environmental Journal for the Global Community, 3*, 1–35.
Murase, M. (2011). The origin and evolution of life by means of endo-exo (or self-nonself) circulation. *Viva Origino, 39*(1), 7–10.
Murase, M. (2018). A self-similar dynamic systems perspective of "living" nature: The self-nonself circulation principle beyond complexity. In Y. Stomu, T. Yagi, & S. Hill (Eds.), *The Kyoto manifesto for global economics the platform of community, humanity, and spirituality* (pp. 257–283). Springer.
Murase, M., & Murase, T. (2020). A grand unified life theory: An extension of the self-nonself circulation theory. *Journal of Integrated Creative Studies*. https://doi.org/10.14989/259183
Murase, M., & Murase, T. (2021). A transdisciplinary study of how to integrate shattered world: The self-nonself circulation principle of "living" wholeness. In S. Hill, T. Yagi, & S. Yamash'ta (Eds.),*Kyoto Manifesto II*. Springer (in press).

Nicolescu, B. (2002). *Manifesto of transdisciplinarity* [Translated by K.-C. Voss], State University of New York Press.
Nicolescu, B. (2017). The bootstrap principle and the uniqueness of the world. *Cybernetics and Human Knowing, 24*(2), 83–88.
Nisbett, R. E. (2003). *The geography of thought: How Asians and westerners think differently ... and why*. Simon and Schuster.
Odling-Smee, F. J., Laland, K. N., & Feldman, M. W. (2013). *Niche construction: The neglected process in evolution*. Princeton University Press.
Prigogine, I. (1977). *Time, structure and fluctuations*. (Nobel Prize Lecture). https://www.nobelprize.org/prizes/chemistry/1977/prigogine/lecture/
Prigogine, I., & Nicolis, G. (1980). *From being to becoming: Time and complexity in the physical sciences*. Freeman.
Raz, N. (2000). Aging of the brain and its impact on cognitive performance: Integration of structural and functional findings. In F. I. M. Craik & T. A. Salthouse (Eds.), *The handbook of aging and cognition* (pp. 1–90). Lawrence Erlbaum Associates.
Rescher, N. (1964). *An introduction to logic*. St. Martin's Press.
Romer, P. M. (1990). Endogenous technological change. *Journal of Political Economy, 98*(5–2), S71–S102 (The University of Chicago Press).
Rossi, F. (2002). An introductory overview of innovation studies. Modena: Universita' di Modena e Reggio Emilia, archived in: Munich Personal RePEc Archive https://mpra.ub.uni-muenchen.de/9106/1/MPRA_paper_9106.pdf
Saijo, T. (Ed.) (2020). *Future design*. Springer.
Scharmer, C. O. (2016). *Theory U: Leading from the future as it emerges*, Berrett-Koehler, Oakland.
Schrödinger, E. (1958). *Mind and matter*. Cambridge University Press.
Schumpeter, J. A. (1942). *Capitalism, socialism and democracy* (2nd ed.) Wilder Publications.
Servigne, P., & Stevens, R. (2015). *Comment tout peut s'effondrer: Petit manuel de collapsologie à l'usage des générations présentes* (How Everything Can Collapse: A Manual for Our Times), Seuil, Paris.
Shiller, R. J. (2019). *Narrative economics: How stories go viral & drive major economic events*. Princeton University Press.
Simondon, G. (1958). *Du mode d'existence des objets techniques* [2nd ed. 1989] Aubier, Paris.
Stammers, S. (2019). Improving knowledge acquisition and dissemination through technological interventions on cognitive biases. *Educational Theory, 68*(6), 585–716 (Special Issue: Symposium: Cheating Education).
Taleb, N. N. (2007). *The Black Swan: The impact of the highly improbable*. Penguin.
Tanabe, A. (2018). Spirituality as the source of human creativity: Insights from India. In Y. Stomu, Y. Tadashi, & S. Hill (Eds.), *The Kyoto Manifesto for global economics the platform of community, humanity and spirituality* (pp. 179–193). Springer.
Thaler, R. H. (2016). *Misbehaving: The making of behavioral economics*. W.W. Norton.
Tversky, A., & Gati, I. (1978). Structures of similarity. In E. Rosch & B.B. Lloyd (Eds.),*Cognition and categorization* (pp. 79–98). Lawrence Erlbaum
van der Leeuw, S. E. (2007). Information processing and its role in the rise of the european world system. In R. Costanza, L. J. Graumlich, & W. Steffen (Eds.), *Sustainability or collapse*? (pp. 213–241). MIT Press (Dahlem Workshop Reports).
van der Leeuw, S. E. (2019a). *Social sustainability, past and future: Undoing unintended consequences for the earth's survival*. Cambridge University Press.
van der Leeuw, S. E. (2019b). The evolution of human information-processing. *Journal of Integrated Creative Studies*. https://doi.org/10.14989/245356 http://www2.yukawa.kyoto-u.ac.jp/~future/icis/wp/wp-content/uploads/2019/12/2019-010-e.pdf
Varela, F. J., Thompson, E., & Rosch, E. (1991). *The embodied mind*. MIT Press.

Sander van der Leeuw an archeologist and historian, pioneered the application of the Complex Adaptive Systems approach to socio-environmental challenges, technology, and innovation. He coordinated the ARCHAEOMEDES research program (1991–2000) using the CAS perspective on sustainability challenges in S. Europe—the first of its kind worldwide. He has held teaching positions in Amsterdam, Leyden, Cambridge, and Paris (Sorbonne), was founding director of Arizona State University's School of Human Evolution and Social Change and dean of its School of Sustainability. He is Fellow of the AAAS, Honorary Fellow of RIHN (Kyoto, Japan), External Fellow of the Santa Fe Institute, and Corresponding Member of the Royal Dutch Academy of Arts and Sciences. He held a Chair at the *Institut Universitaire de France*. In 2012, he was awarded the title "Champion of the Earth for Science and Innovation" by UNEP.

His most important publications are:

1998: S. E. van der Leeuw (ed.), *The Archaeomedes Project—Understanding the natural and anthropogenic causes of land degradation and desertification in the Mediterranean.* Publications of the European Union. Open Access

2009: D. A. Lane, S. E. van der Leeuw, D. Pumain, and G. West (eds), *Complexity Perspectives on Innovation and Social Change*, Springer

2019: S. E. van der Leeuw. *Social Sustainability, Past and FutureUndoing Unintended Consequences for the Earth's Survival*, Cambridge University Press, Open Access: https://www.cambridge.org/core/services/aop-cambridge-core/content/view/811395DC3A8D82EAD39C45657B2FD1AD/9781108498692AR.pdf/Social_Sustainability__Past_and_Future.pdf?event-type=FTLA

Masatoshi Murase received his Ph.D. degree from The University of Tokyo in 1987. Since 1992, he has been an associate professor at the Yukawa Institute for Theoretical Physics, Kyoto University. In 1987 and 1988 he was a visiting scientist at the Duke University Medical Center, Durham, NC, USA. Since 2010, he has been a member of the Cooperation Promotion Committee of the International Research Unit of Integrated Complex System Science, Kyoto University. Between 2015 and 2020 he was a director of the Research Promotion Strategy Office of the International Research Unit of Advanced Future Studies, Kyoto University.

Chapter 22
A Unified Theory and Practice of Creative Complex Systems: Challenging to the Systemic Problems Spanning the Inside and Outside World

Masatoshi Murase, Aine Mori, and Tomoko Murase

Abstract Despite the seemingly incredible triumphs of modern science and technology for advancing our lives and societies, the current COVID-19 pandemic has exploded into our lives because of the emergence of a novel coronavirus. The virus and its impact were first reported in China in scenes that seemed beyond belief, before the scenario then expanded across the globe. We were suddenly thrown into a shattered world in which nothing seemed certain and drastic changes were endured on a daily basis. We have just experienced *constructive destruction*! It is now time to

---

Biological, sociobiological, and sociocultural evolution now appear as linked by *homologous* principles (i.e., principles related through their common origins) and not just by analogous (formally similar) principles. This should not come as a surprise since the whole universe evolved from the same origin.

Erich Jantsch
"The Self-Organizing Universe:
Scientific and Human Implications of the Emerging Paradigm of Evolution" p. 8
Pergamon Press, Oxford, 1980.

M. Murase (✉)
Yukawa Institute for Theoretical Physics (YITP), Kyoto University, Kitashirakawa, Kyoto 606-8502, Japan
e-mail: murase@yukawa.kyoto-u.ac.jp

M. Murase · A. Mori · T. Murase
2021 Project on Challenges of Advanced Future Studies, Kokoro Research Center, Kyoto University, 606-8501 Kyoto, Japan
e-mail: tmurase@rctoyota.ac.jp

M. Murase
International Research Unit of Integrated Complex System Science (IRU-ICSS), Kyoto University, Kyoto, Japan

A. Mori
Shikoku Medical Center for Children and Adults, National Hospital Organization, 169-2 Kinouchi, HandaTsurugiTokushima 779-4403, Mima District, Japan

T. Murase
Japanese Red Cross Toyota College of Nursing, 12-33 Nanamagari, Hakusan-cho, 471-8565 Toyota, Japan

K. Nishimura et al. (eds.), *Creative Complex Systems*, Creative Economy,
https://doi.org/10.1007/978-981-16-4457-3_22

learn from our failures. In doing so, it is necessary for us firstly to realize how we are deeply influenced by the traditional way of thinking that leads to incomplete perceptions; secondly, how we could get rid of such incomplete perceptions to approach ultimate reality; and finally, how we could make our world and ourselves persistent and resilient, together with keeping the dynamic balance between self-interest (or selfishness) and integration (or altruism) of human beings. Persistence, resilience, self-interest, and integration, and even incomplete perceptions are the common characteristics of all living things such as individual cells, species, and huge ecosystems. Now is the best time to act the part of *destructive construction* for inventing the futures. This chapter consists of three parts: Part I (A Grand Unified Theory), Part II (A Mandala Nursing Theory and Practice), and Part III (An Organization Theory and Practice), and discusses the theory and practice of *creative complex systems* toward a new synthesis beyond ever-evolving complexity.

Keywords Constructive destruction · Destructive construction · Ostensive definition · Abduction · Self–nonself circulation theory · Self-consistency principle · Induction · Deduction

Part I: A Grand Unified Theory of Creative Complex Systems

22.1 Introduction to the Self–Nonself Circulation Theory of Life

22.1.1 Overview

Writing as well as reading a book requires us to understand many things at once, such as the meanings of words, grammar, content, context, subjects, components, authors, readers, cultures, histories, future visions, passions, intuitions, and so on. How can we do this? German philosopher Immanuel Kant (1724–1804) introduced a theory that all knowledge is gained via the senses, but this knowledge must be anchored in categories of perception that must be present a priori (i.e., in advance). This a priori concept was not validated in biology, although it has caused an enormous impact on the subsequent history of philosophy.

Thanks to the *recapitulation theory*—that "ontogeny recapitulates phylogeny"—by Ernst Haeckel (1838–1919), it turns out that a seemingly innate form of perception must fit the world for the same reason that a fish's fin fits the water before it slips out of the egg (Lorentz, 1973). Consequently, the *evolutionary epistemology* was proposed to mention that the a priori in ontology can be a posteriori learning products of phylogeny—a product of adaptation (Lorenz, 1973; Riedl, 2019).

As we shall emphasize throughout this chapter, however, learning is not unidirectional, but is bidirectional or even circular in any spatial and temporal domain.

Learning is, therefore, not only accompanied by subliminal pursuits of truths or realities from the past to the future (Riedl, 2019), but also based on a tenacity of purpose together with a vision from the future to the past, just as in physics the anticipated future hypothesis can often guide us to discover facts in the present (Yukawa, 1973).

The same is fundamentally true in contributing to or appreciating nature including creative complex systems in general. In the face of many crises today, we really need to understand "what is nature?" But we do not know how to do so because of a lack of methodology. Nevertheless, we can understand nature based on nature itself as it is one of the most reliable methods.

Of course, we have to keep in mind shared experiences, including especially failures from the past because we can learn from past failures. We should also remind ourselves that future hypotheses create facts and ideas that invent realities in science and technology. In other words, our ideal visions and hypothetical narratives in the future are actually here in the present. Such methodologies are known as scenario planning (Kahane 2012) or Theory U (Scharmer, 2016). We can learn about nature not only from the present view, but also from the past and future perspectives, and even invent a new world from the future view.

In this chapter, we aim at a new synthesis, at letting appear the contours of an emergent unifying paradigm beyond the individual parts mentioned above. The new paradigm is the Self–Nonself Circulation (Murase, 2000). For this purpose, we introduce a guiding principle known as the self-consistency principle (Jantsch, 1980) and three "dynamical" properties characterized by (1) persisting paradox, (2) fractal dynamics, and (3) true analogy.

Interestingly, one principle and three characteristics can be synthesized into the self–nonself circulation theory as a grand unified theory of "living" nature (Murase, 2000, 2018). Indeed, the self–nonself circulation theory provides us with a unifying vision among subject, object, and the circulation dynamics between them. Depending on individual components and/or dynamical balances, however, virtuous and vicious circles appear, which leads to creativity and crises, and even constructive destruction as well as destructive construction.

Without the combinational overviews, we will fail to understand what nature is, who we are, why we study, when we change, where we go, and how we live. Once we start to enjoy the combinational perspectives, as we discuss below, we become creative because of the transfer of learning and intended unlearning based on the self-consistency principle.

In this chapter, let us enjoy the chaotic and unpredictable dynamics typical of creative nature and mind!

22.1.2 The Self–Nonself Circulation Paradigm

Let us consider three dynamic characteristics: (1) paradox, (2) fractals, and (3) analogy. Firstly, nature including creative complex systems is full of *paradox*, as is life or creative mind itself. Within a paradoxical framework, such as the self–nonself circulation theory by Murase (2000), we shall consider life and its environment as a *circular process*, rather than considering them simply as a linear unidirectional process or as an isolated *static thing*.

Secondly, creative complex systems are characterized by *fractal dynamics*, typical of self-nested dynamic hierarchy (Murase, 2018). Within a circular process of life and its environment, each living subsystem involves a recycling "economy" (or circular economy in our modern terminology) instead of linear one-way (throw-away or unidirectional) processes. Whole complex systems are in turn embedded in larger complex systems. As a result, there appear to be self-nested hierarchical complex systems.

Thirdly, exploring the true *analogy* or homology between life and cognition, for example, we can transfer of learning from life to cognitive process. It turns out that cognition or perception is not a linear one-way process, but is a circular process between the cognitive system and its environment (Murase, 2008; Varela et al., 1991). In other words, cognition and even *creative thinking* cannot be making a copy of the object. Understanding is neither a static knowledge nor a linear process, but a circular process between an observer (a self or an endo-system) and an object (a nonself or an exo-world).

This is the starting point where we think of ourselves and our world based on the *self–nonself circulation theory* (Murase, 2000). The reason is simple: as we mentioned above, the self–nonself circulation theory can hold not only *macroscopic* cognitive dynamics, but also *microscopic* living dynamics, and in addition it can also explain the *self-transcendental* dynamics typical of *life's origin* (Murase, 2008). Actually, the origin of life is the transcendental transition from a material world to a living world, and thus it is one of the typical examples of *creativity*. Here, there is a possibility that we can understand the ever-unsolved problem of "What is creativity?".

Our ultimate goal of this chapter is, after understanding how this challenging problem is solvable, to show how we can develop ourselves beyond our past failures and how we can invent our futures according to attractive narratives!

22.1.3 Challenging Problem: What is Creativity?

What on earth is creativity? How can creativity be so crucial in human history? Why can creativity be so important in human thinking? To date, we still have no generally accepted definition of creativity, and thus there is no general agreement on what it is, how to learn it, and how to teach it. Why? Understanding this reason, of course, requires creativity! What we need must be a paradigm shift from a static or a linear process view to a circular view.

Physicist Hideki Yukawa (Figs. 22.1 and 22.2), known for his creative study on meson theory, gave the important remark: *The very attempt to consider objectively such a thing as creativity involves from the very onset a contradiction of a kind. Creativity implies discovering something that nobody has hitherto known about or*

Fig. 22.1 Photograph of Hideki Yukawa (1907–1981)

inventing something new. It would be distinctly odd, in fact, if one could say: this is the nature of creativity, so this is what one must do to display it (Yukawa, 1973, p. 123). It was not only realized that words alone were inadequate to describe creativity itself, but also experienced that words had often prevented creative thinking (Edwards, 1986). Paradoxically, creativity requires silence or emptiness or even nothingness.

To solve the problem of creativity, we have to already understand creativity from the beginning. Without knowing creativity, of course, it is impossible to attack this problem. This seems to be circular causality, but this is the exact point that we emphasize regarding the importance of the *self-consistency principle* (everything must be consistent to itself and everything else). Consider the situation that we are trying to understand a text written in English. Without knowing English grammar, of course, it is impossible to understand the text. But even if we know the grammar, it is still difficult to understand the text without preliminarily expecting what it mentions for whom under what contexts.

Therefore, we need multiple perspectives for understanding creativity. Actually, we have to consider two different approaches simultaneously: one is to think about the more *general* problem of "what is life?", and another is to explore the *fundamental* dynamical processes that are hidden behind creativity. In addition, we have to keep our mind on (1) paradox, (2) fractals, and (3) analogy. Why? Consider the current COVID-19 pandemic. A worldwide systemic destruction of this kind is merely considered as a creative phenomenon in nature! Without learning from our failures, there is little hope of achieving success (Yukawa, 1973).

Fig. 22.2 As a centennial celebration of the late professor Hideki Yukawa's birth, the international symposium on "What is Life?" was organized by Masatoshi Mursae in 2007. See https://www.yukawa.kyoto-u.ac.jp/assets/contents/seminar/archive/2007/ny2007/

22.1.4 Nothingness or Emptiness as the Mother of Infinity

Paradox is the mother of creativity. It has often occurred to us that the subject of our research, its ultimate meaning, was different from what we had originally thought we were studying, as already pointed out by Edwards (1999) and Bateson (2002). Again paradoxically, here is the source of creativity as well as crisis. Whereas no one can predict the future, we can invent our futures if we share attractive narratives. However, no one knows how we can do so and how it does work because of a lack of a unified theory.

Westerners speak of what is unknown primarily in reference to what is known, while most Easterners give honor to the unknown in its own right (Edwards, 1999, p. 153). There appears again the paradox or dilemma that previously learned knowledge—which is useful in some contexts—prevents us from seeing the thing-as-it-is in other contexts, even though it is right there in front of us. We have to believe what we "see," not see what we "believe." Creative individuals in all fields, therefore, benefit not only from seeing both *the trees and forest* (enhanced awareness of part-to-whole relationships), but also from understanding nothingness or emptiness. In a *Zen* sense, the subject must be seen as if never seen before, because nothingness is full of meaning just as there is the knowledge that we do not know what we know.

A creative thinker has to hold in one's mind simultaneously two diametrically opposed ideas: one has to not only consider the problems intensively, but also pay much attention to the empty spaces or seemingly "nothingness" around the problems. *One is aware of some contradiction within oneself. Without some contradiction within, there can be no study; that, indeed, is the essential nature of study* (Yukawa, 1973, p. 131). Contradiction or paradox and creativity must be the different sides of a same coin. Creativity inspired by paradox is therefore necessary for understanding creativity. This is the main reason that creative works or creative individuals often remain unrecognized for a long time.

22.1.5 Unintended Scenario of the World: Constructive Destruction

Thanks to the triumphs of Western science and modern technology (Fig. 22.3 top), on the basis of *human-centric* values, we have rapidly changed our lives, energies, economies, societies, and even the environment of the earth and the universe. On the basis of *nature-centric* or Eastern philosophical values (Fig. 22.3 bottom), however, the seemingly great benefits of these radical changes reveal the counter-effects such as ecosystem collapse, emergent diseases, environmental pollution, and so on. This is an unintended scenario of the world often referred to as trade-off or *constructive destruction*. It is a great pity that our community with so important "Collapsologie" already proposed by Servigne and Stevens (2015) should have paid so little attention to the enormous implications.

When the current COVID-19 pandemic first started to have an impact, we were suddenly thrown into a shattered world where nothing seemed certain, and everything can change radically. It turns out that we, as human beings, appear no longer as the ultimate evolving organisms in the world, and that the world itself is becoming increasingly *alive* and thus evolving. Both are evolving into co-evolution through interplay processes, just like emerging complex life (Jantsch, 1980). What we need is not only a reductionist–physical view but also an emergentist-holistic view for a deep understanding of life itself as a whole. It is possible to possess both views

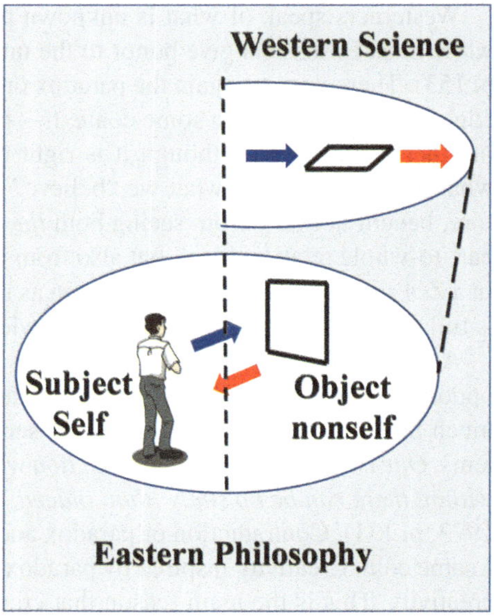

Fig. 22.3 Western science (top) and Eastern philosophy (bottom). The self–nonself circulation theory (bottom) explains the circular interaction between subject and object

alternately through self–nonself circulation theory by Murase (2000), as illustrated in Fig. 22.4.

We began to realize that although modern science based on a reductionist–physical view is true in its own validity, it is not the only way to approach ultimate reality. A new paradigm or new synthesis is urgently needed to understand creative complex systems such as humanity, the mind, and nature, together with life and death (Bateson, 2002; Capra & Luisi, 2014; Murase & Murase 2021a, b).

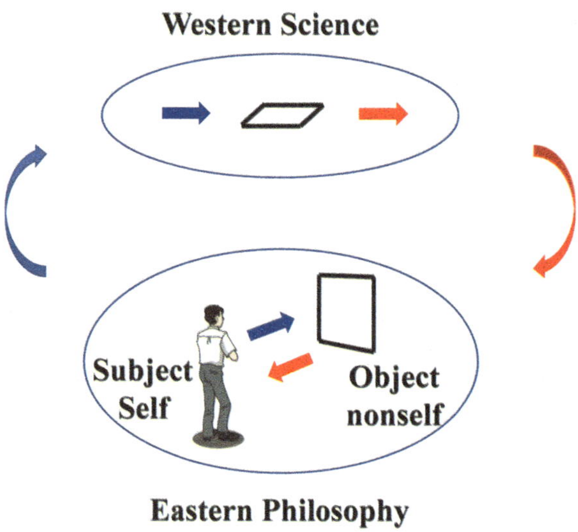

Fig. 22.4 The self–nonself circulation theory (Murase, 2000) as the integration of Western science (top) and Eastern philosophy (bottom). Note that there is the self-consistency principle at the level of the self–nonself circulating process between Western science and Eastern philosophy

Learning from our failures, we have to take the part of *destructive construction* at the current worldwide crises. Ultimately, through a unified theory and practice of creative complex systems, we have to integrate both constructive destruction and destructive construction because they are merely different aspects of the same reality.

22.1.6 Limit of the Traditional Way of Thinking: Ever-Unsolved Problems of What is Life and What is Death?

We have not yet satisfactorily answered the long-lasting problems: "What is life?" and "What is death?" Let us consider the reason why such a situation has happened and still continues. Actually, it turns out that the problem stands not on the unsolved problems themselves, but rather with the approach used toward them. After deep thought, however, it appears that understanding life and death requires an understanding of one's self.

The main interest of Western science has been intended to understand structure, especially exploring fundamental building blocks of matter based on reductionism. Indeed, it has widely been acknowledged that physical reality like non-living things may be understood based on a reduction to composite elements, and even to physical laws. The underlying assumption is that the essential properties of matter may be traced back to these fundamental building blocks, which means that the whole may be completely reduced to its parts. Certainly, material science based on such a reductionist–physical view has been applied to many disciplines involving chemistry, life science, economics, astronomy, and so on.

Concerning the properties of life, of course, material science contributes many detailed findings toward the classification of component molecules as well as the identification of vertically and horizontally transmissible genes. We suddenly realize, however, that it is difficult to provide a scientific definition of life based on traditional reductionism. Why? The reason is very simple: indeed, life is not a *thing* that can be defined in terms of composite molecules; instead, life is an *emergent property* or *process* that always depends on its connectedness, pattern, and context (Murase, 2000, 2018; Gottwald, 2002; Capra & Luisi, 2014; Murase & Murase 2021a, b).

Because there is no universally acknowledged definition of life in terms of traditional reductionism, it may not be so surprising that the same is true for the concept of death. Death or aging has been understood as a progressive process or an intra-individual evolutionary process, corresponding to the destruction of the emergent properties of the various levels characterizing the complexity of the entire organism (Murase, 1996, 2000). The death of a living organism has been thus described in terms of "neg-emergence" (see Capra & Luisi, 2014, p. 139).

As Bortoft (1996) mentioned, nature can manifest in more than one way. Without assuming that one way is more fundamental than another, therefore, there must be different ways toward "ultimate reality" besides traditional reductionism typical of

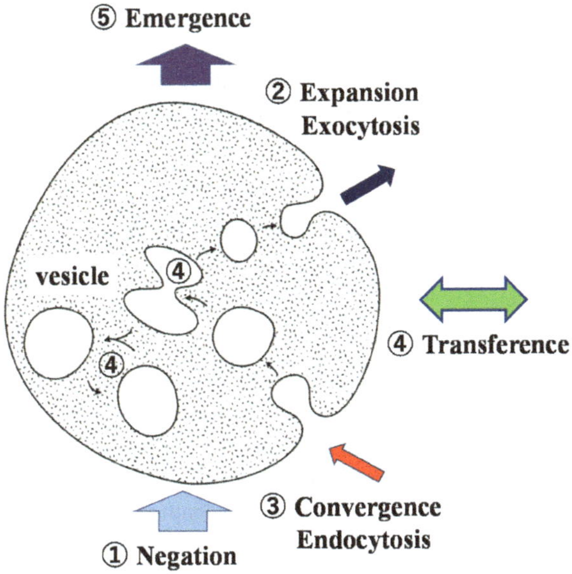

Fig. 22.5 The dynamic characteristics of a living cell. Five different processes are represented: ① Negation, such as mutation or epi-mutation of cellular components intrinsically and/or extrinsically; ② Expansion, leading to the secretion of molecules by exocytosis; ③ Convergence, mediated by the uptake of external molecules by endocytosis; ④ Transference, fusion of different vesicles and the budding of vesicles within the cell, and also occurs between the object and model; and ⑤ Emergence, as the appearance of novel or disappearance of the existing properties

modern science. It is now clear that living processes cannot be defined based on a reduction to nonliving structures (Capra & Luisi, 2014; Murase, 1996, 2000, 2018).

What can we do to understand life itself? Besides a reductionist approach, there is a *phenomenological* approach: it is made based on observations at the level of our experience. This approach is applied to the definition of life in such a way that the complex life phenomena are understood based on the *typical* or *ideal* life (Murase, 1996, 2000). Such a definition is called an *ostensive definition*. As illustrated in Figs. 22.5 and 22.6, through a holistic approach, instead of a reductionistic one, it is possible to provide an ostensive definition—which is relating to *abduction* (Bateson, 2002) or transfer of learning (Edwards, 1986)—of life by raising several instances such as cells, species, and so on (Bortoft, 1996; Murase, 1996, 2000).

22.1.7 Self-Consistency Principle: How to Challenge the Long-Standing Problems

Besides the abovementioned problems of "What is life?" and "What is death?", there are further long-standing problems of "What is the mind?" and "What is nature?"

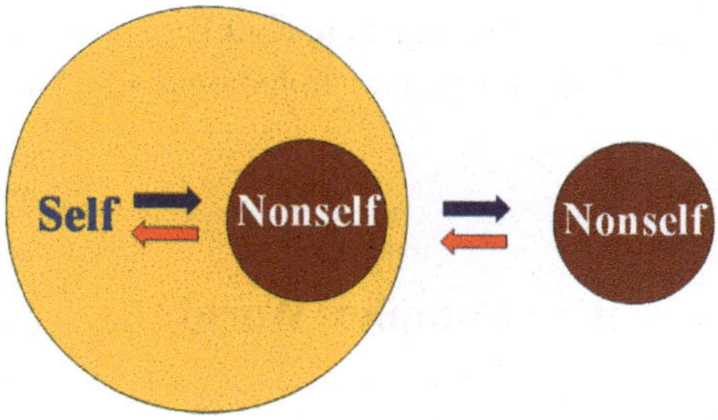

Fig. 22.6 The model of life. The self–nonself circulation theory explains the circular interaction between the subject (or the self or the endo-system) and the object (or the nonself or the exo-world)

As we described, it is difficult to provide a materialistic or scientific definition of life and death because of their emergent properties. The same is true about mind and nature. Instead of exploring the building blocks such as genes as fundamental, there must be an alternative approach how to understand complex life phenomena based on a fundamental principle.

Through the *Copernican revolution* from a structure-centric static to a process-centric dynamic perspective, we can challenge long-standing problems simultaneously on the assumption that, in our unique universe, there is a single fundamental principle, *self-consistency*. That is, there must be a fundamental *homology*—a true relatedness relevant to the common origin—of the evolutionary complex systems at many levels.

Within the self-consistency principle, we no longer need a precise definition of life because life could be considered not only as an elementary process but as a whole, just like fractal dynamics (Murase, 2018). Once we have developed a new conception of life, a sustainable society and resilient individuals could be developed on the condition that we are taking care of the economy, business, structures, technologies, and humanity in order not to interfere with originally emerging "living" nature (Capra & Luisi, 2014).

Therefore, on the one hand, tracing the evolutionary and/or developmental history of life over time back to its origin, we can realize the self-consistency principle at the fundamental level as the homology. Through open evolution towards the future, on the other hand, whatever comes into presence—especially the artifacts of modern science and technology—has to be consistent with itself and with everything else. What we need is, therefore, not the *complete* knowledge of structural components of life, but a paradigm shift from a *reductionist–physical* view to an *emergentist–holistic* or systemic view.

A deep understanding of life, nevertheless, requires an emergentist view, in addition to a traditional reductionist view. This is the reason that we introduced the self–nonself circulation theory. As already illustrated in Figs. 22.3 and 22.4, the theory can make a circular relationship between the two opposed perspectives. Here, we should emphasize the possibility that there could be a different science for understanding life itself, not contradictory but complementary to the science of nonliving things.

22.2 What is a Creative Complex World?

22.2.1 The World is Full of Dichotomies: Too Many Oppositions and Too Few Synergies

It is usually understood that our "world," whether inside or outside, is full of dichotomies such as life and death, oppositions and synergies, awareness and ignorance, construction and destruction, benefits and risks, bottom and top, self and nonself, and so on (see Izutsu, 1973; Murase, 2000). Indeed, as already shown in Fig. 22.3, the Western mode of logical thinking starts from the division of subject and object (Suzuki, 1938).

Even Schrödinger (1958) considered, however, that the Western way of thinking should take into account the Eastern thought process, mainly because with *Zen* all the dichotomies would disappear, and thus a new insight into the nature of life and reality would emerge (Suzuki, 1938). Without mystic or meditative experience, unfortunately, we can hardly access the realization of how we can be enlightened. Again, we are confronted with the serious dichotomy between an enlightened *Zen* master and an unenlightened beginner.

Here, we should realize the difference between two kinds of dichotomies in our current situations, although all the dichotomies should be integrated into the wholeness in the end. When we talk about distinctions or dichotomies, we are apt to think of distinctions at the "visible" level as if we could distinguish one part from the other. However, this is not always the case, because there appears the distinction between what is seen and what is not seen, as we show below.

Figure 22.7 shows an ambiguous figure that has often been used by *gestalt* psychologists. Most of us at first see only random black and white patterns. Some of us will suddenly see a recognizable figure out of the random patterns. It is a giraffe—its head and upper neck can be seen emerging from the complex patterns. Just like turning on a light, the giraffe was switched on. Those who failed to see it at first for themselves could do so sooner or later after reading the above sentence that it is a giraffe. There are no essential changes in our purely sensory experiences before and after this transition!

The sensory stimulus to the eye is the same regardless of whether the giraffe is recognized. As far as we are concerned with sensory experiences, therefore, the difference cannot be explained in terms of sensory stimuli. We have already realized that a slight change of our eye movement or other environmental fluctuations or some conducting information can trigger the sudden appearance of the giraffe. This occurs because we understand the different *meaning* or *value* of the object at once, often

Fig. 22.7 Random black and white patterns? (After Bortoft, 1996)

referred to as *phase transition* typical of complex systems (Murase, 1992, Murase & Murase 2021a, b).

In the case of Fig. 22.7, what we experienced is a sudden emergence of a new meaning. We should, thus, realize that meaning is not an objective thing, but instead an emergent property. This must be an explanation of real difference between an enlightened *Zen* master and an unenlightened beginner, for they are different at the level of meaning or value system in their mind. According to Yukawa (1973, p. 138), *it's something like the man who looks at abstract art and doesn't find it appealing.* Of course, this is not the end of the story!

22.2.2 Systemic Problems Inherent in Creative Complex World Require Systemic Thinking

In addition to many dichotomies, our world is full of systemic problems characterized by complex networks of inseparable patterns of relationships in such a way that the world or the planet as a whole is increasingly considered as a *living* system involving interplay between processes (Lovelock 1979; Capra & Luisi, 2014; Murase & Murase, 2021a, b). The concept that our planet is a nonliving machine-like entity composed of elementary materials is becoming increasingly outdated.

It is now time to realize that the many dichotomies seemingly inherent in our world may be merely the illusion created by contemporary materialistic science, which has tried to understand the world or the universe as a machinery composed of elementary building blocks (Jullien, 1999). Actually, the main interest of Western science has been devoted to recognizing structure, especially on the basis of reductionism, exploring fundamental building blocks of matter on the assumption that the essential properties of matter may be traced back to fundamental building blocks, and also that the whole may be completely reproduced from its parts.

Instead of seeing dichotomies or our shattered world as reality, we need to understand their syntheses or the integrated world as necessary units of mind and nature (Bateson, 2002) or the circles of functions (Gottwald, 2002). We have to realize a new paradigm that considers our world involving systemic problems as “life” or emergent properties or even creative phenomena.

In science, the integrative way of thinking is called “systemic thinking” or “systems thinking” (Capra & Luisi, 2014). Note that the emerging problems within us, around us, and without us are all interconnected and interdependent, characterized as systemic problems. There arises, therefore, a concept that we should apply systemic thinking in our approach to systemic problems, just like the easy case of fighting like with like.

What does the “system” mean? Indeed, a system is no longer considered as a complex structure involving merely interconnected characteristics. Rather, a system itself is evolving through the interplay between differentiation and integration dynamics, which reveals ever-new dimensions of novelty and exchange with the environment. Such a system can be as viewed as creative life. Thus, we call it as a creative complex system. Emerging problems behave like living systems, and it is difficult to predict their behaviors. Consequently, we have to take a systemic and evolutionary approach to address emerging and evolving problems.

From such a complex dynamical systems perspective driven by self–nonself circulation theory (Murase, 2000), a sustainable society, together with resilient individuals, could be fostered whenever we are holding the *self-consistency principle* so that we are taking care of the economy, technologies, humanity, and mind in order not to interfere with originally emerging “living” nature.

22.3 Toward a Grand Unified Theory of Life: What is Creativity?

22.3.1 Reconciling Unidirectional Cause-And-Effect Physiology Typical of Western Science with Circular Hermeneutic Morphology Typical of Eastern Philosophy

Based on a Western way of thinking, human beings acquire the power of rational thought. That is, contradiction or paradox must be avoided so that one cannot both affirm and deny the same proposition at the same time. Consequently, one must reject one or the other. Suppose that, for example, one carries out experiments under different kinds of conditions, collecting special data necessary for the discovery of a general law, known as inductive (from the particular to the general) logic. It is on inductive logic that most scientists rely. We are indeed familiar with such a bottom-up approach.

It is ironic to realize, however, that there appears to be contradiction! As long as we are concerned with successful investigation of individual data in a special sense of usual experiments, we can hardly obtain any "special" revolutionary results in terms of creativity. *What we need is not only the learning of inductive logic, but also the unlearning of itself and even transfer of learning from one subject to the other, and even from one way of thinking to the other.*

Consider the concept of *mathematical induction*: suppose that we have proved that if a certain statement is true of an arbitrary integer N, then we know that it is true of $N + 1$; also, if we have proved that if it is true when $N = 1$, then it is true for all integers N. In every branch of research, therefore, induction provides a way forward because it reveals the preconditions that we accept when we extend our query from the known to the unknown, which leads us to ask "off-limit" questions in a general sense (Riedl, 2019).

Contrary to the name of mathematical induction, however, it seems to be paradoxical not only because there is a kind of operation of *deductive* logic (from the general to the particular, as is the case of $N = 1$), but also because the formula in the case of N has to be known from the beginning. The latter ability may be anchored in a perception of spatial and temporal continuity or intuition (Riedl, 2019, p. 17; Yukawa, 1973).

Induction necessarily contains deductive and speculative elements, because the cognitive process in general is characterized by a natural cycle of alternate synthesis and analysis, referred to as inductive and deductive processes (Riedl, 2019). It appears to be a paradox because, despite even avoiding a paradox in the beginning, the Western way of thinking is confronted with a paradox again!

From the point of view of the self-consistency principle, we should not avoid paradox but instead enjoy it in any space and at any time. Only then, can we understand how we live in the current shattered world. The remaining but still important problem is, of course, that we do not know how we can enjoy too many paradoxes.

22.3.2 *Grammar, Emergent Wholeness, and Passion—Requirements for Creativity*

22.3.2.1 Why Can We not Be Creative? and How Can We Be Creative?

If one hopes to be a great writer, one requires not only a knowledge of grammar, but also an attractive narrative as an emergent wholeness and even passion. The same is true for being a creative individual. There is a simple reason why we struggle with the long-standing problem "What is Creativity?" Ironically, we need the solution to this problem in advance to solve this problem, just like the case of N in mathematical induction mentioned above. In other words, we should "creatively" challenge this problem.

Generally speaking, facts alone are just facts: to invent new values, new meanings, new concepts, or new laws, indeed, we have to know how the facts fit together to form a whole. According to Arieti (1976), products of creativity have strong impacts on us even though the elementary processes of creativity seem to be surprisingly simple. However, he did not show us what the processes are. Then, Betty Edwards (1986, 1999) clearly mentioned that just like drawing, creativity is a global or whole skill requiring only a limited set of basic components.

Here there is a possibility that we understand the "grammar" of creativity—in the sense of a limited set of basic descriptive components—although not all the creative individuals have explicitly defined the "grammar" of creativity.

Based on the discussion in Sect. 3.1, it is true that induction and deduction must be parts of the grammar of creativity. When we think about writing or drawing, however, it is clear that we have to know what we make in advance (i.e., a priori). Likewise, before understanding a new law or narrative, we should know its contents a priori. It must be essential because once we have a preliminary version of a new law or narrative, such a future image can guide us to construct creative products, just as a hypothesis becomes reality in science and technology.

22.3.2.2 The 5-NECTE Principle for Creativity

What are the necessary components for creativity? Based on the considerations by Hideki Yukawa (1973), it is possible to define the following five processes: (1) failure is a source of creativity; (2) knowing a lot of things has an advantage because it serves as a basis for discovering new things; (3) selection of too much information is needed not only to avoid redundancy, but also to reduce a gradual immobilizing side effect; (4) creativity is something that comes from without; and (5) a considerable preparation period is necessary before creativity appears. Note that identical numbers assigned in each process and in Figs. 5 and 8 represent the same process in the following discussions.

Let us introduce the 5-NECTE principle as a grammar of creativity. NECTE means (1) Negation, (2) Expansion, (3) Convergence, (4) Transference, and (5) Emergence. In 2017, when Professor Tae-Soo Chon was a Distinguished Visiting Professor of Kyoto University, he kindly proposed the name "NECTE" as an abbreviation by adopting the first characters of five individual words.

According to Betty Edwards (1986, 1999), after almost 100 years of inquiry, we now realize that there are common processes behind various creative problem solving. As illustrated in Fig. 22.8, the inquiry started in the nineteenth century with Helmholtz, a physicist, who introduced three processes: ②Saturation, ③Incubation, ④Illumination. In the terminology of logical thinking, saturation, incubation,

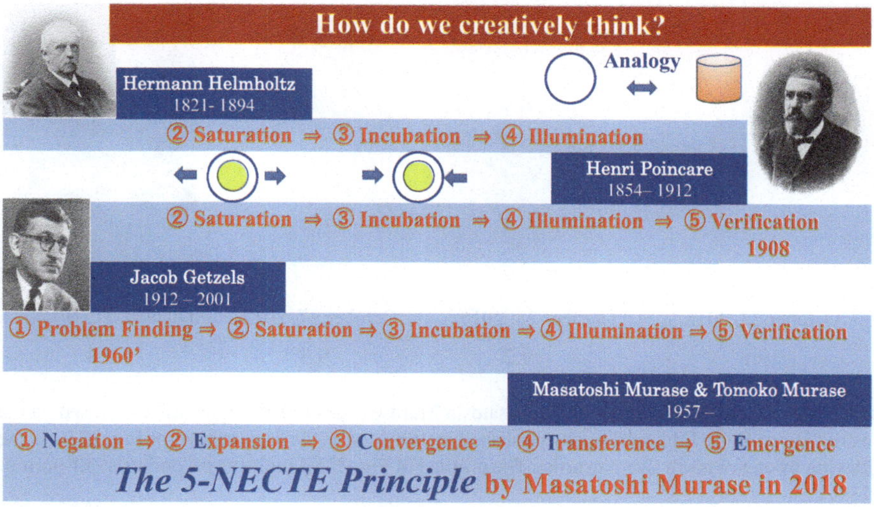

Fig. 22.8 The 5-NECTE principle. Note the analogy (or transference) between creative processes and living processes illustrated in Fig. 22.5

and illumination correspond to deduction, induction, and abduction (or analogy), respectively.

Importantly, the further process of ⑤Verification was introduced in 1908 by Poincare, a mathematician. He introduced this process because he believed that it was necessary to check whether the hypothesis was valid in a mathematical sense. After about half a century, a psychologist called Getzels added another process named ①Problem Finding for consideration before ②Saturation.

The interesting thing is that the essential processes of creativity had been investigated by a physicist, a mathematician, and a psychologist for almost 100 years, and then finally summarized by art educator Edwards (1986), which must be actually the products of creativity!

22.3.2.3 Drawing as Creative Problem Solving: From Negation as Unlearning Preconception to Emergence of Wholeness Through Expansion, Convergence, and Transference

The drawings shown in Fig. 22.9 are artworks completed by Masatoshi Murase (one of the authors) in 2015. According to books written by Betty Edwards (1986, 1999), Murase started to learn how to make drawings. Without *learning* anything special and without *unlearning* any preconceptions or common sense, the result was an unappealing drawing of his left hand (Fig. 22.9a).

Through Edwards' books, Murase actually had an ah-ha! experience with a sense of joyful excitement at seeing everything fall into place. That is, drawing better requires seeing better, which in turn leads to the exciting discovery that there must

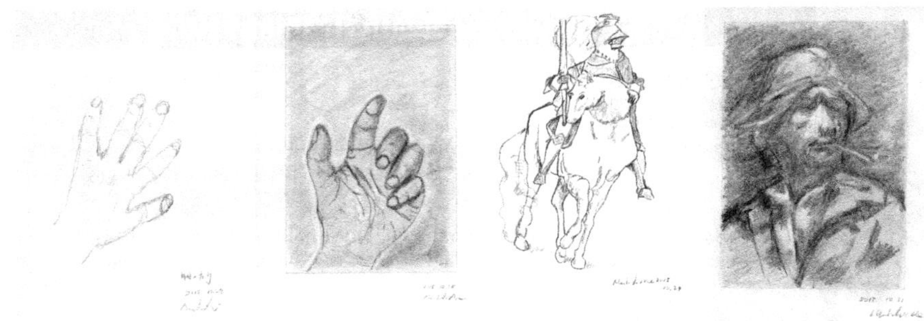

October 27, 2015 9.1 | October 28, 2015 9.2 | October 27, 2015 9.3 | October 31, 2015 9.4

Fig. 22.9 Four different drawings by Masatoshi Murase (one of the Chapter authors) based on the instruction of Betty Edwards' books published in 1986 and 1999. The original of Fig. 22.9c is the drawing by an unknown German artist. The original of Fig. 22.9d is Gustave Courber's self-portrait (1897)

be the essential common elementary processes among various human functions or skills such reading, driving, walking, understanding, enlightening, creative problem solving, and so on. This was exciting because the process revealed a transcendental nature starting from nothingness to infinity, just as the 5-NECTE principle appealed!

It is therefore instructive to observe how Murase's drawings improved remarkably within a very short period of time. Of course, it is generally considered that we should learn harder and harder to achieve a good result in a task like drawing. There appears, however, a paradox because too much knowledge often prevents us from learning radically new things. Unlearning of learned knowledge is a key process, although at schools and even in universities there is little chance to learn the role of *unlearning*.

Murase had the preconception that an eraser should be the only tool for erasing mistakes. Edwards' books, however, showed how he could use the eraser as a special tool for drawing! Fig. 22.9b provides an example of①Negation or unlearning of the preconception about erasers. Following this radical shift on his conceptions, Murase started from a dark background and then drew the edge of his left hand by using an eraser instead of a pencil. It appears very different from his earlier attempt.

Another surprising technique relating to Negation is "turning the original drawing upside down": it looks entirely different from the right side up drawing—simply a pattern of dark and light areas. Murase drew the shapes of light and dark areas, just as he perceived them. After he finished drawing and turned it right side up, as illustrated in Fig. 9.3, the drawing seemed to be so real that it was almost three-dimensional.

As in all the cases in Fig. 22.9, ②Expansion and ③Convergence are also important processes, because it is necessary to see both parts and wholeness at once, and Murase did this to expand and converge his field of view. During this practice, Murase suddenly realized that drawing skills in art require us to see not only *positive forms*, but also *negative spaces* or nothingness, besides the light and dark areas mentioned above. The essential point is that negative spaces are just as important as positive forms.

Here, ④Transference of view plays an important role to see the relationship not only within positive forms, but also between positive forms and negative spaces. In addition, when we transfer the learned skills in art to knowledge in science, we suddenly realize the importance of the radical concept that oppositions must be considered as complementary units, rather than as being mutually exclusive. This is known as the transfer of learning (i.e., an example of ④Transference).

Finally, in art, it is not necessary to draw everything, because the viewer can make a complete image from an incomplete drawing. The key remark is: just draw what one sees, no more, no less. In complex systems terminology, the wholeness is not a *thing*, but just an *emerging property* or *a process*. This is a typical example of ⑤Emergence. In Fig. 9.4, the essence of the drawing can be recognized despite being incomplete.

Let us remember the "transfer of learning" from drawing skills to a specific kind of problem solving. As Edwards (1986) emphasized, it is now clear that there is a striking parallel between drawing and creativity. In other words, creativity can be visualized through how we can learn drawing, which must be very meaningful in education and creative learning.

22.3.2.4 Intuition and Creativity

How do we know the essential features of future products in advance? Hideki Yukawa (1973) emphasized the important role of intuition and tenacity. Indeed, the evolutionary epistemology (Riedl, 2019) mentioned that a priori in ontology can be a posteriori learned products of phylogeny (i.e., adaptation). Besides such a vertically heritable relationship, we can learn creative products from other pioneers through a horizontal relationship like an ostensive definition.

In art schools, it is very instructive for beginners to redraw *only* the great masters' drawings. Redrawing forces one to slow down and really see what the great artist saw. It is therefore possible to understand the artist. The beginners are never asked to copy poor drawings that have little passion. The same is true in other fields including science, religion, and so on. In science, we are recommended to reformulate great theories not only to understand their contents, but also to realize how to invent new theories and even to learn about creative individuals themselves.

In religion, Lama Anagarika Govinda (2012, p. 39) mentioned, *Truths cannot be "taken over," they have to be rediscovered continually. They have ever to be re-formed and transformed, if they are to preserve their meaning, their living value, or their spiritual nutriment.* Besides learning the products of creative individuals, it is also very instructive to learn behaviorally and even psychologically from great people themselves. Fundamentally speaking, although all these examples appear to be different from each other, the essential features behind the creative processes can be understood through the ostensive definition.

22.3.2.5 A Whole World Thinking and a Whole Brain Thinking: Finding the Origins of Curing and Caring

Goldberger et al. (2002) emphasized that there are two different kinds of "world": simple world and complex world, which is illustrated in the upper panel of Fig. 22.10. In the simple world, indeed, simple rules explain simple behaviors so that there is proportionality between input signals and output effects, so that we can expect the outputs with high predictability, but there is no surprise. A whole is easily understood by summing up individual things. Remember the power of mathematical induction discussed in Sect. 3.1. Such a linear world is well understood by Western thinking based on the cause-and-effect relationship.

In the complex world, however, even the simple rules generate complex behaviors, where only small inputs can cause large effects. Just like the current COVID-19 pandemic, no one can predict how the coronavirus evolves through sudden mutations and how we can recover our shattered world, because the whole is not the sum of the parts. The Eastern way of thinking is suitable for such a nonlinear complex world. Of course, the ultimate reality could be realized through the integration of both Western and Eastern thinking, which could be conducted by the self–nonself circulation paradigm.

Why are there two different ways of thinking? This must be a challenging question because different ways of thinking relating to the West and East in the world seem to be, in a horizontal sense, corresponding to left and right hemispheres of our brain. Interestingly, the human brain (Fig. 22.11a) uses two fundamentally different modes of thinking: one is verbal, analytical, and sequential and is often attributed to the dynamics of the left hemisphere of the brain (Fig. 22.11b); the other is visual,

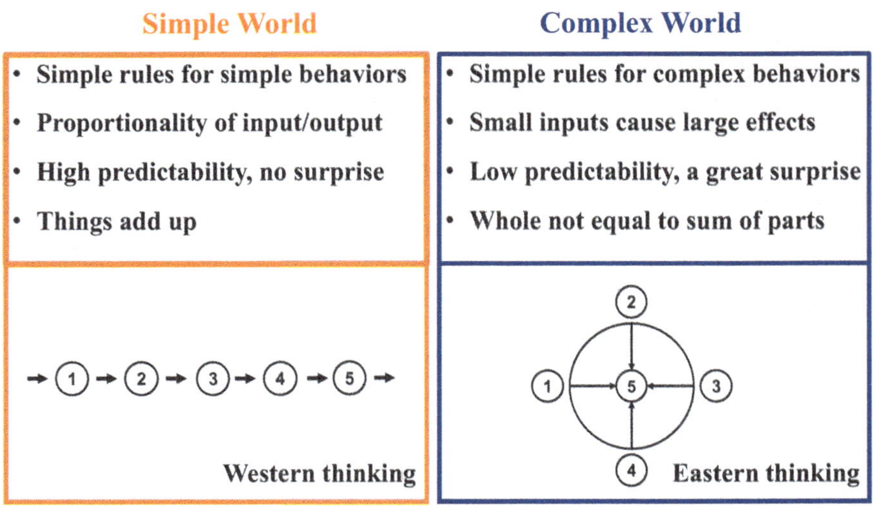

Fig. 22.10 Simple world (left) and complex world (right)

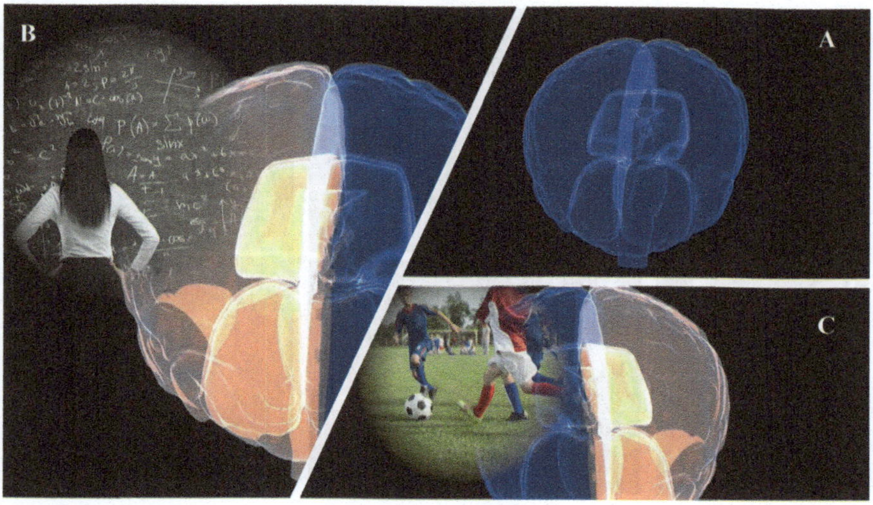

Fig. 22.11 The human brain

behavioral, and simultaneous and is often attributed to the dynamics of the right hemisphere (Fig. 22.11c).

This scenario can deliver us an Ah-ha experience. As we shall show in the latter part of this chapter, we should realize that drawing—although itself is a creative process—is not the *purpose*, but the *means* for becoming creative. Drawing requires us to use a different mode of thinking. Therefore, while drawing, we often find that awareness of the passage of time fades away and words recede from consciousness. According to Edwards (1986, p. 52), *drawing is not locked into linear time and is therefore able to clarify complicated relationships that may include past, present, and future.*

Drawing individuals feel alert and aware yet are relaxed and free of anxiety, experiencing a pleasurable, almost mystical mode of mind. What does this mean? We already know how to draw, but usual habits of living in the "simple" world interfere with that ability, and thus block its unlimited possibilities.

Besides the horizontal relationship within the brain, there are four layers vertically based on the evolution of our brain as illustrated in Fig. 22.12. The lowest layer is the brain stem and spinal cord typical of a "fish brain" (Fig. 22.12a), which is responsible for reflection to environmental stimuli. The second layer appears in the "reptilian brain" as indicated in Fig. 22.12b, which may be relating to instinct. The third one develops in the "mammalian brain," which is responsible for emotion (Fig. 22.12c). Finally, the fourth layer emerges as the "human brain" (Fig. 22.12d). Well-organized society has evolved based on communication through verbal languages. The human brain has a self-nested hierarchical structure that corresponds to the evolutionary history of human beings.

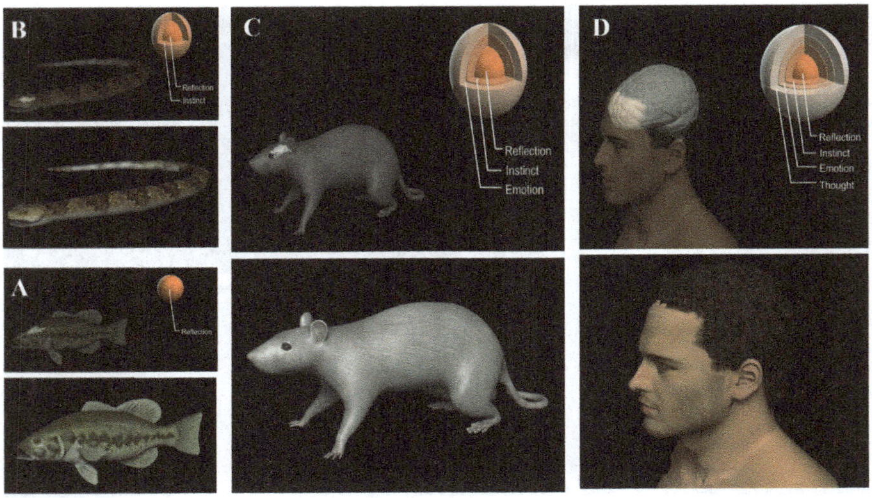

Fig. 22.12 The evolution of the human brain

Let us again consider mathematical induction, because it contains not only induction (from N to $N + 1$), but also deduction (from N to $N = 1$), and even abduction (the statement is valid when $N = N$). This means that vertical bottom-up and top-down approaches as well as horizontal abduction typical of transference play important roles during the construction of the mathematical structure. The same is true in nature. Indeed, it is not usually considered that the structures in the universe were built up one-sidedly, bottom up, from elementary particles and atoms to stars, galaxies, and so on (Jantsch, 1980). It now turns out, however, that there is a mutual co-evolution of the conditions for simultaneous differentiation and complexification along the microscopic and macroscopic branches of co-evolution, as indicated in the previous chapter. Likewise, as our brain has its evolutionary history within its self-nested hierarchy, different modes of thinking can provide us with the simultaneous relationship among the present, past, and future.

How does this occur? In the next section, we briefly mention the self–nonself circulation paradigm.

22.3.3 A Grand Unified Theory

22.3.3.1 The Self–Nonself Circulation Theory and 5-NECTE Principle

As described in the previous sections, we can realize that an object is present in the outside world whatever it is and whenever finding its meaning or value. What happens, however, when there appear different meanings or values like the extended case of ambiguous figures? Then, conflicts of interest appear, which in turn are the causes of different power relationships (Capra & Luisi, 2014).

We are faced with the dichotomy, again! Such situations occur not only within oneself, but between individuals and beyond individuals. They are mutually co-evolved through complex interactions. Therefore, we have to attempt to integrate living individuals and social dimensions simultaneously based on the self-consistency principle.

Here, we should realize the dual dynamics: on the one hand, through the meta-evolution or self-transcendental dynamics, a novel mental system may emerge within an individual person; but on the other hand, the same dynamics challenge us for destruction and the possibility of disintegration appears. Thus, we have to advance the same dynamics once again for destructive construction. Fortunately, the individual person carries the additional dimensions in the form of the macro-world within himself as illustrated in Figs. 22.13 and 22.14.

As van der Leeuw and Murase discuss in Chap. 21, a hypothetical plan can turn into reality through complex evolutionary dynamics, meaning that mental anticipation can pull the future into the present. The direction of causality is reversed, which evokes the deductive process. As time proceeds in inductive ways, we have an example of circular dynamics between the present (or past) and the future, as illustrated in Fig. 22.15.

In other words, there is a parallel between the history of the outside world and that of the inside world. Such a parallel ensures that the *reality* emerges through transcendental co-evolution between complex dynamics *outside* and *inside* us (Murase and Murase 2021a, b). Furthermore, on the basis of the fractal nature of our world, the same arguments can be applied to the various hierarchical systems such as social,

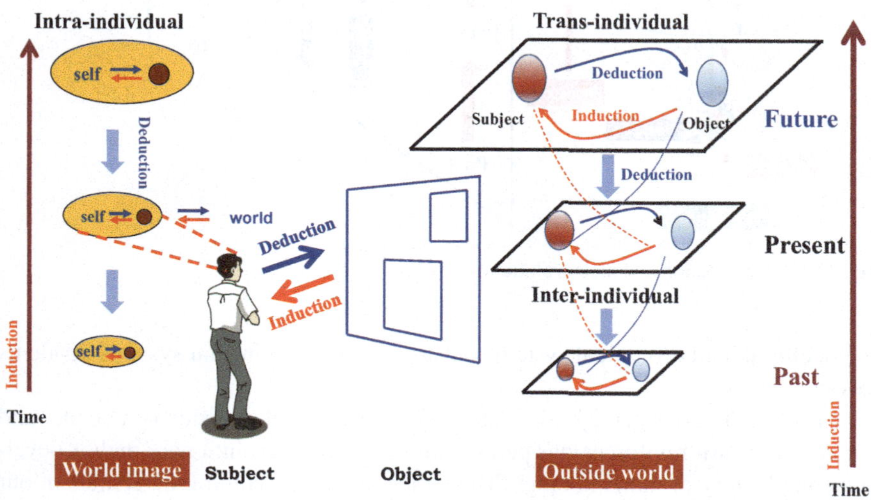

Fig. 22.13 The world image and the outside world

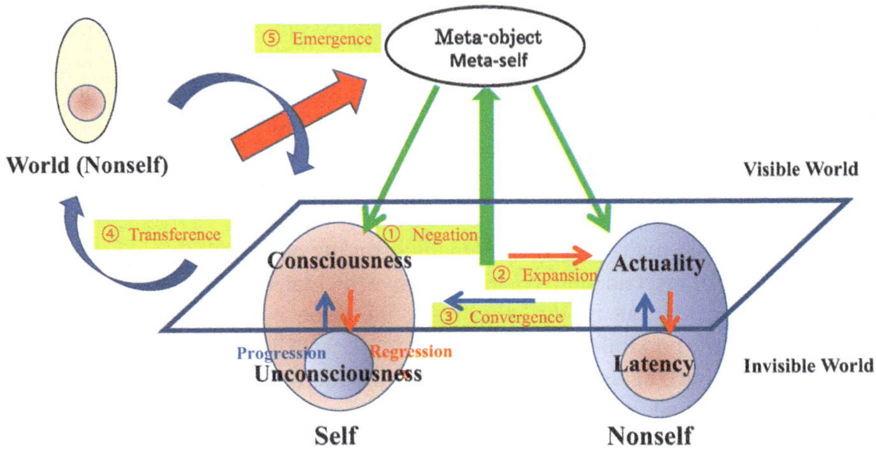

Fig. 22.14 Enlargement of Fig. 22.13 (right). This is the self–nonself circulation theory. Interestingly, the 5-NECTE principle can be found within this diagram

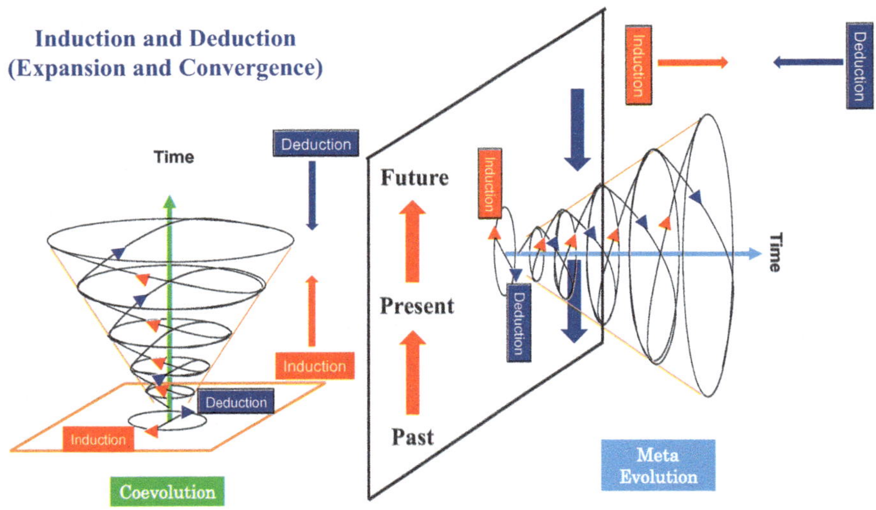

Fig. 22.15 Coevolution and metaevolution

sociocultural, and ecological systems outside us and even mental systems inside us (Murase, 2018).

It is clear that our perception creates our world, whether inside or outside, and our world in turn reinforces our perception through the evolutionary and/or developmental history of human beings. Thus, there are many cultural diversities in our world (Haidt, 2013).

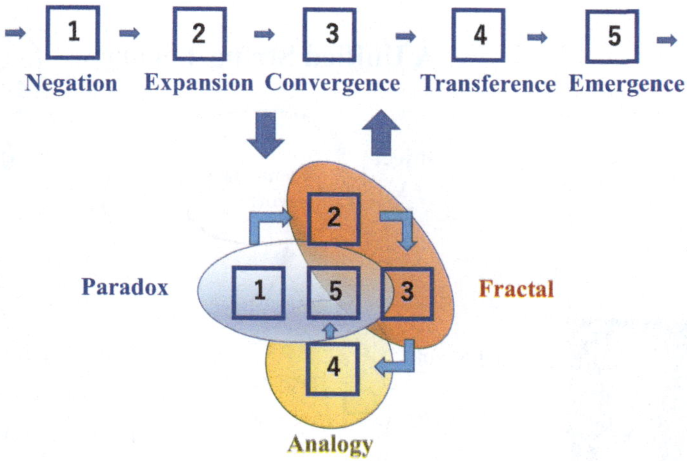

Fig. 22.16 The 5-NECTE processes

22.3.3.2 Transformation Between Simple Chain and Field-Like Shape

Let us consider how the linear chain relationship contributes to the formation of field-like views. In Fig. 22.16, the five numbers represent the 5-NECTE processes. During the learning or evolutionary processes, it is not always necessary to have linear sequences at all; but instead, the different processes often occur simultaneously like the case of mathematical induction (which contains not only induction, but also deduction and even abduction). Such a situation can be understood when the linear sequential chain is considered to be rolled into very different shapes. Actually, such a transformation between a linear chain and a looped shape appears at most composite molecules of a living cell; for example, DNA molecules or amino acid molecules (see Murase, 2018).

Note that the 5-NECTE processes are converted into three characteristics: paradox, fractals, and analogy. When ①Negation and ⑤Emergence occur simultaneously, we observe paradox. If ②Expansion and ③Convergence equally coexist, we will see fractal dynamics. Suppose that ④Transference involves a bidirectional relationship between the two opposing things, we can understand analogy.

Hideki Yukawa considered that the interaction between the two objects would be dictated by the elementary particles and their behaviors in relation to nuclear force. Interestingly, elementary particles are not only the components of each object, but also affect the interactions themselves (upper part of Fig. 22.17).

Yoichiro Nambu introduced the revolutionary idea concerning the elementary particle. He considered that the elementary particle is not a "particle" but rather a "string." Such a string could change its shape from open to closed and vice versa. As a result, the theory could explain not only the gravity but also electromagnetic forces.

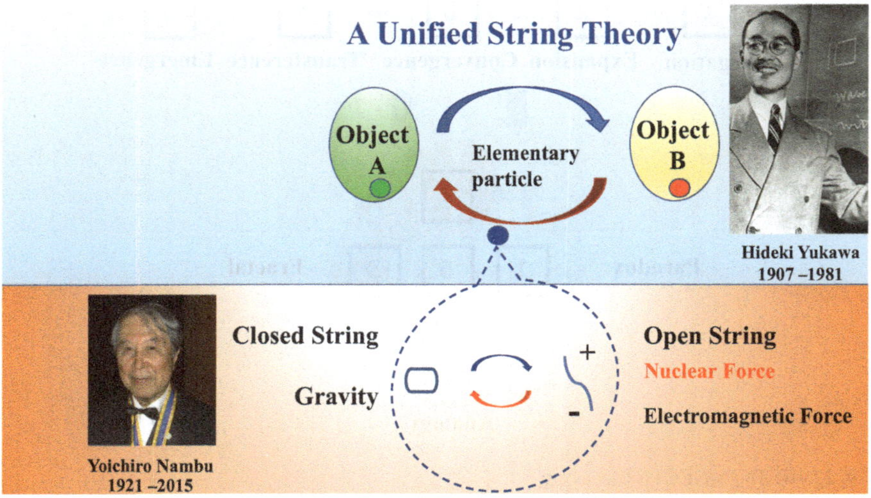

Fig. 22.17 A unified string theory

Similarly, the present consideration suggests that the change between the two different conformations could play important roles in understanding the complex situations.

Nature and life phenomena are so complex, it is difficult to understand what kind of principle is operating among the diverse phenomena. This chapter suggests that there is a scale-invariant principle that could help us to understand such diverse phenomena.

The important guiding perspective is that nature and life phenomena must be conducted by the same principle. In other words, we need the same principle to understand complex phenomena. We should specify the tools to investigate the objects, however, the tools themselves are hidden in the objects. In addition, we (or subjects) ourselves must be operate according to the same principle. It is interesting to realize that subjects, objects, and the dynamical relationships between them are all conducted according to the self–nonself circulation principle.

Part II: The Mandala Nursing Theory and Practice: Based on the Self–Nonself Circulation Principle of "Living" Wholeness.

22.4 Introduction

Nursing is an *art and science*. What is a theory of nursing? First, it proposes structure of philosophy that can explain the mutual relations among nursing phenomena in a unified manner and activity. Second, it focuses on a particular perspective specific to the nurse–patient relationship—as it is the self–nonself circulation—with respect to the health of patients and even nurses. Third, it serves as knowledge for improving nursing practices, guiding research, and nursing curricula. Finally, it provides nurses as well as patients with goals for nursing assessment, diagnosis, and professional education.

The first nursing theory (or philosophy to be precise) was presented by Florence Nightingale in her book *Notes on Nursing: What It Is and What It Is Not* published in 1859 (Nightingale 1859). She provided an answer to the question "What is nursing?" In her philosophy of nursing practice, she focused on the quality of environments around patients in daily life together with her belief in humans' *natural healing vital powers*, because she believed that patients could recover and be protected from diseases when the environment could be defined appropriately.

Indeed, Nightingale considered the importance of the various kinds of environmental conditions such as ventilation, warmth, light, nutrition, medicine, stimulation, room temperature, and activity, avoiding chattering hopes and advice, and so on. From a *strength* perspective, she believed in the equal importance of disease prevention and health promotion. Both would work very well when nurses assist the natural healing vital powers inherent in patients and improve the environmental conditions around the patients. She believed that *all disease is a reparative process.* In this sense, she founded the nursing discipline of today's nursing science.

Since then, there has been a wide variety of nursing theories from Western countries such as philosophies, grand theories, middle-range theories, and so on. Grand nursing theories began to emerge as a movement in the 1950s. Today's critics, however, often find them too abstract because of their vagueness in terms of nursing practice.

Here, we introduce the radically new mandala nursing theory based on the self–nonself circulation theory of life. As we demonstrate in the following sections, the new theory not only provides a philosophical holistic perspective, but also a practical methodology of how to co-create healing of patients and nurses. Health and disease must be emergent phenomena through the interplay of dynamic processes.

22.5 Mandala as a Symbol of Life and Universe

A Japanese philosopher, Kitaro Nishida (1987) placed a strong emphasis on the coexistence of conflicts, oppositions, and contradictions in our everyday experiences and thoughts. Accordingly, as we have just discussed in Part I, our lives, universe, and all emerging problems possess these kinds of properties. More precisely, we introduced three characteristics: (1) persistent paradox, (2) fractal dynamics, and (3) true analogy; and the 5-NECTE principle described by ① Negation, ② Expansion, ③ Convergence, ④ Transference, and ⑤ Emergence. Of course, please note that, as shown in Fig. 22.16, the three characteristics can be integrated into the single 5-NECTE principle.

Now, we introduce the mandala as a powerful tool for understanding ourselves as well as our world. Figure 22.18 gives an example of a mandala. Historically,

Fig. 22.18 Mandala drawn by Masatoshi Murase in 2001

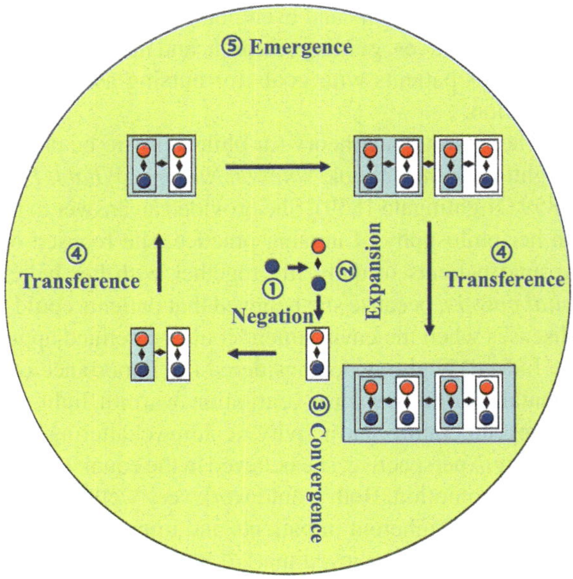

mandalas have been considered as tools or methods for the religious to see the reality behind complex features through meditative experiences. Despite its almost infinite variations, most mandalas have a very universal meaning and structure: it is a symbol of wholeness that represents a synthesis of endless conflicts. Jung (1968) realized that the mandala can be a good tool for psychiatric patients to recover from their mental illnesses.

In 2001, Masatoshi Murase (one of the authors) suddenly started to draw the mandala shown in Fig. 22.18. Even without noticing what is going on, he continued to draw the mandala. Finally, when it was completed, uncertainty and any imbalance of emotion or feeling were suddenly replaced by certainty and satisfaction. As Aine Mori empathizes in Part III of this chapter, it is not only the constructed product but also the process of construction itself that must be important for understanding ourselves as well as our world.

In drawing the mandala, the 5-NECTE principle was working, although at that time no one did understand what was hidden as a guiding principle. ① Negation: a single unit, ○, was placed in the center on the white paper. ② Expansion: the single unit, ○, was placed in the center and was divided into opposing units, ○⇔○. ③ Convergence: the two units in turn build up a higher unit as a synthesis of opposition, [○ ⇔ ○], and ④ Transference: the same process continues. ⑤ Emergence: a whole system is finally emerging. This is driven by the circulation between self (or endo-system) and the nonself (or exo-world).

The targeted whole system is the same as the beginning unit, although there is a difference in size and complexity. So, the mandala shows historical processes of both division (and differentiation) and synthesis (or integration). In other words, it shows its *past*. At the same time, the mandala is also a picture representing spacio-temporal arrangements of different kinds of subsystems at the *present*. As we can see, the mandala has a self-nested hierarchical structure. The origin and the emerging whole are the same, so that it is possible to perceive the dynamics in the *future*.

Put it in a different way, we can consider the emerging problem at the present, not only from the past but from the future! Furthermore, it turns out that the mandala can be a good tool for us to acquire improved understanding of the universe including nature and humans. This is because we can speculate on missing things or hidden dimensions behind complex features. This must be the power of the self-consistency principle of how to achieve creative solutions to complex emerging problems, mainly because the integrated picture shows that such opposed phenomena are merely the two sides of the same coin.

22.6 The Mandala Nursing Theory

22.6.1 Overview of Dynamic Communication Between a Patient and a Nurse

Figure 22.19 illustrates how nursing care takes place when a patient (left side spiral) meets with a nurse (right side spiral). The bottom panel schematically sketches the projected mandala of Fig. 22.18.

When balance is deformed in the self–nonself circulation process, the resulting distortion propagates over the mandala. As a consequence, "illness" sets in and one feels physically and mentally sick. In this case, one identifies the contradiction

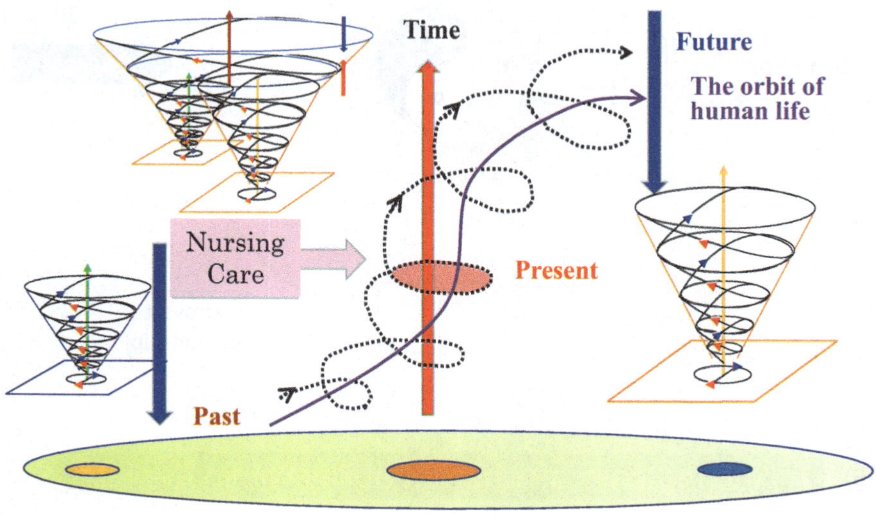

Fig. 22.19 A life process of human beings

causing the distortion of the mandala. Nursing care then begins toward the confrontational coexistence to accommodate the imbalance of the self–nonself circulation process.

22.6.2 A Mandala Nursing Theory

In the mandala nursing theory, there are five circulation processes in the patient's (self) physical–mental characteristics (1 ~ 5) and five circulation processes of nursing care (nonself) patterns (6 ~ 10), as illustrated in Fig. 22.20.

(1) Collapsing Self—Redeveloping Self
This concept is a condition of living self here and now. Collapsing self contains denial of existence and reorganizing self contains affirmation of existence. The concept of "development and collapse" is related to self–nonself circulation theory. The important view for understanding patients is to find positive actions

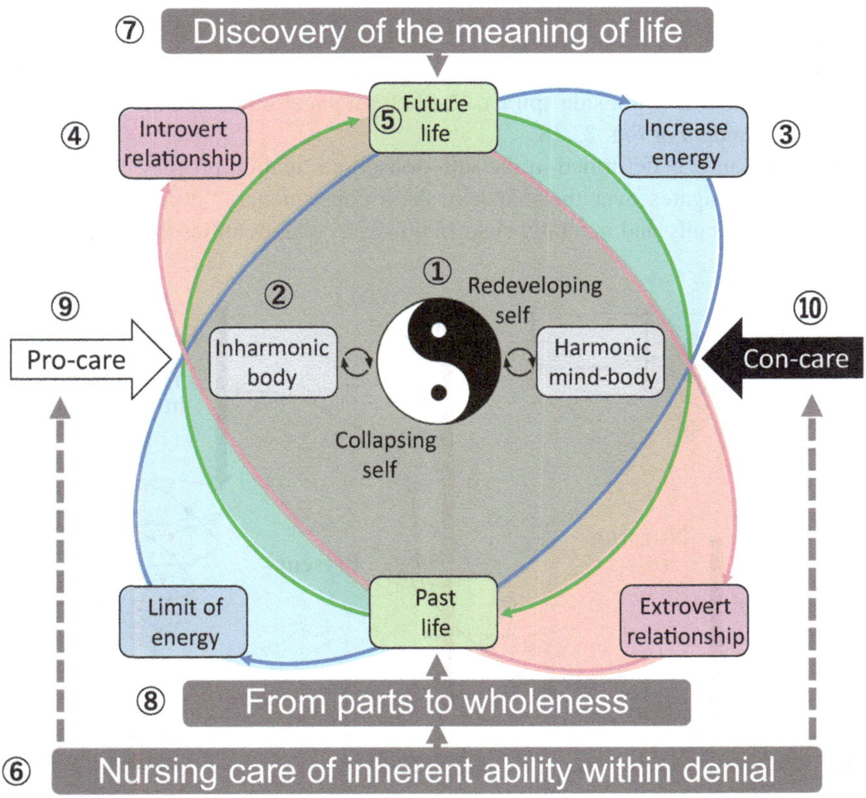

Fig. 22.20 The Mandala nursing theory

in the depth of self-disintegration actions such as illness, self-mutilation, or loss.

(2) Inharmonic Body—Harmonic Mind–Body
This concept is circulating inharmonic body and harmonic mind–body, closely related with the living self here and now, which is swinging between disintegrating self and reorganizing self. Inharmonic body contains pain and the harmonic mind–body contains comfort. For example, depression develops from awareness of a variety of physical disorders. When inharmonic body can be found, the condition of disintegrating self should be accepted. When harmonic mind–body can be found, the condition of reorganizing self should be accepted. The accumulation of small instances of physical care on these kinds of acceptance will also lead to mental care.

(3) Limit of Energy—Increase in Energy
This concept is a circulation factor of limited energy and increased energy that patients acquire during activity, transforming relationships with others from introvert to extrovert, or vice versa. Limit of energy contains pessimism, while increase in energy contains a prospective view. As patients with depression seem to be in limit of energy, their energy of body and soul is exhausted, so their care aims to rest the body and soul until their energy increases from within. Rest is not only relates to the physical state but also to the mental state. To avoid mental threats, distance from others is needed.

(4) Introvert Relationship—Extrovert Relationship
This concept is a circulation of the characteristic cognition for patients that introverted and extroverted relationships are in confrontational coexistence. Introvert relationship involves withdrawal, distrust, solitude, and holding. Extrovert relationship involves association, trust, requests for help, and release. This concept is similar to "extroversion—introversion" in Type Theory by Jung (1921).

(5) Past Life—Future Life
This concept is a history of the patient's life. Past life contains review and future life contains hope for life. This concept has relation to "historicity" of self–nonself circulation theory. History of life has great importance in integrating the identity of the self. Patients with depression are living in emptiness, and do not have affirmation to the living self here and now. Therefore, taking back the history of the self until now will lead to affirmation of living self here and now and future life will open if it is possible. Note that the diagram of Fig. 20 can be represented by this view (Fig. 22.21) when the past–future time axis is taken horizontally.

(6) Inherent Ability Within Denial
The concept of "inherent ability within denial" represents care in which nurses trust the potential power of the patient hidden under his expressions. It includes such assistance that the latent survival and living abilities of the patient are not hampered and the self of the patient denied by himself is

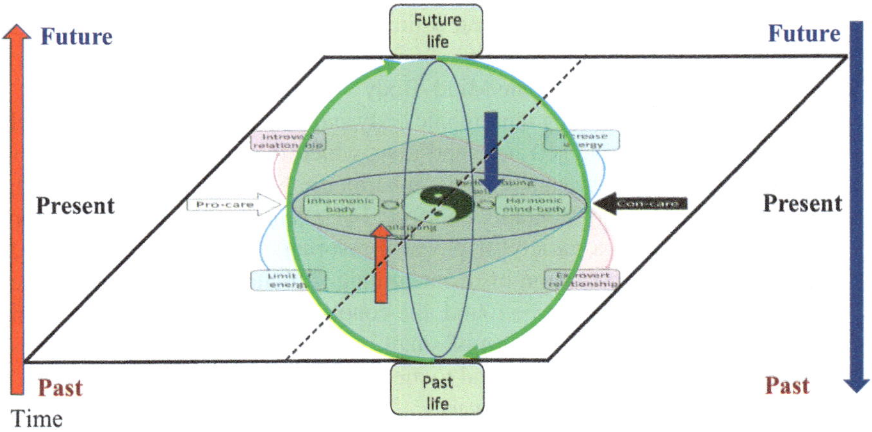

Fig. 22.21 The three-dimensional view of the Mandala Nursing Theory. Note that the diagram of Fig. 22.20 can be represented by this view when the past–future time axis is taken horizontally

approved. It is related to the concepts such as "development and disintegration" and "confrontational coexistence" in self–nonself circulation theory. As "development and collapse" are two sides of the same coin in this model, care should be done with the belief of potential power under negative phenomena. This concept must be similar to resilience. Because developing phenomena and disintegrating phenomena are inextricably linked together in the present theory, the care should be performed with trust in the inherent ability of the patient even though one is faced with negative phenomena.

(7) Discovery of the Meaning of Life

The concept of "discovery of the meaning of life" represents care during which the life of a patient is reconsidered and it is related to the "historical nature of life" in self–nonself circulation theory. In this care, nurses and the patient review the past lifestyle leading to illness and the new lifestyle devised by the patient himself. Rosen (1993) explained that healing the soul through creativity can transform depression. However, if embarrassing experiences of the patient in the past are exposed during the care, there is a risk of causing trauma by threatening the self.

Thus, it is important to create a better "holding environment" and to interrupt conversations when necessary by not missing subtle changes of words and actions of the patient. This kind of care contains risks of psychological trauma by threatening the self because it may touch on traumatic past experiences. When care is undertaken, a "holding environment" and framework of assistance should be set. It is also necessary to stop talking casually and to carefully watch for subtle changes in behavior.

(8) From Parts to Wholeness

The concept of "parts and the whole" represents care with a view to understanding the self from its subtle expressions. It includes "care grasping the entire picture by observing subtle countenances" and "continued physical care." The relevant concept in self–nonself circulation theory is "parts and the whole" and fractals. Because a human being is an indecomposable and holistic existence, parts are contained in the whole and vice versa. Examples of part-whole care occur via the self (the whole) through skin (parts) and through small physical care actions that make patients feel comfortable. It is because parts and the whole are organized into a nested hierarchy that nurses instinctively grasp the entire picture from the information of parts and can start care from the instant they see a patient.

(9) Pro-Care

The concept of "pro-care" represents a care in which the nonself attends the self while agreeing with the cognition of the self. It includes a care accepting and ignoring the anxiety, a care maintaining nonthreatening environment, a care waiting for the energy bubbling up, a care making use of past experience. The relevant concept in self–nonself circulation theory is the endo-system and the exo-world. Because human beings are open systems to the environment, the endo-system and the exo-world appear once patients close their selves. For exhausted patients who withdraw into their endo-systems by closing their selves, nurses should not disturb their withdrawal but perform pro-care by gathering hints from subtle changes in their expressions.

(10) Con-Care

Con-care is more like a "foreign substance" and plays the role of a stimulus perturbing to the equilibrium state. As the energy of the self is bubbling up, a nurse plays the role of the nonself providing the self with a different perspective and perturbs the present state by setting up confrontational coexistence within the cognition of the self. This care is similar to a psychotherapeutic method through dialogue called *yusaburi* (shaking) proposed by Kandabashi (2001).

According to Kandabashi, *yusaburi* induces a movement in the motionlessness state and this occur due to a prominent function of the brain. In addition, medical care workers should play a role to send signals to the brain that makes it easier to cause *yusaburi*, and a "holding environment" would play an important role as preparation. A nurse, although acting as the nonself, needs to recognize the confrontational coexistence between self and nonself and support patients so that they can exert their potential vital force to support themselves. This is the inherent ability within denial extracted from intervention studies.

The quintessence of self–nonself circulation theory lies in its view that the developing and disintegrating processes of the self are inextricably linked together because there is always a risk of disorganization when the nonself in the exo-world is assimilated into the self in the endo-system during the process in which the self and the nonself circulate and maintain their confrontational coexistence. By constructing a nursing theory based on this theory, one is able to recognize that "health and illness," "normality and abnormality," or "life and death" are two sides of the same coin. There are meanings even in depressive behaviors that often trigger suicide and self-mutilation out of pessimism and remorse. The very effort to find such meanings leads to the ability to counter depression or loss of hope.

In such a theory, nurses can show compassion for the patient sensing the deep cognition behind suicide and self-mutilation and support him now believing in the future existence of health, normality, and life despite the current captivity by illness, abnormality, and death. Persons likely to give rise to self-destructive behaviors, on the other hand, have energy for self-development and can exercise their creativity. This gives them the opportunity to transform themselves.

In summary, the theory constructed in this work may be considered as a nursing theory that helps patients discover the fact that there is another aspect in the experience of anguish caused by illness, to find meaning in that experience, and position that experience in their life. As countermeasures to the abovementioned risks, deliberate care strategies are needed in which nurses prepare a "holding environment," have clear frameworks for care, and share information from the viewpoint of teamwork. As a nurse cares for the patient in this way, individual views of nursing care will develop (Fig. 22.20).

In our new nursing theory—the mandala nursing theory—nursing is an involvement as the nonself in the care of a patient who is in the process of self–nonself circulation. Nursing is to help patients give their lives a meaning by recognizing that the current experience of illness is a repetition of past experience in a different guise.

As a patient faces conflicting concepts in experiencing depression, nursing care must be done in such a way that after conflictions are recognized, the patients should not try to resolve them but admit them as they are and seek opposing but coexisting relations. This may be interpreted as a kind of "harmony." We should also mention that in nursing care, there are two opposing but coexisting concepts: pro-care and con-care. We can see the importance of confrontational coexistence from the following points: the conflicting expressions like "affirmative ability within denial" and "the parts and the whole" can coexist, and, in the same sense, we can see the conflicting connotation in sayings such as "believing the affirmative power within denial" and "discovery of the meaning of life by living with illness." This means that nursing care must be done on the premise of opposing and coexisting relations. Thus, the role of nurses as the important nonself to the self who is suffering from illness lies in helping the patient accept the fact that the experience of illness itself has its own meaning. This is achieved by injecting into the patient a view that the current anguish is a necessary experience for the reproduction/reorganization of the self and is a repetition of the past in a different guise.

When the energy of the patient reaches a lower limit and the self narrows down and evades interactions with the nonself, nurses perform a pro-care in which they attend the self while agreeing with the cognition of the self. When the energy of the patient bubbles up and expands itself interacting with the nonself, nurses perform a con-care in which they attend the self while disagreeing with but prompting a change in the cognition of the self so that they can support a new way of life devised by the patient.

In addition, low-level but continued care that the patient feels comfortable with may lead to a significantly positive impact on the whole existence of the patient. Some examples of such care are abdominal massage for constipation, bed baths, keeping watch and showing deep interest, and natural conversations. Because the present theory is based on self–nonself circulation theory, nurses themselves should change as the nonself during the process of nursing care for patients and should live in a circular way.

We have discussed the application of the self–nonself circulation theory for a new nursing care model. The quintessence of the self–nonself circulation theory lies in its view that the reorganizing and disintegrating processes of the self are inextricably linked together. It is because there is always a risk of disorganization when the nonself in the exo-world is assimilated into the self in the endo-world during the process in which the self and the nonself circulate maintaining their confrontational coexistence. By constructing a nursing care model based on this theory, one is able to recognize that "health and illness," "normality and abnormality," and "life and death" are two sides of the same coin. Similarly, there are meanings in depressive behaviors that often trigger suicide and self-mutilation out of pessimism and remorse. The very effort to find such meanings leads to the ability to counter depression or loss of hope.

In our theory, nurses can show compassion for the patient sensing the deep cognition behind suicide and self-mutilation and support patients with depression now believing in the future existence of health, normality, and life despite the current captivity by illness, abnormality, and death. Patients with depression likely to give rise to self-destructive behaviors, but also have the energy for self-reorganizing and can exercise their creativity. This gives them the opportunity to transform themselves.

Part III: The Organization Theory and Practice: Hospital Art at Shikoku Medical Center for Children and Adults.

22.7 The Spirit of Hospital Art

Shikoku Medical Center for Children and Adults is a general hospital located in Zentsuji City, Kagawa Prefecture, in the Shikoku Region of Japan (Fig. 22.22). The hospital has 689 beds in 48 departments. It supports the lives of the people in the community on many levels, from pediatrics to adult medical care, and provides treatment for severely handicapped children.

Art was comprehensively introduced into the center upon its opening in 2013. This does not refer to easily observable art activities, such as pictures being painted on the children's ward walls, artworks being displayed at the entrance and in waiting rooms, and art workshops being held at the center; rather, it refers to the introduction of the spirit of art into the hospital as a whole.

The spirit of this hospital's art aims to face the diverse issues that occur throughout various medical treatments head-on and work to improve them by giving form to unobservable thoughts and feelings through dialogue. In other words, it is a form of

Fig. 22.22 Shikoku medical center for children and adults in Zentsuji City, Kagawa Prefecture in Japan

creative problem-solving and inclusive and participatory hospital-building. We refer to hospital art as not only the display of artworks that are the result of hospital-based art projects, but also as the process behind such work.

22.8 Reasons for the Introduction of Hospital Art

The introduction of hospital art stemmed from a 2008 project to paint a mural on the wall of the children and adolescents' ward of the Kagawa National Children's Hospital, which was the predecessor of the Shikoku Medical Center for Children and Adults. The director at the time wanted to brighten-up the gloomy ward, so art was introduced to address this specific problem. Zentsuji City, where the hospital is located, is the birthplace of the Buddhist monk Kukai and thus, is visited by many pilgrims seeking physical and emotional salvation.

Being responsible for the art direction of the mural, I paid a visit to the Zentsuji Temple in search of motifs to use. At this time, I came into contact with a magnificent 1200-year-old camphor tree that made me feel like I was a part of something much bigger than myself and was being entrusted with something special. People's hearts are easily moved, and concepts of right and wrong can be trifled with by politics and the flow of time. A look can sometimes deeply hurt someone's feelings. However, the gaze of the camphor tree had not changed over the past 1200 years. I had the intuition that such a stable gaze would provide the energy that was needed to soothe the fragile emotional states of the children in the ward.

As such I decided on the camphor tree as a subject and discussed the project with the artist, Hisako Masuda. We decided on a time-intensive process whereby the outline of the tree was divided into finely detailed parts that were then to be filled in gradually, like a coloring-in book. A participatory process was proposed, whereby the mural would be completed by artists, patients, and people from the community. The reason for this decision was that a place such as a children and adolescents' ward is an environment that is created by all who come into contact with it. To do the mural together with people with diverse thoughts and feelings, all of whom have a relationship with the space, was the most natural choice.

After the mural was completed, a number of changes were observed (Fig. 22.23). First, the medical staff noted that they developed an attachment to the old hospital through the process of painting the mural; a medical staff member commented: "I used to feel bad for the patients that our hospital looked so old but since we painted a mural in it, I feel that it is fine to be old." Furthermore, local media and other

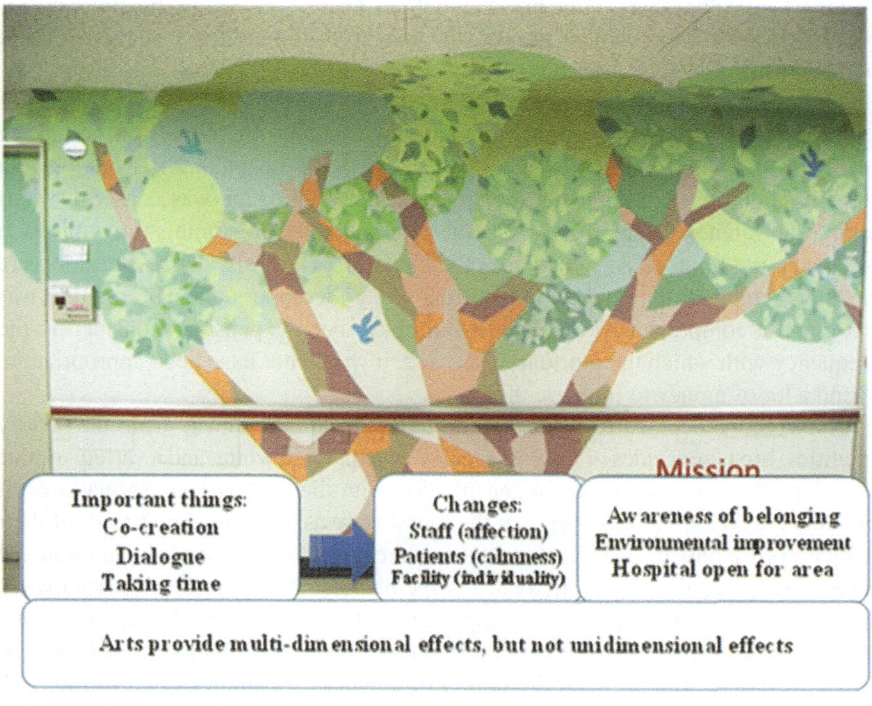

Fig. 22.23 Hospital art

residents became increasingly interested in the hospital when they heard about the project. A possible reason for this is that the field of child psychiatry has a rather closed and negative image, and the act of opening up, going inside together, and creating a mural, served to overturn the common gloomy sense of the hospital.

The biggest change observed was that patients stopped scratching the walls of the ward, resulting in decreased overall maintenance time and costs. The process of improvement through inclusive and participatory mural creation had brought about unexpected benefits on a number of levels, simultaneously. These benefits encouraged the director to introduce hospital art into the new planned hospital, Shikoku Medical Center for Children and Adults, in a comprehensive manner.

22.9 Hospital Art is Not an Art Object, It is Hospital-Building that Involves the Artistic Process

The essence of hospital art at this hospital has remained unchanged from the time of its construction to the present day, 8 years after opening. There is an ongoing process of listening to the current difficulties of the medical staff, engaging in repeated dialogue, and developing creative proposals from new perspectives, with all those involved working toward a better state of affairs. From this process emerges the vision of a hospital that no one has seen before. There is no end to this process.

The corridor improvement project: the area from the mortuary to the car park, which was started soon after the hospital was opened (Fig. 22.24), was inspired by the words of the director of nursing: "It breaks my heart that the corridor where we say goodbye is dreary and smelly." I started addressing this issue by first talking to the people who worked in the relevant areas in the hospital. After talking to a passing cook, it was found that the smell filling the passage was caused by staff taking out kitchen scraps. A request was soon made to the administrative division to change the route taken to dispose of such waste. The walls of the corridor were raw concrete, and many wires were strung along the upper area. The lighting was poor, and a comprehensive overhaul would have been expensive. Considering the frequency with which the mortuary was used, it would not have been appropriate to spend a lot of money to improve the area.

As such, we decided to paint a mural to divert attention away from the nearby facilities. First, both sides of the passageway were painted white, and a variety of blue flower patterns were prepared in consultation with the artist, Reiko Shimada. Staff members involved in the painting of the mural were asked to select a flower of their choice, copy it onto the wall, and paint it as if they were offering it in remembrance of those who have passed away. In total, 177 members of the medical staff participated. In the process, I witnessed an amazing sight; a number of the staff burst into tears while they were painting. They remembered the time their mother passed away or when they were walking with the family of a deceased patient as part of their job as a nurse, for example. One even said that she wanted to paint a flower for her late

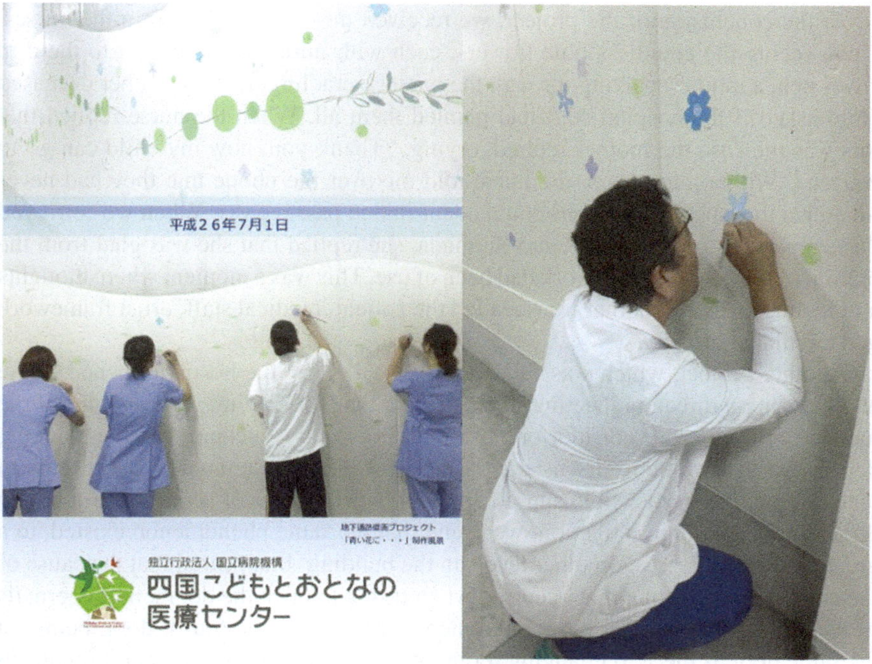

Fig. 22.24 Corridor wall painting

husband, who had died young. One pregnant woman was painting two flowers and said, "Here is one from me and one from my baby." We found that participants were coming face-to-face their own various emotions as they painted these blue flowers. This was definitely an experience that made us think deeply about life. All the people who had shed tears conveyed their feelings of thanks to us.

We came to understand that the act of painting the mural was an act for the patients and their families and also a way for the medical staff to deal with their own emotions. This was the first time that I realized the suffering and pain, not only of the patients and their families, but of the medical staff who look after and support them. In projects since then, I have made conscious efforts to mitigate such pain. Because hospitals are places where the systemization of medical treatment for patients has been optimized, the pain that medical staff experience tends to be overlooked. Despite this fact, medical staff are still able to provide patients with excellent treatment and care. Being aware of, and caring for, the pain that medical staff may experience is directly related to the provision of high-quality medical care for patients. In seeking to acknowledge the medical staff's feelings, the words and figures derived from surveys are insufficient. It is important for medical staff to occupy the same space, to do something together, and be aware of each other's feelings. When we face pain without flinching, the light that we need to walk toward together appears faintly in the distance.

At the conclusion of the project, we received messages of joy from the nurses. Upon seeing the countless blue flowers, each with initials written next to them in silver pen, a mother walking through the passage cradling the body of her deceased child asked if the hospital staff had painted them all. When the nurse replied that this was the case the mother replied, crying, "Thank you, now my child can go to heaven." Withholding tears, the nurse told me over the phone that they had never imagined that such interactions could occur in that passageway. When we conveyed these reactions to the artist, Reiko Shimada, she replied that she was glad from the bottom of her heart that her work had been of use. This was a moment where thoughts and feelings for one another transcended the patient, medical staff, artist framework and circulated broadly.

Another project, which was carried out 7 years after the hospital opened, came about during a directors meeting. It was found that the coating was coming off the metal door to the staff corridor on the fourth floor and someone from the planning department had come to discuss the situation. I went to the area in question with the staff member responsible and we found a large number of scratches around the door handle. Upon investigation, we found that the same phenomenon existed, to a greater or lesser degree, on other levels in the building. It appeared that the cause of the issue was scratching that occurred when trying to open the doors, by fingernails or when holding books or other equipment. The door to the fourth floor, being on the same floor as the staff changing room, was in a particularly bad condition due to its high frequency of use. As metal doors are finished with baked enamel, it takes a lot of time and money to remove and repair them. In this light, we decided to place stickers with graphic designs over the damaged areas (Fig. 22.25). At this stage, it was important to stay in line with the existing signage rules; each floor of the hospital had been assigned its own color. The colors for the stickers were selected to match the colors of the individual floors, and the designs were created with a level of versatility that would enable them to be used on other floors when such an issue occurred in the future. In this way, we respected past decisions and anticipated the future while deciding on what needed to be done in the present. This could be done precisely because we understand hospital art as a process and from a long-term perspective.

Florence Nightingale (1860) once states in her book, *Notes on Nursing*:"All disease, at some period or other of its course, is more or less a reparative process, not necessarily accompanied with suffering: an effort of nature to remedy a process of poisoning or of decay, which has taken place weeks, months, sometimes years beforehand, unnoticed."In the case of problems that occur in hospitals, it is important to view such issues as phenomena that occur in the process of recovery. If, instead of constantly abhorring problems, passing them on to others, and trying to suppress them as quickly as possible, we face and accept them as part of the process of making our hospital even better, the actions toward recovery will become apparent. To do so, is not simply to suppress the symptoms, but to seek the cause of the problem as true participants and to search for hope, that we must together strive for. Problems should be viewed as an important resource from which to create improvement. The spirit of hospital art creates the circulation of energy by making the switch to comprehend pain as hope.

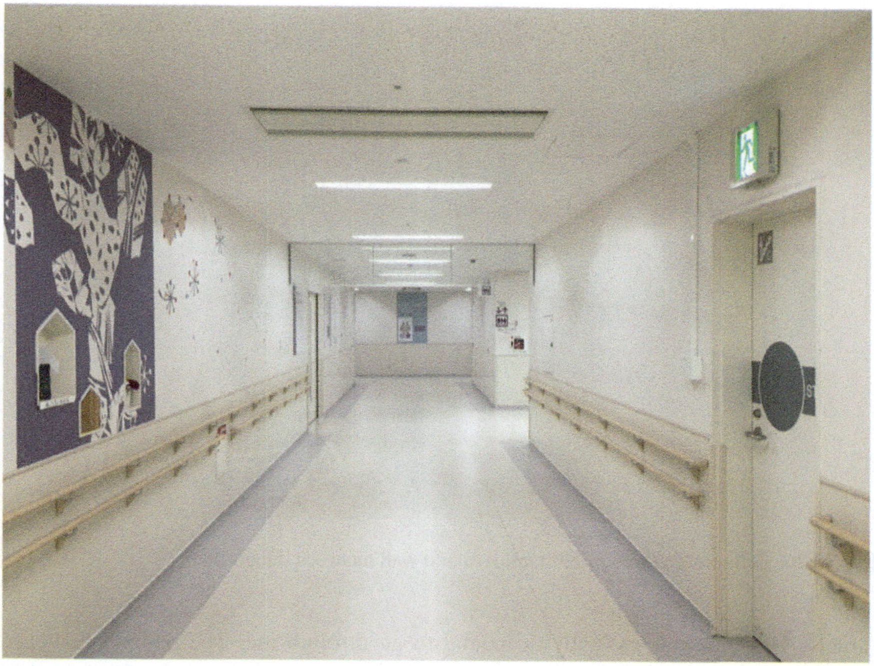

Fig. 22.25 Placing stickers with graphic designs on the metal door (right)

22.10 Designing a System for the Circulation of Feelings and Thoughts

Since the start of the coronavirus pandemic in 2020, our hospital has received approximately 200 gifts from members of our volunteer group who live all over Japan. These gifts were placed, along with a message, in 19 small recessed wall niches at the hospital. (Fig. 22.26). They can be taken freely by anyone who happens to notice them. Many people have been heartened by these anonymous presents since the opening of the hospital. The small holes were built at the time of the hospital's construction after we made a sudden request to change the design. This idea was born out of dialogues with patients with psychiatric disorders.

During the 2008 mural project at Kagawa National Children's Hospital, I became friendly with a number of these patients when I visited the children and adolescents' ward. At this time, I found that they were making beautiful handicrafts while in the hospital. Over a long period of hospitalization, the objects began to pile up.

One day, a patient entrusted me with some handicrafts and asked me to give them to children who were scheduled to undergo surgery. When I conveyed this to the nurses, they were extremely happy and soon put the plan into action. After repeating this several times, I suggested to the patient that she pass the objects to the nurses herself. The reason for this was that I thought she would be happy to receive thanks

Fig. 22.26 Gifts were placed in 19 small recessed wall niches at the hospital

directly from the nurses, without having to go through me. The patient, however, felt differently; she was intent on staying anonymous. The reason for this was that, due to her condition, she was unsure how long she would be able to meet people's expectations of her and the feelings of obligation would put pressure on her. While in hospital she felt like everything was being done for her, making her feel unsure if it was acceptable for her to keep on living. It was enough for her that someone enjoyed something that she had made. She did not require anything else.

What the patient desired was not evaluation but to be in the position of being able to give something to someone. The reason we built small holes into the wall of the hospital was to enable doors to be opened for the smooth circulation of the modest and sincere feelings of the patients. During construction, no one had any idea about what might be put into those small holes, but we deeply understood the role of those holes for circulating genuine feelings without any unnecessary bias.

22.11 The Role of the Hospital Art Director

Since the opening of the hospital, I have been in charge of the operations of the volunteer room. In hospitals, every space has a purpose and is named accordingly, for example, operating theater, conference room, or treatment room. The volunteer room, however, is the only place where its role changes by the day or hour. It can be used by patients and doctors for arts and crafts one moment, and by the planning department for a project meeting the next. Sometimes exhausted nurses come to seek personal advice and at other times local senior citizens come to participate in volunteer activities. As such, the volunteer room functions as a blank canvas where the role can change according to what is required at a particular time. It is a place within the internal organization of the hospital that is also connected to external stakeholders and thus maintains a high degree of open communication.

The objectives of the work that I have undertaken up to this point can be categorized as making ideas tangible, improving operations, and social inclusion. Even though they can be categorized in this way, each project may involve one or all three of these objectives. Making ideas tangible refers to expressing what the hospital wants to convey in the form of shapes, colors, and words. Mural production and public relations activities fit into this category. Improving operations refers to developing a deep understanding of the perspectives of the medical staff and working to improve the medical environment in terms of functionality and efficiency. The signage project, for example, always kept medical safety in mind. Social inclusion refers to sharing the spirit of hospital art with medical staff and alienated patients and families and welcoming them into its implementation, facilitating mutual support between internal and external stakeholders, circulating the desire to make the hospital a good place, and nurturing the growth of the hospital, together.

The role of the hospital art director is to meet these three objectives, in dialogue and in coordination with all departments (Fig. 22.27).

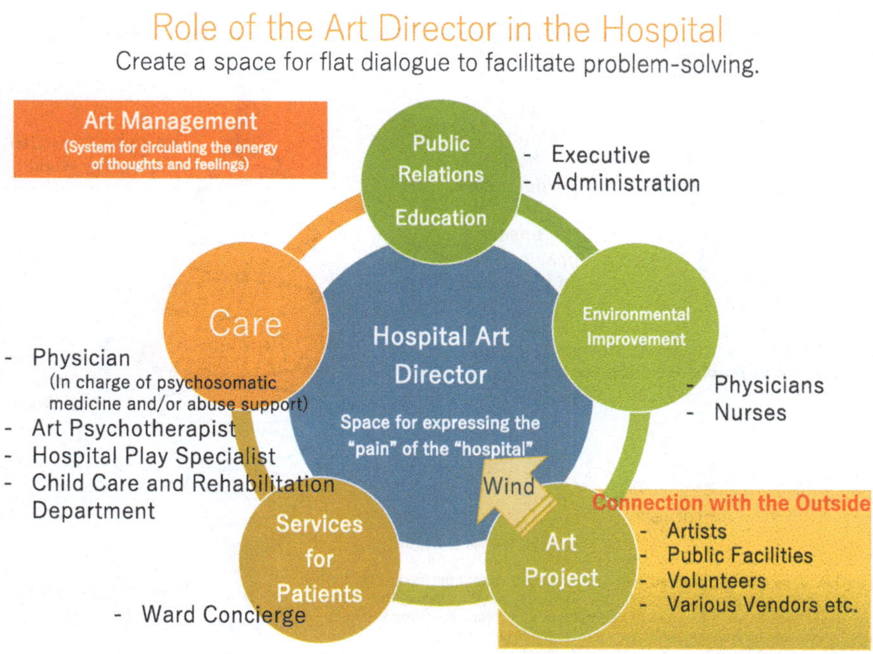

Fig. 22.27 The role of the hospital art director

22.12 The Energy Cycle of the Camphor Tree and the Energy Cycle of the Hospital

A hospital organizational chart, in most cases, resembles Fig. 22.28. A pyramid is the most logical and effective structure for delivering good medical treatment in a quick and accurate manner. But even though commands from the top may flow smoothly, what about the transmission of problems at the site? Significant time and effort are required to raise them even one level in the hierarchy. Furthermore, the greater the distance from the site to the issue, the more the raw sensations of the experience get chipped away, eventually leaving only standardized expressions to tell the story. By the time they reach the top, the true nature of the problem has totally changed.

In this context, the role of the art director is to collect a wide range of issues across all departments, as unadulterated as possible, and share them with the department heads while they are still fresh and relevant. Rather than favoring art from a top-down perspective, this is a system for circulating information from the bottom up as fast as possible. The information referred to here is not just quantified facts or objective and logical opinions; it also includes the emotions of the people directly involved. Imagine this as a flow of energy that travels in a spherical, three-dimensional way while maintaining balance between horizontal and vertical circulations. What is

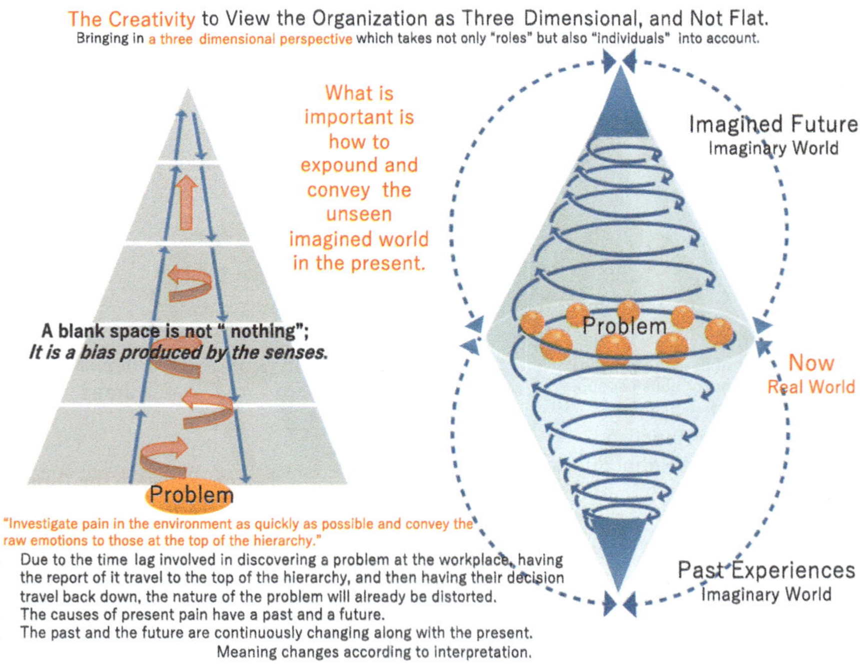

Fig. 22.28 A hospital organizational chart

essential for the improvement of this circulation is direct, everyday communication that transcends departmental boundaries.

The energy cycle of hospital art can find suggestions from the 1200-year-old camphor tree (Fig. 22.29). The fact that it has lived for 1200 years in this world means that the tree possesses a magnificent life support system. When I am feeling lost in my everyday life, I visit the 1200-year-old camphor tree and return to the origins of the mural we painted of it. Each time, I discover and learn new things about the life support system of the tree. The breadth of the camphor tree's canopy is approximately the same as that of its roots, which have spread throughout the ground. It continuously draws water and nutrients up through its trachea from its roots and photosynthesizes light, by way of its leaves, in a complete energy cycle that involves both its trachea and its sieve tubes.

In other words, the camphor tree expertly balances that which can be gained from its roots, which are the traces of its past growth, and that which can be gained from the light, which comes from the place it is reaching toward, and grows horizontally and vertically in a spherical form. Just as any plant will wither and die if it is suddenly planted in a large pot, it grows by balancing the observable ground with the unobservable soil below in a cycle of energy. Similarly, it is important to comprehend the

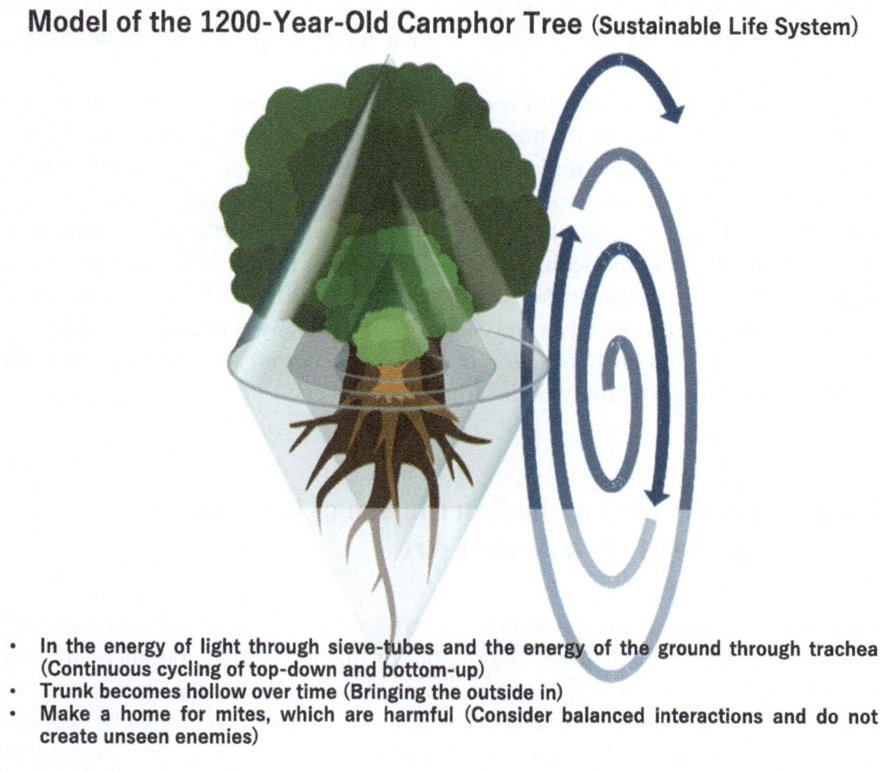

Fig. 22.29 The energy cycle of hospital art inspired by the 1200-year-old camphor tree

organization of a hospital as a three-dimensional ever-growing process consisting of unseen elements, past experiences, and future goals, and to understand the "now" as the only thing that we can see, touch, and choose for ourselves. How can we edit the energy flowing through unseen elements, share it, and select the optimal "now"? This requires transdisciplinary dialogue.

In hospitals, humans do not exist on reason and duty alone. Our lives also involve unique emotions that bubble up at every moment, a subconscious that cannot be fully controlled, and uncertainties that cannot be seen. In order that we do not forget that it is not machines, but humans, each with their own emotions and creativity, that make up the organization, it is important that we understand the organization as an ever-changing and growing living organism. To do this, observational power with respect to the present and the creativity to link the past and the future is essential. We need to closely observe what is happening right now, shed light on it through past experiences and data, and engage in repeated transdisciplinary dialogue in the pursuit of creating a vision that cannot be imagined by data alone. Then, every individual needs to prepare themselves as true participants and construct the present.

Hospital art may be a circulatory system for thoughts and feelings that can draw out the powers of observation and creativity that all humans possess. Hospital art is not something that is entrusted to artists with special abilities in hospital-building; rather, it is a process whereby all people who work in the hospital become able to use their creative and observational powers to their fullest extent and create the future, as artists themselves.

Having a treatment system that is expertly managed and controlled is important, but that alone is insufficient for hospitals as places that are living organisms for the regeneration of life. People from all disciplines need to spontaneously and naturally exercise their powers of observation and creativity through repeated cycles of success and failure and work together to create hospitals that no one has seen before. It is my belief that the field of medicine will progress even further if this perspective is introduced as a complement to the current system. This will result in an organization that is a beautiful living organism and cannot be recreated by machines. As expressed by Kenji Miyazawa in his book, *An Introduction to Peasant Art*, "Everyone should strive to achieve the sensitivity of the artist."

22.13 Complementing the Paternalism of the Medical with the Maternalism of Art

The camphor tree intentionally creates a place for mites, which are its natural enemy, to live in the veins of its leaves. The reason for this is not clear, but there are two theories. One is that plant mites are kept to a fixed number by having them live there. The other is that the number of mites in the entire tree can be controlled by keeping them in the domatium of the leaves and dropping all their leaves at once. Regardless of whichever is the correct theory, it is important to focus on the fact that something that at first glance looks disadvantageous, and possibly life-threatening, is being proactively included in the tree's life support system. This is a perspective whereby rather than trying to constantly eliminate one's enemies, one welcomes and coexists with them. It is to include rather than separate us from that which is different from ourselves. For all of us who live in an environment that involves a wide variety of lifecycles, this way of being has significant implications for prosperous co-existence. What should not be forgotten is that the camphor tree, which has clearly been able to maintain its existence for a much longer period than humanity, has selected this as its best strategy for survival.

When art was comprehensively introduced in the hospital, a large number of medical staff understood art only as: "a luxury we can indulge in only because we have an ample budget," or "a window dressing to help cheer people up." Some even viewed it as "a complication and a risk which will hinder the genuine medical system." In response, we made a clear statement regarding the meaning of art activity in hospitals. The meaning of introducing art into hospitals is to temper the paternalistic results-focused approach of medicine, which employs scientific theory in its fight against disease, with a maternalism that sometimes moves away from conflict and accepts everything as it is; in other words, a process-focused perspective that respects the workings of life.

It is true that it is a risk to introduce the unquantifiable, vague, and uncertain maternalism into a reproduceable evidence-based theoretical treatment system for defeating illness. It can even appear as wasteful or futile. But what if we step back from medical treatment and view the hospital inclusively from a higher dimension, as an environment and space that fosters powers of natural healing and as a place where people live? What gently functioning environment, which accepts and circulates the scattered elements that cannot be handled by paternalism alone, comes to light? It is impossible to fight off disease forever. We should not forget that we will one day die eventually, and that we exist as part of a larger cycle of life. No matter what wonderful surgical procedure one undergoes, or what newly developed medicine one take medication are offered, we must ultimately entrust ourselves to our own individual powers of natural healing. If we think in this way, then what hospital art expresses, in a place that regenerates life, is the process-focused maternalism that complements the paternalism of results-focused medicine. It is the development of an environment that increases one's individual powers of natural healing for an improved effectiveness and quality of medical treatment, similar to the way of nature, which reaches for the light in pursuit of happiness (Figs. 22.30 and 22.31).

22.14 Concluding Remarks: Let Us Invent Our Futures

Have you imagined the paradoxical story that there is a book on "ultimate reality" written in *known* language, but the contents are still *hidden*? This story suggests that even if we know the language and its grammar, it is impossible to understand the contents of the book without the necessary key information. What's that? Of course,

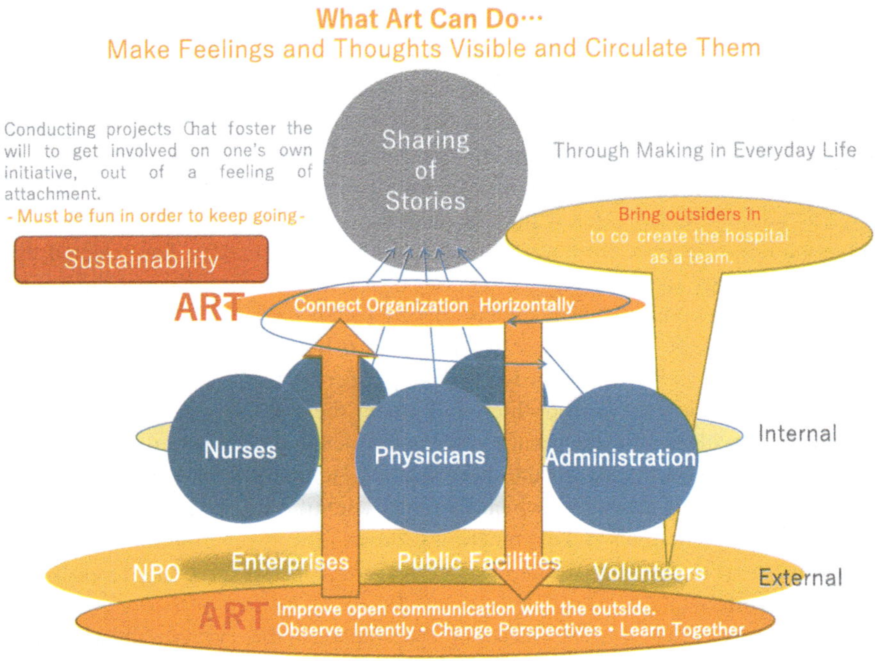

Fig. 22.30 The development of an environment

it must be very complex as usual. It now turns out, however, that one of the key pieces of information is *the knowledge that we do not know what we know*. We have not understood the above paradoxical story, simply because we have forgotten that we ourselves are paradoxical beings!

The same must be true in our everyday lives. We can realize that there are too *many* examples such as learning of martial arts, of drawing, of living, and even of learning itself, which must have provided us the same, only *one*, story mentioned above. We begin to understand that *one* is not always in *many*, *many* is in turn fitting together to form *one* wholeness! Likewise, we start to find the power of paradox as the key to understand the book. Besides paradox, in the present chapter, we emphasized two other important characteristics such as fractal dynamics and true analogy.

Now let us turn to our whole planet. Despite the huge diversities of our cultural, habitual, educational, and spiritual situations just like today, through the advanced development of modern science and technology, globalization has been strongly enhanced through interconnected webs across the planet.

As Zolli and Healy (2012) mentioned, globalization is the mother of all coupled systems, and for all its benefits and wonders. Today's COVID-19 pandemic has revealed how globalization leads to so many difficult systemic problems not only in our outside world such as bankruptcies and job losses but also in our inside world such as depression symptoms and suicide. It is reported that prevalence of depression symptoms in the USA was more than threefold higher during COVID-19 than before

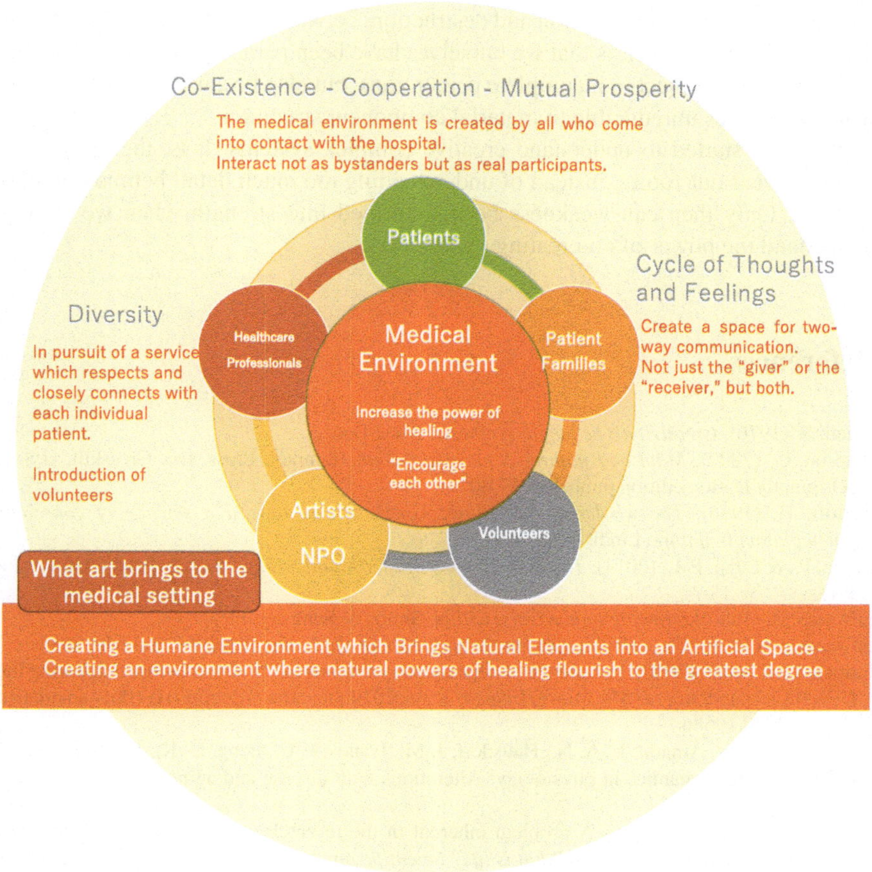

Fig. 22.31 An improved effectiveness and quality of medical treatment

the COVID-19 pandemic (Ettman et al., 2020). It occurs because of lower social resources, lower economic resources, and greater exposure to stressors such as job loss.

Although there always seems to be too many complex systemic problems, most of them could be ascribed to a single origin: that is, our *incomplete perceptions* (Capra & Luisi, 2014; Murase & Murase 2021a, b). Indeed, it is proved mathematically by Kurt Gödel in terms of "Incompleteness Theorems" that there is always incompleteness in any *axiomatic* system of propositions and rules. It is not universally understood in our everyday lives, however, how we are subject to incomplete perceptions inherent in ourselves (Gottwald, 2002). This is the reason that we are always surprised to see novelty or creativity or even abrupt disappearance of being, which cannot be deduced, defined, or predicted beforehand.

Gödel's incomplete theorems can be applied to the objective world as well. It is true that we have been confronted with many paradoxes between change and constancy, between construction and destruction, between life and matter, and so on. What is ironic, however, is that we ourselves have been paradoxical beings not only because we are subject to incomplete perceptions, but also because we are emergent whole processes unreducible to material component structures.

We have started to understand creative complex systems. It is, therefore, now time to invent our future, instead of understanding too much detail behind complex systems. Only then can weakness be transformed into strength when we start to understand the power of co-creating dynamics.

References

Arieti, S. (1976). *Creativity: The magic synthesis*. Basic Book.

Bateson, G. (2002). *Mind and nature: A necessary unit*. Hampton Press, Inc. Cresskill, U.S.A. (Originally Dutton edition published 1979).

Bortoft, H. (1996). *The wholeness of nature: Goethe's way toward a science of conscious participation in nature*. Lindisfarne Press.

Capra, F., & Luisi, P. L. (2014). *The systems view of life: A unifying vision*. Cambridge University Press.

Edwards, B. (1986). *Drawing on the artist within*. Simon & Schuster, Inc.

Edwards, B. (1999). *The new drawing on the right side of the brain: The 1999*, 3rd edn, Tarcher.

Ettman, C. K., et al. (2020). Prevalence of depression symptoms in US adults before and during the COVID-19 pandemic. *JAMA Network Open, 3*(9), e2019686. https://doi.org/10.1001/jamanetworkopen.2020.19686

Goldberger, A. L., Amaral, L. A. N., Hausdort, J. M., Ivanov, P. C., Peng, C.-K., & Stanley, H. E. (2002). Fractal dynamics in physiology: Alterations with disease and aging. *PNAS, 99*, 2466–2472.

Gottwald, F.-T. (2002). Life—A problem inherent in the research context. In H.-P. Dürr, F.-A. Popp, & W. Schommers (Eds.), *What is life? Scientific approaches and philosophical positions* (pp. 25–37). World Scientific.

Govinda, L. A. (2012). *Foundations of tibetan mysticism*. Martino Publishing.

Haidt, J. (2013). *The righteous mind: Why good people are divided by politics and religion*. Penguin Books.

Izutsu, T. (1973). "The interior and exterior in Zen Buddhism" In Portmann, A. & Ritsena, R. (Eds) *Correspondence of man and world*, Vol. 42. Eranos Yearbook.

Jantsch, E. (1980). *The self-organizing universe: scientific and human implications of the emerging paradigm of evolution*. Pergamon Press, Oxford.

Jullien, F. (1999). *The propensity of things: Toward a history of efficacy in China (Translated by Lloyd J)*. Zone Books.

Jung, C. G. (1921). Type theory, Japanese ver. Misuzu Book.

Jung, C. G. (1968). Psychology and alchemy, the collected works of C. G. Jung. In Read, H., Fordham, M., Adler, G. & Mcguire, W., (Eds.) Vol. 12. Princeton University Press.

Kahane, A. (2012). *Transformative scenario planning: Working together to change the future*. Berrett-Koehler Publishers, Inc.

Lorenz, K. (1973). *Behind the mirror: A search for a natural history of human knowledge*. (translated by Taylor, R.) Methuen and Co Ltd.

Lovelock, J. (1979). *Gaia*. Oxford University Press.

Murase, M. (1992). *The dynamics of cellular motility*. Wiley.

Murase, M. (1996). Alzheimer's disease as subcellular 'Cancer': The scale-invariant principles underlying the mechanisms of aging. *Progress of Theoretical Physics, 95*(1), 1–36.

Murase, M. (2000). *Life as history: The construction of self-nonself circulation theory (in Japanese)* (pp. 369–376). Kyoto University Press, (in English). https://repository.kulib.kyoto-u.ac.jp/dspace/bitstream/2433/49765/1/Murase2000b2.pdf
Murase, M. (2008). Environmental pollution and health: an interdisciplinary study of the bioeffects of electromagnetic fields, SANSAI. *An Environmental Journal for the Global Community*, (3), 1–35.
Murase, M. (2018). "A self-similar dynamic systems perspective of "Living" nature: The self-nonself circulation principle beyond complexity" In Yamash'ta, S., Yagi, T., Hill, S. (Eds.) *The Kyoto manifesto for global economics the platform of community, humanity, and spirituality*, (pp. 257–283). Springer.
Murase, M., & Mursae, T. (2021a) Transdisciplinary study of how to integrate shattered world: the self-nonself circulation principle of "Living" wholeness. In Hill, S., Yagi, T., & Yamash'ta, S. (Eds.) *The Kyoto manifesto II*. Springer
Murase, M., & Mursae, T. (2021b) The self-nonself circulation principle of "Living" nature: How to survive shattered world. In Hill, S., Yagi, T., & Yamash'ta, S. (Eds.) *The Kyoto Manifesto II*. Springer
Nightingale, F. (1859). Notes on nursing: What it is, and What it is not, Facsimile. (1st edn) J.B. Lippincott Co.
Nishida, K. (1987). An inquiry into the good, (translated by Abe, M. & Ives, C) Yale University Press.
Riedl, R. (2019). *Structures of complexity: A morphology of recognition and explanation*. Springer.
Rosen, D. (1993) T*ransforming depression: Healing the soul through creativity*, Japanese ver.
Scharmer, O. (2016). Theory U: Leading from the future as it emerges, Berrett-Koehler Publishers, Inc.
Schrödinger, E. (1958). *Mind and matter*. Cambridge University Press.
Servigne, P., & Stevens, R. (2015). Comment tout peut s'effondrer: Petit manuel de collapsologie à l'usage des générations présentes.
Suzuki, D. T. (1938). Lectures on Zen Buddhism. In Fromm, E., Suzuki, D. T. & De Martino, R. (Eds.) *Zen Buddhism and psychoanalysis*. Harper & Row, Publishers.
Varela, F. J., Thompson, E., & Rosch, E. (1991). *The embodied mind*. The MIT Press.
Yukawa, H. (1973). Creativity and intuition: A physicist looks at East and West, Kodansha International. New York.
Zolli, A., & Healy, A. M. (2012). *Resilience: Why things bounce back*. Simon and Schuster.

Masatoshi Murase received his Ph.D. degree from The University of Tokyo in 1987. Since 1992, he has been an associate professor at the Yukawa Institute for Theoretical Physics, Kyoto University. In 1987 and 1988 he was a visiting scientist at the Duke University Medical Center, Durham, NC, USA. Since 2010, he has been a member of the Cooperation Promotion Committee of the International Research Unit of Integrated Complex System Science, Kyoto University. Between 2015 and 2020 he was a director of the Research Promotion Strategy Office of the International Research Unit of Advanced Future Studies, Kyoto University.

Aine Mori graduated from the Department of Photography at the Osaka University of Arts, Japan, in 1995. In 2005, she received Fuji Photo Salon 'the New Face Prize', Tokyo. After that, Mori worked as a photographer. In 2009, after producing a mural at the National Hospital Organization Kagawa Children's Hospital, Mori was engaged as art director at the hospital. In 2012, she worked as a director to establish the Shikoku Medical Center for Children and Adults. She now works as art director at the hospital. Mori was awarded second prize in the Medical Welfare Architectural Awards in 2017, and won first place in the Public Architectural Awards in 2019.

Tomoko Murase graduated from the School of Nursing at Chiba University, Japan, in 1985 and received a Master's degree from Chiba University in 1987. In the same year, she became an assistant professor at the Japanese Red Cross College in Tokyo, Japan. She was a lecturer at the Japanese Red Cross College in Tokyo from 1989 to 1991. Between 1991 and 1993, she was a visiting researcher at the University of Tokyo (Division of Health Science and Nursing, Department of Medicine). After spending almost 10 years caring for children, Murase entered a doctoral course at the Graduate School of Nursing at Chiba University and was awarded her Ph.D. in 2010. In 2011, she became an associate professor at the Faculty of Nursing, Himeji University, Japan, and in 2012, became a professor and Dean of Nursing at Himeji University. In 2013, she was appointed professor at the School of Nursing at Japanese Red Cross Toyota College of Nursing, Japan, and has been Dean of the School of Nursing since 2016.

Correction to: David Pines and Me

Kazuo Nishimura

Correction to:
Chapter 1 in: K. Nishimura et al. (eds.), *Creative Complex Systems*, Creative Economy, https://doi.org/10.1007/978-981-16-4457-3_1

The chapter "David Pines and Me", written by Kazuo Nishimura, was originally published electronically on the publisher's internet portal without open access. With the author's decision to opt for Open Choice the copyright of the chapter changed on 27-Jan-2024 to "© The Authors 2024" and the chapter is forthwith distributed under a Creative Commons Attribution 4.0 International License (http://creativecommons.org/licenses/by/4.0/), which permits use, sharing, adaptation, distribution and reproduction in any medium or format, as long as you give appropriate credit to the original author(s) and the source, provide a link to the Creative Commons license and indicate if changes were made.

The images or other third party material in this chapter are included in the chapter's Creative Commons license unless indicated otherwise in a credit line to the material. If material is not included in the chapter's Creative Commons license and your intended use is not permitted by statutory regulation or exceeds the permitted use, you will need to obtain permission directly from the copyright holder.

The updated version of this chapter can be found at
https://doi.org/10.1007/978-981-16-4457-3_1

K. Nishimura et al. (eds.), *Creative Complex Systems*, Creative Economy,
https://doi.org/10.1007/978-981-16-4457-3_23

Index

K. Nishimura et al. (eds.), *Creative Complex Systems*, Creative Economy, https://doi.org/10.1007/978-981-16-4457-3

The manufacturer's authorised representative in the EU is Springer Nature Customer Service Centre GmbH, Europaplatz 3, 69115 Heidelberg, Germany. If you have any concerns regarding our products, please contact ProductSafety@springernature.com

Printed and bound by CPI Group (UK) Ltd, Croydon, CR0 4YY
30/09/2026
02225254-0016